PREFACE

These Proceedings contain most of the papers presented at the 8th IFIP Conference on Optimization Techniques held in Würzburg, September 5-9,1977.
The Conference was sponsored by the IFIP Technical Committee on System Modelling and Optimization (TC 7) with the cooperation of

- European Research Office (ERO), London
- Gesellschaft für Angewandte Mathematik und Mechanik (GAMM)
- Bayerisches Staatsministerium für Unterricht und Kultus
- Bundesministerium für Forschung und Technologie
- Deutsche Forschungsgemeinschaft.

The Conference was attended by 241 scientists from 28 countries. The program offered a broad view of optimization techniques currently in use and under investigation. Major emphasis was on recent advances in optimal control and mathematical programming and their application to modelling, identification and control of large systems, in particular, recent applications in areas such as biological, environmental and socio-economic systems.

The Proceedings are divided into two volumes: In the first are mainly collected the papers dealing with optimal control, in the second those dealing with mathematical programming and various application areas.

The international Program Committee of the Conference consisted of:

A.V. Balakrishnan (Chairman, USA), L.V. Kantorovich (USSR),
W.J. Karplus (USA), R. Kluge (GDR), H.W. Knobloch (GER),
J.L. Lions (France), G.I. Marchuk (USSR), C. Olech (Poland),
L.S. Pontryagin (USSR), A. Prekopa (Hungary), E. Rofman (Argentina),
A. Ruberti (Italy), B.F. de Veubeke † (Belgium), K. Yajima (Japan).

TABLE OF CONTENTS

INVITED SPEAKERS

ROUND TABLE DISCUSSION ON WORLD MODELS

COMPUTATIONAL TECHNIQUES IN OPTIMAL CONTROL

*paper not received

DIFFERENTIAL GAMES

OPTIMAL CONTROL OF PARTIAL DIFFERENTIAL EQUATIONS

STOCHASTIC OPTIMAL CONTROL

PART 2

TABLE OF CONTENTS

MATHEMATICAL PROGRAMMING, THEORY

ECONOMICS

OPERATIONS RESEARCH

STOCHASTIC OPTIMIZATION: TIME-CONTINUOUS DATA MODELS

A. V. Balakrishnan*
Department of System Science
UCLA, Los Angeles, California

Introduction By stochastic optimization we shall mean optimization problems based on real data ('observed' data) -- in which the operation on the data is to be optimized. It is essential then that account be taken of errors in the observation. There is always an unavoidable error which can only be modelled as an additive random process labelled 'noise' in engineering practice. Hence the qualification 'stochastic' in the title.

We shall be concerned with the case where the data is indexed by time -- this should come as no surprise as we are talking about handling real data -- where furthermore the time parameter is not discrete but 'continuos'. It is not our purpose here to go into the question of why we choose the 'time-continuous' model. [See [1] for more on this point]. The subtelties involved on the discrete-versus-continuous choice can only be appreciated if one has had occasion to handle real data. On the other hand our choice of the time-continuous model can be accepted and studied on its own merits, even if only theoretical. We also hasten to point to the tremendous volume of literature both engineering and mathematical in which time-continuous observation models are used.

We may describe then the observation as a 'stochastic process', $y(t)$, $0 < t \leq T < \infty$, with range in R_n. At time t, any decision or action can only be based on data observed upto that time. We are interested primarily in the limitation due to random errors in the data. Hence we may write

$$y(t) = S(t) + N(t) \quad 0 < t < T \tag{1.1}$$

where $N(t)$ is the random process corresponding to the 'noise' and $S(t)$ is the true information-bearing 'signal' or 'system response'. The latter is also allowed to be a random process, (assumed, naturally, to be independent of the noise process). In identification problems we write $S(t)$ as $S(\theta;t)$ where θ is the unknown entity (usually a finite dimensional vector) to be identified, but $S(\theta;t)$ is completely specifiable as a stochastic process once θ is specified. We shall for the moment not introduce θ since it will clutter up our notation. A basic assumption is that for each finite T, $0 < T < \infty$:

$$\int_0^T E||S(t)||^2 dt < \infty \tag{1.2}$$

where E denotes expected value.

Much depends on how the noise process $N(t)$ is modelled. Usually not much is known of a precise nature about it. We may assume it is stationary and Gaussian. (Indeed we may assume whatever we wish; the hitch is to obtain useful answers in practice.) For instance we may disdain the Gaussianness; but if we do so we are faced with having to specify the precise 'non-Gaussianness' -- a hopeless task, in general. Even if we assume stationarity and Gaussianness we cannot hope to specify the exact spectral density. About all we can say is that the bandwidth of the noise (i.e. where the spectral density is non-zero) being the bandwidth of the instrument must be large compared to that of the signal and that over that region it is a constant since otherwise the instrument would 'distort' the signal. The only sensible way to translate the qualitative specification 'large' is to allow the bandwidth to be infinite -- leading to "white noise". However, for white noise the variance ($E||N(t)||^2$) must be infinite and this brings in difficulties in theory as well. But for many problems in which operations on the data are 'linear' this does not create much of a stumbling block. However, serious difficulties arise if we wish to consider non-linear operations (as we must, for instance, in Identification problems). This was the situation around 1960, when a different model was brought in as "more rigorous", [see [2]]. This was based on the recent advances in the theory of stochastic differential equations due primarily to the work of Ito, crystallised in the integral bearing his name. First we write (1.1) in the "integrated" version:

$$Y(t) = \int_0^t y(s)ds = \int_0^t S(\sigma)d\sigma + \int_0^t N(s)ds. \tag{1.3}$$

If N(t) is white noise, then

$$E(\int_0^t N(s)ds \quad (\int_0^t N(s)ds)^* = tD \tag{1.4}$$

where D is the 'spectral density' matrix of the white noise. Next these authors make the transition to

$$Y(t) = \int_0^t S(\sigma)d\sigma \quad + W(t) \tag{1.5}$$

where W(t) is a Wiener process which of course preserves (1.4) and (1.5) is certainly a mathematically valid formulation. Moreover once this is done, one can lift the complicated but well-developed apparatus of the Ito theory of stochastic processes: of Ito integrals and martingales. One result, in particular, of importance to us in identification problems is the 'likelihood ratio'. For readers unfamiliar with this, it is given by

$$f(Y(\cdot))$$

where f(·) is the Radon-Nikdoym derivative of the measure induced by Y(·) on C[0,T] with respect to the Wiener measure. It is the correct extension of the notion of probability density of the process y(·) that we can write down in case 't' was discrete (and finite). A triumph of the Ito theory is that we can write: [see [2,7]]:

$$f(Y(\cdot)) = \mathrm{Exp} - \frac{1}{2}\{\int_0^T [\hat{S}(t),\hat{S}(t)]dt - 2\int_0^T [\hat{S}(t),dY(t)]\} \tag{1.6}$$

where

$$\hat{S}(t) = E[S(t)|B_Y(t)]$$

where $B_Y(t)$ is the σ-algebra generated by Y(s), $0 \leq s \leq t$. But the most salient feature of this formula is the 2nd term which is an Ito-integral. In particular, its value is undefined on all sample-paths $Y(\cdot)$ which are absolutely continuous with respect to Lebesgue measure, since this set has zero Wiener measure.

This is a remarkable formula in mathematics. But unfortunately it is totally unusable in practice for the very simple reason that (1.5) (the basis for it) was obtained from (1.3) and hence we can only replace dY by y(t)dt, and this would be wrong. In other words the formula simply does not have an operational meaning on real data since it was derived on the basis of a different model in the first place! It is remarkable that the promulgators of this formula in the engineering literature (even proferring engineering interpretation of the Ito integral!) had never bothered to actually use it on any real problem. But then of course this is a typical instance of the 'intellectualisation' of engineering†; perhaps it should be labelled 'academic engineering', by now an established discipline!

It turns out that there is an alternate theory (which is no more than exploiting the interpretation of 'white noise' as the 'asymptotic' case of increasing bandwidth) which is far simpler in contract to the Wiener-process theory and which moreover leads to operationally meaningful answers. Unfortunately space does not permit a detailed exposition of the white noise theory. We shall present the main ideas winding up with the likelihood ratio formula.

2. White Noise: Basic Notions

Let $W = L_2[0,T;R_n]$. By 'white noise' we mean the elements of W with 'Gauss' measure μ on W (also called 'weak distribution') such that

$$\int_W \mathrm{Exp}\ i[\omega,h]d\mu = \mathrm{Exp} - \frac{1}{2}[h,h] \tag{2.1}$$

where ω denotes points in W and h is a fixed element in W, and [,] denotes innerproduct in W. The right-hand-side of (2.1) defines the 'characteristic function' of the measure which is finitely additive on the class of cylinder sets with Borel bases. For more explanation see [3]. The measure is simply <u>not</u> countably additive on the Borel sets of W. Note that (2.1) implies that

$$\int_0^T [\omega(t),h(t)]dt$$

†The term (alas!) is not due to me. I borrowed it from Professor N. DeClaris.

is Gaussian with mean zero and variance

$$[h,h]$$

consistent with the 'process' $\omega(t)$, $0 < t < T$ being 'white noise'. In fact we shall adopt this as the 'precise' definition of white noise.

Let $f(\cdot)$ be any Borel measurable function mapping W into another Hilbert space H_r, real separable. Suppose there is a finite dimensional projection P on W into W such that $f(\omega) = f(P\omega)$ -- a so-called 'tame' function. Then $f(P\omega)$ is a random variable in the sense that the probability of inverse images of Borel sets in H_r is well-defined. The class of all tame functions is a linear space and we make it a linear metric space by defining the metric

$$d(f,g) = \int_W \frac{||f-g||}{1 + ||f-g||} \, d\mu$$

Completing this space, we obtain the class of all random variables -- a Frechet space. More important for us however is a subclass of these random variables. Given any Borel measurable function $f(\cdot)$ mapping W into H_r, we shall call $f(\cdot)$ itself a random variable, if H_r is finite dimensional (otherwise 'weak' random variable) provided $f(\cdot)$ has the following property: for any sequence of finite dimensional projections $\{P_n\}$, strongly convergent to the Identity:

$$E[e^{i[f(P_n\omega),h]}] = C_n(h), \; h \in H_r$$

converges for each h in H_r to a function C(h) which is independent of the particular chosen. Then C(h) is a characteristic function and the corresponding distribution (weak, if H_r is not finite dimensional) will be defined to be that of the 'random variable' $f(\cdot)$. For example if

$$f(\omega) = L\omega$$

where L is linear bounded, $f(\cdot)$ is a random variable if and only if L is Hilbert-Schmidt; see [4] for other examples. We are not so much interested in characterising this class by a set of necessary and sufficient conditions because there would still remain the problem of verifying these conditions in any particular case anyway. We shall be content rather to confine our attention to functions that we need in the likelihood-ratio formula.

3. Main Results:

Let us now go back to (1.1). We allow $N(\cdot)$ to be white noise and rewrite it as

$$y = S + N \tag{3.1}$$

where N is white noise in W and S a process that induces a countably probability measure on W which is countably additive by virtue of our assumption (1.2). See [5]. Because the two processes are independent we may write down for the characteristic function of the weak distribution induced:

$$C_y(h) = C_s(h) \text{ Exp } - \frac{[h,h]}{2}$$

where

$$C_s(h) = E[\exp i\, [s,h]]$$

Our first result is that the distribution induced by y is absolutely continuous with respect to the Gauss measure (induced by N) in the following sense [see [5] for more explanation]: there exists a Borel measurable function $f(\cdot)$ mapping W into R_1 such that it is a random variable and moreover for any cylinder set C (with Borel base) in W:

$$\mu_y(C) = \underset{n}{\text{limit}} \int_C f(P_n\omega)d\mu_G$$

where μ_G is Gauss measure, μ_y the measure induced by y and P_n is any sequence of finite dimensional converging strongly to the identity. The function $f(\cdot)$, which we shall naturally call the Radon-Nikodym derivative, is given by:

$$f(\omega) = E[\exp - \frac{1}{2} \{ \int_0^T ||S(t)||^2 dt - 2 \int_0^T [S(t),\omega(t)]dt \}] \tag{3.2}$$

See [5] for proofs. The likelihood ratio is obtained by replacing $\omega(\cdot)$ in (3.2) by the observed process $y(\cdot)$. To go further, we need to introduce the notion of conditional expectation in the white noise theory. Let P(t) denote the projection W into $W(t) = L_2[[0,t], R_n]$, $0 < t \leq T$. Let ζ be a random variable [i.e. ζ has its range in H_r which is finite-dimensional] with finite expectation. By

$$E[\zeta|P(t)y]$$

we shall mean a function $f(\cdot)$ defined on W(t) with range in H_r which is a random variable such that

$$f(P_nP(t)y) = E[\zeta|P_nP(t)y]$$

where P_n is any finite dimensional projection on W(t) into itself, the right-hand side being then well-defined, and for any sequence of finite dimensional projections approximating (strongly) the identity, $f(P_nP(t)y)$ converges and is by definition the random variable $f(P(t)y)$. In [6] we show that

$$E[S(t)|P(t)y] \quad \text{a.e. } 0 < t < T$$

and

$$E[||S(t)||^2|P(t)y] \quad \text{a.e. } 0 < t < T$$

can be defined as random variables. Finally [see [6]], by elementary manipulations, in comparison with the elaborate machinery in the Wiener process version via Girsanov theorem etc., see [7], we obtain that the likelihood ratio [a random variable] can be expressed:

$$\text{Exp} - \frac{1}{2}\left\{ \int_0^T ||\hat{S}(t)||^2 dt - 2 \int_0^T [\hat{S}(t), y(t)]dt + \int_0^T P(t)dt \right\} \qquad (3.3)$$

where

$$P(t) = E[\widehat{||S(t)||^2}] - E[\,||\hat{S}(t)||^2]$$

where

$$\widehat{||S(t)||^2} = E[\,||S(t)||^2 | P(t)y]$$

$$\hat{S}(t) = E[S(t)|P(t)y]$$

both 'random variable' functions defined on W. Unlike the Wiener process version (1.3), we can see that (3.3) is directly instrumentable. The 3rd term in (3.3) can be interpreted as a "correction-term" to (1.3). But it is clearly ridiculous to create an elaborate machinery and then to 'correct' it ad hoc, even when it is possible -- which is not always so even in our case when the condition (1.2) is not satisfied. Whereas the white noise theory gives you directly the anser in the first place with far less effort.

Of course the problem of calculating $\hat{S}(t)$ and P(t) still remains, in both theories. On the other hand in the case where S(t) is Gaussian it can fortunately be carried thru -- see [5]. Moreover it is still by far the most important case of practical interest [see the application to aircraft identification problems in [8]]. For a non-Gaussian example -- the random telegraph signal -- see [6].

References

1. A. V. Balakrishnan: Likelihood Ratios for Time Continuous Data Models. The White Noise Approach, IRIA Symposium on New Trends in Systems Analysis, December 1976.
2. T. E. Duncan: Evaluation of Likelihood Functions, Information and Control, 13, 1968.
3. A. V. Balakrishnan: Applied Functional Analysis, Springer-Verlag, New York 1976.
4. A. V. Balakrishnan: A White Noise Version of the Girsanov Formula, Proceedings of the Symposium on Stochastic Differential Equations, Kyoto 1976, edited by k. Ito.
5. A. V. Balakrishnan: Radon-Nikodym Derivatives of a Class of Weak Distributions on a Hilbert Space, Journal Applied Mathematics and Optimization, Vol. 3, No. 2/3, 1977.
6. A. V. Balakrishnan: Likelihood Ratios for Signals in White Guassian Noise, Journal of Applied Mathematics and Optimization, Vol. 3, 1977.
7. R. S. Liptser and A. N. Shiryayev: Statistics of Random Processes, Nauka, Moscow, 1975.
8. A. V. Balakrishnan: Parameter Estimation in Stochastic Differential Systems, Theory and Application in Advances in Statistics, 5th Edition, Edited by P. R. Krishnaniah, Academic Press 1977.

* Research supported in part under Grant No. 73-2492, Applied Mathematics Divn., AFOSR, USAF.

CONJUGATE DIRECTION METHODS IN OPTIMIZATION

Magnus R. Hestenes

Professor of Mathematics, University of California, Los Angeles

1. Introduction. Conjugate direction methods have become well established techniques for obtaining extreme points of a real valued function of n real variables. They are also applicable to the infinite dimensional case. The concept of conjugacy is an old concept which arises in the study of quadric surfaces. Algebraically conjugacy is associated with the polar form $Q(x,y)$ of a quadratic form $Q(x)$. The surfaces, $Q(x)$ = constant, are quadric surfaces. For a fixed vector x_1 the hyperplanes, $Q(x_1,x)$ = constant are conjugate to the vector x_1. Two vectors x and y are conjugate if $Q(x,y) = 0$. Consequently when $Q(x,y)$ is an inner product, conjugacy is synonymous to orthogonality. The author has used the concept of conjugacy extensively in the study of variational problems and in the study of quadratic forms in Hilbert space. Orginally conjugacy was used primarily for geometrical and analytical purposes and played no siginificant role in the theory of computation. With the advent of high speed computers it became clear to some of us that conjugacy could play an important role in computation. The first paper which significantly exploits the concept of conjugacy for computational purposes appears to be the joint paper by Stiefel and the author.* Undoubtedly the concept of conjugacy has been used earlier at least implicitly. Stiefel and I independently developed the method of conjugate gradients. At a conference at UCLA we discovered that we had devised this method simultaneously and accordingly presented our results in a joint paper. My original paper was stimulated by a seminar on linear systems in which J. B. Rosser, G. Forsythe, C. Lancjos, L. J. Paige, and others were active participants. In the earlier period we were concerned primarily with solutions of linear systems and hence with optimization of quadratic functions. We gave some thought to nonquadratic functions but did not give them serious consideration. It must be remembered that during this period high speed computers were in their infancy and techniques for using them effectively had not been developed. The handling of linear problems was a major task. We even had difficulties with Gaussian elimination and it was not until later that effective techniques for using Gaussian elimination were developed. As will be noted one of

---

*M. R. Hestenes and E. Stiefel, Methods of conjugate gradients for solving linear systems, Journal of Research of the National Bureau of Standards, Vol. 49 (1952), pp. 409-436.

See also, E. Stiefel, Ueber einige Methoden der Relaxationsrechnung, Z. ange. Math. Physik (3) (1952); M. R. Hestenes, Iterative methods for solving linear equations, NAML Report 52-9, National Bureau of Standards (1951), reproduced in Journal of Optimization Theory and Applications, Vol. 11 (1973), pp. 323-334.

the standard Gaussian elimination methods can be viewed as a conjugate direction method. Conjugate gradient methods complement Gaussian elimination methods. Neither displaces the other.

An important feature of conjugate direction methods for minimizing a quadratic function is that they are easily extensible to minimization of a nonquadratic function. Extensions of this type have been carried out by Davidon, Fletcher, Powell, and many other researchers. An account of these developments has been given by Fletcher.* Recently Wolfe has shown that conjugate gradient techniques can be extended even to the optimization of nondifferentiable functions.

It is my purpose in this paper to describe the geometrical and analytical bases for conjugate direction methods. It is my hope that this will lead to a better understanding if these techniques and perhaps to new versions of these methods. I shall leave the historical account of these methods to others.

2. Fundamental concepts. Minimization of quadratic functions plays a significant role in Optimization Theory. It is the basis of Newton's method for finding a minimum point x_0 of a real valued function f. Newton's algorithm consists of obtaining a new estimate x_2 of x_0 from a previous estimate x_1 of x_0 by minimizing the quadratic part

$$F(x) = f(x_1) + f'(x_1, x - x_1) + (1/2)f''(x_1, x - x_1)$$

of the second order Taylor expansion

$$f(x) = f(x_1) + f'(x_1, x - x_1) + (1/2)f''(x_1, x - x_1) + R(x_1, x - x_1)$$

of f about x_1. Here $f'(x,z)$ and $f''(x,z)$ are the first and second differentials

$$f'(x,z) = f'(x)^*z \quad , \quad f''(x,z) = z^*f''(x)z$$

of f at x. The vector $f'(x)$ is the gradient of f at x and the matrix $f''(x)$ is the Hessian of f at x. All vectors, such as x, z, $f'(x)$, are considered to be column vectors and $f'(x)^*$, z^* are the transposes of $f'(x)$, z. Thus Newton's method consists of successive minimizations of suitably chosen quadratic functions. This leads to the following principle.

An efficient method for minimizing a quadratic function

$$F(x) = (1/2)x^*Ax - h^*x + c \tag{2.1}$$

*Fletcher, R., A review of unconstrained minimization, Optimization, Academic Press, London and New York, 1969.

can be extended so as to obtain an efficient method for minimizing a nonquadratic function f. Of course, a similar principle holds for obtaining maximum points and saddle points of f.

In view of this principle we shall consider in some detail the problem of minimizing a quadratic function F given by Formula (2.1). We assume that the Hessian $F''(x) = A$ of F is positive definite. The gradient

$$F'(x) = Ax - h$$

of F vanishes at the minimum point x_0 of F so that x_0 solves the linear equation

$$Ax = h$$

and is given by the formula $x_0 = A^{-1}h$. A level surface, $F(x) = \text{const.}$, of F is an $(n - 1)$-dimensional ellipsoid having x_0 as its center. The problem of minimizing a quadratic function F is therefore equivalent to that of finding the center of an ellipsoid.

The fundamental property of a quadratic function, upon which our computation algorithms are based, is the following.

The minimum point of F on parallel lines lie on an $(n - 1)$-plane π_{n-1} containing the mimimum point x_0 of F. Analytically, the minimum point x_2 of F on the line $x = x_1 + \alpha p$ through x_1 in direction p lies on the $(n - 1)$-plane

$$\pi_{n-1}: \qquad p^*(Ax - h) = 0$$

having Ap as its normal. The $(n - 1)$-plane π_{n-1} is said to be conjugate to these lines and to the vector p.

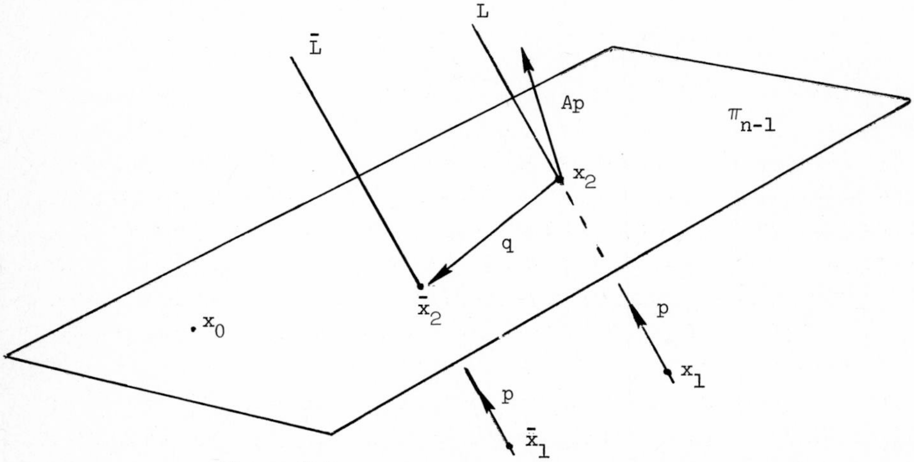

Figure 2.1

The situation is illustrated schematically in Figure 2.1. At the minimum point x_2 of F on the line $x = x_1 + \alpha p$, the gradient $F'(x_2) = Ax_2 - h$ is perpendicular to p. Hence

$$p^*(Ax_2 - h) = 0 \quad ,$$

that is, the point x_2 lies in the $(n - 1)$-plane π_{n-1}. Consequently, π_{n-1} has the alternative representation

$$p^*A(x - x_2) = 0 \quad . \tag{2.2}$$

Inasmuch as $Ax_0 = h$, the minimum point x_0 of F is in π_{n-1}. If $x = \bar{x}_1 + \alpha p$ is a second line in direction p, the minimum point $\bar{x}_2$ of F on this line is also in π_{n-1} so that

$$p^*A(\bar{x}_2 - x_2) = 0 \quad .$$

The vector $q = \bar{x}_2 - x_2$ joins the minimum points x_2 and $\bar{x}_2$ of F on parallel lines with direction p acoordingly satisfies the relation

$$p^*Aq = 0 \quad . \tag{2.3}$$

When this relation holds, the vectors p and q are said to be <u>conjugate</u>.

The term "conjugacy" has its origin in Euclidean geometry. In the two dimensional case the midpoints of parallel chords of an

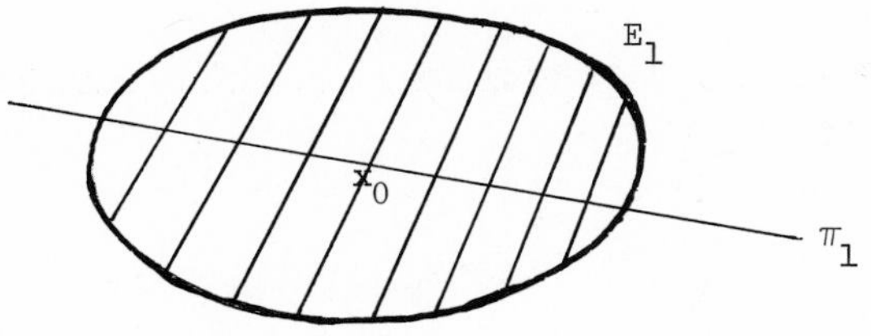

ellipse E_1 lie on a line π_1 conjugate to these chords. The line π_1 passes through the center x_0 of E_1. In the n-dimensional case the midpoints of parallel chords of an $(n - 1)$-dimensional ellipsoid E_{n-1} lie on an $(n - 1)$-plane π_{n-1} conjugate to these chords and passing through the center x_0 of E_{n-1}. Of course, the midpoint of a chord minimizes on this chord the quadratic function F defining E_{n-1}. A study of poles and polars of an ellipsoid yields further insight of the concept of conjugacy.

The basic property of a positive definite quadratic function F on $\mathcal{E}^n$ described above leads us to the following computational procedure. Select a direction vector p_1 in $\pi_n = \mathcal{E}^n$ and minimize F on parallel lines having direction p_1.

The minimum points of F on these lines determine an $(n - 1)$-plane π_{n-1} which contains the minimum point x_0 of F. Henceforth we restrict our space of search to π_{n-1} so that we have diminished the dimension of our space of search by one. We now repeat our process and select a direction p_2 in π_{n-1}. We then minimize F on parallel lines in π_{n-1} having direction p_2. We obtain thereby an $(n - 2)$-plane π_{n-2} in π_{n-1} which contains x_0. Again we have diminished our space of search by one. A further repetition of this procedure on π_{n-2} with a vector p_3 in π_{n-2} yields an $(n - 3)$-plane π_{n-3} containing x_0. Continuing in this manner we locate x_0 in n steps. The vectors $p_1, p_2, \ldots, p_n$ generated by this algorithm are <u>mutually conjugate</u>, that is

(2.4) $$p_j^*Ap_k = 0 \qquad (j \neq k; \quad j,k = 1,\ldots,n) \quad .$$

To carry out this procedure we need not determine the planes $\pi_{n-1}, \pi_{n-2}, \ldots$ explicitly. All we need are the mutually conjugate vector $p_1, p_2, \ldots, p_n$. Starting at a point x_1 we select a direction p_1 and minimize F on the line $x = x_1 + \alpha p_1$ to obtain $x_2 = x_1 + a_1 p_1$. We then select p_2 conjugate to P_1 and find the minimum point $x_3 = x_2 + a_2 p_2$ of F on the line $x = x_2 + \alpha p_2$. Having obtained x_k we select p_k conjugate to $p_1, \ldots, p_{k-1}$ and find the minimum point $x_{k+1} = x_k + a_k p_k$ on the line $x = x_k + \alpha p_k$. The final point

$$x_{n+1} = x_n + a_n p_n = x_1 + a_1 p_1 + \cdots + a_n p_n$$

is the minimum point x_0 of F. Observe that the $(n - k)$-plane referred to above is given by the equations

$$p_j^*(Ax - h) = 0 \qquad (j = 1,\ldots,k)$$

or equivalently by

$$p_j^*A(x - x_{k+1}) = 0 \qquad (j = 1,\ldots,k) \quad .$$

The procedure just described is called the <u>method of conjugate directions</u>.

3. <u>Method of parallel displacements, Elimination</u>. The procedure described in the last section can be implemented as follows. Start with an n-simplex having $x_1, x_{11}, x_{12}, \ldots, x_{1n}$ as vertices, or equivalently, start with an initial point x_1 and n linearly independent vectors $u_1, u_2, \ldots, u_n$ and set

$$x_{1j} = x_1 + u_j \qquad (j = 1,\ldots,n) \quad .$$

Select $p_1 = x_{11} - x_1$ and minimize F on the parallel lines

$$x = x_{1j} + \alpha p_1 \qquad (j = 1,\ldots,n) \quad .$$

This yields n points $x_2 = x_{21}, x_{22}, x_{23}, \ldots, x_{2n}$ which determine an $(n - 1)$-plane π_{n-1} containing the minimum point x_0 of F. Next select $p_2 = x_{22} - x_2$ and minimize F along the parallel lines

$$x = x_{2j} + \alpha p_2 \qquad (j = 2,\ldots,n)$$

in π_{n-1} to obtain $n - 1$ points $x_3 = x_{33}, x_{33}, \ldots, x_{3n}$ which determine an $(n - 2)$-plane π_{n-2} in π_{n-1} containing x_0. Repeating, set $p_3 = x_{33} - x_3$ and minimize F on the parallel lines

$$x = x_{3j} + \alpha p_3 \qquad (j = 3,\ldots,n)$$

in π_{n-2} to obtain points $x_4 = x_{43}, x_{44}, \ldots, x_{4n}$ which define an $(n - 3)$-plane π_{n-3} containing x_0. At the n-th repition we obtain x_0 as the point $x_{n+1} = x_{n+1,n}$. We call this procedure the method of parallel displacements. The case $n = 3$ is illustrated in Figure 3.1.

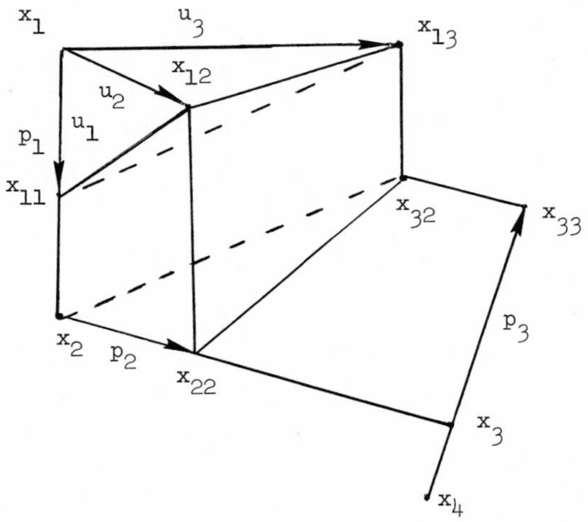

Figure 3.1

If $x_1 = 0$ and $u_1 = (1,0,\ldots,0)$, $u_2 = (0,1,0,\ldots,0),\ldots,$ $u_n = (0,\ldots,0,1)$, then this method is a geometric version of a standard Gaussian elimination method for solving the linear system $Ax = h$. A reordering of vertices in any step is a type of pivoting.

Obviously this procedure can be applied to a nonquadratic function f. Of course in the nonquadratic case a cycle on n steps will not yield the minimum point of f but an improved estimate of the minimum point x_0 of f. Normally one cycle of this type approximates one Newton step in the minimization of f. Repetition of these cycles with smaller initial n-simplices usually leads us to a satisfactory estimate of x_0. In applying this algorithm to a nonquadratic function f we prefer not to minimize f along a line. Instead we take one linear Newton step. A linear Newton step along a line $x = x_1 + \alpha p$ can be found by an application of the formula

$$x_2 = x_1 + ap \quad , \quad a = c/d \quad ,$$

where

$$c = - p^*f'(x_1) \quad , \quad d = p^*f''(x_1)p$$

if first and second derivatives are used,

$$c = - p^*f'(x_1) \quad , \quad d = p^*s \quad , \quad s = \frac{f'(x_1 + bp) - f(x_1)}{b}$$

with a small positive number b when only first derivatives are used,

$$c = \frac{f(x_1 - bp) - f(x_1 + bp)}{2b} \quad , \quad d = \frac{f(x_1 - bp) - 2f(x_1) + f(x_1 + bp)}{b^2}$$

when functional values are used.

4. Method of parallel planes. The computations described in the last section need not be done in parallel. When they are done serially in the proper order we obtain the following algorithm which we call the method of parallel planes. It is based on the following property of a positive definite function F as illustrated schematically in Figure 4.1.

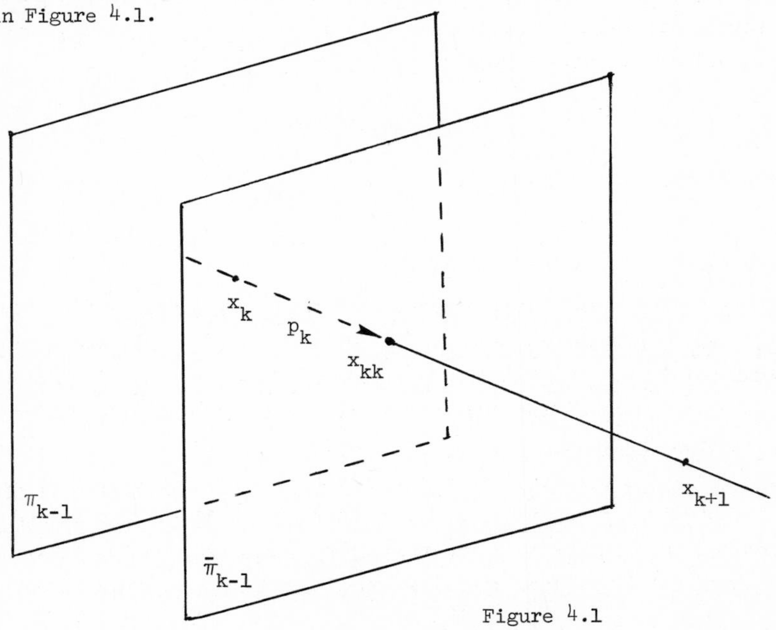

Figure 4.1

Let x_k and x_{kk} be the minimum points of F on two distinct parallel $(k - 1)$-planes π_{k-1} and $\bar{\pi}_{k-1}$. The vector $p_k = x_{kk} - x_k$ is conjugate to these $(k - 1)$-planes. The minimum point x_{k+1} of F on the line $x = x_k + \alpha p_k$ through x_k and x_{kk} affords a minimum to F on the k-plane $\pi_k = \pi_{k-1} + \alpha p_k$ spanning π_{k-1} and $\bar{\pi}_{k-1}$.

This property leads us to the following procedure. Select an initial point x_1 and a line π_1 through x_1 in a direction $p_1 = u_1$ and find the minimum point x_2 of F on π_1. Select a vector u_2 not in π_1 and obtain the minimum point x_{22} of F on the translate $\bar{\pi}_1 = \pi_1 + u_2$ of π_1 by u_2. The line $\bar{\pi}_1$ is parallel to π_1. Next set $p_2 = x_{22} - x_2$ and find the minimum point x_3 of F on the line $x = x_2 + \alpha p_2$ through x_2 and x_{22}. The point x_3 minimizes F on the 2-plane $\pi_2 = \pi_1 + \alpha p_2$ spanning π_1 and $\bar{\pi}_1$. For $k = 3, \ldots, n$ proceed as follows. Having obtained the minimum point x_k of F on the $(k - 1)$-plane π_{k-1} determined previously, select a vector u_k not in π_{k-1} and find the minimum point x_{kk} on the translate $\bar{\pi}_{k-1} = \pi_{k-1} + u_k$ of π_{k-1} by u_k. Clearly $\bar{\pi}_{k-1}$ is parallel to π_{k-1}. Set $p_k = x_{kk} - x_k$ and determine the minimum point $x_{k+1} = x_k + a_k p_k$ of F on the line $x = x_k + \alpha p_k$ through x_k and x_{kk}. The point x_{k+1} minimizes F on the span $\pi_k = \pi_{k-1} + \alpha p_k$ of π_{k-1} and $\bar{\pi}_{k-1}$. The point x_{n+1} obtained in this manner is the minimum point x_0 of F.

Of course in our computational procedure we do not find the k-planes π_k and $\bar{\pi}_k$ explicitly. Our computation proceeds as follows. Select an initial point x_1, an initial direction $p_1 = u_1$, and obtain the minimum point $x_2 = x_1 + a_1 p_1$ of F on the line $x = x_1 + \alpha p_1$. For $k = 2, \ldots, n$ select u_k so that $u_1, u_2, \ldots, u_k$ are linearly independent. Set $x_{1k} = x_1 + u_k$ (or $x_{1k} = x_k + u_k$) and successively obtain the minimum points

$$x_{j+1,k} = x_{jk} + a_{jk} p_j$$

of F on the lines

$$x = x_{jk} + \alpha p_j \qquad (j = 1, \ldots, k) \quad .$$

Put $p_k = x_{kk} - x_k$ and find the minimum point $x_{k+1} = x_k + a_k p_k$ of F on the line $x = x_k + \alpha p_k$ through x_k and x_{kk}. The point x_{n+1} obtained in this manner is the minimum point x_0 of F. When the choice $x_{1k} = x_1 + u_k$ is made we reproduce the points generated by the parallel displacement routine with the same initial point x_1 and translation vectors $u_1, \ldots, u_n$.

The algorithm just described is easily modified so as to be applicable to nonquadratic functions. When this is done we obtain a variant of a method introduced by Fletcher for minimizing a function using functional values only. Again we find it convenient to replace minimization along lines by single linear Newton steps.

5. Conjugate Gram Schmidt processes. The algorithms given in Sections 3 and 4 can be simplified. The points x_{jk} were introduced to obtain the mutually conjugate vectors $p_1, \ldots, p_n$ and are not of particular interest in themselves. By modifying our procedure we can avoid the computation of the vectors x_{jk}. Observe the formulas

$$x_k = x_1 + a_1 p_1 + \cdots + a_{k-1} p_{k-1}$$

$$x_{kk} = x_{1k} + a_{1k} p_1 + \cdots + a_{k-1,k} p_{k-1}$$

for the points x_k and x_{kk} in the parallel planes algorithm. If we select $x_{1k} = x_1 + u_k$ we have

$$p_k = x_{kk} - x_k = u_k + (a_{1k} - a_1)p_1 + \cdots + (a_{k-1,k} - a_{k-1})p_{k-1} \quad .$$

A somewhat simpler formula is obtained when the formula $x_{1k} = x_k + u_k$ is used. In either case we see that the mutually conjugate vectors $p_1, \ldots, p_n$ are obtained from the vectors $u_1, \ldots, u_n$ by the Gram Schmidt process.

(5.1)
$$\begin{aligned} p_1 &= u_1 \\ p_2 &= u_2 - b_{21}p_1 \\ p_3 &= u_3 - b_{31}p_1 - b_{32}p_2 \\ &\text{-----------------------------} \\ p_k &= u_k - b_{k1}p_1 - b_{k2}p_2 - \cdots - b_{k,k-1}p_{k-1} \\ &\text{---} \end{aligned}$$

where the scalars b_{kj} $(j = 1,\ldots,k-1)$ are chosen so that p_k is conjugate to the vectors $p_1, \ldots, p_{k-1}$. We have accordingly the formulas

(5.2)
$$b_{kj} = \frac{u_k{}^*Ap_j}{d_j} \quad , \quad d_j = p_j{}^*Ap_j \qquad (j = 1,\ldots,k-1) \quad ,$$

where A is the Hessian of our quadratic function F. When formulas (5.1) and (5.2) are combined with the relations

(5.3)
$$\begin{aligned} &x_{k+1} = x_k + a_k p_k \\ &a_k = c_k/d_k \quad , \quad c_k = -F'(x_k)^*p_k \quad , \quad d_k = p_k{}^*Ap_k \end{aligned}$$

we obtain a conjugate direction algorithm for minimizing F which we call a

Conjugate Gram Schmidt algorithm or simply a CGS-algorithm.

This algorithm is also extensible to the nonquadratic case. Differences are used when computation of derivatives is to be avoided. When properly carried out n steps of the CGS-algorithm is basically equivalent to a single n-dimensional Newton Step.

It is of interest to note that Formula (5.1) for p_k can be put in the form

$$p_k = C_{k-1}^* u_k \quad ,$$

where the matrices $C_0, C_1, \ldots$ are generated by the recursion formulas

$$C_0 = I \quad , \quad C_j = C_{j-1} - \frac{s_j p_j^*}{d_j} \quad , \quad s_j = Ap_j \quad , \quad d_j = p_j^* s_j \quad .$$

We also have the matrices

$$B_0 = 0 \quad , \quad B_j = B_{j-1} + \frac{p_j p_j^*}{d_j}$$

so that $C_k = I - AB_k$, $B_n = A^{-1}$, and $C_n = 0$. Of particular interest is the case in which the vectors $u_1, \ldots, u_n$ are mutually orthogonal, or more generally $u_k = Hv_k (k = 1,\ldots,n)$, where

$$v_j^* H v_k = 0 \qquad (j \neq k) \quad ,$$

and H is a positive definite matrix. In this event the matrices

$$G_k = C_k^* H C_k + B_k \qquad (k = 0,\ldots,n)$$

are generated by the algorithm

$$p_k = G_{k-1} v_k \ , \ s_k = Ap_k \ , \ q_k = G_{k-1} s_k \ , \ d_k = p_k^* s_k \ , \ \delta_k = q_k^* s_k$$

(5.4)

$$G_0 = H \ , \ G_k = G_{k-1} - \frac{p_k q_k^* + q_k p_k^*}{d_k} + \left(\frac{\delta_k}{d_k} + 1\right) \frac{p_k p_k^*}{d_k} \quad .$$

Clearly $G_n = A^{-1}$. Formulas (5.3) and (5.4) give a matrix version of the CGS-algorithm. When $v_k = - F'(x_k)$ we obtain a variant of the Davidon-Fletcher-Powell algorithm. Again this technique can be extended to nonquadratic functions.

If in the CGS-algorithm (5.1) and (5.2), we select $u_k = F'(x_k)$ $(k = 1,2,3,\ldots)$ we obtain an algorithm, called the CGS-cg-algorithm, which is mathematically equivalent to the conjugate gradient algorithm (cg-algorithm) described in the next section. It is somewhat more complicated than the cg-algorithm in that the coefficients b_{kj} $(j < k - 1)$ are computed whereas in the cg-algorithm these coefficients are presumed to be zero, as would be the case if no roundoff errors occur. The CGS-cg-routin accordingly can be viewed as a devise for correcting roundoff errors in a cg-routine.

6. Conjugate Gradient routines, cg-routines. The conjugate gradient method is the simplest conjugate direction routine in the sense that it involves the computation of the fewest number of inner products. The conjugate gradient routine, cg-routine, for minimizing a positive definite quadratic function

$$F(x) = (1/2)x^*Ax - h^*x + c$$

proceeds as follows.

After selecting an initial point x_1, we compute the steepest descent vector $p_1 = -F'(x_1)$ of F at x_1 and obtain the minimum point x_2 of F on the line L_1 through x_1 in direction p_1. As noted in Section 2 the minimum point x_0 of F is in the $(n-1)$-plane π_{n-1} through x_2 conjugate to p_1 so that we can restrict our search for x_0 to the $(n-1)$-plane π_{n-1}. We repeat the process in π_{n-1}. We select a steepest descent vector p_2 of F at x_2 in π_{n-1} and find the minimum point x_3 of F on the line L_2 through x_2 in direction p_2. Now the minimum point x_0 of F is in the $(n-2)$-plane π_{n-2} in π_{n-1} through x_3 conjugate to p_2, thereby diminishing our space of search by one dimension again. Proceeding in this manner, in the k-th step we select a steepest descent vector p_k of F at x_k in the $(n-k+1)$-plane π_{n-k+1} through x_k conjugate to $p_1, \ldots, p_{k-1}$ and determine the minimum point x_{k+1} of F on the line L_k through x_k in direction p_k. After $m \leq n$ steps we obtain a point x_{m+1} which is the minimum point x_0 of F.

This description of the conjugate gradient algorithm (cg-algorithm) is somewhat involved. However, it does exhibit the reason for the terminology "conjugate gradients" for the vectors $p_1, p_2, \ldots$. Fortunately, we do not need to determine the planes $\pi_{n-1}, \pi_{n-2}, \ldots$ explicitly. All that is needed are the formulas

$$(6.1) \qquad p_k = r_k + b_{k-1}p_{k-1}, \quad r_k = -F'(x_k), \quad b_{k-1} = \frac{|r_k|^2}{|r_{k-1}|^2}$$

for the conjugate gradient p_k $(k > 1)$ of F at x_k in the $(n-k+1)$-plane π_{n-k+1} conjugate to $p_1, \ldots, p_{k-1}$. As noted by Stiefel, Formula (6.1) for p_k can also be obtained by a relaxation technique. Observe that the 2-plane $x = x_k + \alpha r_k + \beta p_{k-1}$ cuts the ellipsoid $F(x) = F(x_k)$ in an ellipse E_2, as indicated in Figure 6.1. The standard gradient technique would require us to move from x_k in the steepest descent direction r_k. Instead we relax this condition and move in the direction p_k to the center x_{k+1} of the ellipse E_2. Observe that the vector p_k joins x_k to the midpoint of the chord $z'z''$ of E_2 whose direction is p_{k-1}.

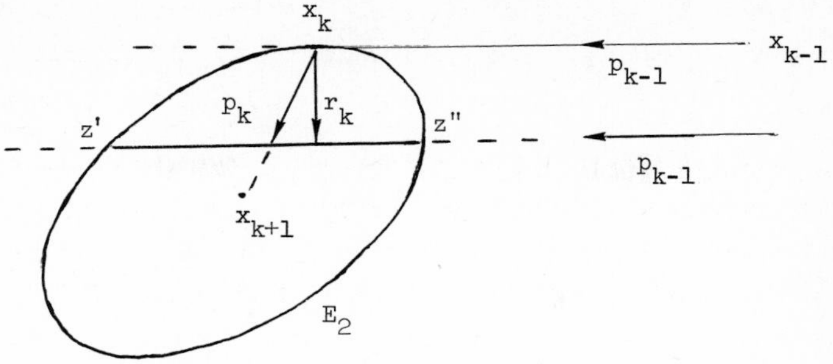

Figure 6.1

Perhaps the simplest description of the conjugate gradient algorithm (cg-algorithm) is the following. Initially select a point x_1 and set $p_1 = r_1 = -F'(x_1)$. For $k = 1, 2, \ldots$ determine x_{k+1} and p_{k+1} from x_k and p_k by the rules:

(i) Select the minimizer $\alpha = a_k$ of $\phi_k(\alpha) = F(x_k + \alpha p_k)$ and obtain $x_{k+1} = x_k + a_k p_k$;

(ii) Compute p_{k+1} by the formulas

$$p_{k+1} = r_{k+1} + b_k p_k \quad , \quad r_{k+1} = -F'(x_{k+1}) \quad , \quad b_k = \frac{|r_{k+1}|^2}{|r_k|^2} \quad .$$

This algorithm terminates in $m \leq n$ steps if no roundoff errors occur. When high precision arithmetic is used, the point x_{n+1} is normally a satisfactory estimate of the minimum point x_0 of F.

This cg-algorithm is applicable without change to a nonquadratic function if we restart the algorithm after n or $n + 1$ steps. The nonquadratic routine is known as the Fletcher-Reeves algorithm. In this algorithm we can replace the minimization of $\phi_k(\alpha)$ for obtaining a_k by the one-step linear Newton formula $a_k = \phi_k'(0)/\phi_k''(0)$. Again we can approximate $\phi_k'(0)$ and $\phi_k''(0)$ by difference formulas.

When the cg-algorithm is expressed algebraically without reference to minimization, it is defined by the relations.

$$x_1 \text{ arbitrary} \quad , \quad r_1 = -F'(x_1) = h - Ax_1 \quad , \quad p_1 = r_1$$

$$s_k = Ap_k \quad , \quad d_k = p_k{}^* s_k \quad , \quad c_k = p_k{}^* r_k \quad , \quad a_k = c_k/d_k \tag{6.2}$$

$$x_{k+1} = x_k + a_k p_k \quad , \quad r_{k+1} = r_k - a_k s_k$$

$$p_{k+1} = r_{k+1} + b_k p_k \quad , \quad b_k = - \frac{s_k{}^* r_{k+1}}{d_k} \quad \text{or} \quad b_k = \frac{|r_{k+1}|^2}{c_k}$$

The scalars a_k, b_k, c_k, d_k are positive and the formulas for these quantities are unaltered if we scale p_k, that is, if we replace p_k by $\rho_k p_k$, where ρ_k is a positive scale factor. If we do not scale p_k we can use the alternative formula $c_k = |r_k|^2$ for c_k. When a scale factor ρ_k for p_k is introduced we have $c_k = \rho_k |r_k|^2$.

The residuals $r_1, r_2, \ldots$ generated by (6.2) are mutually orthogonal and the vectors $p_1, p_2, \ldots$ are mutually conjugate. The residual r_{k+1} is orthogonal to the vectors $p_1, \ldots, p_k$, signifying that the point x_{k+1} minimizes F on the k-plnne given parametrically by $x = x_1 + \alpha_1 p_1 + \cdots + \alpha_k p_k$.

The formula

$$p_{k+1} = r_{k+1} + b_k p_k$$

has various interpretations. We have seen already that it is a steepest descent vector for F at x_{k+1} on the $(n - k)$-plane π_{n-k} conjugate to $p_1, \ldots, p_k$. It was also obtained by a relaxation of r_{k+1}, as explained above. This property can be restated as follows. Translate the line $L_k : x = x_{k+1} + \alpha p_k$ by r_{k+1} to obtain a parallel line $\bar{L}_k : x = x_{k+1} + r_{k+1} + \alpha p_k$. Let $\bar{x}_{k+1}$ be the minimum point of F on $\bar{L}_k$. Then p_{k+1} is given by the formula

$$p_{k+1} = \bar{x}_{k+1} - x_{k+1} = r_{k+1} + b_k p_k \quad .$$

A further interpretation is obtained as follows. For $k = 1, 2, \ldots$ rescale p_{k+1} so that

$$p_{k+1} = \frac{r_{k+1} + b_k p_k}{1 + b_k}$$

is the weighted average of r_{k+1} and p_k of shortest length. Then p_{k+1} is the shortest vector p of the form

$$p = \alpha r_{k+1} + \beta p_k \quad , \quad \alpha + \beta = 1 \quad .$$

In fact p_{k+1} is the shortest vector p of the form

$$p = \beta_1 r_1 + \cdots + \beta_{k+1} r_{k+1} \quad , \quad \beta_1 + \cdots + \beta_{k+1} = 1 \quad .$$

Equivalently, $p_{k+1} = -F'(\hat{x}_{k+1})$, where $\hat{x}_{k+1}$ minimizes

$$\hat{F}(x) = (1/2)|F'(x)|^2$$

on the k-plane through the points $x_1, \ldots, x_{k+1}$. This property was used by P. Wolfe to obtain a method of conjugate subgradients for minimizing convex functions which may be nondifferentiable. Wolfe invented his procedure before he became aware of its connection with conjugate gradients. The relations between F and $\hat{F}$ described above lead us to the following version of the conjugate gradient algorithm.

Choose an initial point x_1 and set $\hat{x}_1 = x_1$. For $k = 1, 2, \ldots$ select $p_k = -F'(\hat{x}_k)$ and obtain the minimum point x_{k+1} of F on the line $x = x_k + \alpha p_k$. Then obtain the minimum point $\hat{x}_{k+1}$ of F on the line joining $\hat{x}_k$ to x_{k+1}. The algorithm terminates when $\hat{x}_{m+1} = x_{m+1}$. This version of the cg-algorithm is not as efficient as the original algorithm.

A variant of cg-algorithm (6.2), in which the residuals $r_1, r_2, \ldots$ are not computed explicitly, is given by the relations.

$$x_1 \text{ arbitrary}, \quad p_1 = -F'(x_1), \quad q_1 = Ap_1, \quad d_1 = p_1{}^*Ap_1$$

$$x_{k+1} = x_k + a_k p_k, \quad a_k = c_k/d_k, \quad c_k = |p_k|^2$$

(6.3) $$p_{k+1} = p_k - \hat{a}_k q_k, \quad \hat{a}_k = d_k/\delta_k, \quad \delta_k = |q_k|^2 \text{ or } \delta_k = p_k{}^*Aq_k$$

$$q_{k+1} = Ap_{k+1} + \hat{b}_k q_k, \quad d_{k+1} = p_{k+1}{}^*Ap_{k+1}$$

$$\hat{b}_k = d_{k+1}/d_k \text{ or } \hat{b}_k = -q_k{}^*Ap_{k+1}/\delta_k .$$

The vectors $p_1, p_2, \ldots$ are mutually conjugate and the vectors $q_1, q_2, \ldots$ are mutually orthogonal. The vector p_{k+1} is orthogonal to the vectors $q_1, q_2, \ldots$.

The residuals $r_1, r_2, \ldots$ generated by Algorithm (6.2) satisfy the relations

$$r_2 = r_1 - a_1 Ar_1$$

$$r_{k+1} = (1 + \tilde{b}_{k-1})r_k - a_k Ar_k - \tilde{b}_{k-1}r_{k-1}, \tilde{b}_{k-1} = \frac{a_k b_{k-1}}{a_{k-1}} .$$

The corresponding relations for $x_1, x_2, \ldots$ are

$$x_2 = x_1 + a_1 r_1$$

$$x_{k+1} = (1 + \tilde{b}_{k-1})x_k + a_k r_k - \tilde{b}_{k-1}x_{k-1} \quad .$$

By setting

$$\alpha_1 = a_1 \quad , \quad \alpha_k = \frac{a_k}{1 + \tilde{b}_{k-1}} \quad , \quad \beta_{k-1} = \frac{\tilde{b}_{k-1}}{1 + \tilde{b}_{k-1}} \quad ,$$

these equations yield the algorithm

$$x_1 \text{ arbitrary, } r_1 = - F'(x_1)$$

$$x_2 = x_1 + \alpha_1 r_1 \quad , \quad r_2 = r_1 - \alpha_1 A r_1$$

(6.4) $$\tilde{x}_{k+1} = x_k + \alpha_k r_k \quad , \quad \tilde{r}_{k+1} = r_k - \alpha_k A r_k$$

$$x_{k+1} = \frac{\tilde{x}_{k+1} - \beta_{k-1}x_{k-1}}{1 - \beta_{k-1}} \quad , \quad r_{k+1} = \frac{\tilde{r}_{k+1} - \beta_{k-1}r_{k-1}}{1 - \beta_{k-1}}$$

$$\alpha_k = \frac{|r_k|^2}{r_k^* A r_k} \quad , \quad \beta_{k-1} = \frac{r_{k-1}^* r_{k+1}}{|r_{k-1}|^2} \quad .$$

This version of the cg-algorithm is known as <u>Gradient-Partan</u>. At the k-th step $(k > 1)$ we first minimize F in the direction r_k of steepest descent to obtain $\tilde{x}_{k+1}$ and obtain the minimum point x_{k+1} of F on the line joining x_{k-1} to $\tilde{x}_{k+1}$.

In a cg-algorithm we can replace the gradient $F'(x)$ by the generalized gradient $HF'(x_k)$, where H is a positive definite symmetric matrix. This is equivalent to performing a linear transformation $y = U^{-1}x$, where $UU^* = H$, and applying the cg-algorithm to the transformed function $G(y) = F(Uy)$.

As noted by Davidon, Fletcher, Powell, and others, the cg-algorithm can be put in matrix form. Starting with a positive definite symmetric matrix H we have the following algorithm.

$$x_1 \text{ arbitrary, } H_0 = H$$

$$r_k = - F'(x_k) \quad , \quad p_k = H_{k-1} r_k$$

$$x_{k+1} = x_k + a_k p_k \quad \text{minimizes } F \text{ on } x = x_k + \alpha p_k$$

$$(6.5)\qquad s_k = \frac{F'(x_{k+1}) - F'(x_k)}{a_k} \;,\quad q_k = H_{k-1}s_k$$

$$d_k = p_k^*s_k \;,\quad \delta_k = q_k^*s_k \;,\quad \alpha_k = d_k/\delta_k$$

$$H_i = H_{k-1} - \frac{q_kq_k^*}{\delta_k} + \frac{p_kp_k^*}{d_k} + (\alpha_k - \omega_k)\frac{\bar{p}_{k+1}\bar{p}_{k+1}^*}{d_k}$$

$$\bar{p}_{k+1} = \frac{p_k - \alpha_kq_k}{\alpha_k}$$

and ω_k is restricted so that H_k is positive definite.

If in (6.5) we replace the minimization of F by a linear Newton step we compute x_{k+1} and x_k by the formulas

$$s_k = \frac{F'(x_k + \sigma_kp_k) - F'(x_k)}{\sigma_k} \;,\quad d_k = p_k^*s_k$$

$$(6.6)$$

$$x_{k+1} = x_k + a_kp_k \;,\quad a_k = c_k/d_k \;,\quad c_k = p_k^*r_k$$

where σ_k is a small positive number. Variations in ω_k alters the length of p_{k+1} but not its direction. The direction of q_{k+1}, however, is changed. The choices $\omega_k = \alpha_k$ and $\omega_k = 0$ yield two of the standard routines of the form (6.5).

Algorithm (6.5) or algorithm (6.5) modified by (6.6) is immediately applicable to the nonquadratic case. It is not necessary to restart after n steps. Again n steps approximates one n-dimensional Newton step.

7. A planar cg-algorithm. In the preceding pages we assumed that the Hessian A of the quadratic function

$$F(x) = (1/2)x^*Ax - h^*x + c$$

is positive definite. This assumption guarantees that the divisor

$$d_k = p_k^*Ap_k$$

in cg-algorithm (6.2) is not zero unless $p_k = 0$. If A is indefinite the cg-algorithm can be applied without change if $d_k \neq 0$ at each step. However, occasionally we encounter the situation in which $d_k = 0$ and $p_k \neq 0$, thereby terminating the algorithm prematurely. Of course, when d_k is small, considerable roundoff errors are introduced. Inasmuch as saddle points play a significant role in applications, it is desirable to obtain a modification of the cg-algorithm which is applicable even when A is indefinite. This is accomplished by obtaining critical points of F on 2-planes as well as on lines in an appropriate manner. In particular if we choose a linear combination $\bar{p}_{k-1}$ of $p_1, \ldots, p_{k-1}$ so that the vector $q_k = Ap_k + \bar{p}_{k-1}$ is conjugate to $p_1, \ldots, p_{k-1}$, then we obtain the critical point of F on the 2-plane

$$x = x_k + \alpha p_k + \beta q_k$$

whenever p_k and q_k are almost orthogonal. Otherwise we find the critical point of F on the line $x = x_k + \alpha p_k$. A specific rule for carrying out these computations is the following. Select an initial point x_1 and obtain

$$r_1 = -F'(x_1) \quad , \quad p_1 = r_1 \quad , \quad q_1 = Ap_1 \quad .$$

Having found x_k, r_k, p_k, q_k compute

$$d_k = p_k{}^*Ap_k \quad , \quad \delta_k = p_k{}^*Aq_k = |Ap_k|^2 \quad , \quad e_k = q_k{}^*Aq_k \quad ,$$

$$\Delta_k = d_k e_k - \delta_k^2 \quad , \quad c_k = p_k{}^*r_k \quad .$$

If $|\Delta_k| < \delta_k^2/2$ find $x_{k+1}, r_{k+1}, p_{k+1}, q_{k+1}$ by the formulas

$$x_{k+1} = x_k + a_k p_k \quad , \quad r_{k+1} = r_k - a_k Ap_k \quad , \quad a_k = c_k/d_k$$

$$p_{k+1} = r_{k+1} + b_k p_k \quad , \quad b_k = |r_{k+1}|^2/c_k$$

$$q_{k+1} = Ap_{k+1} + \beta_k p_k \quad , \quad \beta_k = -q_k{}^*Ap_{k+1}/d_k$$

Otherwise we obtain $x_{k+2}, r_{k+2}, p_{k+2}, q_{k+2}$ by the formulas

$$a_k = \frac{c_k e_k - d_k \delta_k}{\Delta_k} \quad , \quad a_{k+1} = \frac{d_k^2 - c_k \delta_k}{\Delta_k}$$

$$x_{k+2} = x_k + a_k p_k + a_{k+1} q_k \quad , \quad r_{k+2} = r_k - a_k Ap_k - a_{k+1} Aq_k$$

$$\bar{p}_{k+1} = p_k - \alpha_k q_k \quad , \quad \alpha_k = d_k/\delta_k$$

$$p_{k+2} = r_{k+2} + b_{k+1}\bar{p}_{k+1} \quad , \quad b_{k+1} = \frac{\delta_k q_k{}^*Ar_{k+2}}{\Delta_k}$$

$$q_{k+2} = Ap_{k+2} + \beta_{k+1}\bar{p}_{k+1} \quad , \quad \beta_{k+1} = \frac{\delta_k q_k{}^* A^2 p_{k+2}}{\Delta_k} \quad .$$

As before alternative formulas can be given for the scalars in this algorithm. This algorithm can also be put in matrix form giving us an algorithm of the Davidon-Fletcher-Powell type. The algorithm can be extended so as to be applicable to non-quadratic functions in the usual manner.

The planar cg-algorithm can be used to solve the constrained minimum problem

$$f(x) = \min. \quad , \quad g(x) = 0 \quad ,$$

where $g(x)$ is a m-vector whose Jacobian matrix $g'(x)$ has rank m. The solution x_0 of this problem is given by the critical point (x_0,y_0) of the Lagrangian

$$L(x,y) = f(x) + g(x)^*y$$

with y_0 as the corresponding Lagrange multiplier. The Hessian of L is indefinite.

8. Clustered eigenvalues. The conjugate gradient routine for minimizing a quadratic function

$$F(x) = (1/2)x^*Ax - h^*x + c$$

has the desirable property that, if the eigenvalues of A are clustered about N values $\mu_1, \ldots, \mu_N$, then N steps of the cg-algorithm will yield a good estimate x_{N+1} of the minimum point x_0 of F. This follows because if we set

$$\bar{x} = x_1 + P(A)r_1 \quad , \quad r_1 = - F'(x_1) \quad ,$$

where $P(\lambda)$ is the polynomial defined by the formulas

$$R(\lambda) = \frac{1 - R(\lambda)}{\lambda}$$

$$R(\lambda) = (1 - \lambda/\mu_1)(1 - \lambda/\mu_2) \ldots (1 - \lambda/\mu_N) \quad ,$$

then

$$\bar{r} = - F'(\bar{x}) = R(A)r_1 \quad , \quad F(\bar{x}) - F(x_0) = (1/2)r_1{}^*R(A)^2A^{-1}r_1 \quad .$$

Consequently

$$\frac{|r|}{|r_1|} \leq \|R(A)\| \quad , \quad \frac{F(\bar{x}) - F(x_0)}{F(x_1) - F(x_0)} \leq \|R(A)\|^2 \quad .$$

Inasmuch as

$$\|R(A)\| = \max[|R(\lambda_1)|,\dots,|R(\lambda_n)|] \quad ,$$

where $\lambda_1, \dots, \lambda_n$ are the eigenvalues of A, it follows that, because these eigenvalues are clustered about $\mu_1, \dots, \mu_N$, the values $R(\lambda_1), \dots, R(\lambda_n)$ are very small. In view of (8.1), $\bar{x}$ is a good estimate of x_0. Because $\bar{x}$ is on the N-plane

$$x = x_1 + \alpha_1 p_1 + \dots + \alpha_N p_N = x_1 + \beta_1 r_1 + \dots + \beta_N A^{N-1} r_1$$

on which F is minimized by x_{N+1}, we have $F(\bar{x}) \geq F(x_{N+1})$ so that X_{N+1} is also a good estimate of x_0.

We illustrate this result by an example. Let H_k be the Hilbert Matrix

$$H_k = \left(\frac{1}{i + j - 1}\right) \quad (i,j = 1,\dots,k) \quad .$$

For $n = 16$ let

$$A_1 = I + H_{16} \quad , \quad A_2 = I + \begin{pmatrix} H_8 & 0 \\ 0 & H_8 \end{pmatrix} \quad .$$

These matrices a very well conditioned. The matrix A_1 has 16 distinct eigenvalues, most of which are close to $\mu = 1$. The Matrix A_2 has 8 distinct double eigenvalues, three of which are close to $\mu = 1$. We select $h = Ax_0$ so that x_0 is a known minimum point of F. Choosing $x_0 = (1,1,\dots,1)^*$ we apply the cg-algorithm to F with $x_1 = 0$ and observe the values of

$$\rho_k = \frac{|r_{k+1}|}{|r_1|} \quad , \quad \delta_k = \frac{|x_{k+1} - x_0|}{|x_1 - x_0|} \quad , \quad \varepsilon_k = \frac{F(x_{k+1}) - F(x_0)}{F(x_1) - F(x_0)}$$

first with $A = A_1$ and then with $A = A_2$. A repetition with $x_0 = (1,2,\dots,16)^*$ does not alter these ratios significantly. The orders of magnitude of these ratios are given in the following table.

Step	A_1			A_2		
k	ρ_k	δ_k	ε_k	ρ_k	δ_k	ε_k
4	10^{-6}	10^{-6}	10^{-12}	10^{-7}	10^{-7}	10^{-14}
5	10^{-10}	10^{-10}	10^{-20}	10^{-12}	10^{-12}	10^{-25}
6	10^{-15}	10^{-15}	10^{-27}	10^{-19}	10^{-19}	10^{-27}

Observe that the point x_5 obtained at the end of 4 steps of the cg-algorithm is already a good estimate of the minimum point x_0 of F. The CGS-algorithm (elimination) required the full 16 steps to obtain even a reasonable estimate of the solution.

REMARKS ON THE RELATIONSHIPS BETWEEN FREE SURFACES AND OPTIMAL CONTROL OF DISTRIBUTED SYSTEMS

Jacques-Louis LIONS
IRIA-LABORIA
Domaine de Voluceau
B.P. 105
F-78150 Le Chesnay, France

INTRODUCTION

Boundary value problems are called problems of *free surfaces* when *one of the unknowns* (and generally the most important one) is a *surface*.

Such problems arise in very many questions of mathematical physics : jets, cavities, evaporation, elasto-plasticity, etc..

In this survey paper, we want to briefly indicate how problems of this type are *closely connected with problems of optimal control*.

The relationship between these questions appear for the following reasons :

(i) in problems of *optimization with constraints* of functionals in a space of functions, the *interfaces* between regions where constraints are *"saturated"* or *"not saturated"* are free surfaces ;

(ii) in problems of *optimal stopping times*, the surfaces *where the process has to be stopped* (in order to achieve optimality) are free surfaces ;

(iii) optimization of functionals depending on *functions* and on *geometrical arguments* leads to free surface problems ; one of the first results along these lines is classical in hydrodynamics and is due to Riabouchinsky [1] ;

(iv) problems of optimization *with respect to domains (optimum design)* lead to problems of free surface and reciprocally, *one can transform problems of free surfaces into problems of optimum design.*

In what follows, we shall give some indications on these questions, together with the *numerical consequences* of these remarks.

The plan is as follows :

1. <u>Optimization with constraints</u>. <u>Variational Inequalities</u>.
   1.1. The obstacle problem.
   1.2. Interpretation of the V.I.
   1.3. Extensions and Numerical Approximation.
   1.4. Transformation of problems of free surfaces into V.I.
2. <u>Optimal stopping time</u>.
   2.1. Minimum cost function.
   2.2. Characterization of the minimum cost function.
3. <u>Calculus of variations with respect to test functions and to free surfaces</u>.
   3.1. An Example.
   3.2. A problem in the Calculus of Variations.
4. <u>Optimum Design and Free Boundary Problems</u>.
   4.1. Setting of the problem.
   4.2. Problem of Optimum Design.
   4.3. Numerical Method.
5. <u>Concluding Remarks</u>.

1. OPTIMIZATION WITH CONSTRAINTS. VARIATIONAL INEQUALITIES.

1.1. The obstacle problem.

Let $\mathcal{O}$ be an open set in $\mathbb{R}^n$ bounded, with (smooth) boundary Γ.
We denote by $H^1(\mathcal{O})$ the Sobolev's space

(1.1) $$H^1(\mathcal{O}) = \left\{v \mid v, \frac{\partial v}{\partial x_1}, \ldots, \frac{\partial v}{\partial x_n} \in L^2(\mathcal{O})\right\}$$

provided with the norm given by

$$\left(\int_{\mathcal{O}}\left[|v|^2 + |\mathrm{grad}\, v|^2\right]dx\right)^{\frac{1}{2}} = \|v\|$$

for which $H^1(\mathcal{O})$ is a Hilbert space.
We define ([1])

(1.2) $$H_0^1(\mathcal{O}) = \{v \mid v \in H^1(\mathcal{O}),\quad v=0 \text{ on } \Gamma\} ;$$

$H_0^1(\mathcal{O})$ is a closed subspace of $H^1(\mathcal{O})$.
We consider a function ψ (the "obstacle") which is, say, *continuous* in $\bar{\mathcal{O}} = \mathcal{O} \cup \Gamma$ and which is such that

(1.3) $$\psi \geq 0 \quad \text{on } \Gamma.$$

We then introduce

(1.4) $$K = \{v \mid v \in H_0^1(\mathcal{O}),\quad v \leq \psi \quad \text{a.e. in } \mathcal{O}\}$$

(condition (1.3) insures the compatibility between the condition $v \leq \psi$ and $v = 0$ on Γ).
For $u, v \in H_0^1(\mathcal{O})$ we set

(1.5) $$a(u,v) = \int_{\mathcal{O}} \mathrm{grad}\, u . \mathrm{grad}\, v \, dx$$

and

(1.6) $$J(v) = \tfrac{1}{2} a(v,v) - (f,v),$$

where $$(f,v) = \int_{\mathcal{O}} fv \, dx, \quad f \in L^2(\mathcal{O}).$$

([1]) We refer to S.L. Sobolev [1], J.L.Lions and E. Magenes [1], J. Necas [1] and to the bibliographies therein for technical details.

We consider the problem of *optimization with constraints*

(1.7) $$\inf J(v), \quad v \in K.$$

It is a simple matter to verify that problem (1.7) admits a *unique solution* u *which is characterized by*

(1.8) $$\begin{cases} u \in K, \\ a(u,v-u) \geq (f,v-u) \quad \forall v \in K \quad ; \end{cases}$$

(1.8) is what is called a *Variational Inequality* (V.I. in short) (cf. Lions-Stampacchia [1] for the case - useful in application - where a(u,v) is *not* symmetric in u and v).

We give now an interpretation of (1.8) which explains the terminology : "obstacle problem", and which shows the character of free surface problem of (1.8).

1.2. Interpretation of the V.I.

It can be proven (cf. Brézis-Stampacchia [1], Brézis [1]) that - assuming ψ smooth enough - the solution u of (1.8) satisfies

(1.9) $$\frac{\partial^2 u}{\partial x_i \partial x_j} \in L^2(\mathcal{O}) \quad \forall i, j \ (^1) \quad .$$

Then (1.8) is equivalent to

(1.10) $$\int_{\mathcal{O}} (-\Delta u-f)(v-u)dx \geq 0 \quad \forall v \in K, \quad u \in K$$

and one easily verifies that (1.10) is in turn equivalent to

(1.11) $$\begin{cases} u-\psi \leq 0, \quad -\Delta u - f \leq 0, \\ (u-\psi)(-\Delta u-f) = 0 \quad \text{in } \mathcal{O} \end{cases}$$

with the boundary condition

(1.12) $$u = 0 \quad \text{on } \Gamma.$$

One can think of u as describing the equilibrium of a membrane (in case n = 2) subject *to stay below the obstacle* represented by ψ.

According to (1.11) there are *two regions in* $\mathcal{O}$; in one region (schematically represented by the shaded part in Fig. 1) one has $u = \psi$; it is the *contact region* ; in

(1) In short $u \in H^2(\mathcal{O})$.

the other part of $\mathcal{O}$ one has the usual equilibrium equation : $-\Delta u = f$.
Since $u = \psi$ on the contact region $\mathcal{C}$ and since $u \in H^2(\mathcal{O})$, one has (at least if $\mathcal{C}$ is "not too small")

$$u = \psi \text{ and } \frac{\partial u}{\partial \nu} = \frac{\partial \psi}{\partial \nu} \text{ on the interface } S$$

(1.13) between the *contact region* and the *equilibrium region.*

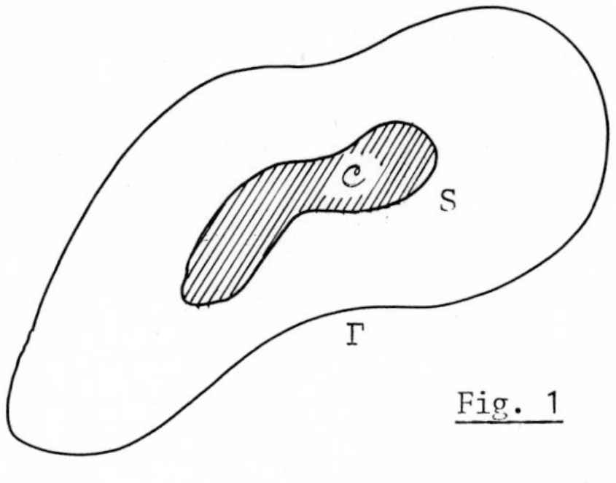

Fig. 1

(in (1.13) ν denotes the normal to S - assuming it exists).
One can think of (1.13) as a *free boundary condition* where u and S are both unknown but they are subject to *two* conditions ($u = \psi$ and $\frac{\partial u}{\partial \nu} = \frac{\partial \psi}{\partial \nu}$).

1.3. Extensions and Numerical Approximation.

Many other V.I. arise in applications in Mechanics and in Physics (cf. G. Duvaut and J.L. Lions [1]). Systematic and simple tools are now available for *the numerical solution of V.I.* We refer to the books Glowinski, Lions Trémolières [1] and to the bibliography therein.

1.4. Transformation of problems of free surfaces into V.I.

Starting with a problem of free surface arising in Hydrodynamics (seepage in a earth dam), C. Baiocchi [1] has observed that one can - by a (non trivial) change of unknown function - *transform some problems of free surface into problems of V.I.* (1).
We refer to the exposition given by E. Magenes [1] and to the bibliography therein - without pursuing this matter here. Cf. also J.L. Lions [1].

2. OPTIMAL STOPPING TIME

2.1. Minimum cost function.

Let us consider in $\mathbb{R}^n$ a standard Wiener process $w(s)$ and let us consider the (trivial) Ito's differential equation

$$dy(s) = \sqrt{2}\, dw(s), \quad s > 0 \text{ }^{(2)}, \quad y(o) = x \quad (x \in \mathcal{O}). \tag{2.1}$$

(1) Or into *Quasi Variational Inequalities* (Q.V.I.), a tool introduced in A.Bensoussan and J.L. Lions [1] for the study of problems of *Impulse Control.*

(2) The factor $\sqrt{2}$ is introduced just to find exactly the same problem than the one considered in Section 1.2.

We denote by $y_x(s)$ the solution of (2.1) and by τ_x the exit time from $\mathcal{O}$ of $y_x(s)$.
Let f and ψ be given functions in $\mathcal{O}$.
For any stopping time $\theta \leq \tau_x$, we define the *cost function*

(2.2) $$J_x(\theta) = E\left[\int_0^\theta f(y_x(s))ds + \psi(y_x(\theta))\,\chi_{\theta<\tau}\right]$$

(where $\chi_{\theta<\tau} = 1$ if $\theta<\tau$, 0 otherwise).

The problem of *optimal stopping time* consists in finding

(2.3) $$u(x) = \inf J_x(\theta) \quad , \quad \theta \leq \tau_x$$

and the θ (which exists) that minimizes (2.3).

2.2. Characterization of the minimum cost function.

One can prove (cf. A. Bensoussan and J.L. Lions [2] [3]) that $u(x)$ is *characterized* by

(2.4) $$u - \psi \leq 0, \quad -\Delta u - f \leq 0 \;, \quad (u-\psi)(-\Delta u - f) = 0 \quad \text{in } \mathcal{O} \;,$$

(2.5) $$u = 0 \quad \text{on } \Gamma,$$

i.e. the *obstacle problem* considered in Section 1.2. But here the interpretation of the free surface S (with notations of Section 1.2) is entirely different :

> the equilibrium region is the *continuation set*, i.e. the set where one does not stop the process before reaching the *stopping set* (the contact region) where $u = \psi$.

We refer to the book [3] of A. Bensoussan and J.L. Lions for the proof and for the study of much more general situations (including non stationary problems).

3. CALCULUS OF VARIATIONS WITH RESPECT TO TEST FUNCTIONS AND TO SURFACES.

3.1. An Example.

Let us consider the following classical problem in Hydrodynamics.
Let ψ be the stream function of an ideal flow with free surface S under gravity (Fig. 2). Let $\mathcal{O}$ be the (shaded) domain occupied by the flow ; the function $\psi = \psi(x_1,x_2)$ satisfies :

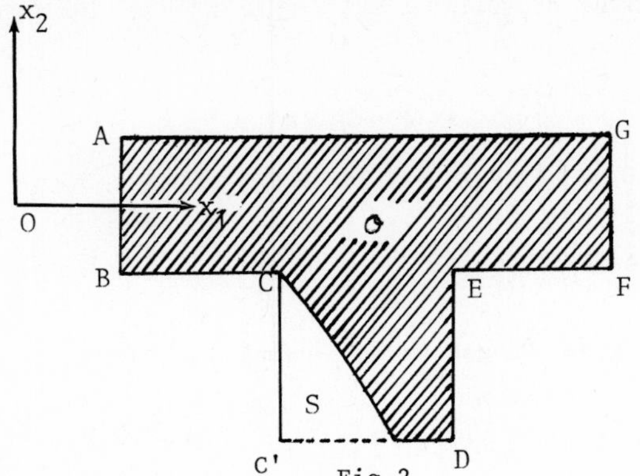

Fig.2

(3.1) $$\Delta\psi = 0 \text{ in } \mathcal{O}.$$

The *free surface* is denoted by S. On S one has

(3.2) $$\begin{cases} \psi = 0 \\ \frac{1}{2}|\text{grad}\psi|^2 + gx_2 = 0 \end{cases}$$

and on the other part of the boundary one has

(3.3) $$\psi = g \quad \text{on AG, DEF}$$

and

(3.4) $$\frac{\partial\psi}{\partial\nu} = 0 \text{ on the remaining part of the boundary of } \mathcal{O}.$$

Remark 3.1.

The view point is here different from the one in Sections 1 and 2 ; we *start* here from a free boundary problem and we want to study this problem by *variational methods*[1]. It is not known if this problem can be transformed into a V.I. but we are going to verify (this is well known and goes back to Riabouchinsky [1] ; cf. also Birkhoff and Zarantonello [1] and the bibliography therein) that this problem is equivalent to a problem in the calculus of variations.

([1]) Of course there are *other* methods to treat free surface problems (use of fixed point theory, of functions of complex variables, etc..). They are not considered here.

3.2. A problem in the calculus of variations.

We define on $\mathcal{O} = \mathcal{O}_S$ ([1])

(3.5) $$K_S = \{v \mid v \in H^1(\mathcal{O}_S),\ v = g \text{ on } AG, BC, DEF,\ v=0 \text{ on } S\}$$

and for $v \in H^1(\mathcal{O}_S)$, we define

(3.6) $$J(v,S) = \int_{\mathcal{O}_S} (\tfrac{1}{2}|\text{grad}\, v|^2 - gx_2)\, dx.$$

We now consider the problem

(3.7) $$\inf J(v,S)\ ,\quad v \in K_S\ ,\quad S \in \mathcal{S}$$

where

(3.8) $$\mathcal{S} = \text{set of "all" curves joining } C \text{ to some point on } C'D.$$

Of course the definition (3.8) is somewhat ambiguous. This ambiguity disappears in the *numerical treatments* of this type of problem, where S is defined through the nodes of triangulations having a finite number of parameters of freedom.

We now verify that (3.7) is *equivalent* to (3.1)...(3.4). Let ψ, S be a couple realizing the minimum in (3.7) ; we assume this couple to exist, and that S is smooth enough - so that, in particular, we can use the normal ν to S (directed toward the exterior of $\mathcal{O}_S$ to fix ideas).

We define variations S_λ of S as follows ; let α be a continuous function given on S and let us define

(3.9) $$S_\lambda = \{x + \lambda\alpha(x)\nu(x) \mid x \in S\};$$

let v be a given function with compact support in a neighborhood of S ; we assume that v will be defined in $\mathcal{O}_S$ and that ψ is extended (smoothly) in a neighborhood of S ; we introduce

(3.10) $$w = \psi + \lambda v \quad (\textit{same } \lambda \text{ than in } (3.9))\ ;$$

w will be an element of K_{S_λ} iff

$$w(x+\lambda\alpha(x)\nu(x)) = 0\ ,\quad x \in S$$

hence

$$\psi(x)+\lambda\alpha(x)\frac{\partial\psi}{\partial\nu}(x)+\lambda v(x)+o(\lambda) = 0 \quad \text{on } S$$

([1]) We use $\mathcal{O}_S$ instead of $\mathcal{O}$ to emphasize the fact *that* $\mathcal{O}$ *will be variable.*

and since $\psi = 0$ on S, it remains (up to higher order terms)

$$(3.11) \qquad v(x) + \alpha(x) \frac{\partial\psi}{\partial\nu}(x) = 0 \quad , \quad x \in S.$$

We must have then

$$\frac{d}{d\lambda} J(w,S_\lambda) = 0 \quad \text{for } \lambda = 0 \, ,$$

i.e.

$$(3.12) \qquad \int_{\mathcal{O}} \operatorname{grad}\psi \operatorname{grad} v \, dx + \int_S \left[\tfrac{1}{2}|\operatorname{grad}\psi|^2 - gx_2\right] \alpha \, dS = 0 \, ,$$

for all v with compact support in a neighborhood of S and which satisfy (3.11). Therefore (3.12) becomes

$$(3.13) \qquad \int_{\mathcal{O}} (-\Delta\psi) v \, dx + \int_S \frac{\partial\psi}{\partial\nu} v \, dS + \int_S \left[\tfrac{1}{2}|\operatorname{grad}\psi|^2 - gx_2\right] \alpha \, dS = 0.$$

Using (3.11),(3.13) becomes :

$$\int_{\mathcal{O}} (-\Delta\psi) v \, dx - \int_S \left(\tfrac{1}{2}|\operatorname{grad}\psi|^2 + gx_2\right) \alpha \, dS = 0$$

which should be true $\forall$ v and $\forall\alpha$, hence (3.1) and the second condition (3.2) follow.

For the numerical implementation (*net moving method*) of this technique and for other examples, we refer to O'Carroll [1] and O'Carroll and H.T. Harrison [1] and to the bibliography therein.

4. OPTIMUM DESIGN AND FREE BOUNDARY PROBLEMS

4.1. Setting of the problem.

Let $\mathcal{O}$ be a bounded open set of $\mathbb{R}^n$; its boundary consists of a givent part Γ_o and of an unknown part S (cf. Fig. 3). Let g_o and g_1 be given functions in the whole space. We look for a function u in $\mathcal{O}$ such that

$$(4.1) \qquad \Delta u = 0 \text{ in } \mathcal{O} \, ,$$

$$(4.2) \qquad u = h \text{ (h = a given function) on } \Gamma_o,$$

$$(4.3) \qquad u = g_o, \ \frac{\partial u}{\partial\nu} = g_1 \text{ on } S$$

(where ν denotes the normal to S oriented toward the exterior of $\mathcal{O}$).

We show now how *this free boundary problem can be transformed into a problem of optimum design.*

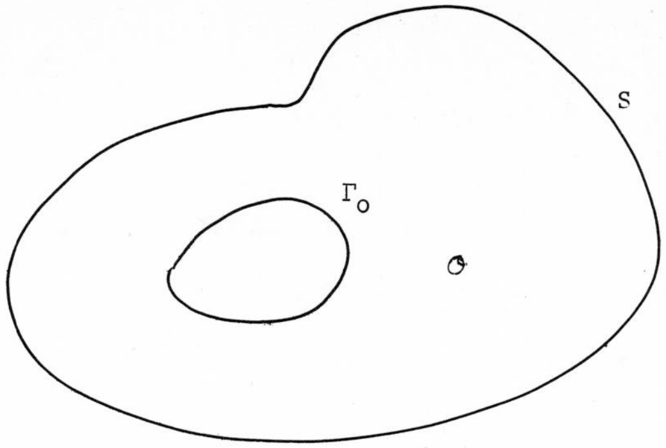

Fig. 3

4.2. Problem of optimum design.

We set $\mathcal{O} = \mathcal{O}_S$ in order to emphasize the fact that $\mathcal{O}_S$ is variable with respect to the "*control variable*" S.
Given S we define the *state* of the system as $y(S) = \{y_o(S), y_1(S)\}$ where

$$(4.4) \qquad \begin{cases} \Delta y_o(x,S) = 0 \quad \text{in } \mathcal{O}_S, \\ y_o(S) = h \quad \text{on } \Gamma_o, \\ y_o(S) = g_o \quad \text{on } S \end{cases}$$

and

$$(4.5) \qquad \begin{cases} \Delta y_1(x,S) = 0 \quad \text{in } \mathcal{O}_S, \\ y_1(S) = h \text{ on } \Gamma_o, \\ \dfrac{\partial y_1}{\partial \nu}(S) = g_1 \quad \text{on } S. \end{cases}$$

Problem (4.1)(4.2)(4.3) will be solved (as far as existence is concerned) *if we find* S *such that* $y_o(S) = y_1(S)$. Therefore one has to introduce a *cost function* J(S) which equals some norm of $y_o - y_1$. There are many possible choices. In order to have a functional with a "not too complicated" gradient, one is led (following O.Pironneau [1]) to the cost function given by

$$(4.6) \qquad J(S) = \int_{\mathcal{O}_S} |\operatorname{grad}(y_o(x,S) - y_1(x,S))|^2 \, dx.$$

We are now looking for S (among "all" surfaces which contain Γ_o in their interior) *minimizing* J(S) (and inf J(S) = 0 if (4.1)(4.2)(4.3) admits a solution).

4.3. Numerical method.

For the numerical solution of problems of *the type* of (4.6) one uses (cf. in particular Marrocco and Pironneau [1]) the method of finite elements with *moving triangulations*.
Basic formulas which express the "variations" of finite elements (of any order) and of integrals when *one node* of the triangulation is modified are given in Marrocco and Pironneau [1].

5. CONCLUDING REMARKS.

Remark 5.1.

We have indicated *some* connections between Optimal Control and Free Surfaces. In this respect one should add the remark that "*switching*" *lines or surfaces* can be also thought of as free surfaces. Cf. Lions [2].

Remark 5.2.

Problems of optimum design have a great interest in themselves for many applications. Cf. E.R. Barnes [1], J. Céa [1], J. Céa and K. Malanowski [1], N.V. Banichuk and A.A.Mironov [1][2], K.A. Lurié [1], O. Pironneau [1][2][3]and the bibliography therein (no attempt is made for giving a complete bibliography on this subject).

Remark 5.3.

Boundary value problems which are not necessarily of the free boundary type can be transformed into problems of optimal control for distributed systems. This technique has proven to be useful for the numerical analysis of transonic flows ; cf. M.O.Bristeau, R. Glowinski and O. Pironneau [1] , R. Glowinski, J. Périaux and O. Pironneau [1].

Remark 5.4.

Another method which can be used in problems of free surfaces is to transform the domain (unknown) into a fixed domain (using the solution) ; this method has permitted to D. Kinderlehrer and L. Nirenberg [1] to prove the *regularity* of free boundaries in several questions.

Remark 5.5.

Methods of optimal control of distributed systems are useful in problems of *identification*. We refer to G. Chavent and G. Cohen. [1].

REFERENCES

E.A. BARNES [1] Some Max-Min problems arising in optimal design studies, in Control Theory of Systems governed by Partial Differential Equations, ed. by A.K. Aziz J.W. Wingate and M.J. Ballas, Acad. Press, 1977, pp. 177-208.

C. BAIOCCHI [1] Sur un problème de frontière libre de l'hydraulique. C.R. Acad. Sci. Paris, 273 (1971), pp. 1215-1217.

N.V. BANICHUK and A.A. MIRONOV [1] Optimization of vibration frequencies of an elas - tic plate in an ideal fluid. P.M.M. 39(5), 1975, pp. 889-899.

[2] Optimization problems for plate oscillating in an ideal fluid. P.M.M. 40(3), 1976, pp. 520-527.

A. BENSOUSSAN and J.L. LIONS [1] Nouvelles méthodes en Contrôle Impulsionnel. Applied Mathematics and Optimization 1 (1975), pp. 289-312.

[2] Problèmes de temps d'arrêt optimal et Inéquations Variationnelles paraboliques. Applicable Analysis 3 (1973), pp. 267-294.

[3] Problèmes de temps d'arrêt optimal, Paris, Dunod (1978).

G. BIRKHOFF and E.H. ZARANTONELLO [1] Jets, wakes and cavities. Acad. Press, 1957.

H. BREZIS [1] Problèmes unilatéraux. J. Math. Pures et Appli. 51 (1972),p. 1-168.

M.O. BRISTEAU, R. GLOWINSKI and O. PIRONNEAU [1] Numerical solution of the transonic equation by the finite element method via optimal control, in Control Theory of Systems governed by Partial Differential Equations, ed. by A.K. Aziz, J.W. Wingate and M.J. Balas, Acad. Press, 1977, pp. 265-178.

J. CEA [1] Une méthode numérique pour la recherche d'un domaine optimal. IN Computing Methods in Applied Sciences and Engineering. Lecture Notes in Economics and Mathematical Systems. Springer, 134, 1976.

J. CEA and K. MALANOVSKI [1] An example of a max-min problem in partial differential eqiations. SIAM J. on Control, 8 (3), 1970, pp. 305-316.

G. CHAVENT and G. COHEN [1] These proceedings.

R. GLOWINSKI, J.L. LIONS and R. TREMOLIERES [1] Analyse numérique des Inéquations Variationnelles. Vol. 1 and 2, Dunod, Paris, 1976.

R. GLOWINSKI, J. PERIAUX and O.PIRONNEAU [1] Transonic flow computation by the finite element method via optimal control. Proceedings Second Symposium on Finite Element Methods in Flow Problems. I.C.C.A.D., Genoa, 1976.

D. KINDERLEHRER and L. NIRENBERG [1] Regularity in Free Boundary Problems. Annali Scuola Normale Sup. di Pisa, IV, 1977, pp. 373-391.

J.L. LIONS [1] Sur la théorie du contrôle. Actes Congrès International des Mathématiciens, Vancouver, 1974, pp. 139-154. (Canadian Math. Congress, 1975).

[2] Sur le contrôle optimal des systèmes gouvernés par des équations aux dérivées partielles. Paris, Dunod, Gauthier Villars. 1968. (English translation by S.K. Mitter, Springer, 1971).

J.L. LIONS and E. MAGENES [1] Problèmes aux limites non homogènes et Applications. Vol. 1, Paris, Dunod, 1968 ; (translated in English by J. Kenneth, Springer 181, 1972).

J.L. LIONS and G. STAMPACCHIA [1] Variational Inequalities. C.P.A.M. XX (1967) pp. 493-519.

K.A. LURIE [1] Optimal control of problems of mathematical physics. Moscow, 1975, in Russian.

E. MAGENES [1] Topics in parabolic equations : some typical free boundary problems, in Boundary Value Problems for Linear Evolution Partial Differential Equations. Ed. H.G. Garnir. NATO Adv. Study Inst. Series, Reidel, 1977.

A. MARROCCO and O. PIRONNEAU [1] Report LABORIA, 1977.

J. NECAS [1] Les Méthodes directes en théorie des équations elliptiques. Masson, 1967.

M.J. O'CARROLL [1] Variational principles for two dimensional open channel flows. Proc. Second INt. Symposium on Finite Elements Methods in Flow Systems. I.C.C.A.D., Genoa, 1976.

M.J. O'CARROLL and H.T. HARRISON [1] Variational techniques for free streamline problems. Proc. Second. Int. Symposium on Finite Elements Methods in Flow Problems. I.C.C.A.D., Genoa, 1976.

O. PIRONNEAU [1] Thesis, Paris, 1976.

[2] Optimisation de structures. Application à la Mécanique des Fluides Springer Verlag. Lecture Notes in Economics and Mathematical Systems. 107 (1976), pp. 610-624.

[3] Variational methods for the numerical solutions of free boundary problems and optimum design problems, in Control Theory of Systems governed by Partial Differential Equations, ed. by A.K. Aziz, J.W. Wingate, and M.J. Balas, Acad. Press, 1977, pp. 209-225.

D. RIABOUCHINSKY [1] Sur un problème de variations. C.R. Acad. Sci. Paris, 185 (1927), pp. 840-841.

S.L. SOBOLEV [1] Some Applications of Functional Analysis to Mathematical Physics. Leningrad, 1950.

SOME MATHEMATICAL MODELS IN IMMUNOLOGY

G.I. Marchuk
Computing Center, Novosibirsk, USSR

At the present time immunology has achieved such a success that mathematical simulation is naturally becoming a basic tool in the study of complex processes that take place in a living organism.

Bernet's outstanding investigations of clonal selection stimulated development of theoretical and experimental immunology. His theory has been further refined and specified and now it constitutes a solid basis for mathematical simulation of immune processes.

Today we can say that the immune processes taking place in an organism can be regarded from the viewpoint of description of vital activity of a very complicated system in which optimization processes are included as a natural component. Studies of immune processes in an organism from the viewpoint of the theory of complex systems will make it possible to realize control of these processes particularly in cases when pathologic changes occur in an organism. We can predict that methods of medical treatment of various diseases will be more and more based on profound investigation of mathematical simulation.

Problems of mathematical simulation of immune processes have already been substantially studied by many authors. Here I would like to mention first investigations by I. Hege and L. Cole (1966), M. Jilek, G. Bell, G. Bruni and colleagues. Most important results were recently obtained by R. Mohler, C. Barton and others and reported on at the Symposium on Optimization Techniques in Nice. The present paper deals with some immunological models devoted to describe virus diseases which have been developed at the Computing Center of the Siberian Branch of the USSR Academy of Sciences in the last five years. It should be noted that the $T-B$ model is based on Feldmann's $T-B$ cooperation hypothesis.

I. A Simple Model of a Virus Disease

Let us assume that a small population of viruses (antigens) has penetrated a human organism and after a time reached an organ, which they are able to affect.

Let t_0 be the average time of penetration. The viruses begin to multiply and infect the cell. This process goes on until the albuminous reserve of the cell is exhausted.

Having injured the cell, the viruses hit blood and lymphatic nodes. Part of them again penetrates healthy cells and continues multiplying. The other part hits lymphatic nodes where there are concentrated lymphocytes specific with respect to different antigens. The antigens that have penetrated into the lymphatic nodes have some probability to meet with lymphocytes that react to antigen of a given kind.

As a result of binding and the catalytic reaction that follows it the lymphocyte divides and transforms into an antibody-producing plasmacyte. The time of formation of plasmacytes is about 12 to 18 hours. In this manner, in principle, there goes on a development of defensive mechanisms of the organism on the basis of active stimulation by antigens.

In accordance with our model we assume that the following are basic factors of a virus disease:

V the number of viruses in an organism,

C the number of plasma cells producing antibodies F ,

m relative characteristics of the damaged part of the tissue.

Let us construct model equations. The first equation will describe variation of the number of viruses in the organism

$$dV = \beta V dt - \gamma F V dt . \tag{I.1}$$

The first right-hand term of the equation represents the increase of the number of viruses dV for the time dt as a result of their division. Naturally, this number is proportional to V and to some quantity β which is called a coefficient of virus multiplication. The term $\gamma F V dt$ denotes the number of viruses neutralized by antibodies F for the interval dt . Indeed, the above number is apparently proportional to both the number of antibodies in the organism and that of viruses $\gamma F V dt$,

γ is the coefficient associated with probable neutralization of virus by antibody during their encounter. In this model the coefficient is considered a constant value.

Having divided (I.1) by dt we obtain

$$\frac{dV}{dt} = (\beta - \gamma F) V . \tag{I.2}$$

Let us form a second equation devoted to describe the growth of plasmacytes C. To this aim we make use of the simple hypothesis on formation of plasmacytes cascade populations from lymphocytes in their interaction with the antigen. The lymphocyte interacts with the antigen which it has been programmed to recognize. Then it starts a cascade process of formation of cells which synthesize the antibodies. The latter neutralize the antigens of the given kind. It should be noted that the process of plasmacyte formation apparently consists in that it takes into account a probability of a bound virus to stimulate the immune system. The number of lymphocytes thus estimated is, evidently, proportional to VF. Hence, we arrive at

$$dC = Q(t-\tau)\,dt\ , \tag{I.3}$$

where

$$Q(t) = pFV\,. \tag{I.4}$$

Here τ is the time of formation of a plasmacytes cascade. The more complete equation of the plasmacytes generation is as follows:

$$dC = Q(t-\tau)\,dt - \mu_c(C-C^*)\,dt\,. \tag{I.5}$$

The first right-hand term in (I.5) represents the plasmacytes generation, τ the period of the cell cascade formation and the beginning of a mass antibody synthesis, p constant.

The second term in the formula describes the decrease of the number of plasma cells due to aging. Here C^* is the immunological level of plasmacytes in the organism and μ_c the coefficient equal to the inverse of the cells lifetime. Usually it is equal to several days.

Having divided (I.5) by dt we arrive at

$$\frac{dC}{dt} = Q(t-\tau) - \mu_c(C-C^*)\,. \tag{I.6}$$

Let us calculate the balance of the number of antibodies of a given structure that react with antigens of the virus. We have the formula

$$dF = \rho C dt - \eta\gamma FV dt - \mu_f F dt \tag{I.7}$$

which consists of three components.

The first term, $\rho C dt$, represents the antibodies generation by plasmacytes in the interval dt, ρ the antibodies reproduction coefficient per unit time calculated for a single cell. The second term $\eta\gamma FVdt$ denotes the decrease of the antibodies number in the interval dt at the expense of the binding to the antigens and the elimination from an active fight. Indeed, as indicated above, in deriving equation (I.2) the number of viruses, eliminated in the interval dt as a result of neutralization by antibodies, was $\gamma FVdt$. If neutralization of a single virus takes η antibodies, we obtain the above-mentioned term. The third term, $\mu_f Fdt$, represents the decrease of antibodies due to aging. Here μ_f is the coefficient inversely proportional to the time of the antibody decay. It is approximately equal to several days.

Having divided (I.7) by dt we arrive at

$$\frac{dF}{dt} = \rho C - (\mu_f + \eta\gamma V)F. \qquad (I.8)$$

The final model equation is obtained for the balance of the damaged tissue of the organism. If m is relative characteristics of the damaged part of the tissue, we obtain the relation of balance

$$dm = \sigma V dt - \mu_m m dt \qquad (I.9)$$

which describes a change of the damaged tissue dm for the time dt caused by the damage of the tissue by viruses $\sigma V dt$ and its recovery in the organism $\mu_m m dt$. The coefficient σ depends on the extent of damage per unit time, and μ_m is the inverse of recovery of the organ's damaged part.

Having divided (I.9) by dt we arrive at the fourth, and final equation

$$\frac{dm}{dt} = \sigma V - \mu_m m. \qquad (I.10)$$

Up to this moment our model equations have not taken into account the effect of the damaged tissue on the dynamics of the disease. It will be noted that if vital organs are seriously damaged, the efficiency of antibody production falls. This fact may be fatal for the organism. The factor of damage of such organs can be considered in equation (I.6), if we substitute the coefficient $\rho \cdot \xi(m)$ for ρ. Here ξ is the function, taking into account the efficiency of plasmacytes due to the change of m. $m = 1 - \frac{M'}{M}$, M is characteristics of a healthy organ (its mass or area) and M' is, respectively, characteristics of a healthy part of the damaged tissue. Fig. 1 shows that in the interval $0 \leqslant m \leqslant m^*$ the variation of the organ's mass does not practically affect the plasmacytes production.

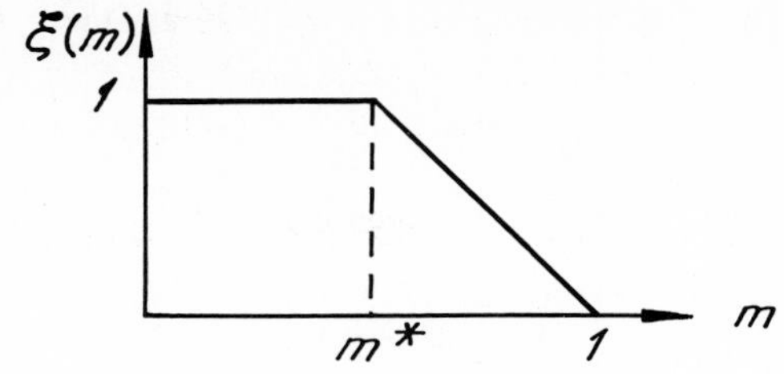

Fig. 1

In the interval $m^* \leqslant m \leqslant 1$ there is a rapid drop in plasmacyte productivity. This form of the simple dependence can be specified on the basis of clinical data.

Thus, we arrive at the following system of nonlinear differential equations describing the dynamics of a virus disease

$$\begin{aligned} \frac{dV}{dt} &= (\beta - \gamma F)V , \\ \frac{dC}{dt} &= \xi(m)Q(t-\tau) - \mu_c(C - C^*) , \\ \frac{dF}{dt} &= \rho C - (\mu_f + \eta\gamma F)V , \\ \frac{dm}{dt} &= \sigma V - \mu_m m . \end{aligned} \qquad (I.11)$$

We add to (I.11) the initial data

$$\begin{aligned} V &= V^0 \\ C &= C^0 \qquad\qquad t = 0 \\ F &= F^0 \\ m &= m^0 . \end{aligned} \qquad (I.12)$$

Let us note that in the absence of viruses the system of equations has the solutions

$$V = 0; \quad C = C^*, \quad F = \frac{\rho}{\mu_f} C^*, \qquad m = 0 . \qquad (I.13)$$

Here $F = F^* = \frac{\rho}{\mu_f} C^*$ is the immunological barrier of antibodies which is maintained by productivity $C = C^*$.

Thus, if the initial conditions (I.12) satisfy equations (I.13), problem

(I.11), (I.12) has the stationary solution (I.13), which represents a state of a healthy organism.

A Qualitative Analysis of Solutions

Let us analyze equations (I.11). There immediately follows the Theorem: Whatever the nonzero original number of viruses, that have penetrated the cells of the damaged tissue, a disease progresses, if only the immunological antibody level F^* satisfies the inequality

$$0 \leqslant F^* \leqslant \frac{\beta}{\gamma} . \qquad (I.14)$$

The proof of the statement is based on the analysis of the first equation of system (I.11). Indeed,

$$\text{if } F^* < \frac{\beta}{\gamma} , \quad \text{then } \left.\frac{dV}{dt}\right|_{t=0} = (\beta - \gamma F^*) V_0 > 0 . \qquad (I.15)$$

Hence, the number of viruses grows with time.

$$\text{If } F^* > \frac{\beta}{\gamma} , \quad \text{then } \left.\frac{dV}{dt}\right|_{t=0} = (\beta - \gamma F^*) V_0 < 0 \qquad (I.16)$$

and the number of viruses decreases with time.

$$\text{If } F^* = \frac{\beta}{\gamma} , \quad \text{then } \left.\frac{dV}{dt}\right|_{t=0} = 0, \qquad (I.17)$$

that is $V = const$ and a stationary solution is possible.

The theorem indicates that in this mathematical model the virus disease is independent of the number of viruses which have penetrated the damaged organ.

Of special interest for the disease dynamics is the acute period, when the number of viruses in the organism is growing and one can assume $\xi = 1$. This period lasts several tens of hours and the disease dynamics can be represented by the simple formula

$$\frac{dV}{dt} = (\beta - \gamma F) V , \qquad (I.18)$$

$$\frac{dC}{dt} = \rho \cdot F(t-\tau) \cdot V(t-\tau) ,$$

$$\frac{dF}{dt} = \rho C - \eta \gamma V F , \qquad (I.18)$$

$$\frac{dm}{dt} = \sigma V .$$

To this we add the initial data

$$V = \varepsilon , \quad C = C^* , \qquad F = F^* , \quad m = 0 . \qquad (I.19)$$

Let $\bar{\beta} = \beta - \gamma F^* > 0 , \quad \alpha = \rho F^* .$

Here we consider a simple case, when $F = F^*$.

If the time τ of formation of plasmacytes, producing the necessary antibodies, is fixed the solution of (I.18), (I.19) in the interval $0 \leqslant t \leqslant \tau$ can be written as follows

$$V = \varepsilon e^{\bar{\beta} t} , \quad C = C^* , \quad F = F^* , \quad m = \frac{\sigma \varepsilon}{\bar{\beta}} (e^{\bar{\beta} t} - 1) . \qquad (I.20)$$

Using this solution one can obtain an approximate solution to problem (I.18), (I.19) on the interval $\tau \leqslant t \leqslant 2\tau$ in the form:

$$V = \varepsilon \exp \left\{ \bar{\beta} t - \alpha \rho \frac{\gamma \varepsilon}{\bar{\beta}^3} e^{\bar{\beta}(t-\tau)} \right\} . \qquad (I.21)$$

One can see that in the second interval $\tau \leqslant t \leqslant 2\tau$ the exponential increase of viruses is followed by an exponential drop and a rapid approach of V to zero.

In our analysis we put $\xi(m) = 1$. In fact the critical damage of the organ may cause a failure of the organism's vital activity before the viruses concentration has reached zero. Therefore only a direct numerical calculation will provide a true description of the disease dynamics in a more or less perfect formulation.

Stationary Solutions

The basic system of equations has a whole class of stationary solutions. The latter can be obtained if we assume that the functions V, C, F, m are independent of time. As a result we arrive at the system of nonlinear equations

$$\begin{aligned}
&(\beta - \gamma F)V = 0, \\
&\rho F V - \mu_c (C - C^*) = 0, \\
&\rho C - (\mu_f + \eta\gamma V)F = 0, \\
&\sigma V - \mu_m m = 0 .
\end{aligned} \qquad (I.22)$$

For simplicity we consider the case when $0 \leqslant m \leqslant m^*$ which corresponds to $\xi(m) = 1$.

The solution of equations (I.22) is as follows

$$V = \mu_c \frac{\rho C^* - \frac{\mu_f \beta}{\gamma}}{\eta\beta\mu_c - \frac{\rho\beta}{\gamma}\rho},$$

$$F = \frac{\beta}{\gamma}, \qquad (I.23)$$

$$C = \frac{\mu_c \eta C^* - \frac{\rho\beta}{\gamma}\frac{\mu_f}{\gamma}}{\eta\mu_c - \frac{\rho}{\gamma}\rho}, \quad m = \frac{\sigma}{\mu_m} V .$$

If the above solutions are positive they can be interpreted as a chronic or a persistent disease. It will be noted that the stationary solutions for a healthy organism obtained before are also included in this class of stationary solutions.

A Principal Scheme of the Virus Disease Dynamics

On the basis of the above stated mathematical model one can get a typical picture of the virus disease dynamics. It can be presented as follows.

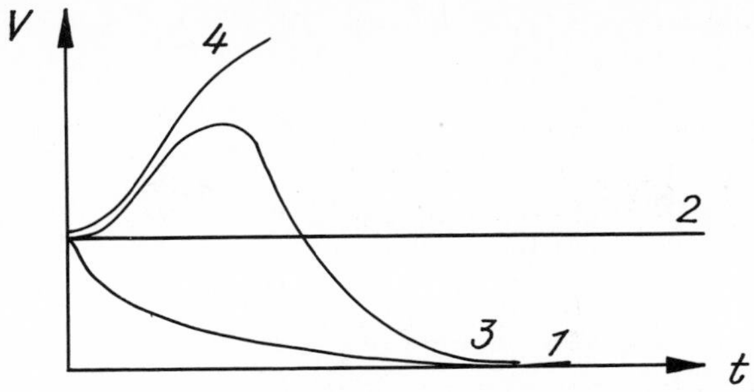

Fig. 2

If there is a sufficient number of functioning antibodies with respect to their antigen, the viruses that penetrate the organism will meet with a powerful response and their concentration will decrease and approach zero. This is a mild case of a disease (curve 1).

It may so happen, however, that in the organism there goes a process of a viruses multiplication. The viruses bind with the antibodies, present in the blood plasma. Thus there establishes a balance between the number of viruses generated every second and those captured by antibodies. Here we deal with a stationary process which can be interpreted as a chronic or a persistent disease (curve 2).

If the number of viruses grows more rapidly than reproduction of antibodies neutralizing the latter the curve of virus concentration begins to grow exponentially. However, after plasmacytes have formed and begun mass antibody production, the growth of the virus concentration decelerates and some time later it rapidly falls. At the same time there goes on a reproduction of new antibodies whose total number decelerates exponentially until the normal immunological level is reached. The damaged part of the organ, in which there went an evolution of the virus population, begins to recover exponentially (curve 3).

Finally, it may happen that the damage of the virus-affected organ is essential. In that case the vital activity of the organs responsible for antibody formation is seriously upset. Then the number of viruses in the organism will continuously grow, which results in a lethal outcome (curve 4).

Discussion of Results: the Simple Model of a Virus Disease

Let us discuss now some results of mathematical simulation of a virus disease using the basic system of equations. Fig. 3 shows the dynamics of all the components of the immune process under the condition that the immune barrier is not passed $\left(F^* > \frac{\beta}{\gamma}\right)$.

In this case the concentration of viruses begins immediately to fall and after a time it approaches zero. This is a latent, lightest form of a disease for which the above-mentioned theorem about the immune barrier holds.

Fig. 4 illustrates a normal response of another type. In this case the concentration of viruses grows and infects the mass of a tissue. In addition the viruses stimulate the immune system and over the delay time τ there appear plasmacytes. The latter start an intensive antibody production. On combining with the viruses the antibodies form an antigen-antibody complex which is eliminated from the organism. The time during which the virus affects the organism is equal to several days. This process is characterized by a fall in the virus concentration from its maximum to zero. After the viruses are eliminated from the organism antibody and plasmacyte concentrations tend to their normal values and the mass of the affected tissue regenerates. So the system tends to a state of a healthy organism.

Fig. 5 shows an immune process corresponding to another stationary solution at $V(t) = const > 0$, which is called a second stationary solution. In this case all variables are constant and positive. Plasmacyte and antibody concentrations are higher as against their norms. A dynamic balance establishes when the number of viruses generated every second is exactly the same as that of viruses eliminated from the organism.

It is also found out that the system has periodic solutions depending on a chronic form of a disease (Fig. 6). This process is due to the fact that the number of the antibodies produced is enough to eliminate only part of viruses rather than all of them. Therefore a virus may have a possibility to multiply again and stimulate the immune system. In the course of investigation it has been found out that such processes arise if the ratio of antibody to plasmacyte lifetimes is great. The process normalizes if this ratio is made less (Fig. 4).

Fig . 7 demonstrates the effect of the mass of the damaged tissue on the immune response. Fig. 4 represents the immune response when the effect of the mass is neglected, that is $\xi(m) \equiv 1$. Fig. 7 shows the same when the effect of the mass is taken into account. One can see that in the first case the end result is recovery. In the second case the mass of the affected tissue causes reduction in antibody production up to the level which is not sufficient for a complete elimination of viruses.

2. An Immunological Model with B-, T-antibodies

A simple mathematical model of a virus disease was derived in Sec. I.

The model was based on the concepts of nineteen sixties [2]. It consisted of the three important components of the process: antigens, antibodies, antibody-producing plasmacytes.

The object of this part of the study is to formulate a more complicated model of a virus disease based on up-to-date facts and the laws of immunology [3].

We will begin with a construction of an equation for the dynamics of an antigen V concentration. Let the increase of antigens per time interval dt be defined by the formula

$$dV = \beta V dt - \gamma_1 F V dt - \gamma_2 T V dt - \gamma_3 B V . \tag{2.1}$$

Here the first term on the right represents the increase of antigens due to multiplication. The second, third and fourth terms represent the decrease of antigens due to: interaction with antibodies F, the attack of killer T-lymphocytes and the binding with immunocompetent B-lymphocytes, respectively. β is the coefficient of antigen multiplication, $\gamma_1, \gamma_2, \gamma_3$ are the constants describing the number of antigens neutralized by antibodies, killer- and B--lymphocytes.

The structure of the second, third and fourth right-hand terms in (2.1) in a form of binary products expresses a probability of an antibody-antigen, a killer lymphocyte-antigen and a B-lymphocyte-antigen "encounter". Obviously the probability of such an encounter is proportional to the product of concentrations of interacting populations.

Let us derive an equation for the dynamics of concentration of F- antibodies of a given kind, able to combine with antigens. Write the relation

$$dF = \mu_B C_B(t) dt - \eta_1 \gamma_1 F V dt - \alpha_F (F - F^*) dt . \tag{2.2}$$

Here the first term on the right is the number of antibodies produced per time interval dt due to the formed plasmacytes B producing antibodies. An equation for the function $C_B(t)$ is derived below. The second term on the right is the same as in (2.1). It represents the number of antigens neutralized per time interval dt. Taking into the account the fact that neutralization requires η_1 antibodies we obtain a tern to describe the decrease of antibodies due to the binding of antigens. The last term represents the antigen decrease due to aging with the period inverse to the coefficient α_F.

To determine $C_B(t)$, we make up the equation of balance

$$dC_B = P_B(t-\tau_B)dt - \alpha_{C_B}(C_B - C_B^*)dt\,, \qquad (2.3)$$

where

$$P_B(t) = \gamma_4 VT\Lambda B\,.$$

The equation is based on Feldmann's $T-B$ cooperation hypothesis [11]. The first term on the right in (2.3) represents the increase of B-cells producing antibodies specific for a given antigen. To within the coefficient γ_4 this term is proportional to the product of the factors $VT\Lambda B$. Here besides the values V and T there are present: concentration of macrophages Λ and concentration of B- lymphocytes. It will be noted that the term P_B in the equation has $t-\tau_B$ as its argument. It corresponds to the delay time beginning from the moment when a B- lymphocyte reacts with antigen up to the moment a B- cell cascade is formed. The following scheme of the immune process of the antibody formation is a base for this construction of the B-cell source. First, it is assumed that an initial phase of this process is a reaction of T- lymphocytes with V-antigen. As a result of the reaction the antigen binds with a T- lymphocyte receptor. Then with some probability it "tears off" the receptor from the lymphocyte. On the interval dt there are about $VTdt$ of such interactions. This means that the more concentration of V and T the more probability for V to come across and interact with a T- lymphocyte which has a given receptor. The antigens labelled in this manner (with T- lymphocyte receptors) are recognized by macrophages Λ and the macrophage collects them on its membrane. In this way the macrophage becomes a mediator of a great number of antigens. The number of macrophage-bound antigens per time interval dt is proportional to $VT\Lambda dt$. Finally, such a macrophage-antigen complex on contacting with a B- lymphocyte forms a multideterminant B- lymphocyte-antigen bond. As a result of division antibody-producing plasma B- cells begin to form.

The coefficient γ_4 can be found by analyzing empirical data.

The second term on the right in (2.3) defines the decrease of B- cells due to aging. The term C_B^* denotes the normal amount of B- cells in a healthy organism (it can be equal to zero).

Now we will describe the dynamics of T- lymphocytes. Assuming T to be their concentration we write

$$dT = \mu_T C_T(t)dt - \eta_2\gamma_2 TVdt - \alpha_T(T-T^*)dT\,. \qquad (2.4)$$

This equation for killer T- lymphocytes is similar to equation (2.2).

Let us derive an equation for the number of plasma T- cells. Consider the relation

$$dC_T = P_T(t-\tau_T)dT - \alpha_{C_T}(C_T - C_T^*)dT, \qquad (2.5)$$

where

$$P_T(t) = \gamma_5 VT.$$

Eq. (2.5) is similar to (2.3). Let us consider the structure of $P_T(t)$. This relation shows that part of T- lymphocytes on combining with antigens generates T- plasmacytes which develop an increase of killer lymphocytes.

Let us construct an equation of balance of B- lymphocytes. Consider the expression

$$dB = \gamma_6 BVdT - \gamma_7 BVdT - \gamma_8 VTB\Lambda dT - \\ - \alpha_B(B-B^*)dT. \qquad (2.6)$$

The first term on the right represents a source of formation of B- lymphocytes whose number will increase in proportion to the available BV- complexes. It is assumed that free complexes $D_B V$ (where D_B are receptors of respective immunocompetent lymphocytes), torn off from B- lymphocytes, stimulate the activity of lymphocyte-producing organs. The next term represents a loss of B- lymphocytes whose receptors are either torn off by antigens or completely bound by them. The third term is the number of B- lymphocytes lost to form B-plasmacytes. It is proportional to $VT\Lambda Bdt$. Thus the total loss of B-lymphocytes is $\gamma_8 VT\Lambda Bdt$. The last term in the formula represents dissipation of B-lymphocytes. Let us construct an equation for balance of macrophages. Assume that it consists of the four components:

$d\Lambda =$ [generation] — [utilization] — [loss of lymphocytes on formation of plasmacytes] — [aging].

Let generation of new leucocytes take place in corresponding organs and be stimulated by $D_T V$ and $D_B V$ free complexes. Taking account of the fact that these complexes are proportional to TV and BV respectively, we obtain

$$\gamma_9 TV + \gamma_{10} BV.$$

Utilization of different complexes by macrophages results in their quantitative decrease, equal to $TV\Lambda$, $BV\Lambda$, $FV\Lambda$. Besides, there occurs utilization of the used plasmacytes C_B and C_T. Using $\gamma_{16}VT\Lambda B$ to denote the loss of macrophages on formation of C_B plasmacytes and $\alpha_\Lambda(\Lambda-\Lambda^*)$ to denote aging, we write the relation

$$d\Lambda=(\gamma_9 TV+\gamma_{10}BV)dt-\Big[\gamma_{11}TV+\gamma_{12}BV+\gamma_{13}VF+\gamma_{14}(C_B-C_B^*)+$$
$$+\gamma_{15}(C_T-C_T^*)\Big]\Lambda dt-\gamma_{16}VT\Lambda B dt-\alpha_\Lambda(\Lambda-\Lambda^*)dt\,. \quad (2.7)$$

Like in the simple model we add to this system the equation for relative characteristics of the damaged tissue.

The final system of equations becomes

$$\begin{aligned}
\frac{dV}{dt}&=(\beta-\gamma_1F-\gamma_2T-\gamma_3B)V,\\
\frac{dF}{dt}&=\mu_B\xi(m)C_B-\eta_1\gamma_1FV-\alpha_F(F-F^*),\\
\frac{dC_B}{dt}&=P_B(t-\tau_B)-\alpha_{C_B}(C_B-C_B^*),\\
\frac{dT}{dt}&=\mu_T\xi(m)C_T-\eta_2\gamma_2TV-\alpha_T(T-T^*),\\
\frac{dC_T}{dt}&=P_T(t-\tau_T)-\alpha_{C_T}(C_T-C_T^*),\\
\frac{dB}{dt}&=\gamma_6\xi(m)BV-\gamma_7BV-\gamma_8VTB\Lambda-\alpha_B(B-B^*),\\
\frac{d\Lambda}{dt}&=(\gamma_9TV+\gamma_{10}BV)\xi(m)-\Big[\gamma_{11}VT+\gamma_{12}BV+\gamma_{13}FV+\\
&\quad+\gamma_{14}(C_B-C_B^*)+\gamma_{15}(C_T-C_T^*)\Big]\Lambda-\gamma_{16}VT\Lambda B-\alpha_\Lambda(\Lambda-\Lambda^*),\\
\frac{dm}{dt}&=\sigma V-\alpha_m m,
\end{aligned} \quad (2.8)$$

where

$$P_B(t)=\gamma_4VT\Lambda B\,,\quad P_T(t)=\gamma_5VT\,.$$

It is solved under the initial conditions

$$V=V^0,\ F=F^*,\ C_B=C_B^*,\ T=T^*,\ C_T=C_T^*,\ \Lambda=\Lambda^*,\ B=B^*,$$
$$m=0,\ \ t=0\ . \tag{2.9}$$

Thus we arrive at a nonlinear system of equations of eight functions to describe the dynamics of a virus disease. This system of equations together with the initial data will be taken as a basis for mathematical simulation of a virus disease.

We next discuss a simpler model. Assume that there are no B- lymphocytes in an organism, that is $F=0$. In this case the organism defenses itself only by means of killer T- lymphocytes. Indeed, from (2.8) follows

$$\frac{dV}{dt}=(\beta-\gamma_2 T)V,$$

$$\frac{dT}{dt}=\mu_T\xi(m)C_T-\eta_2\gamma_2 TV-\alpha_T(T-T^*),$$

$$\frac{dC_T}{dt}=P_T(t-\tau_T)-\alpha_{C_T}(C_T-C_T^*)\ , \tag{2.10}$$

$$\frac{dm}{dt}=\sigma V-\alpha_m m,$$

$$P_T=\gamma_4 VT\ .$$

To this we can add the equation for concentration of macrophages

$$\frac{d\Lambda}{dt}=\xi(m)\gamma_9 TV-\left[\gamma_{11}TV+\gamma_{15}(C_T-C_T^*)\right]\Lambda-\alpha_\Lambda(\Lambda-\Lambda^*)\,. \quad (2.11)$$

Thus we arrive at the simple model of a disease considered above. This model can be used to describe the disease dynamics by means of T- lymphocytes only.

We will now consider another limiting case when specific T- lymphocytes are absent and the immune process is based on the antigen-antibody interaction.

We have the following system of equations:

$$\frac{dV}{dt}=(\beta-\gamma_3 B)V\,,$$

$$\frac{dB}{dt}=\xi(m)\gamma_6 BV-\gamma_8 BV-\alpha_{13}(B-B^*)\,, \quad (2.12)$$

$$\frac{dm}{dt}=\sigma V-\alpha_m m\,.$$

To this, again, we add the equation for macrophages

$$\frac{d\Lambda}{dt}=\xi(m)\gamma_{10}BV-\gamma_{12}BV\Lambda-\alpha_\Lambda(\Lambda-\Lambda^*)\,. \quad (2.13)$$

In this way we obtain the second limiting case. It should be noted that in both limiting cases antibodies are not produced. Therefore the unique opportunity for a quick suppression of viruses by means of antibodies is lost. These limiting cases are usually referred to as B- and T- failures or primary immunodeficits.

In studying the new immunological model of a virus disease, we obtained the forms of the immune response similar to those mentioned above. As before if the immunological barrier is not overcome $(\beta<\gamma_1 F^*+\gamma_2 T^*+\gamma_3 B^*)$ the infection ceases to develop whatever the initial values of the virus V^0 concentration.

To conclude with, we will note that the simple model of a virus disease allowed us to obtain and investigate basic forms of the immune response. The most important result is the theorem about the immunological barrier and the study of conditions under which chronic forms of the immune response arise.

A logic development of the simple model was the model based on immunological discoveries, particularly on the data about cooperation of T- and B-cells and macrophages. With the help of the new model we studied the interaction of the immunity T- and B-systems. It appears that the T-system plays a major role in virus diseases.

In the immunological models presented in this study it is assumed that antibodies have the same immunological basis. Usually this is I_gG. However it is known that in the antigen-antibody reaction there participate different classes of immunoglobulins. Therefore it is possible to develop more complete models including at least two different B - antibodies on the basis of I_gM and I_gG. It appears that these two groups bear special responsibility for the development of immunological means to fight with antigens.

At the present time at the Computing Center of the Siberian Branch of the USSR Academy of Sciences research is carried out with mathematical models of the immune reaction. In these models the immunoglobulins are represented by the classes I_gM and I_gG.

Finally it will be emphasized that our model by no means describe the whole spectrum of immune processes. We did not try to simulate any specific antigen. Nevertheless the models reflect the laws that govern development of the immune response in the course of a disease. Therefore, it is hoped that they are adequate to describe a specific disease or antigen, etc.

The author expresses a hope that cooperative efforts by physicians, immunologists and mathematicians will make it possible to find solutions to immunological problems, problems of predicting the dynamics of virus diseases and eventually to design effective methods of medical treatment.

References

1. БЕРНЕТ Ф. Целостность организма и иммунитет. М., "Мир", 1964.
2. НОССЕЛ Г. Антитела и иммунитет. М., "Медицина", 1973.
3. ПЕТРОВ Р.В. Иммунология и иммуногенетика. М., "Медицина", 1976.
4. МАРЧУК Г.И. Простейшая модель вирусного заболевания. Сб. Применение математических методов в клинической практике. ВЦ СО АН СССР, Новосибирск, 1976.
5. МАРЧУК Г.И. Иммунологическая модель вирусного заболевания. Препринт, ВЦ СО АН СССР, Новосибирск, 1977.
6. BELL, G.J. Mathem. Biosciences 16, 291 (1973).
7. MOHLER, R.R. et al. T-and B-Cell Models in the Immune Response.
8. BRUNI, C. et al. A Dynamical Model of Humoral Immune Response. Universita di Roma, Istituto di Automatica, R. 74-22, Luglio 1974.
9. JILEK, M. Folia microbiologica 16, No 1, No 2, No 3 (1971).
10. HEGE, I.S., COLE, L.I. J. Immunology 97, No 1, 34 (1966).
11. FELDMANN, M., P. Erb. Cell. Immunol. 19, 356 (1975).

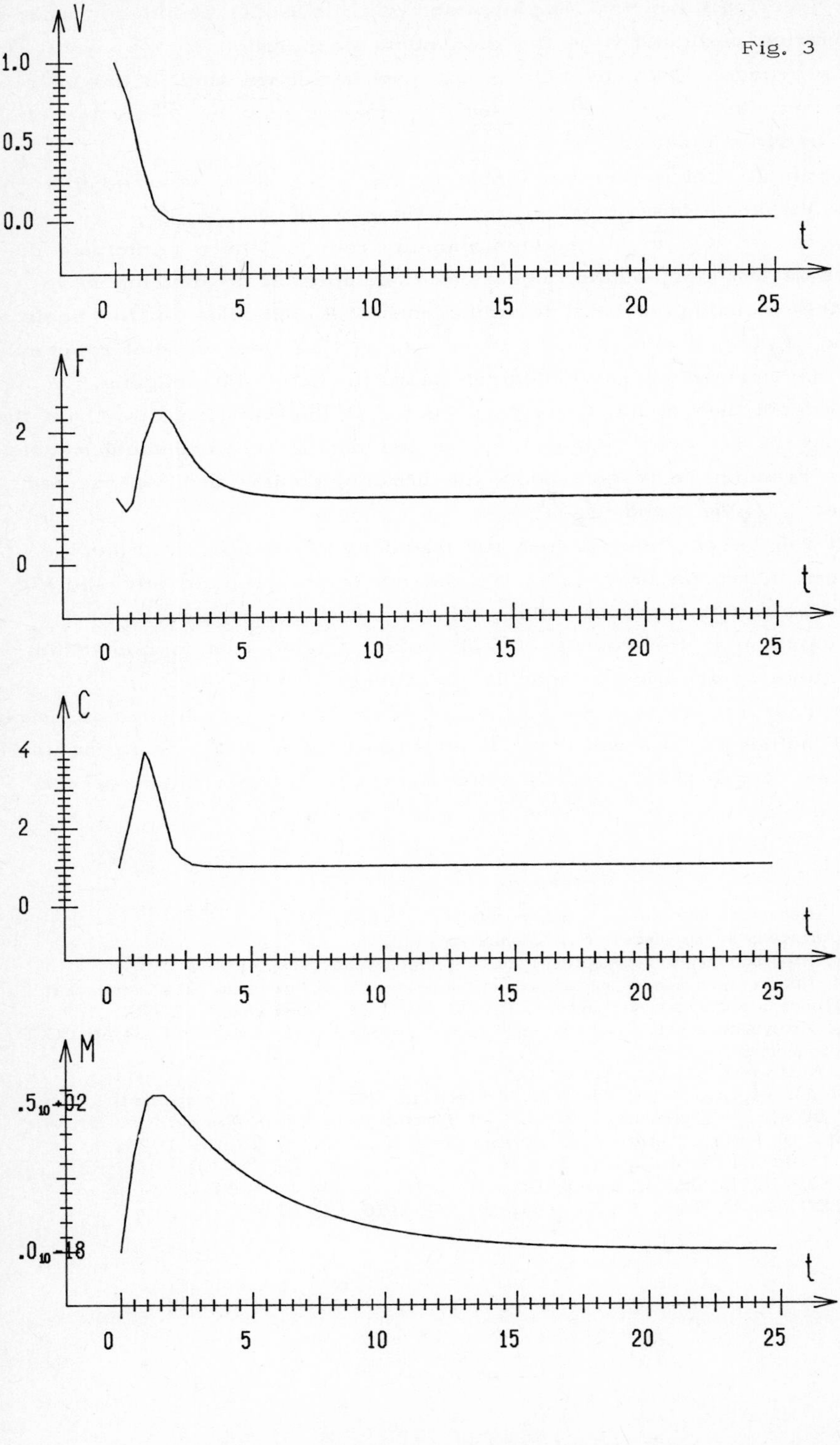
V
1.0
0.5
0.0
t
0
5
10
15
20
25
F
2
0
t
0
5
10
15
20
25
C
4
2
0
t
0
5
10
15
20
25
M
.5₁₀-02
.0₁₀-18
t
0
5
10
15
20
25

Fig. 3

Fig. 4

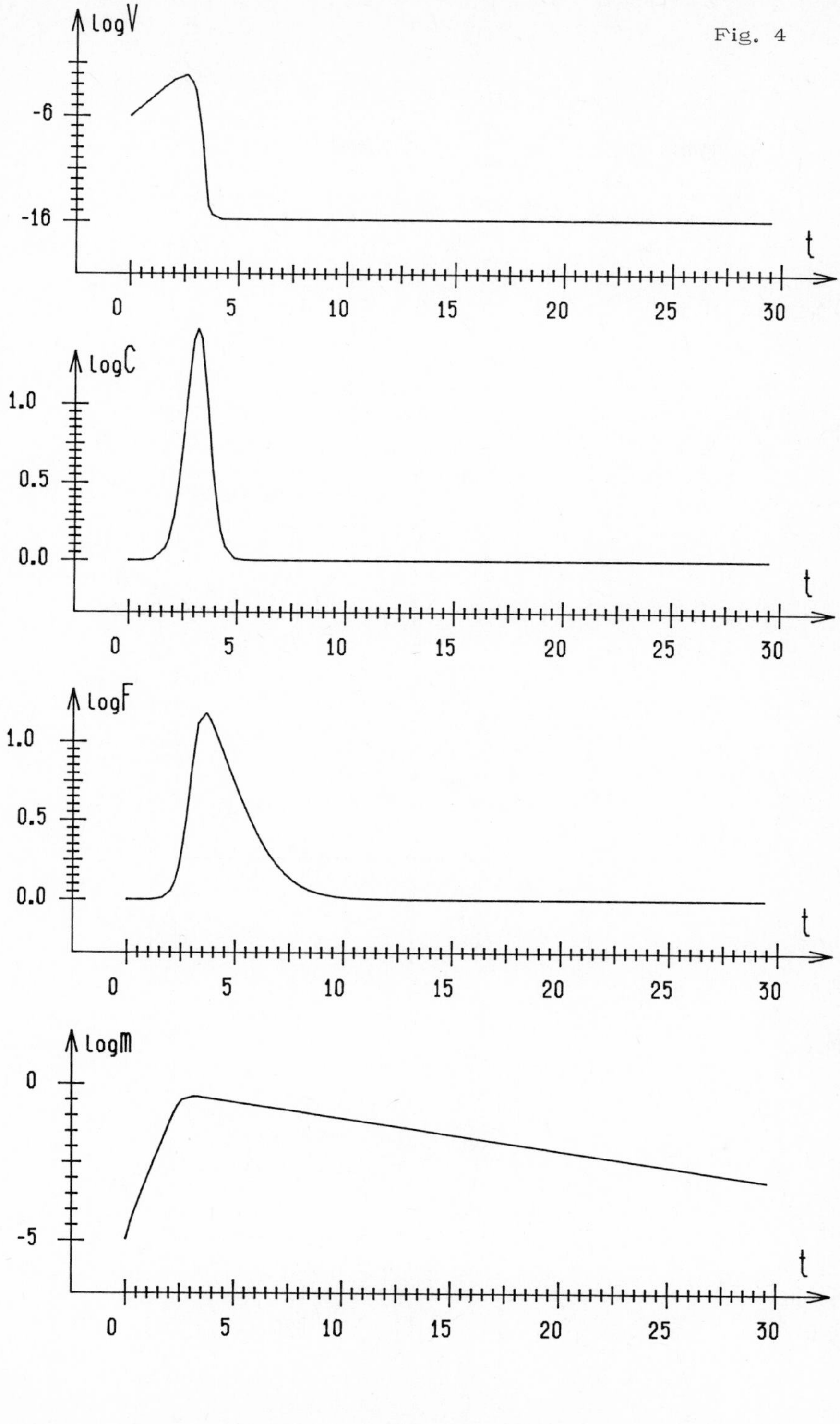

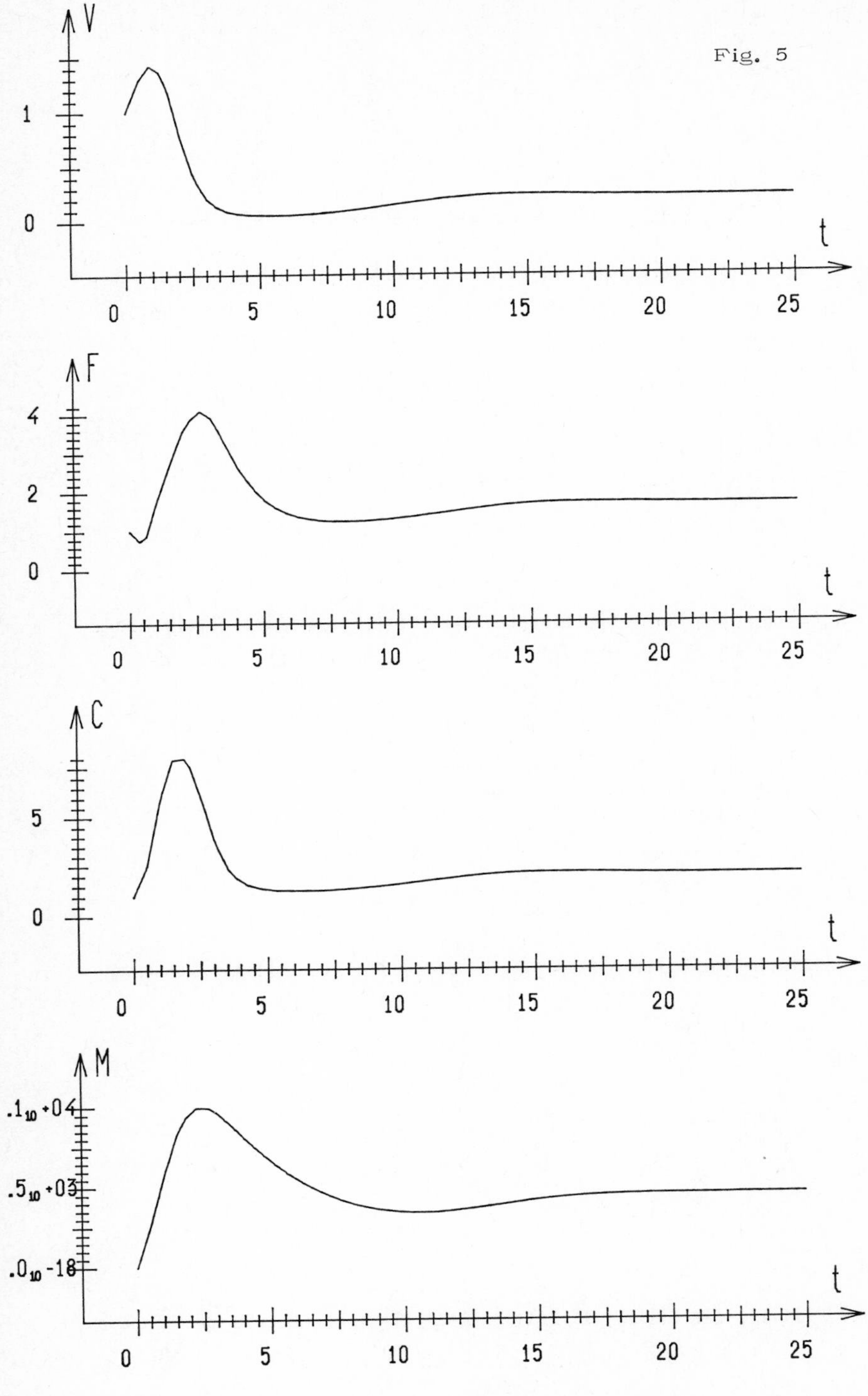
V
1
0
t
0
5
10
15
20
25
F
4
2
0
t
0
5
10
15
20
25
C
5
0
t
0
5
10
15
20
25
M
.1₁₀+04
.5₁₀+03
.0₁₀-18
t
0
5
10
15
20
25

Fig. 5

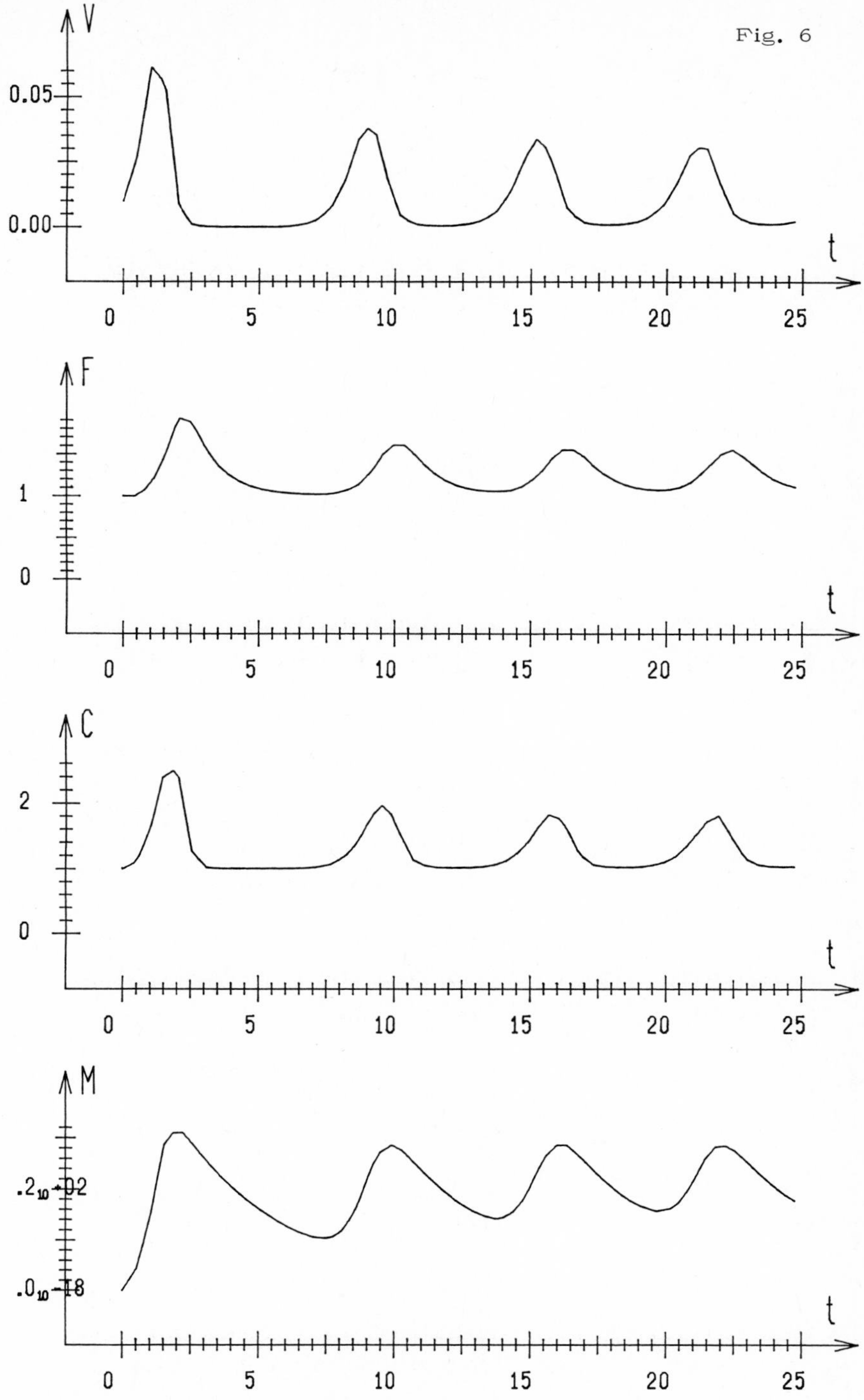
V
0.05
0.00
t
0
5
10
15
20
25
F
1
0
t
0
5
10
15
20
25
C
2
0
t
0
5
10
15
20
25
M
.2$_{10}$-02
.0$_{10}$-18
t
0
5
10
15
20
25

Fig. 6

Fig. 7

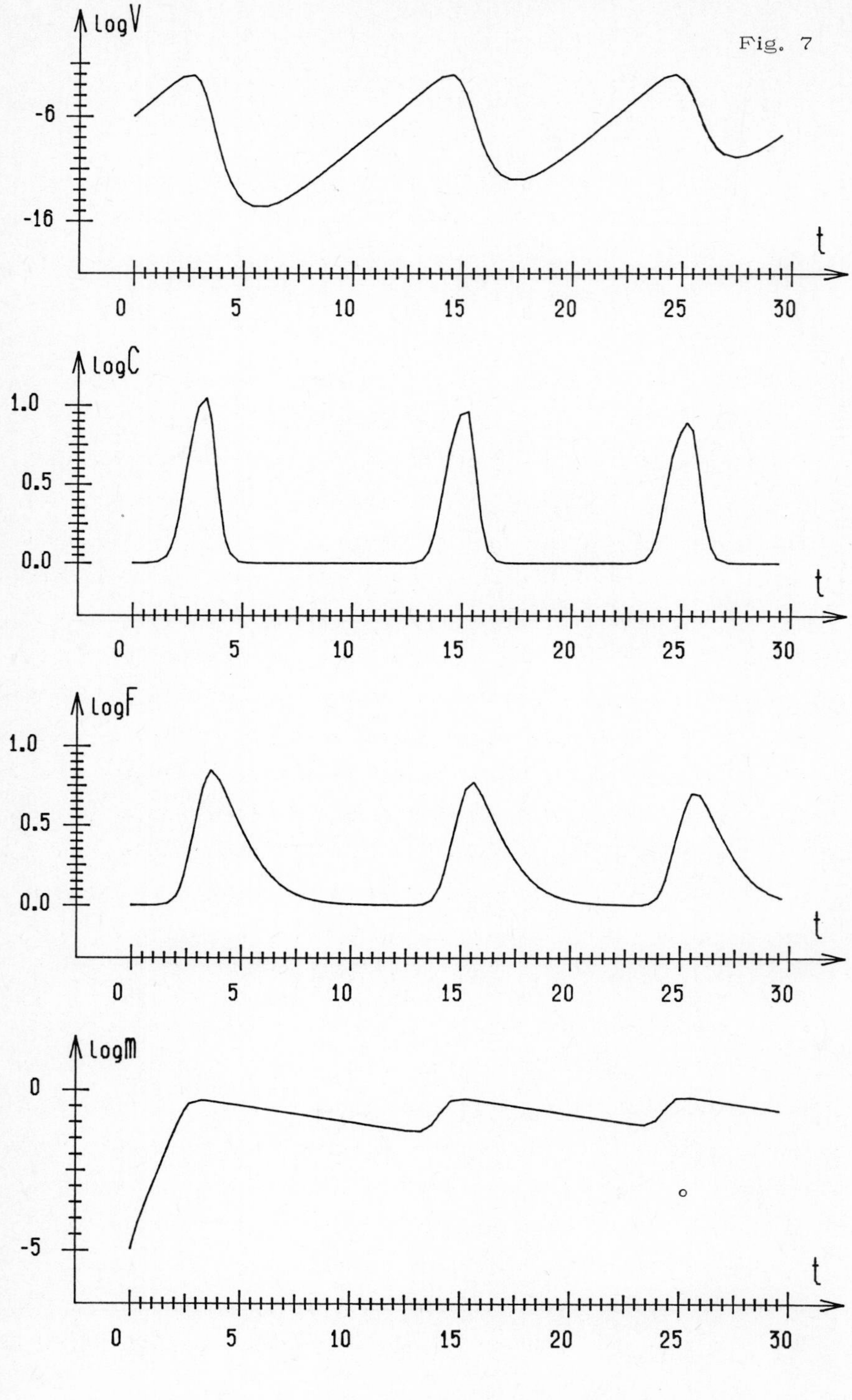
logV
-6
-16
t
0
5
10
15
20
25
30
logC
1.0
0.5
0.0
t
0
5
10
15
20
25
30
logF
1.0
0.5
0.0
t
0
5
10
15
20
25
30
logM
0
-5
t
0
5
10
15
20
25
30

ON SOME SELF-IMMUNIZATION MECHANISMS OF APPLIED MATHEMATICS: THE CASE OF CATASTROPHE THEORY

H. J. Sussmann
Department of Mathematics
Rutgers University
New Brunswick, NJ 08903/USA

1. Introduction

The Catastrophe Theory (CT) phenomenon deserves a close study both by applied mathematicians and by philosophers of science. The questions it raises lead straight into a discussion of some basic issues about scientific method and the sociology of science. The main aim of this lecture is to give a brief sketch — the available time permits no more — of these issues. We feel it is not a coincidence that our lecture will be followed by a panel discussion on World Models. The fact that two plenary sessions of this conference are devoted to the critical analysis of certain proposed applications of mathematics is evidence of a growing awareness by scientists that mathematical theories, if they claim to talk about physical, or biological, or social phenomena, must be subjected to the same kind of critical discussion that is common in the natural, social and biological sciences. One aim of this paper is to show that the attempts to undertake a critical discussion of applied mathematics involve certain peculiar methodological difficulties which derive from the specific nature of this intellectual discipline. Catastrophe Theory will serve as a good illustration of these difficulties.

Most scientists view scientific research as a "trial and error" process. Theories are proposed, and they are discussed critically. The critical discussion of a scientific theory involves its confrontation with experimental facts, with other theories, and with itself (i. e. the search for internal contradictions within the theory). A large amount of work by philosophers of science has attempted to throw light on how this is done. The naive view that the truth of a theory can somehow be established beyond doubt by experiment occurs much less frequently nowadays than the idea —forcefully defended by K. Popper, cf., e. g., [8] and [9] — that theories can be refuted by experiments that contradict them, but can never be proved correct by an accumulation of successful experimental predictions. An attempt to verify a prediction made by a theory constitutes a "challenge" to it. If the theory withstands the challenge then an extra reason has been added for trusting it. A good, reliable theory is one that has successfully overcome many challenges. But it is always possible that such a theory may fail us in the future, when confronted with some new challenge. Hence no theory can be said to be known with absolute certainty.

The view sketched above has more recently been subject to discussion, especially by those who have pointed out that the distinction between "fact" and "theory" is not as clear as it may seem at first, and that no single theory can ever be said to be refuted by an experimental "fact" (e.g. the Michelson-Morley experiment refutes either Galilean relativity or the theories about optics and measurement instruments that are involved in our interpretation of the observations). However, these important problems need not be touched upon here. All we need is an agreement on the most basic point, namely, that the extent to which a theory constitutes a valuable addition to our knowledge depends very strongly on its ability to be challenged and to successfully withstand challenges. It follows from this view that criticism is an integral part of scientific activity, and that efforts to challenge theories are necessary for scientific progress. It is possible to make a case in favor of almost any proposition by a careful one-sided selection of evidence. Only if the search for evidence relevant to a theory is not restricted to those that are a priori inclined in its favor can the outcome of such a search be meaningful. Moreover, it also follows from this view that it is essential that the theory offer the possibility of being challenged. A theory which avoids challenges is like a weather forecaster that refuses to make even a single forecast. It will never be shown wrong, but it would be absurd to regard this fact as evidence for the theory. Now, it is a fact that, in the past, many intellectual activities have operated in such a way as to make it hard or even impossible for anybody to challenge them. This was achieved by procedures ranging from the outright persecution of critics to the use of unnecessarily obscure language to hide the lack of real content. Although the most extreme of such procedures are rare nowadays, milder mechanisms replace them sometimes in today's academic world. We shall use the name self-immunization mechanism to refer to any factor which contributes to grant a proposed scientific theory some a priori immunity from critical challenges. The expression immunization strategy is often used in the philosophical literature with a similar meaning (cf. Albert [1]), but we prefer to speak about "mechanisms" rather than "strategies" in order to avoid the idea — which the latter word inevitably conveys — that these mechanisms are deliberately used by the proponents of the theory in order to immunize it from criticism. Although deliberation may be present in some cases, it is quite likely that in many other situations the existence of the immunization mechanism is an unintended byproduct of other factors.

The main contention of this paper is that applied mathematics, in addition to the immunization mechanisms which are found in other disciplines, is endowed with some peculiar ones that arise from its unique position, half-way between mathematics and the other sciences, and that even those mechanisms that it shares with other sciences take on new characteristics when they operate to protect applied mathematics.

Why concentrate on this topic, rather than on a straightforward discussion of CT? The controversy about CT has already given rise to several publications, and we feel that it is not necessary to repeat here what has already been discussed in a much more detailed form in our previous work (cf. [11] , [20], and [21]). However, we feel that the reaction to the criticism of CT has brought the methodological problem to the forefront, and that it is now time to begin the discussion of this problem. CT proponents have brought this about by their peculiar way of answering criticism. When an assertion A is made about CT (e.g. that CT does not lead to predictive models), they normally reply by acknowledging the truth of A but denying that A should be considered as a valid criticism that can be made of a theory (e.g. by saying: why should a theory have to lead to predictive models?). But, since the Catastrophists themselves do not seem to accept any methodological constraints whatsoever, this puts them in a position which no criticism can possibly affect, since any criticism can be answered by rejecting the critic's methodological presuppositions. So the combination of lack of methodological commitments on the part of CT together with its readiness to argue against critics by objecting to their methodological commitments operates as an immunization mechanism. The question of how this and other mechanisms work and interact thus appears immediately in the discussion, and it is clear that it leads into problems that transcend CT and concern the broader issue of the application of mathematics. We feel that these problems are much more interesting than the particular story of CT. However, since it is our intention to illustrate our discussion with the example of CT, we must first present some facts about it.

2. Some facts about Catastrophe Theory

Catastrophe Theory is the creation of one of the greatest mathematicians of today, René Thom. His book ([13]) was hailed by some as a true revolution in science, and even compared with Newton's Principia(cf., e. g., [4]). Although the book is primarily about biology, it has the subtitle "A general theory of models", indicating the author's aspiration to cover a much broader range. And, indeed, the book itself has sections on other topics, such as: animal in quest of an ego, dreaming, play, organs and tools, the double origin of language, the origin of geometry, art, delirium, human play, basic types of society (there are two, according to Thom: the "military society" and the "fluid society"), money, the mind of a society, a model for memory. Thom's all-encompassing interests have lead him to continue the pursuit of some of these subject by writing articles on symbolism, crises, and other "catastrophic" events.(cf. [14] and [15]).

However, the recent popularity of CT is primarily due to the work of E. C. Zeeman. In a number of articles, he has attempted to apply the theory to biology, economics, sociology, psychology, politics and other fields. An article by Zeeman in Scientific American ([21]), another one by I. N. Stewart in the 1977 Encyclopedia Britannica Book of the Year, and various other non-scholarly publications have disseminated the belief that, indeed, CT is a scientific achievement of outstanding magnitude.

We have argued that this belief is completely unsubstantiated (cf. [11]), and there is no room here for details. However, we shall briefly present a few facts. First, we must make an important distinction ; CT is both the study of certain mathematical questions and the search for applications of the results of this study to other fields of science. The search is undertaken because of the Catastrophists' strong conviction that, due to the nature of their mathematical results, it is to be expected that many applications will indeed be found. Moreover, the Catastrophists claim to actually have found them. For instance, Stewart writes in [10] that CT "...can boast of an enormous variety of applications to broad areas of human concern, including many that had hitherto resisted mathematical description". E.C. Zeeman claims in [21] that the theory "has the potential for describing the evolution of forms in all aspects of nature". One would expect the "boasting" and the fantastic claims to be supported by some evidence that at least some the applications that the Catastrophists have proposed actually work. However, any attempt to look for such evidence is made difficult ab initio because of a semantical problem. The Catastrophists do not seem to distinguish between "application" and "successful application". Any paper in which a model of some phenomenon is <u>proposed</u> is regarded by them as an "application". Since the number of such papers is indeed "enormous", this may provide some rationale for Stewart's boasting. However, the acceptance of such a boasting as legitimate on this basis would be tantamount to admitting that the mere existence of a theory suffices to provide evidence for its correctness, provided only that its creators write sufficiently many papers. This belief may indeed be held by Stewart or Zeeman and, if it is, we know of no argument against it except for the observation that, if accepted, then one has to accept as well that many other theories, such as Velikovsky's cosmic catastrophes, Von Daniken's chariots of the Gods, and even Middle Age demonology have at least as good a claim to legitimacy as CT.

We shall proceed from a different perspective, and take it for granted that "boasting" is legitimate only when substantiated with some evidence that goes beyond the publication of many papers in (mostly unrefereed) journals and books. And then we find the difficulty that this evidence is hard to trace, and that the repeated claims by the Catastrophists that it actually exists do not even resist the confrontation with the sources that they themselves quote. For instance, the claim has been made for quite a few years that Zeeman's biological applications were the most convincing successes. Zeeman himself writes, in April 1976, in [21] : "I have constructed catastrophe models of the heartbeat, the propagation of nerve impulses and the formation of gastrula and of somites in the embryo. Recent experiments by J. Cooke and

T. Elsdale appear to confirm some of my predictions". An independent inquiry carried out by R. Zahler and this author revealed that: a)no experiments exist that confirm any of Zeeman's predictions on nerve impulses; b) the only experiment that is supposed to exist on the heartbeat is one made by Zeeman himself in 1972, the details of which are not available; c)both Cooke and Elsdale fail to agree with Zeeman's evaluation of their experiments.

The fate of the social science applications of CT seems similar. Stewart acknowledges that "most such models still lack precise data", but he claims that "an interesting exception is a study by Zeeman and several collaborators of how tension and alienation among prison inmates influence disorder. A cusp catastrophe fits the data very well". Here, again, it suffices to look at the original paper by Zeeman and his collaborators to see that Stewart's assertion is false (cf. [22]). They do not even assert that one cusp fits the data. They claim that two cusps are needed (the cusp must have moved during the process, they say). And, even for the fitting of two cusps, no attempt is made to evaluate its goodnes by means of some statistical technique. The authors limit themselves to plotting the data points in the plane, and then drawing the cusp curves. A skeptic can achieve an equally good fit by means of many other kinds of curves, and even the very crude agreement shown by the authors between their data points and their cusp curves should not surprise us, given that a)they had a family of curves depending on infinitely parameters to choose from and that b) they granted themselves the additional freedom to change cusps if necessary.

So, the evidence of success in predicting testable results seems not to exist. This is particularly damning to the Catastrophists because of their repeated claims to the contrary, some of which seem to justify the labelling of Zeeman's Scientific American article by M. Kac as "the height of scientific irresponsibility" (cf. [5]). Moreover, the size of the gap between the Catastrophists' claims and the actual truth makes it natural to expect that an even wider gap will be found in the writings by nonprofessionals. And, indeed, Nesweek magazine prints the assertion that "the theory has been hailed as an intellectual revolution in mathematics —the most important development since calculus—", and that one can use it to "make precise forecasts of behavior or events which Zeeman says are superior to any that can be achieved with the best statistical techniques known". And, in a similar vein, the brochure that advertises the volume where Stewart's essay [10] appears states that CT "is being applied to help predict accurately a number of situations (earthquakes, and floods, the ups and downs of the stock market)". These statements go only a bit farther than what Zeeman or Stewart themselves have written, and constitute a natural consequence of their own extravagant expository style.

However, the excesses of some extremists do not in themselves prove a theory wrong. Moreover, in the case of CT it is particularly important to go beyond the criticism of the exaggerated claims by Zeeman and his school, because there are at least two different ways of viewing CT — Zeeman's and Thom's —, and Thom's version of CT is not affected by the criticism of the preceding paragraphs. Whereas Zeeman has devoted most of his effort to the use of CT to make what he considers to be "predictive models", Thom has consistently argued that this is precisley what CT cannot do. In [16], p. 387, he warns that "many people, understandably eager to find for CT an experimental confirmation (?), may embark into precarious quantitavive modelling" (the question mark is his), and he prophetically adds that, among "positivist-minded scientists", this may cause a reaction against CT. Three years later, in April 1977, Thom maintains that "the 'practical' results of CT are, up to now, not very striking; evaluated by the strict-positivist-criterion of the discovery of 'new phenomena', they reduce to a few (not too surprising) facts in geometric optics elaborated by M. Berry at Bristol in his work on caustics" (cf [17], p. 190).

Thom's view of what CT is seems a lot less clear than his description of what it is not. In [17], p. 189, he claims for CT "a novel epistemological status". He denies that CT is a mathematical theory, and writes that it is "a 'body of ideas', I daresay a 'state of mind' ". In p. 193, he indicates that "CT is fundamentally qualitative, and has as its foundamental aim the explanation of an empirical morphology. Its epistemological status is the one of an interpretative-hermeneutic theory. Hence it is not obvious that it will necessarily develop into new pragmatic developments". In [16], p. 388, he tells us that "In social sciences, still more than in exact sciences, the hope of finding quantitative modelling of catastrophes is very slight. Granted that CT leads to basically qualitative modelling, what may be the interest of such models? A first answer, I think, is as follows: CT is —quite likely— the first coherent attempt (since Aristotelian Logic) to give a theory on analogy. When narrow-minded scientists object to CT that it gives no more than analogies, or metaphors, they do not realise that they are stating the proper aim of CT, which is to classify all possible types of analogous situations".

The philosophers in the audience will recognize the language as part of a very old intellectual tradition, (and will see that the "novel epistemological status" claimed by Thom for CT is the same that others in that tradition have claimed). It is the most extreme version of idealistic philosophical rationalism found, for instance, in Hegel, on in contemporary German idealism. The salient features of this tradition are: a) the belief that knowledge acquired by pure reason is the only knowledge that matters, and that "so-called facts" are of lesser importance, and can be disgarded when they conflict with reason; and b) the belief that it is possible to acquire a significantly large amount of such knowledge. Now, traditionally, rationalists of all persuasions have been confronted with the difficulty that pure thought is unable to establish with certainty anything at all about the world external to the thinking subject, not even its existence. In order to hold beliefs a) and b) in a consistent fashion, one has to build a "bridge" other than perception through the senses, for reason to reach out to the external world and get to know something about it. No satisfactory bridge seems to have been found (e. g. Descartes' idea, of proving the existence of God via the ontological argument and then inferring the existence of the external world from the fact that God would not so perverse as to deceive him all the time, does not seem convincing). So this philosophical tradition has developed a number of "interpretative-hermeneutic" approaches which ultimately are based on some incommunicable intuition which somehow allow the philosopher to apprehend the essential aspects of reality. And it has always found it very hard to deal with those who fail to experience the same mystical extrasensory contact with reality as the philosopher. (See, e. g., Kolakowsky's study on Husserl [6]).

The radical novelty of Thom's idealism lies in the fact that he boldly draws from belief a) the conclusion which his predecessors refused to draw. Since the only knowledge that can be acquired by pure thought is that of mathematics (or no knowledge at all, if one wishes to question even this one, which Thom does not) then it is not surprising to find Thom writing that, in his "optique de l'explication scientifique", only mathematical theorising would exist, and that "seul le mathématicien, qui sait charactériser et engendrer les formes durables à longue portée, a le droit d'utiliser les concepts (mathématiques); seul, au fond, il a le droit d'être intelligent" (in Ref. 12, p. 373). Nonmathematicians and experimentalists are regarded as incapable of reaching the heights of knowledge, and only considered suited for the performance of menial tasks (e.g. "You may leave that to professional biologists like Wolpert who are unable to conceive of anything else", in [16], p. 387 and, two paragraphs later :"I agree with P. Antonelli when he states that theoretical biology should be done in mathematical departments; we have to let the biologists busy themselves with their very concrete, but almost meaningless, experiments; in developmental Biology, how could they hope to solve a problem they cannot even formulate?"

So Thom thinks that mathematicians like himself are in possession of a special kind of knowledge, a "body of ideas" which members of other, empirically oriented professions, cannot easily penetrate. What is this body of ideas? It consists, primarily, of a series of results and methods that have arisen from the study of singularities of smooth maps, an area to which Thom himself made fundamental contributions and which he now regards —perhaps not too surprisingly— as the key to the understanding of the world. In order to convey the flavor, we briefly describe some of these ideas.

3. Some mathematics

Suppose that you are interested in the solutions of an equation of the form $f(a, x) = 0$, where f is a real-valued function of the variables a and x. Suppose that you regard a as a parameter and want, for each a, to find the corresponding value, or values, of x, and want to know how x changes as a changes, i.e. you want to study x as a "function" of a (with quotation marks because, for a given a, the set of values of x that satisfy the equation may be empty, or may have more than one element). Then, many things can happen. For instance, it may be the case that for each a there is a unique x (e.g. if $f(a, x)=a-x$), or two (e.g. if $f(a, x)= a^2+1-x^2$), or none (e.g. if $f(a, x)=a^2+1+x^2$). Also, there can be cases where the situation is "qualitatively different" for different values of a, as shown by the example $f(a, x)=x^2-a$, which has the property that for $a<0$ the equation has no solutions, and for $a>0$ it has two. At $a=0$ a drastic, "catastrophic" change occurs. A different example is provided by the function $k(a, x)=x^3+ax$, , which is such that for $a<0$ there are three solutions, and for $a>0$ only one. These are two examples of "catastrophes", and one may ask the question: what types of catastrophes are possible for functions $f(a, x)$? The word "types" indicates that one wants to classify the catastrophes into classes, and then one wants a description of these classes. The classification must be specified by an equivalence relation which somehow corresponds to our intuitive idea of what it means for two catastrophes to be "qualitatively of the same kind" (thus, for instance, the catastrophe that occurs at $a=0$ for $f(a, x)=x^2-a$, and the catastrophe at $a=1$ for $g(a, x)=(x-3)^2-(a-1)^2$ are clearly "the same type of phenomenon", although they differ from each other by a translation of the a and x coordinates. On the other hand, the function $h(a, x)=a-x^2$, which is obtained from f by a rotation of the axes, does not involve this catastrophe). These examples give some indication of what the definition of equivalence should be like. We should declare two catastrophes to be equivalent if one can obtain one from the other by means of a change of coordinates, but we should not allow arbitrary changes, since this would force us to regard f and h as equivalent, which we do not want. The precise definitions are as follows. Two sets S, S' of points in the a, x plane are <u>equivalent</u> if there is a C^∞ diffeomorphism T of the plane which maps S onto S' and takes vertical lines (i.e. lines a=constant) to vertical lines. A similar definition can be given for <u>local equivalence</u> (i.e. for equivalence of germs of sets), by only requiring that the transformation be defined in some neighbourhood of the given base point. Also, naturally, one can define equivalence in a similar way when a and x are vectors rather than scalars.

<u>Thom's theorem for one control parameter</u> asserts that, <u>for "most" families of functions F of one variable x depending on one parameter a, the critical set of F (i.e. the set of solutions of $\partial F/\partial x=0$) is equivalent to the set of solutions of $x^2-a=0$ (i.e. :if C is the critical set of F, and p is an arbitrary point of C, then there is a point q in the set of solu-</u>

tions of $x^2-a=0$ and a transformation T of some neighbourhood U of p to some neighbourhood V of q which is C^∞, invertible, maps verticals to verticals, takes p to q, and is such that, for r in U, then r is in C if and only if and only if $T(r)$ is a solution of $x^2-a=0$).

The word "most" means the following: the set of those F that satisfy this property is open and dense in the space $C^\infty(\mathbb{R}^2, \mathbb{R})$ of all C^∞ real-valued functions of two variables, endowed with the C^∞ Whitney topology (cf., e.g., [2]). The intuition behind this theorem is as follows: certain kinds of "catastrophes" appear in a "stable way", i.e. cannot be removed by means of a small perturbation of F. Other catastrophes are of a more "special" kind in that, when an F exhibits one of them, then one can find, arbitrarily close to F, families of functions F' for which this catastrophe does not occur. As an example, the situation that arises from the family F_1 given by $F_1(a,x)=x^3/3-ax$ (whose critical set C_1 is the solution set of $x^2-a=0$) is stable in this sense. The critical set C_1 is a curve in $\mathbb{R}^2$ which is almost like the graph of a function $x=X(a)$, except for the fact that it folds over itself at the point $(0,0)$. Any critical set C corresponding to a family of functions F which is sufficiently close to F_1 will have to exhibit this same feature. The point where C folds over itself will not necessarily be $(0,0)$, but a folding will have to occur at some nearby point. On the other hand, the catastrophe that occurs at $a=0$ for the function $K(a,x)= x^4/4 + ax^2/2$ is of the "unstable" kind. To see this, observe that the critical set of K is the solution set of $x^3+ax=0$. As was pointed out before, its salient feature is that, for each negative a, the vertical through $(a,0)$ contains three points in the set, and that these three points collapse into one at $a=0$ (the catastrophe point). Think of a as "time", and of K as a function of x which is "changing with time". Then K has three critical points for negative a and those three collapse into one at $a=0$. Now, being somewhat vague, we can also describe this by saying that two of the three critical points collapse against each other and annihilate each other (exactly as in the "fold" case discussed above) and that this collapsing happens to take place exactly at the place where the third critical point is located. The fact that the collapsing occurs precisely there is "very special situation", and a small perturbation of K should be able to eliminate this by changing the location of the collapsing to a point which is close, but not identical, to the other critical point. Now this is, naturally, not very precise. But the reader should be able to verify by himself that, if $b \neq 0$, the family of functions K_b defined by $K_b(a,x) = x^4/4 + ax^2/2 + bx$ has the property that its critical set only contains "catastrophes" of the fold type.

More generally, let us define the critical set $C(F)$ of a family of functions $F(a_1, \ldots, a_p; x_1, \ldots, x_n)$ (i.e. a function of $p+n$ variables, thought of as a function of the last n variables which depends on the first p as parameters) to be the set of solutions of the system of equations

$$\partial F/\partial x_1 = 0 \quad , \quad \ldots \quad , \quad \partial F/\partial x_n = 0 \quad .$$

Equivalence of critical sets and of germs of critical sets is defined as before. An ordinary point of $C(F)$ is a point P in $C(F)$ with the property that, at P, the Hessian determinant $\det(\partial^2 F/\partial x_i \partial x_j)$ does not vanish. Equivalently, P is a critical point if, and only if, it is possible to apply the implicit function theorem to conclude that, in some neighbourhood of P, the equations defining $C(F)$ can be solved for the x variables in terms of the a's. A point in $C(F)$ which is not ordinary will be called singular. If (a,x) is a singular point, then we call a a catastrophe point of F, and we say that a catastrophe occurs at a. If F, F' are families of functions, with the same p,n and if P, P' are ordinary points of $C(F)$, $C(F')$, then the germs of $C(F)$ at P and of $C(F')$ at P' are equivalent. For $p=1$, $n=1$, a precise statement of Thom's theorem is: there is an open dense subset U of $C^\infty(\mathbb{R}^2, \mathbb{R})$ such that, if F is in U, then the critical set $C(F)$ consists only of ordinary points and of fold points. (Here, a fold point of $C(F)$ is a point P such that the germ of $C(F)$ at P is equivalent to the germ at $(0,0)$ of the solution set of $x^2 - a = 0$).

For $p=2$, $n=1$, there is a similar result, which we now state. First, let $F_1(a,b,x) = x^3/3 - ax$. The singular points of $C(F_1)$ are, clearly, the points $(0,b,0)$, with b arbitrary It is easy to see that the germs at any two of these points of the set $C(F_1)$ are equivalent. If C is a subset of $\mathbb{R}^3$, , we shall say that C has a fold singularity at a point P of C if the germ of C at P is equivalent to the germ at $(0,b,0)$ of $C(F_1)$ for some (and hence all) b. Now let $F_2(a,b,x) = x^4/4 - ax^2/2 + bx$. The critical set of F_2, or any subset of $\mathbb{R}^3$ which is equi alent to it, will be called a cusp surface. The singular points of $C(F_2)$ are all folds, with the exception of $(0,0,0)$. If F is a C^∞ function in $\mathbb{R}^3$, let us say that F has a cusp singularity at a point P if the germ of $C(F)$ at P is equivalent to the germ of $C(F_2)$ at $(0,0,0)$. Then Thom's theorem for $n=1$, $p=2$, says: there is an open dense subset U of $C^\infty(\mathbb{R}^3, \mathbb{R})$ such that, if F is in U , then every point of $C(F)$ is either an ordinary point, or a fold, or a cusp. So, for $n=1, p=2$, there are two kinds of catastrophes that are "unavoidable", namely, folds and cusps. Other catastrophes can occur (e.g. if $F(a,b,x)=ax$, then $C(F)$ is the vertical plane $a=0$) , but they are "avoidable", or "unstable", in the sense that a small perturbation can remove them.

A similar theorem is true for $p=3$, and for $p=4$;. In the case $p=4$ there are seven "unavoidable catastrophes", the famous "elementary seven". There is no need to describe them here. (See, e.g. [18]). Also, one still finds seven catastrophes independently of the number n of internal variables, so that n can be arbitrarily large. However, the same kind of theorem cannot be proved for a larger number of parameters, since , as soon as $p=6$, there appear infinitely many nonequivalent types of catastrophes. (For $p=5$ there are eleven, as shown by Siersma in [23]).

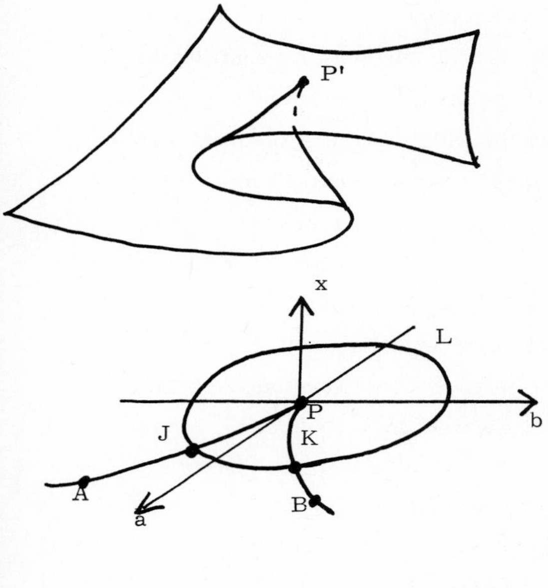

Figure 1. A cusp surface. Point P' is a cusp singularity, and its projection P a cusp catastrophe point. The curve through A, J, P, K, B is the bifurcation curve, or catastrophe set. The a, b plane is the control plane, and x is the internal, or state variable . The significance of the circle in the control plane that passes through J, K, and L is explained in Section 4

4. The Bridge

So far, we have presented some results of pure mathematics. No statement has been made about any relationship with the external world. In the view of the Catastrophists, the mathematics that we have described ought to be expected to be of help to build explanatory theories about other phenomena. Why? Because a large number of situations should, p sumably, be describable in terms of a state space (the space of the x variables) on which a potential function F is defined, which depends on some other variables (the a's) as parameters. When those parameters change, the internal state x given by the equilibrium equations $\partial F/\partial x_i = 0$ will change, and the change will not necessarily be continuous, i.e. "catastrophes" will occur. The claim made by Thom and his followers is that his theory tell us how this will happen, i.e. that we can deduce from the theory which types of catastrophic behavior will occur. In an overenthusiastic description of these ideas, Zeeman writes (in [21], p.65), that "the underlying forces in nature can be described by smooth surfaces of equilibrium, it is when the equilibrium breaks down that catastrophes occur. The problem for CT is therefore to describe the shapes of all possible equilibrium surfaces. Thom has solved this problem in terms of a few archetypal forms, which he calls the elementary catastrophes". This statement involves a number of unstated assumptions, as well as a distortion of Thom's results. Some of the unstated assumptions are: 1) That the underlying forces in nature can be described by smooth surfaces of equilibrium; 2) That "all possible equilibrium surfaces" means the same as "the surfaces that arise in the very particular framework to which Thom's theorem applies", and 3) that "describing the shape of a surface" —i.e. what Thom's theorem does, according to Zeeman— is the same as "listing the types of singularities (up to equivalence) that the surface contains" —which is what Thom's theorem really does— Now, the first of these is essentially meaningless (how does one "describe forces" by a "smooth surface of equilibrium"?). The third one is clearly false, since the information given by the conclusion of Thom's theorem deals exclusively with the local properties of the surfaces, whereas the concept of shape is a global one. In Thom's theory, a critical set in the a, x plane which is a horizontal line, and another one which is the graph of a wildly oscillating function $x(a)$ are equivalent, i.e. have the same "shape".

Now, this is either false, if Zeeman means by "shape" what most people usually do, or misleading, if he means something else, since he does not say what he means, thereby leading the reader to believe he is using the word in its ordinary sense. Finally, unstated assumption 2) amounts to believing that "all possible equilibrium surfaces" arise as critical sets of families of functions, and no substantiation of this is given by Zeeman.

As we said before, there is also a distortion of Thom's theorem. The theorem does not give us information about all possible critical sets, but only about generic ones (i.e. the critical sets C(F), for F in some open dense set of the space of all families of functions). So, a particular "equilibrium surface" that appears in a specific situation need not be one that has the singularities allowed by Thom's theorem, even if it arises as a critical set. Thom is clearly aware of this, as shown by his assertion, in [16] p. 387, that "in no case has mathematics any right to dictate anything to reality. The only thing one might say is that, due to such and such theorem, one has to expect that the empirical morphology will take such and such form. If reality does not obey the theorem —that may happen— this proves that some unexpected constraints cause some lack of transversality, which makes the situation all the more interesting". So, in Thom's view, if a critical set is known to be of importance for some phenomenon, and if one knows nothing about it, one ought to expect it to be in that open set where Thom's theorem applies, because it would be quite exceptional if it did not. If it is, then this is good. If it is not, then this is also good, because we are now confronted with the very interesting problem of finding out why it was not. Perhaps a parallel may be drawn with the following hypothetical situation: suppose you have proved that a number, randomly chosen in the interval $[0, 10]$, is, with probability one, transcendental. Then you can say, if you wish, that, whenever you are confronted with a number in that interval (say π, or Euler's constant C), it is natural to expect that it will be transcendental. It may not be but, if it not, it is interesting to ask why. You can then claim the proof by Hermite and Lindemann that π is transcendental as evidence for your approach. You will not be very helpful to those who are trying to answer the same question about C, but you know you will win no matter what. If C is transcendental, you can say "I told you so" and, if it is not you can say "I never asserted it was".

An even more fundamental difficulty exists. When we talk about a "randomly chosen" number, we have a particular process in mind for generating it. The theorem of the preceding paragraph only applies if the numbers are distributed according to the probability distribution used in the theorem, i.e; the uniform law. Now, one can easily generate numbers "at random" in ways that will never lead to a transcendental number (e.g. by asking children to write numbers with no more than ten decimal figures) and, if this is done, then it is no longer reasonable to expect, when confronted with a number so generated, that it will be transcendental. In a similar fashion, the process by which Nature confronts us with phenomena in need of explanation may be random, but we have no reason to believe that its randomness is of the particular kind that will make it reasonable to expect that Thom's theorem will apply. If any evidence at all exists, it points against rather than for Thom's belief. Indeed, the basic laws of physics are clearly nongeneric, various physical quantities are conserved, and, in a different vein, the catastrophists have not yet been too successful in finding all those cusps which they thought should be so easy to find. Thom would reply that this raises the interesting problem of explaining why this is so. But the difficulty here is that the problem of explaining why something does not happen is interesting only if there is the a priori belief that it ought to happen. Since Thom presents no evidence why we should believe that his catastrophes should be found everywhere, the fact that they are found so rarely does not appear to be of particular significance. The interest of CT becomes circumscribed to a subjective problem of the Catastrophists, namely, their difficulty in explaining to their own satisfaction why their expectations are not fulfilled in nature. Any set of propositions can be presented in a similar way, and a similar "epistemological status" can be claimed for it. The belief in a Flat Earth is also a "state of mind" which leads to expect certain things, such as that, if you fly a plane in a straight line, you will not come back to the starting point. If you find you did come back, you can either regard that as evidence against your belief, or as proof that your belief raises interesting new problems, such as that of finding an explanation for your return which is compatible with a flat earth. If you do the latter, your mental universe becomes disjoint from that of believers in a round earth, since the feelings that make you entertain your belief are incommunicable. In a similar way, the world of the Catastrophists operates on the basis of reasons which seem hardly intelligible to an outsider. Communication seems difficult, and the only alternative left is to look from outside to see how it works.

A simple example will show us how. Let us recall that the cusp surface shown in Figure 1 is, in a certain sense, a kind of critical set that arises "naturally". If a circle centered at the origin is drawn in the a, b plane, then the circle will cut the catastrophe set in two points , J and K . If we move counterclockwise along this circle, then , for a point such as L , the function F_2 will have a unique minimum. As we reach J , a point of inflexion appears which, after crossing J , becomes a new maximum-minimum pair. At K , the maximum and the "old" minimum come together and disappear. This bit of mathematics is turned by Thom into the following: "...For an animal, feeding, that is restoring its reserves in chemical energy, is the most fundamental regulative process. This also is a periodic process, hence it is described by a loop, which we call the predation loop. Here we meet with a fundamental difficulty: predation implies the presence of a prey, that is, a being external to the animal itself. Feeding is, fundamentally, engulfing a prey in the organism (as seen very clearly in phagocitosis for the unicellulars). Hence we have, to describe the the predation loop, to use the simplest of the capture catastrophes, the Riemann-Hugoniot catastrophe [i. e. the cusp]. The predation loop is the circle in the unit plane ... This circle meets the bifurcation curve in two points J , K . In J appears a new minimum, an actor. In K the newly appeared actor captures the old one, K is the catastrophe point for capture. But if we continue to describe the unit circle we see that, after a turn, the predator, in a hungry state, becomes its prey. This apparently paradoxical statement may involve in fact the explanation of many facts in Mythology (The Werwolf), in Ethnology (Hunting rituals involve in general simulation of the prey by the hunters), in magic thinking in general". ([24]). This type of reasoning is representative of the logic used by Thom in his book (where the same picture is used to talk about father and son, and where captures of preys are repeatedly described in similar terms). His other biological models are very ellaborate, but the pattern is always the same, namely, to observe the existence of some vague analogy between something observed and something known to happen in the theory of singularities , and then to declare that this analogy somehow "explains" , or "may explain", or"gives an insight into" the phenomenon. The insight is usually incommunicable, as it leaves most readers asking how and why the phenomenon is "explained". Moreover, the steps are incommunicable as well. Why is predation a periodic phenomenon? It certainly has some periodic aspects—e. g. a wolf may eat rabbits periodically—as well as nonperiodic ones —it is a different rabbit every time—, but why does Thom ignore the latter? How does it follow that predation must be represented by a loop, and what does this mean? Why does the fact that the prey is a being external to the predator imply that predation must be described by a cusp catastrophe? And what does it mean to say that the cusp "describes" predation? Is anything more being claimed other than the obvious fact that there is a vague analogy between the cusp and predation, in that in both cases there is something that appears and later something else that disappears? If more than this is being claimed, then what is it? If no more is being claimed, then what is the value of this particular analogy? Why isn't any other phenomenon where an appearance is followed by a disappearance equally good as a "description" of predation? Why not, for instance, say that predation is "described" or "represented" by the arrival of train A to a station, followed by the departure of train B ? It may be argued that train A does not "engulf" train B , but neither does the minimum born at J "engulf" the old one in any reasonable sense of the word, since, after all, the disappearance of the old minimum takes place far away from the place where the new one is located .

It seems hard to answer these questions except by telling the skeptic that we are dealing with an "insight" that may or may not be experienced, but cannot be communicated to those who do not experience it. This is the fate of all extreme forms of idealistic rationalism: the thinker immerses himself in his own thoughts so deeply and for so long that the products of his imagination acquire for him the status of objective, self-evident truths. He then believes that his thoughts have succeeding in building a bridge to the outside world, whose pillars are built with the solid material of his reason, rather than the uncertain data conveyed by the senses. But, sadly, only he can see the pillars. What seems self-evident to him, is unable to withstand the questions by the onlookers. He then needs to protect himself, and it is to the protective mechanisms that we now turn.

5. The immunization mechanisms

a) The methodological double standard. We have already described this mechanism in the introduction, so we shall deal with it very briefly. It consists of not accepting any methodological constraints, while turning all criticism into a discussion about the methodological presuppositions of the critic. Its typical manifestation is a propensity to explain what the proposed theory is not, and to be vague about what it is. Any objection can then be taken to be evidence that the critic does not understand "the proper aims of the theory". But the "proper aims" are never stated, or stated only in vague general terms (e. g. "the classification of all types of analogous situations").

b) Lack of commitment. A theory which commits itself to an assertion that can independently be examined by others must accept the consequences when it is found wrong. Einstein, some years after he had introduced the cosmological constant, decided that he had been wrong, and publicly called the constant the biggest error he had made in his life. Catastrophists "boast" about the applicability of their theory, but refuse to commit themselves to the defense of any particular example. At each moment in the history of CT, it has been true that the only applications on which a defense of CT was based were those that had been proposed so recently that critics had not yet had time to examine them. As of today (11/23/77), the "boasting" by Stewart about applications to "broad areas of human concern" seems to have been replaced by a more modest request that the scientific status of CT be evaluated on the basis of a few applications to physics.(I. Stewart, letter to The Sciences). If these are found faulty then, presumably, the critic will be told that a set of even newer applications, just developed three days ago, is the one that should really be used. The consequence of this is to turn the critical examination of CT into a velocity contest in which CT wins provided that it succeeds in producing papers faster than the critics can study them.

c) Justification by mathematics. Einstein never attempted to defend the cosmological constant by claiming that he had never said that the Universe was that way, and was only playing a purely mathematical game. In today's applied mathematics, one finds often theories about other phenomena that are perceived by everybody as talking about these phenomena but, when objected to on the basis of their poor agreement with reality, are defended by claiming that they were only meant as mathematics, or as an exploration of how one might deal with the phenomenon if certain things (which are known to be false) were true. In [3], Isnard and Zeeman claim to study the influence of public opinion on an administration. They state several "assumptions", and assert that, using "the deep theorems of catastrophe theory", they can obtain some conclusions. "It is not immediately obvious that the sociological hypotheses imply the sociological conclusions, and that is the purpose of using CT". As shown in [1], the sociological hypotheses are too vague for anything to be provable from them and, depending on how they are interpreted, the conclusions either do not follow or follow trivially (i.e. without "deep Mathematics"). But our main point here is a different one: Isnard and Zeeman create the presumption that they are making a theory about the phenomenon (e.g. "To test the theory, the social scientist would have to test the hypotheses by experiment, or by interpretation of historical data", or "From the model one can explain, predict, and relate a variety of phenomena that previously may not have appeared to be related".) But they never really say they are. The critic who argues that administrations do not behave that way can then be dismissed with a "So what? We only wanted to show the use of the mathematical technique to determine what would be the case if the hypotheses were true, and this is interesting even if the hypotheses are false!"

d) Justification by application. If the critic of the Isnard-Zeeman paper then points out that the preceding justification is only acceptable (if at all) if the mathematics is rigorous (since no claim is really being made other than the usefulness of the theory as a tool to connect hypotheses with conclusions) and that the mathematics of Isnard and Zeeman is not rigorous, nor even intelligible (e.g. near the beginning of their paper, they announce that they will "introduce sociological hypotheses, and translate them into mathematics". A typical sociological hypothesis is" if the cost of the war is low, the opinion will be unified". Contrary to the promise, no mathematical translation is given. Now, "low" is not a precise mathematical word), then the reply will be: we are doing applied mathematics, and writing for nonmathematicians, so there is no need to be rigorous. This reply has often been used in public discussions on CT and, if combined with c), provides perfect immunization.

e) Justification by existence. As shown before, Catastrophists argue that CT has an enormous number of applications, and produce as evidence the many papers where applications are proposed. The existence of all those papers is hard to deny, but its value as evidence is null.

f) Technical use of nontechnical words Thom insists that, in CT, the word "Catastrophe" is used in the precise technical sense that was explained above. Statements on earthquakes and floods are based on a misconception by the non-mathematical audience. One wonders if the same is true about werewolves.

g) The intimidatory power of mathematics. Its existence can hardly be denied. It contributes to make criticism hard, because the —mostly nonmathematical— public presumes mathematicians to be always right when they talk mathematics, and the critic can only make his points by explaining the issues, and demanding that the audience exercise its own judgement.

h) Illusory precision. The mathematical concept of catastrophe may be a precise one, but the nature of its link with the real world is not at all clear from the writings by the Catastrophists. By giving precise definitions of mathematical entities (even if called with nonmathematical words, cf. f) above), and by proving rigorous theorems about them, the illusion is created that the theory is precise. This makes it harder to see that precision is totally absent precisely where it is most needed, i.e. in the justification for the claim of a link with the nonmathematical phenomena that the theory supposedly describes.

Conclusion

The eight mechanisms listed above are far from being a complete list. However they suffice to illustrate our main point, as announced in the introduction, i.e. that the question of immunization mechanisms acquires distinctively peculiar characteristics in the case of Applied Mathematics. The last six of them seem to us to be unique to it. As for the first two, they clearly are not, but they assume a new form in Applied Mathematics. Because of the certainty of its theorems, mathematics has traditionally stayed away from methodological problems. No clear criteria are agreed upon as to how to evaluate works in Applied Mathematics. This makes it an extremely fertile ground for theories which do not commit themselves to specific statements about facts, nor even to some methodological constraints. What is needed is the acceptance of the legitimacy and the need for critical activity. We look forward to it.

REFERENCES

1 Albert, H., Traktat über kritische Vernunft, Tubingen, Mohr, 1969.

2 Golubitsky, M., and Guillemin, V., Stable maps and their singularities, Springer-Verlag, Graduate Texts in Math. 14, 1973.

3 Isnard, C. A., and Zeeman, E. C., Some models from catastrophe theory in the social sciences, in The use of models in the social sciences (L. Collins Ed.) London, 1976.

4 Kilmister, C., review of Thom [13], quoted in the back cover of [13].

5 Kolata, G., Catastrophe Theory, the Emperor has no clothes, in Science, April 15, 1977.

6 Kolakowski, L., Husserl and the search for certitude, Yale Univ. Press, 1975.

7 Panati, C., Catastrophe Theory, in Newsweek, January 19, 1976.

8 Popper, K., The logic of scientific discovery, Basic Books, New York, 1959.

9 Popper, K., Conjectures and refutations, Basic Books, New York, 1963.

10 Stewart, I., Catastrophe theory, in Encyclopedia Britannica book of the year 1977.

11 Sussmann, H. J., and Zahler, R., Catastrophe theory as applied to the social and biological sciences, a critique, Synthese, to appear.

12 Thom, R., D'un modèle de la science à une science des modèles, Synthese 31(1975), 359-374.

13 Thom, R., Structural stability and morphogenesis, Benjamin, New York, 1975.

14 Thom, R., De l'icone au symbole, Cahiers Internationaux de Symbolisme 1973, 85-106.

15 Thom, R., Crise at catastrophe, Communications, 25, 1976, 34-38.

16 Thom, R., Answer to Christopher Zeeman's reply, in Dynamical Systems Warwick 1974, Springer-Verlag, Lecture Notes in Math. No. 468, 384-389.

17 Thom, R., Structural stability, catastrophe theory and applied mathematics, SIAM Review 19, No. 2, 189-201.

18 Wasserman, G., Stability of unfoldings, Springer-Verlag, Lecture Notes in Math. No. 393, 1974.

19 Zahler, R., and Sussmann, H. J., Catastrophe theory: exaggerated claims, meager accomplishments, Nature, to appear.

20 Zahler, R., and Sussmann, H. J., Mathematics misused, the case of catastrophe theory, The Sciences, Vol. 17, No. 6, Oct. 1977, 20-23.

21 Zeeman, E. C., Catstrophe Theory, in Scientific American, April 1976, p. 65-83.

22 Zeeman, E. C., and four coauthors, A model for institutional disturbances, British Journal of Mathematical and Statistical Psychology, Vol. 29, Part 1, 1976, p. 66-80.

23 Siersma, C., The singularities of C^∞ functions of right codimension lesser or equal than eight, Indag. Math. 35 (1973) p. 31-37.

24 Thom, R., A global dynamical scheme for vertebrate embriology, Lectures on Mathematics in the Life Sciences, Amer. Math. Society, Providence, Rhode Island, 1973.

WORLD MODELS

D.C.J. de Jongh
National Research Institute for Mathematical Sciences
CSIR, P.O. Box 395, Pretoria, South Africa

The problems discussed under the heading 'world models' are important issues invol= ving mankind as a whole. A recent description of this complex of problems was given by Alexander King [1]. The following problems are listed.

1. The present very rapid increase in the world population.
2. The concomitant increased demand for food, energy and materials.
3. The excessive disparities, both within and between countries.
4. The rise of expectations, both within and between countries.
5. Uncertainty as to the basic value systems of society.
6. The subordination of longer-term issues to matters of political expediency and the gratification of immediate desires.

The report also suggests that the darkest shadow lies on the problems interrelating population increase, food production and energy.

Our own study in this area involved an analysis of some of the mathematical aspects of the 'Limits to Growth' world model [2]. The techniques described by Tomovic [3] were used in an attempt at a comprehensive sensitivity analysis of the Meadows model. This has been found to be very sensitive to small perturbations of certain parameters [4]. If only three constant parameters in the capital sector of the model are each given a 10% change in the year 1975, the disastrous population collapse, which is an important conclusion of the Meadows study, is postponed to beyond the simulation interval, i.e. to beyond the year 2100 [5]. These changes are well within the limits of accuracy of the data, as described in the Meadows technical report [6]. The idea of parameter sensitivity has subsequently been generalized to the concept of so-called structural parameter sensitivity analysis, with resulting implications for the Meadows world model and modelling in general [7].

The results described above imply that a combination of small changes of parameters may be sufficient to produce a qualitatively different predicted outcome for the evolution of the world. Far-reaching philosophical arguments and policy recommen= dations based on such a flexible mathematical model may therefore be invalid and misleading.

On the other hand, let us suppose for argument's sake that the Meadows model gives a reasonable portrayal of the relationship between, in particular, the economic and population subsectors. Then the above analysis has uncovered the sensitive pressure points of the system. In order to obtain a desirable evolution of the world system, parameter optimization procedures may be employed involving a suitable cost function and the use of the indicated sensitive parameters. Thus the same result could pos= sibly be achieved by small pressures in the appropriate directions instead of the drastic changes proposed by Meadows [2].

A very important word of warning comes from economist Milton Friedman in his Nobel lecture [8]. 'In order to recommend a course of action to achieve an objective, we must first know whether that course of action will in fact promote the objective. Positive scientific knowledge that enables us to predict the consequences of a possible course of action is clearly a prerequisite for the normative judgment whether that course of action is desirable'. The term 'positive scientific know= ledge' in the context of world models can be interpreted as referring to knowledge of the dynamics of the system. In the Meadows case it means the mathematical struc= ture of the difference equations describing the model. In our opinion it is pre= cisely this aspect of world modelling which is still in its infancy. Adequate knowledge of the dynamics of a system as complex as the economic-demographic-poli= tical system of the world is perhaps impossible to gain.

References

[1] King, Alexander. Report on the state of the planet. *International Federation of Institutes for Advanced Study. Stockholm*, 1976.

[2] Meadows, D.H., Meadows, D.L., Randers, J. and Behrens, W.W. The limits to growth. *Potomac Associate/Universe Books, New York*, 1972.

[3] Tomovic, R. and Vukobratovic, R. General sensitivity theory. *Elsevier, New York*, 1972.

[4] Vermeulen, P.J. and De Jongh, D.C.J. Growth in a finite world - a comprehen= sive sensitivity analysis. *Automatica* 13 (1977), *no.* 1, 77-84.

[5] Vermeulen, P.J. and De Jongh, D.C.J. Parameter sensitivity of the 'Limits to Growth' world model. *App. Math. Modelling*, 1 (1976), *no.* 1, 29-32.

[6] Meadows, D.L. *et al.* The Dynamics of growth in a finite world. *Wright-Allen Press, Cambridge, Mass.* 1973.

[7] De Jongh, D.C.J. and Vermeulen, P.J. Structure and prediction in 'The Limits to Growth'. *Proceedings of the international conference on cybernetics and society*, 1976. *The Institute of Electrical and Electronics Engineers, Inc., New York*, 1976, 135-142.

[8] Friedman, Milton. Nobel lecture: inflation and unemployment. *J. of Political Economy* 85 (1977), *no.* 3, 451-472.

World Modeling

Hugo D. Scolnik
Candido Mendes University and Federal University of Rio de Janeiro, Brazil.

Although there exists the tendency to talk about world models as if they were all alike, in fact the existing ones are very different from each other. The Forrester - Meadows {1,2} models are simple systems of linear differential equations with initial conditions solved by crude numerical methods. Thus, the "future" is completely determined by the structural equations and the initial values, without including adaptative capabilities for facing changing conditions. The authors of the second report to the Club of Rome {3} claim they have used a "multilevel hierarchical approach". In fact, the computer listings show that nothing different from a standard numerical simulation model had been implemented {4}. Another type of models are those based upon input-output techniques like the one developed by V. Leontief for the United Nations {5}. These are "accounting" models useful for studying the possible consequences of exogenously determined policies.

The last category I would like to talk about corresponds to models employing optimization techniques like the one we have developed in Latinamerica from 1971 to 1976 {6}. This model was built with the aim of studying the feasibility of achieving the satisfaction of the basic needs (defined in terms of food, housing, education and health) in different regions of the world. This "target oriented or goal seeking model" tries to calculate the optimal allocation of resources (capital and labour) among the different economic sectors in order to achieve the goals as fast as possible. From the mathematical viewpoint the problem can be characterized as a "non-conventional" non-linear discrete time control one, because very often it is not possible to satisfy all the constraints. Hence, a hierarchical order of the restrictions is defined, and the optimization procedure violates the less important ones if no feasible solution exists. Moreover, there is no analytical representation of the whole problem since, for instance, linear programming is used for the internal allocation of resources with in the agricultural submodel. In other words, one is forced to use derivative free optimization techniques for solving a non-linear programming problem in which not all the involved functions are differentiable. I think this sort of models are of special interest for the IFIP audience because they present a great variety of unsolved problems in sensitivity analysis, long-term optimization, etc.

On the other hand, I do believe scientific organizations like IFIP should help to enlighten the public's opinion regarding world models since they deal with some of the key problems faced by mankind. Naturally, the different conceptual and political positions should be respected, but another matter is to push for drastic actions to be taken at global level based upon shaky foundations. As an example, the conclusions arrived at in the book "Limits to Growth" called for the detention of the demographic and economic growth (hence condemning the poor masses of the world to remain in miserable conditions forever) as the "only" solution to avoid a global disaster. A sensitivity analysis made in 1972 {7,8}, showed that by changing in less than 5% some of the parameters (like the depreciation rates of the capital and consumer goods which, by the way, were chosen as the ones of the United States for the whole planet) no disaster appeared up to the year 2300. Similar results were later obtained by de Jongh {9} and Gelovani {10}. This does not mean that technically correct world models cannot be used for studying global problems. In fact, global optimizing models are very useful for deriving macro-policies and, linked with detailed sectorial models, can be effectively used for real planning. Along this direction, UNESCO is organizing different activities (national applications, courses for planners from developing countries, etc) using the Latinamerican model as a tool. But, of course, a lot of serious interdisciplinary research must be done in this fascinating field which is a crossroad of many branches of the scientific knowledge.

{1} Forrester, J.W., World Dynamics, Wright-Allen Press, Cambridge, Mass. 1971.

{2} Meadows, D.H., et al, The Limits to Growth, Universe Books, New York, 1972.

{3} Mesarovic, M. and Pestel, E., Mankind at the Turning Point, E.P. Dutton, New York, 1974.

{4} Saavedra, N, Pericchi, L., and Sagalousky, B., Analisis critico del modelo mundial de M. Mesarovic y E. Pestel, Simon Bolivar University, Caracas, 1976. To appear also in the Journal of Applied Mathematical Modelling.

{5} Leontief, V., Structure of the World Economy, American Economic Review, 64 (6), Dec. 1974.

{6} Herrera, A., Scolnik, H., et al, Catastrophe or New Society? A Latinamerican World Model, International Development Research Centre, Ottawa, 1976.

{7} Scolnik, H., Critica Metodologica al Modelo World 3, Ciencia Nueva, Nº 25, Buenos Aires, 1973.

{8} Scolnik, H., and Talavera, L., Mathematical and Computational aspects of the construction of self-optimizing dynamic models, Proceedings of the Second International Seminar on Trends in Mathematical Modelling, UNESCO, Jablonna, 1974.

{9} De Jongh, D.C.J., and Vermeulen, P.J., parameter sensitivity of the Limits to Growth World Model, Journal of Applied Mathematical Modelling, Vol. 1, Nº 1, 1976.

{10} Gelovani, V., et al, paper published in russian in the proceedings of the UNESCO Conference on Global Modelling, Moscow, 1976

LIMITATIONS OF WORLD MODELS

Rajko Tomović

To begin with, one has to point out of which type of world models we are talking about: input-output relations, optimization models, dynamic models, etc. Taking into account the nature of this meeting, I shall restrict my remarks to state space approaches and multilevel models of world dynamics as initiated by efforts of Club of Rome.

Evidently, the complexity of the world modeling is such that not one round table discussion but many extended seminars are needed in order to evaluate fully the available results. Therefore, my discussion will deal with just a few basic questions of world modeling which are of much concern to those involved in this area of research. I shall also emphasize the point of view of a multidisciplinary research group in my country engaged in the analysis of the impact of the transfer of technology to developing countries.

The greatest difficulty in assessing properly the value of formal approaches in world modeling is the nature of the object itself. The formal description of world dynamics requires a multivariable, multilevel model. This is in itself a great challenge to computer modeling because there are very few cases in practice where multivariable, multilevel models of "soft" large systems have produced reliable results for far reaching conclusions. To apply these techniques to predict the world future and thus derive conclusions about certain aspects of world politics is not an easy matter.

All attempts to develop World Models will run into a fundamental difficulty. Namely, the world dynamics as a whole cannot be described by just measurable attributes. A mixture of quantitative and qualitative attributes must be used to understand properly the world evolution. Therefore, each formal model in this case carries from its very beginning a great uncertainty as to the validity of its results due to the fact that important system attributes cannot be involved in computer simulation or have been transformed into the measurable form in a highly subjective way.

The above difficulty, as known, may be relieved only by long range checking of computer and real life results. Even this method is of questionable value in the case of evolutionary soft systems to which the concept of initial state, or rather the equivalence of initial states, must be applied with much caution. Only in the case of those objects, mainly of mechanical nature, where it is possible to derive formal models by deductive reasoning instead of induction, the accuracy of computer simulations, may be more objectively assessed without extended periods of experimentation. Therefore, it would have been much better if the great publicity of first computer simulation results has been avoided and much less pretentious claims were expressed.

Further tests of the available World Models have only proved that their predictive value is at present highly questionable. You will hear in this discussion the results of extensive sensitivity analysis of some World Models which are conclusive but very pessimistic. Only one percent changes of some parameters produced qualitatively different responses. Other attempts to check the validity of World Models by time reversal have also produced very poor results.

Let us mention another unresolved basic problem of World Models. As known, the first attempts to represent the world dynamic without horizontal and vertical decomposition have been immediately abandoned, since this is clearly an unjustified simplification. However, there is no operational scientific procedure available by which a large system such as the world can be decomposed into subsystems. All what can be done then is to use subjective decomposition criteria and undertake extensive experimentation before generally accaptable decomposition approaches will appear.

The above remarks are given in order to point out to the vulnerability of formal models of the world dynamics. There are many more remarks which could support this statement. Rather than doing so, I shall outline at the end a constructive approach which could revive interest in World Models, specially in developing countries. First of all, the methodology of world modeling must be embedded as an unseparable part of much more complex considerations such as the transfer of technology, the gap between industrialized and nonindustrialized regions, general goals of world development, etc. Secondly, such an undertaking must become an international project in the true sense of the world. By this I imply that parallel but coordinated multidisciplinary groups should work on World Models in different regions of the world. They should look at this problem from various angles taking into account regional interests and different socio-economic goals. Finally, an adequate theoretical effort should be developed by such international organizations like IFIP in order to explore deeply certain fundamental problems of large scale system modeling.

THE USE OF NONLINEAR PROGRAMMING IN A DIRECT/ INDIRECT METHOD FOR OPTIMAL CONTROL PROBLEMS

M.C. Bartholomew-Biggs
Numerical Optimisation Centre
Hatfield Polytechnic
Hatfield, England.

Abstract

This paper describes an alternative to the classical indirect approach to optimal control problems. The idea is discussed with particular reference to spacecraft orbital manoeuvres and the motivation is to avoid some of the well-known difficulties associated with switching functions (for determining the starting and stopping of thrusting) and with boundary value problems for determining the initial adjoint multipliers.

Introduction

In this paper we shall chiefly be concerned with the problem of optimising spacecraft orbital manoeuvres. Typically we shall be considering the minimum-fuel transfer of a low-thrust vehicle from a given initial orbit to a desired final orbit. This will involve the calculation of optimum times for starting and stopping the thrusters and also the calculation of optimum thust directions. We shall also take account of constraints on the thrust directions and constraints on the allowable times for thrusting.

Since the approach presented in this paper may contain features relevant to other kinds of control problem we first state the problem, and some fundamental optimality conditions in general terms. We shall then use these in subsequent sections in connection with specific examples.

Let $X_1(t) \ldots X_N(t)$ be state variables describing the system at time t. (In our application these variables define spacecraft position and velocity.) Let $u_1(t) \ldots u_M(t)$ be the continuous control variables to be optimised. (For our application these variables define thrust directions.) Let the time interval (t_{2k-1}, t_{2k}) define the k-th thrust period (k=1 ... T) and let (t_o, t_F) denote the whole period within which the desired manoeuvre is to be performed.

We can state the optimisation problem as

Minimise $$F(\underline{x}(t_F)) \tag{1}$$

subject to $$\frac{d}{dt}(x_i(t)) = \eta_i(\underline{x}(t), \underline{u}(t), t) \qquad i = 1 \ldots N \tag{2}$$

$$\xi_\ell(\underline{x}(t), \underline{u}(t)) \geqslant 0 \qquad \ell = 1 \ldots L \tag{3}$$

$$\phi_r(\underline{x}(t_F)) = 0 \qquad r = 1 \ldots R \tag{4}$$

$$\psi_s(\underline{x}(t_1), t_1, \ldots \underline{x}(t_{2T}), t_{2T}) \geqslant 0 \qquad s = 1 \ldots S \tag{5}$$

For our application, the function (1) would be a measure of performance such as fuel used up to time t_F. The constraints (2) would be the equations of motion. Inequalities (3) would express limitations on the choice of thrust directions while inequalities (5) would be restrictions on the permitted times for thrusting. The equations (4) would represent the desired orbital conditions at the end of the manoeuvre.

In order to write down the optimality conditions for the problem (1) - (5) we introduce adjoint multipliers $p_1(t) \ldots p_N(t)$ associated with the equations of motion (2). We also define the Hamiltonian function H by

$$H = \sum_{i=1}^{N} p_i(t)\, \eta_i(\underline{x}(t), \underline{u}(t), t) \tag{6}$$

The following conditions (7) - (11) can be shown to hold at the solution of the optimisation problem. For brevity we drop the explicit dependence on t. Between switching times, the adjoint multipliers satisfy the <u>adjoint equations</u>

$$\frac{dp_k}{dt} = -\frac{\partial H}{\partial x_k} - \sum_{\ell=1}^{L} \pi_\ell \frac{\partial \xi_\ell}{\partial x_k} \qquad k = 1 \ldots N \tag{7}$$

where $\pi_\ell(t)$ (ℓ=1 ... L) are multipliers associated with the constraints (3). At every instant, $\underline{u}(t)$ solves the problem

$$\underset{\underline{u}}{\text{Max}} \quad H \qquad \text{s.t.} \quad \xi_\ell(\underline{x}, \underline{u}) \geq 0 \qquad \ell = 1 \ldots L. \tag{8}$$

At each switching time the adjoint multipliers are subject to jumps

$$\Delta p_k = -\sum_{s=1}^{S} \mu_s \frac{\partial \psi_s}{\partial x_k} \tag{9}$$

where μ_s(s=1 ... S) are multipliers on constraints (5). Similarly, at each switching time, the Hamiltonian function experiences a jump

$$\Delta H = -\sum_{s=1}^{S} \mu_s \frac{\partial \psi_s}{\partial t} \tag{10}$$

At the final time t_F the adjoint multipliers satisfy the <u>transversality conditions</u>

$$p_k(t_F) = -\frac{\partial F}{\partial x_k} - \sum_{r=1}^{R} \lambda_r \frac{\partial \phi_r}{\partial x_k} \qquad k = 1 \ldots N \tag{11}$$

where λ_r(r=1 ... R) are multipliers associated with the equality constraints (4). These relationships are fundamental in indirect methods of solving problem (1) - (5). We shall illustrate this after we have described the optimal orbit control problem more precisely in the next section.

<u>Notation & definitions for the orbital control problem</u>

In the rest of the paper we shall use the following notation for orbital elements.

a = semi-major axis; e = eccentricity; i = inclination ;

Ω = argument of ascending node; ω = argument of perigee ;

ν = true anomaly of spacecraft; θ = range angle of spacecraft ;

$\phi = \nu + \omega - \theta$

The state variables used in the optimisation calculations are defined as follows.

$$\left.\begin{array}{l} x_1 = \sqrt{a(1-e^2)/\mu} \quad \text{where } \mu \text{ is the gravitational constant} \; ; \\ x_2 = e\cos(\omega - \phi) \quad ; \quad x_3 = e\sin(\omega - \phi) \quad ; \\ x_4 = (\text{fuel used})/M_o \quad \text{where } M_o \text{ is initial mass of spacecraft} \; ; \\ x_5 = \text{time} \quad ; \end{array}\right\} \tag{12}$$

$$X_6 = \cos\frac{i}{2}\cos\frac{\Omega+\phi}{2} \quad ; \quad X_7 = \sin\frac{i}{2}\cos\frac{\Omega-\phi}{2} \quad ;$$
$$X_8 = \sin\frac{i}{2}\sin\frac{\Omega-\phi}{2} \quad ; \quad X_9 = \cos\frac{i}{2}\sin\frac{\Omega+\phi}{2} \qquad (13)$$

These variables are introduced to avoid singularities in the equations of motion in the case of circular or zero-inclination orbits.

We also define τ as the thrust magnitude; V_e as the exhaust velocity; f_1 as the perturbing force on the spacecraft in the orbit plane and perpendicular to the radius vector; f_2 as the perturbing force on the spacecraft in the orbit plane and along the radius vector; and f_3 as the perturbing force on the spacecraft perpendicular to the orbit plane.

The thrust pitch and yaw angles are defined in figure 1 below.

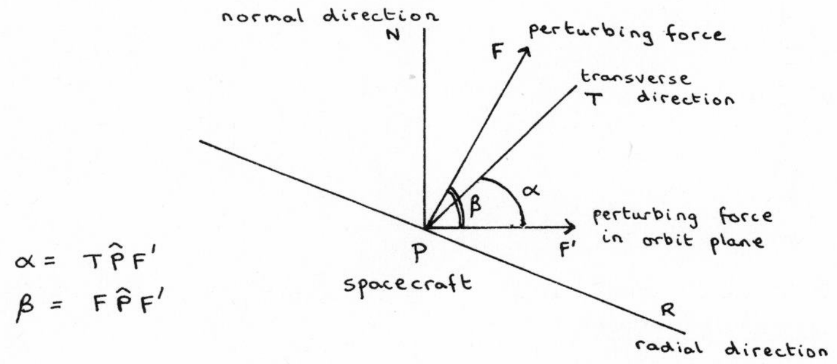

Figure 1; Thrust pitch and yaw angles

It follows that

$$f_1 = \tau\cos\alpha\cos\beta \quad ; \quad f_2 = \tau\sin\alpha\cos\beta \quad ; \quad f_3 = \tau\sin\beta \qquad (14)$$

It will be convenient for numerical integration purposes to take the range angle θ as the independent variable.

If we define

$$\gamma = 1 + X_2\cos\theta + X_3\sin\theta \quad ; \quad \sigma = \mu X_1^4/(\gamma^3 M_0(1-X_4)) \qquad (15)$$

then we shall use the following equations of motion.

$$\frac{dX_1}{d\theta} = \eta_1 = \sigma X_1 f_1 \qquad (16)$$

$$\frac{dX_2}{d\theta} = \eta_2 = \sigma\{[(\gamma+1)\cos\theta + X_2]f_1 + \gamma f_2\sin\theta\} \qquad (17)$$

$$\frac{dX_3}{d\theta} = \eta_3 = \sigma\{[(\gamma+1)\sin\theta + X_3]f_1 - \gamma f_2\cos\theta\} \qquad (18)$$

$$\frac{dX_4}{d\theta} = q_4 = \sigma \gamma \tau (1 - X_4) / X_1 V_e \tag{19}$$

$$\frac{dX_5}{d\theta} = q_5 = \sigma \gamma (1 - X_4) M_o / X_1 \tag{20}$$

$$\frac{dX_6}{d\theta} = q_6 = -\frac{1}{2} \sigma f_3 (X_7 \cos\theta + X_8 \sin\theta) \tag{21}$$

$$\frac{dX_7}{d\theta} = q_7 = \frac{1}{2} \sigma f_3 (X_6 \cos\theta - X_9 \sin\theta) \tag{22}$$

$$\frac{dX_8}{d\theta} = q_8 = \frac{1}{2} \sigma f_3 (X_9 \cos\theta + X_6 \sin\theta) \tag{23}$$

$$\frac{dX_9}{d\theta} = q_9 = \frac{1}{2} \sigma f_3 (X_7 \sin\theta - X_8 \cos\theta) \tag{24}$$

Discussion of a two-dimensional example

For simplicity we first consider a problem involving only two-dimensional motion. Hence we suppose that the constraints (4) define some in-plane orbital correction which does not depend on the final time. We suppose also that only one thrust period is allowed for completing the manoeuvre but that there are no restrictions on the position of this thrust period or on the thrust directions. The optimisation variables will therefore be the range angles at the start and finish of the thrust period and the pitch angle during thrusting.

Since we are concerned only with an in-plane manoeuvre independent of time we need only the first four equations of motion (16) - (19). Hence the Hamiltonian can be written

$$H = p_1 q_1 + p_2 q_2 + p_3 q_3 + p_4 q_4$$

In the absence of constraints the adjoint equations can be shown to be

$$\frac{dp_1}{d\theta} = -(4H + p_1 q_1 - p_4 q_4) / X_1 \tag{25}$$

$$\frac{dp_2}{d\theta} = \frac{\cos\theta}{\gamma}(3H - p_4 q_4) - \sigma\{p_2 f_1 + \cos\theta\,[\cos\theta\,(p_2 f_1 - p_3 f_2) + \sin\theta\,(p_2 f_2 + p_3 f_1)]\} \tag{26}$$

$$\frac{dp_3}{d\theta} = \frac{\sin\theta}{\gamma}(3H - p_4\eta_4) - \sigma\{p_3 f_1 + \sin\theta[\sin\theta(p_2 f_2 + p_3 f_1) + \cos\theta(p_2 f_1 - p_3 f_2)]\} \qquad (27)$$

$$\frac{dp_4}{d\theta} = -(H - p_4\eta_4)/M_0(1 - X_4) \qquad (28).$$

Notice that when there is no thrusting, $\eta_i = 0$, $H = 0$ and $dp_i/d\theta = 0$. Also, as there are no constraints on starting and stopping the thrusts, ΔH given by (10) is zero. Hence at the start and end of thrusting

$$H = p_1\eta_1 + p_2\eta_2 + p_3\eta_3 + p_4\eta_4 = 0 \qquad (29).$$

The subproblem (8) for calculating the continuous control variable α can be written as

$$\left.\begin{aligned} &\max_{\alpha} \left\{ \sum_{k=1}^{4} p_k\eta_k = q_1\cos\alpha + q_2\sin\alpha \right\} \\ &\text{where} \\ &q_1 = p_1X_1 + p_2[(\gamma+1)\cos\theta + X_2] + p_3[(\gamma+1)\sin\theta + X_3] \\ &q_2 = p_2\gamma\sin\theta - p_3\gamma\cos\theta \end{aligned}\right\} \qquad (30)$$

This is solved by using

$$\sum_{k=1}^{4} p_k \frac{\partial \eta_k}{\partial\alpha} = 0 \qquad (31)$$

and obtaining

$$\sin\alpha = \frac{+q_2}{\sqrt{q_1^2 + q_2^2}} \quad ; \quad \cos\alpha = \frac{+q_1}{\sqrt{q_1^2 + q_2^2}} \qquad (32)$$

The relationship (29) allows us to define a <u>switching function</u>

$$\tilde{H} = q_1\cos\alpha + q_2\sin\alpha - \gamma/X_1V_e \qquad (33).$$

It is easy to show that $\tilde{H}$ passes through zero at θ_1 the start of thrusting and θ_2 the end of thrusting.

<u>Algorithms for solving the optimisation problem</u>

We can now outline the <u>classical indirect approach</u> to the optimisation problem by means of the following steps.

STEP 0. Guess some initial values $p_1(0)$ ..., $p_4(0)$ for the adjoint multipliers at $\theta = 0$.

STEP 1. Integrate the equations of motion (16) - (19) and the adjoint equations (25) - (28). At each step compute α to satisfy (32). Start thrusting when $\tilde{H}(\alpha) = 0$ and continue thrusting until $\tilde{H}(\alpha)$ becomes zero again.

STEP 2. Stop integrating at $\theta = \theta_F$. Solve equation (11) in the least-squares sense for λ_r (r = 1 ..., R) and let $\hat{\phi}_k$ (k = 1 ..., N) be the corresponding residuals.

STEP 3. If $\lambda_r = 0$ (r = 1 ..., R) and $\hat{\phi}_k = 0$ (k = 1 ... N) then we have a feasible trajectory which satisfies all the first order optimality conditions (7) - (11) and the method has converged.

STEP 4. If some $\lambda_r \neq 0$ or $\hat{\phi}_k \neq 0$ then $p_1(0) \ldots, p_4(0)$ are adjusted so as to reduce

$$\sum_{r=1}^{R} \lambda_r^2 + \sum_{k=1}^{N} \hat{\phi}_k^2$$

(using a non-linear least-squares algorithm for instance) and the method continues from step 1.

As a comment on this approach we mention that it can be difficult to find the zeros of $\tilde{H}(\alpha)$ very precisely during the integration in step 1. Moreover the switching function idea becomes more complicated if we wish to introduce constraints on the switching points. Therefore we would prefer to optimise the switching points explicitly. Another fact that has often been noted is that the adjoint multipliers $p_1(0) \ldots, p_4(0)$ are not well suited to iterative procedures of the kind sketched above. Since the multipliers have no obvious physical significance a good initial guess is seldom available. Furthermore the transversality conditions (11) needed in step 2 are not easy to compute and the sensitivity of the integration to small changes in the adjoint multipliers can be such that the iteration is badly-conditioned. We would therefore like to replace the adjoint variables by some equivalent physical quantities in the hope of obtaining a better behaved problem.

We note first of all that if θ_1 is the start of thrusting then $p_i(\theta_1) = p_i(0)$ $i = 1 \ldots, 4$ since the adjoints remain constant up to the start of thrusting. Now if α_1 is the initial pitch angle (32) implies

$$\tan \alpha_1 = q_2(\theta_1) / q_1(\theta_1) \qquad (34)$$

and, repeating (33),

$$\tilde{H}(\theta_1) = 0 \qquad (33).$$

We can obtain a further relationship similar to (34) if we suppose that $\tilde{\alpha}_1$ is the pitch angle that would have been implied by (34) <u>if the start of thrusting had been delayed by some prescribed amount $\Delta\theta$</u>. Hence the equation involving $\tilde{\alpha}_1$ would be

$$\tan \tilde{\alpha}_1 = q_2(\theta_1 + \Delta\theta) / q_1(\theta_1 + \Delta\theta) \qquad (35).$$

Now (34) is a linear equation in $p_1(\theta_1)$, $p_2(\theta_1)$ and $p_3(\theta_1)$; (33) is a linear equation in $p_1(\theta_1) \ldots, p_4(\cdot\theta_1)$; and (35) is linear in $p_1(\theta_1{+}\Delta\theta) \ldots p_4(\theta_1{+}\Delta\theta)$.

The adjoint equations imply that $p_i(\theta_1+\Delta\theta) = p_i(\theta_1)$ if no thrusting occurs between θ_1 and $\theta_1+\Delta\theta$. Hence (35) is also a linear equation in $p_1(\theta_1)\ldots p_4(\theta_1)$. Now the homogeneity of the adjoint system means that, without loss of generality, we can normalise one of the initial adjoint multipliers to ± 1. Therefore if we guess values for the physical variables θ_1, α_1, $\tilde{\alpha}_1$ then we have enough information in equations (33) - (35) to calculate the three remaining multipliers at θ_1. To be more specific, it can be shown that the arbitrary choice $p_4(\theta_1) = -1$ is consistent with the fact that subproblem (30) involves maximisation.

This kind of device for relating the initial adjoint multipliers to some more manageable and more easily estimated physical variables has been called an 'adjoint-control transformation' (ref. 1).

We now propose an alternative to the classical approach to problem (1) - (5). It will be called the hybrid approach because it includes features of both indirect and direct optimisation, and it consists of the following steps.

STEP 0. Guess some initial values of θ_1, θ_2 (starting and stopping of thrust) and α_1, $\tilde{\alpha}_1$ (actual and fictitious pitch angles, discussed above).

STEP 1. Calculate $p_1(\theta_1)$ ..., $p_4(\theta_1)$ from equations (33) - (35). Integrate the equations of motion and the adjoint equations from θ_1 to θ_2, choosing α to satisfy (32) throughout.

STEP 2. Obtain new estimates of θ_1, θ_2, α_1, $\tilde{\alpha}_1$ by performing one iteration of a nonlinear programming algorithm applied to the problem

$$\{\text{Min } F \quad \text{s.t. } \phi_r = 0, \; r = 1 \ldots R\} \tag{36}$$

where F and ϕ_r are calculated during the integration in step 1.

STEP 3. If the current values of θ_1, θ_2, α_1, $\tilde{\alpha}_1$ solve (36) then they also solve the original problem because of the optimality conditions satisfied during the integration. Therefore the method has converged.

STEP 4. If the values of θ_1, θ_2, α_1, $\tilde{\alpha}_1$ are changed by the subproblem (36) then the algorithm continues from step 1.

This approach does not need switching functions and uses physical variables rather than adjoint variables as the basis of the iterative scheme. Moreover the subproblem (36) is both closer to the original form of the problem and also eliminates the need to work with the transversality conditions explicitly. We can show that any termination point for the classical algorithm will also be a termination point for the hybrid algorithm. For simplicity, the proof given here relates to the unconstrained single-thrust-arc case currently under discussion. We first observe that the values of $\underline{X}$, $\underline{p}$ and α at any point are completely determined by the values of θ_1, θ_2, α_1, $\tilde{\alpha}_1$ and by the integration process. We can therefore regard X_k, p_k (k=1 ..., 4) and α as functions of θ_1, θ_2, α_1, $\tilde{\alpha}_1$.

We observe also that both the hybrid and the classical approach satisfy the optimality conditions (7) and (8). We need therefore to show that the conditions (10) and (11) are equivalent to those obtained by the hybrid algorithm. Consider the transversality conditions, rewritten here to show θ as the independent variable.

$$p_k(\theta_F) = -\frac{\partial F}{\partial x_k} - \sum_{r=1}^{R} \lambda_r \frac{\partial \phi_r}{\partial x_k} \qquad k = 1 \ldots 4$$

Since $\theta_F \geq \theta_2$ and since the adjoint and state variables do not change in a coast arc it follows that

$$p_k(\theta_2) = -\frac{\partial F}{\partial x_k} - \sum_{r=1}^{R} \lambda_r \frac{\partial \phi_r}{\partial x_k} \qquad k = 1 \ldots 4 \tag{37}$$

where the right hand side is evaluated at $\theta = \theta_2$.

Now let ρ denote any one of θ_1, θ_2, α_1, $\tilde{\alpha}_1$. Multiplying (37) by $\frac{\partial x_k(\theta_2)}{\partial \rho}$ for k=1 ..., 4 and adding the results gives

$$\sum_{k=1}^{4} p_k(\theta_2) \frac{\partial x_k(\theta_2)}{\partial \rho} = -\frac{\partial F}{\partial \rho} - \sum_{r=1}^{R} \lambda_r \frac{\partial \phi_r}{\partial \rho} \tag{38}$$

Now from the definition of $X_k(\hat{\theta})$

$$\frac{\partial x_k(\hat{\theta})}{\partial \rho} = \int_{\theta_1}^{\hat{\theta}} \left[\sum_{j=1}^{4} \frac{\partial \eta_k}{\partial x_j} \frac{\partial x_j}{\partial \rho} + \frac{\partial \eta_k}{\partial \alpha} \frac{\partial \alpha}{\partial \rho} \right] d\theta$$

Define

$$I(\hat{\theta}) = \sum_{k=1}^{4} p_k(\hat{\theta}) \frac{\partial x_k(\hat{\theta})}{\partial \rho}$$

$$= \sum_{k=1}^{4} p_k(\hat{\theta}) \left\{ \int_{\theta_1}^{\hat{\theta}} \left[\sum_{j=1}^{4} \frac{\partial \eta_k}{\partial x_j} \frac{\partial x_j}{\partial \rho} + \frac{\partial \eta_k}{\partial \alpha} \frac{\partial \alpha}{\partial \rho} \right] d\theta \right\}$$

Then

$$\left.\begin{aligned} \frac{d}{d\hat{\theta}} I(\hat{\theta}) = \sum_{k=1}^{4} \frac{dp_k(\hat{\theta})}{d\hat{\theta}} \frac{\partial x_k(\hat{\theta})}{\partial \rho} &+ \sum_{k=1}^{4} p_k(\hat{\theta}) \sum_{j=1}^{4} \frac{\partial \eta_k}{\partial x_j} \frac{\partial x_j(\hat{\theta})}{\partial \rho} \\ &+ \sum_{k=1}^{4} p_k(\hat{\theta}) \frac{\partial \eta_k}{\partial \alpha} \frac{\partial \alpha}{\partial \rho} \end{aligned}\right\} \tag{39}$$

The third term on the right hand side of (39) is zero by (31). Moreover the first term can be rewritten, using (7), as

$$-\sum_{k=1}^{4} \frac{\partial x_k}{\partial \rho}(\hat{\theta}) \left[\sum_{j=1}^{4} p_j(\hat{\theta}) \frac{\partial \eta_j(\hat{\theta})}{\partial x_k} \right]$$

This is the negative of the second term, and hence $dI/d\hat{\theta}$ is identically zero and so $I(\hat{\theta})$ is constant for $\theta_1 \leq \hat{\theta} \leq \theta_2$. Therefore (38) can be rewritten as

$$\sum_{k=1}^{4} p_k(\theta_1) \frac{\partial x_k(\theta_1)}{\partial \rho} = -\frac{\partial F}{\partial \rho} - \sum_{r=1}^{R} \lambda_r \frac{\partial \phi_r}{\partial \rho} \tag{40}$$

When $\rho = \theta_1$, (40) becomes

$$\left.\begin{aligned} \frac{\partial F}{\partial \theta_1} + \sum_{r=1}^{R} \lambda_r \frac{\partial \phi_r}{\partial \theta_1} &= \sum_{k=1}^{4} p_k(\theta_1)\, \eta_k(\theta_1) \\ &= 0 \text{ at a stopping point of the classical algorithm by (29).} \end{aligned}\right\} \tag{41}$$

When $\rho = \theta_2$, (38) becomes

$$\frac{\partial F}{\partial \theta_2} + \sum_{r=1}^{R} \lambda_r \frac{\partial \phi_r}{\partial \theta_2} = \sum_{k=1}^{4} p_k(\theta_2)\, \eta_k(\theta_2) \tag{42}$$

= 0 at a stopping point of the classical algorithm by (29).

When $\varsigma = \alpha_1$, (40) becomes

$$\frac{\partial F}{\partial \alpha_1} + \sum_{r=1}^{R} \lambda_r \frac{\partial \phi_r}{\partial \alpha_1} = \sum_{k=1}^{4} p_k(\theta_1) \frac{\partial X_k(\theta_1)}{\partial \alpha_1} \tag{43}$$

= 0 because the state variables $X_k(\theta_1)$ do not depend on α_1.

When $\varsigma = \tilde{\alpha}_1$, (40) becomes

$$\frac{\partial F}{\partial \tilde{\alpha}_1} + \sum_{r=1}^{R} \lambda_r \frac{\partial \phi_r}{\partial \tilde{\alpha}_1} = \sum_{k=1}^{4} p_k(\theta_1) \frac{\partial X_k(\theta_1)}{\partial \tilde{\alpha}_1} \tag{44}$$

= 0 because the state variables $X_k(\theta_1)$ do not depend on $\tilde{\alpha}_1$.

The equations (41) - (44) are simply the stationarity conditions which hold at the solution of (36). Hence we have shown an equivalence between the optimality conditions used by the classical algorithm and those used by the hybrid algorithm.

Computational experience

The hybrid approach has, for simplicity, been developed and justified in the context of a particular, rather simple application. The ideas have in fact been extended in practice to more complicated problems. For instance, if there is more than one thrust arc then we simply introduce as extra optimisation variables the start and end of each additional thrust arc. Equation (9) shows that adjoint multipliers are unchanged from one thrust arc to the next if there are no constraints of the form (5). If, however, constraints (5) are present then one more optimisation variable is needed per thrust arc. For the first thrust arc we introduce a (fictitious) range angle θ^* (say) representing the position of the unconstrained optimum start. For subsequent thrust arcs we need the initial pitch angle to be an optimisation variable which allows the jumps (9) in the adjoint multiplier to be calculated. The adjoint-control transformations for calculating the values of $p_1 \dots, p_4$ from these optimisation variables are discussed in ref. 2. Adjoint control transformations are also described in ref. 2 for the three dimensional case. For an unconstrained single-thrust-arc problem the optimisation variables are θ_1, θ_2, α_1, $\tilde{\alpha}_1$ as before, together with yaw angles β_1, $\tilde{\beta}_1$ at θ_1 and $\theta_1 + \Delta\theta$. The particular form of the state variables and the differential equation enables the initial adjoint variables to be calculated just from these six variables and the given initial conditions on the problem.

Several different kinds of problem have been dealt with by the methods described in this paper. As a 2-dimensional manoeuvre we have considered a semi major axis change for a low thrust vehicle in a highly eccentric orbit. This problem has been solved for a single unconstrained thrust-arc and also for a single thrust arc with a constraint forbidding thrusting in certain regions. A solution has also been found for the case when two separate thrust arcs are allowed. We have also dealt with a three dimensional manoeuvre involving a semi major axis and inclination change for a low thrust vehicle in near-circular orbit. As a complete contrast we have also considered, as a high-thrust example, the optimisation of a third stage launch vehicle trajectory into a desired tranfer orbit. For all these problems it appears that the approach suggested in this paper was successful.

References

1. Dixon, L.C.W. & Biggs, M.C.
The advantages of adjoint-control transformations when determining optimal trajectories by Pontryagin's Maximum Principle. J.R.Ae.S. March 1972.

2. Hersom, S.E., Dixon, L.C.W., Bartholomew-Biggs, M.C. & Pocha, J.J.
The optimisation of satellite trajectories. Hawker Siddeley Dynamics Report HSD TP 7633, 1977.

APPROXIMATION OF FUNCTIONAL-DIFFERENTIAL EQUATIONS BY ORDINARY DIFFERENTIAL EQUATIONS AND HEREDITARY CONTROL PROBLEMS

F. Kappel
Institut für Mathematik
Universität Graz
Steyrergasse 17, A 8010 Graz
AUSTRIA

If one deals with optimal control problems where the dynamics of the system is governed by a functional-differential equation one encounters a problem with infinite dimensional state space. In [6] H.T.Banks and A. Manitius proposed an approximation technique for the numerical treatment of such control problems with linear autonomous state equations using projection of the system onto eigenspaces (which in this case are finite dimensional). The general idea is to project the given system onto a sequence of finite dimensional subspaces in such a way that the dynamics of each of the projected systems is governed by a system of ordinary differential equations. Next one calculates the solutions of the projected problems and finally one has to show that these solutions converge to the solution of the original problem. In this paper we give a short survey regarding approximation of functional differential equations by ordinary differential equations. With respect to control problems we refer to [3,4,6].

In [6] the linear system

$$\dot{x}(t) = L(x_t) + f(t) \tag{1}$$

is considered, where L is a bounded linear functional on $C = C(-r,0;R^n)$ and f is locally integrable. As usual for a function $x(\cdot)$ the symbol x_t denotes the function in C defined by $x_t(s) = x(t+s)$, $s \in [-r,0]$. As a special case of (1) we have difference differential equations

$$\dot{x}(t) = \sum_{j=0}^{m} A_j x(t-h_j) + f(t), \tag{2}$$

$$0 = h_0 < \ldots < h_m = r.$$

It is well known that the eigenspaces corresponding to eigenvalues of the homogeneous equation are finite dimensional and that the projection of a solution of (1) into an eigenspace is solution of an ordinary differential equation (see for instance [11] with respect to the general theory for equation (1)). There are two difficulties which one has to deal with. The first is of principal importance and consists in the fact that the system of eigenfunctions in general is not complete in C without rather severe restrictions for L. The second difficulty is a more numerical one: the eigenvalues of the equation are zeros of an entire function of exponential type which in case of equation (2) is an exponential-polynomial.

In order to avoid these difficulties H.T. Banks and J.A. Burns considered in [3] equation (1) in the state-space $Z = R^n \times L^2(-r,0;R^n)$ under conditions given by Borisovic and Turbabin in [7]. These conditions hold for all equations of type (1) which are important in applications. The use of the state-space Z in connection with hereditary control problems goes back to Delfour and Mitter [10]. We give the main ideas of the approach in [3] and refer the reader to this paper for details and an extensive bibliography. For simplicity we deal with the homogeneous equation

$$\dot{x}(t) = L(x_t). \tag{3}$$

For initial data $x(0) = \eta \in R^n$ and $x(s) = \varphi(s)$, $s \in [-r,0)$, $\varphi \in L^2(-r,0;R^n)$ the corresponding solution is denoted by $x(t) = x(t;\eta,\varphi)$. We define the operators

$$T_t: Z \to Z,\ t \geq 0,$$
$$(\eta,\varphi) \to (x(t),x_t).$$

T_t, $t \geq 0$, is a C_o-semigroup with infinitesimal generator $\mathcal{A}$:

$$D(\mathcal{A}) = \{(\varphi(0),\varphi) \mid \varphi \in W^{1,2}(-r,0;R^n)\},$$
$$\mathcal{A}(\varphi(0),\varphi) = (L(\varphi),\dot{\varphi}),\quad (\varphi(0),\varphi) \in D(\mathcal{A}).$$

Instead of taking eigenspaces one takes subspaces of step-functions. Let $t_j^N = -j\frac{r}{N}$, $j = 0,\ldots,N$, be a subdivision of $[-r,0]$ and χ_j^N be the characteristic function of $[t_j^N, t_{j-1}^N)$. Then

$$Z^N = \{(\eta,\varphi) \mid \eta \in R^n,\ \varphi = \sum_{j=1}^N v_j^N \chi_j^N,\ v_j^N \in R^n\}.$$

The projection $P^N: Z \to Z^N$ is given by

$$P^N(\eta,\varphi) = (\eta,\varphi^N),\ \varphi^N = \sum_{j=1}^N \varphi_j^N \chi_j^N,\ \varphi_j^N = \frac{N}{r}\int_{t_j^N}^{t_{j-1}^N} \varphi(s)ds.$$

Next one defines an approximation $\mathcal{A}^N$ of $\mathcal{A}$ by

$$\mathcal{A}^N(\eta,\varphi) = (L(\eta,\varphi), \sum_{j=1}^N \frac{N}{r}(\varphi_{j-1}^N - \varphi_j^N)\chi_j^N),\ \varphi_o^N = \eta,\ \varphi = \sum_{j=1}^N \varphi_j^N \chi_j^N,$$

$$\mathcal{A}^N(\eta,\varphi) = \mathcal{A}^N P^N(\eta,\varphi) \text{ for } (\eta,\varphi)\in Z.$$

$\mathcal{A}^N$ is a bounded linear operator on Z and infinitesimal generator of $T_t^N = \exp(\mathcal{A}^N t)$. For an obvious choice of a basis for Z^N, $\mathcal{A}^N$ has a very simple matrix representation A^N. Let $w^N(t)$ denote the solution of

$$\dot{w}^N = A^N w^N, \quad w^N(0) = w_o^N,$$

where w_o^N is the coordinate vector of $P^N(\eta,\varphi)$. Then by an application of the Trotter-Kato-Theorem [14] Banks and Burns prove

<u>Theorem 1</u> ([3]). For any $(\eta,\varphi)\in Z$ we have

$$\lim_{N\to\infty} (w_o^N(t), \sum_{j=1}^{N} w_j^N(t)\chi_j^N) = T_t(\eta,\varphi)$$

uniformly on bounded intervals. For sufficiently smooth initial data the convergence is linear.

If instead of equation (3) one deals with the nonlinear autonomous equation

$$\dot{x}(t) = f(x(t),x_t) \tag{4}$$

the situation is much more complicated. The main difficulty is that even in the case where we have global existence of solutions the nonlinear solution semigroup T_t, $t \geq 0$, corresponding to equation (4) is not a semigroup with a dissipative generator and therefore does not fit into the existing theory of semigroups of nonlinear transformations. In [16,17] Webb considered applications of nonlinear semigroup theory to functional differential equations in the C-space and Z-space settings. In [12] the authors considered a rather general type of equations (4) assuming a local type of Lipschitz condition for f. Equation (4) is approximated by a family of equations which can be treated in the existing framework of nonlinear semigroup theory. An important feature of the approach in [12] is that equation (4) is considered as a perturbation of the linear equation $\dot{x}(t) \equiv 0$. Using the nonlinear version of the Trotter-Kato-Theorem given in [8] the nonlinear analogon of Theorem 1 was established (i.e. convergence of the approximating ODE-solutions obtained by averaging projections) for initial data $(\varphi(0),\varphi)\in Z$ with $\varphi\in W^{1,\infty}(-r,0;R^n)$.

Using the approach of [12] K. Kunisch in his thesis [13] proved the same result for functional differential equations of neutral type,

$$\frac{d}{dt}\left(\sum_{j=0}^{m} A_j x(t-h_j)\right) = f(x(t),x_t) \tag{5}$$

assuming $\sum_{j=1}^{m} |A_j| < 1$ and f globally Lipschitzean. The most difficult aspect of the work done in [13] is to develop a reasonable concept of a solution of (5) with initial data in Z.

In order to avoid the difficulties with respect to nonlinear semi-group theory H. T. Banks made a more direct approach and was able to obtain analogous results to Theorem 1 for nonlinear time-dependent equations under the expense of more restrictive assumptions on the right-hand side f (globally Lipschitzean and continuous on Z) [1,2].

Quite recently H.T. Banks and the author obtained a general approximation scheme for autonomous linear functional differential equations in the same spirit as in [3], i.e. using the state space Z and the Trotter-Kato-Theorem. Here we just give the idea of the approach. A complete representation together with examples will appear elsewhere [5]. Consider again equation (3) under the same assumptions on the right-hand side as in the case of averaging projections. Let (Z^N) be a sequence of subspaces of Z and $P^N: Z \to Z^N$ be the corresponding orthogonal projections. We assume that

$$Z^N \subset D(\mathcal{A})$$

for all N and define the operators $\mathcal{A}^N: Z \to Z^N$ by

$$\mathcal{A}^N = P^N \mathcal{A} P^N.$$

Let T_t^N, $t \leq 0$, be the C_o-semigroup on Z with infinitesimal generator $\mathcal{A}^N$. The subspace Z^N is invariant with respect to T_t^N. If we assume in addition that dim $Z^N < \infty$ then as in the case of averaging projections the restriction of $\mathcal{A}^N$ to Z^N has a matrix representation A^N with respect to a given basis of Z^N. For each element $z_o^N \in Z^N$ the coordinate vector $w^N(t)$ of $T_t^N z_o^N$ is solution of

$$\dot{w}^N(t) = A^N w^N(t), \quad w^N(0) = w_o^N, \tag{6}$$

where w_o^N is the coordinate vector of z_o^N.

For an element $(\eta,\varphi) \in Z$ we put $P^N(\eta,\varphi) = (\varphi^N(0),\varphi^N) \in D(\mathcal{A})$. The assumptions with respect to the projections P^N are: There is a natural number k such that

$$P^N(\varphi(0),\varphi) = (\varphi^N(0),\varphi^N) \overset{Z}{\to} (\varphi(0),\varphi),$$
$$\text{and } \dot\varphi^N \overset{L_2}{\to} \dot\varphi \quad \text{as } N\to\infty \text{ for all } \varphi\in C^k(-r,0;R^n). \tag{7}$$

Then an application of the Trotter-Kato-Theorem gives

Theorem 2 (H.T. Banks and F. Kappel [5]). Under assumption (7) we have

$$\lim_{N\to\infty} T_t^N z_o = T_t z_o, \; z_o\in Z,$$

uniformly on bounded intervals.

Corollary ([5]). Suppose $\dim Z^N < \infty$ for all N and for $z_o\in Z$ let $w^N(t)$ be the solution of (6) with $z_o^N = P^N z_o$. Then

$$\lim_{N\to\infty} z^N(t) = T_t z_o,$$

where $z^N(t)$ denotes the element in Z with coordinate vector $w^N(t)$.

The Corollary shows that again the solution semigroup of equation (3) is approximated by solution semigroups of ordinary differential equations of increasing dimensions.

As an example to this general scheme we mention approximation by cubic splines (or more general, splines of some order k) [5] where the subspaces Z^N are given by

$$Z^N = \{(\varphi(0),\varphi) \mid \varphi \text{ is a cubic spline on } [-r,0] \text{ with knots at } t_j^N = -j\tfrac{r}{N},\; j=0,\dots,N\}.$$

In this case we have for $z_o = (\varphi(0),\varphi)$, $\varphi\in W^{4,2}(-r,0;R^n)$ and $T\geq 0$ the estimate

$$\| z^N(t) - T_t z_o \| \leq \frac{K}{N^3}, \quad t\in[0,T],$$

with some positive constant K.

As a final remark we want to draw attention to other possible approaches which are related to that described here by the fact that semigroups or abstract Cauchy-problems are associated with the given equation, but are different in an important aspect, because the system is also discretized with respect to t. As representative papers we cite [9,15].

References.

[1] Banks, H.T.: Approximation methods for optimal control problems with delay-differential systems, Séminaires IRIA: analyse et contrôle de systémes, 1976, Rocquencourt, France

[2] ———— : Approximation of nonlinear functional differential equation control systems, Manuscript, June 1977

[3] Banks, H.T., and J.A. Burns: Hereditary control problems: numerical methods based on averaging approximation, to appear in SIAM J. Control and Optimization

[4] Banks, H.T., J.A. Burns, E.M. Cliff and P.R. Thrift: Numerical solution of hereditary control problems via an approximation technique, Brown University LCDS Techn. Rep. 65-6, Providence, R.I., October 1975

[5] Banks, H.T., and F. Kappel: Spline approximation and linear functional differential equations, Manuscript August 1977

[6] Banks, H.T., and A. Manitius: Projection series for retarded functional differential equations with applications to optimal control problems, J. differential Eqs. 18(1975), 296 - 332

[7] Borisovic, J.G., and A.S. Turbabin: On the Cauchy problem for linear nonhomogeneous differential equations with retarded argument, Soviet Math., Dokl. 10(1969), 401 - 405

[8] Brezis, H., and A. Pazy: Convergence and approximation of semigroups of nonlinear operators in Banach spaces, J. Functional Analysis 9(1972), 63 - 74

[9] Delfour, M.C.: The linear quadratic optimal control problem for hereditary differential systems: theory and numerical solution, Appl. Math. Optimization 3(1977), 101 - 162

[10] Delfour, M.C., and S.K. Mitter: Controllability, observability and optimal control of affine hereditary differential systems, SIAM J. Control 10(1972), 298 - 328

[11] Hale, J.K.: Theory of Functional Differential Equations, Appl. Math. Sci. Vol. 3, Springer 1977

[12] Kappel, F., and W. Schappacher: Autonomous nonlinear functional differential equations and avergaing approximations, to appear in J. nonl. Analysis

[13] Kunisch, K.: Neutrale Funktional-Differentialgleichungen in L^p-Räumen, Ph.D. Thesis, Technical University of Graz, November 1977

[14] Pazy, A.: Semi-groups of linear operators and applications to partial differential equations, University of Maryland, Lecture Note No. 10(1974)

[15] Reddien, G.W., and G.F. Webb: Numerical approximation of nonlinear functional differential equations with L^2 initial functions, to appear

[16] Webb, G.F.: Autonomous nonlinear functional differential equations and nonlinear semigroups, J. math. Analysis Appl. 46(1974), 1 - 12

[17] ————: Functional differential equations and nonlinear semigroups in L^p-spaces, J. differential Eqs. 20(1976), 71 - 89

AN ALGORITHM TO OBTAIN THE MAXIMUM SOLUTION OF THE HAMILTON-JACOBI EQUATION.

Roberto GONZALEZ(*) *and Edmundo ROFMAN*(**)
Instituto de Matematica "Beppo Levi"
Universidad Nacional de Rosario
ARGENTINA

(work included in the cooperation program with I.R.I.A.- Rocquencourt- FRANCE)

ABSTRACT :

The analysis of the optimal control problem with initial data as parameters leads us to the study of the associated Hamilton-Jacobi equation.

In the first part of this paper we prove for the stationary problem that the optimal cost function $v(x) = \text{Inf } J(x,u(\cdot))$ is an upper bound of a suitable set W of subsolutions $w(x)$ of the H.J. equation. In particular, for $v(x)$ lipschitzien, it will follow $v \in W$, v being the maximum element of W for the partial order $w_1 \leq w_2$ iff $w_1(x) \leq w_2(x)$, $\forall x \in \Omega$.

In the second part of the paper we obtain approximate values of v after introducing an auxiliary problem (maximization in a convex set of a functional space); to solve it, we use internal approximations of the space in such a form that the final numerical problem is of linear programming type.

Several extensions (with some examples) of the above results will be published in [4], [5].

§1. STATEMENT OF THE PROBLEM.

We consider a system guided by the (matricial) differential equation :

$$(1.1)\qquad \begin{cases} y(t) = f(y,u) \\ y(o) = x \end{cases} \qquad x \in \Omega \subset R^n \text{ , } \Omega \text{ open bounded domain}$$

$$u \in \mathcal{U}_{ad} \subset R^m$$

and the cost functional

$$(1.2)\qquad J(x,u(.)) = \int_o^{\tau} e^{\int_o^s k(y(r),u(r))dr} \ell(y(s),u(s))ds + g(y(\tau))\, e^{\int_o^{\tau} k(y(r),u(r))dr}$$

(*) Researcher of the "Consejo de Investigaciones de la Universidad Nacional de Rosario" for the project : "Optimization and Control. Theory and Applications".

(**) Director of the above referred project.

(1.3) with $\tau = \inf\{s \,/\, y(s) \in \Omega\}$; if $\forall s,\ y(s) \in \Omega$ we put $\tau = \infty$

and the second term of (1.2) will vanish (after (1.6)).

We assume

(1.4) f, ℓ, g, k lipschitzian respect of y, L_1 being the constant of Lipschitz (independent of u)

(1.5) $\|f(y.u)\| \leq M$, $|\ell(y.u)| \leq M$, $|g(y)| \leq M$, $\forall y \in \Omega$, $\forall u \in \mathcal{U}_{ad}$

(1.6) $k(y.u) \leq -(L_1 + \gamma)$, $\gamma > 0$, $\forall y \in \Omega$, $\forall u \in \mathcal{U}_{ad}$.

We define :

(1.7) $$\bar{v}(x) = \inf_{u \in \mathcal{U}_{ad}} J(x,u(.))$$

and we can prove under one of the two following hypothesis a_1) or b_1)

a_1) $\langle f(y.u),\bar{n}\rangle < 0$ $\forall y \in \partial\Omega$, $\forall u \in \mathcal{U}_{ad}$, $\bar{n}$ normal to $\partial\Omega$ pointing to $\complement\Omega$,

b_1) $\langle f(y.u),\bar{n}\rangle \geq \beta > 0$ $\forall y \in \partial\Omega$, $\forall u \in \mathcal{U}_{ad}$

that $\bar{v}(x)$ is lipschitzian. Then, from Rodemacher's Theorem and the dynamic programming theory we know that $\bar{v}(x)$ satisfies

(1.8) $$\inf_{u \in \mathcal{U}_{ad}} \left(\frac{\partial \bar{v}}{\partial x} f(x,u) + \ell(x,u) + k(x,u)\bar{v}(x)\right) = 0 \quad \text{a.e. in } \Omega$$

Our aim is to obtain $\bar{v}(x)$; for it we give the following

§2. CHARACTERIZATION OF $\bar{v}$.

Let us introduce the set

$$W = \{w : \Omega \to R \,/\, a_2),\ b_2),\ c_2)\} \quad \text{with}$$

a_2) $\exists\, L(w) \,/\, |w(x_1) - w(x_2)| \leq L(w) \|x_1 - x_2\| \ \forall x_1, x_2 \in \Omega$

b_2) $w(x) \leq g(x)$ $\forall x \in \partial\Omega^+$, $\partial\Omega^+ = \{x \in \partial\Omega \,/\, \exists u \,/\, \langle f(x\ ,u),\bar{n}\rangle > 0\}$,

c_2) $\inf_{u \in \mathcal{U}_{ad}} \left(\frac{\partial w}{\partial x} f(x,u) + \ell(x,u) + k(x,u)w(x)\right) \geq 0$ a.e. in Ω

We can prove now the

<u>Theorem 1</u> : If (1.4), (1.5), (1.6) are satisfied we have (adding a_1 or b_1) :

(2.1) $\bar{v}(x) \geq w(x)$ $\forall x \in \Omega$, $\forall w \in W$

Proof : Let be $x \in \Omega$, $u(.)$ an admissible control and τ the corresponding exit time.

We take $T < \tau$ and we define

$$\Gamma_T = \{y(s) \ / \ 0 \le s \le T\} \tag{2.2}$$

From the definition of τ it follows

$$d(\Gamma_T, \partial\Omega) = \rho > 0 . \tag{2.3}$$

So we can define for $x \in \Gamma_T$

$$w_\rho(x) = (w * \varphi_\rho)(x) = \int_{B(0,\rho)} w(x-y)\varphi_\rho(y)dy \tag{2.4}$$

with

$$\varphi_\rho \in C_0^\infty(R^n), \ \operatorname{supp} \varphi_\rho \subset B(0,\rho) = \{x \ / \ \|x\| < \rho\} , \ \int_{B(0,\rho)} \varphi_\rho(y)dy = 1.$$

It is known that $w_\rho \to w$ uniformly on the compact sets of Ω. As Γ_T is compact we obtain :

$$\frac{\partial w_\rho}{\partial x} . f(x,u) + \ell(x,u) + k(x,u) . w_\rho(x) \ge \xi(x,u,\rho) \tag{2.5}$$

with
$$\xi(x,u,\rho) = \int_{B(0,\rho)} (\ell(x,u) - \ell(x-y,u))\varphi_\rho(y)dy +$$

$$+ \int_{B(0,\rho)} w(x-y)(k(x,u)-k(x-y,u))\varphi_\rho(y)dy +$$

$$+ \int_{B(0,\rho)} \frac{\partial w}{\partial x}(x-y)(f(x,u) - f(x-y,u))\varphi_\rho(y)dy .$$

From (1.4), (1.6) and a_2) it follows :

$$\xi(x,u,\rho) \ge \xi(\rho) = -\rho(1 + L(w) + \sup_{x\in\Omega}|w(x)|)L_1 ;$$

so, we have :

$$\int_0^T e^{\int_0^s k(y(r),u(r))dr} \ell(y(s),u(s))ds \ge \int_0^T \xi(\rho) \, e^{\int_0^s k(y(r),u(r))dr} ds + \tag{2.6}$$

$$+ \int_0^T e^{\int_0^s k(y(r),u(r))dr} \left(- \frac{\partial w_\rho}{\partial x} f(y(s),u(s)) - k(y(s),u(s))w_\rho(y(s))\right)ds \ge$$

$$\ge \xi(\rho).T + w_\rho(x) - w_\rho(y(T)) \, e^{\int_0^T k(y(s),u(s))ds}$$

Now, we take limit in (2.6) for $\rho \to 0$ and we obtain :

$$\int_0^T e^{\int_0^s k(y(r),u(r))ds} \ell(y(s),u(s))ds \ge w(x) - w(y(T)) e^{\int_0^s k(y(s),u(s))ds} \tag{2.7}$$

The next step is to take limit for $T \to \tau$.

If $\tau < +\infty$ we obtain :

$$(2.8)\quad \int_0^\tau e^{\int_0^s k(y(r),u(r))dr}\,\ell(y(s),u(s))ds \geq w(x) - w(y(\tau))\,e^{\int_0^\tau k(y(s),u(s))ds}\ ;$$

taking into account that $y(\tau) \in \partial\Omega^+$ it follows $w(y(\tau)) \leq g(y(\tau))$; so (2.8) becomes :

$$(2.9)\quad \int_0^\tau e^{\int_0^s k(y(r),u(r))dr}\,\ell(y(s),u(s))ds + g(y(\tau))\,e^{\int_0^s k(y(s),u(s))ds} \geq w(x)$$

On the other hand, if $\tau = +\infty$, after taking limit in (2.7) as $T \to \infty$ (we recall (1.6)) we obtain :

$$(2.10)\quad \int_0^\infty e^{\int_0^s k(y(r),u(r))dr}\,\ell(y(s),u(s))ds \geq w(x)$$

In both cases (2.9) and (2.10) we can conclude

$$J(x,u(.)) \geq w(x)\ ,$$

then,

$$\inf_{u(\cdot)} J(x,u(.)) \geq w(x)\ ,\ \text{i.e.}$$

$$(2.11)\qquad \bar{v}(x) \geq w(x)\qquad \forall x \in \Omega\ . \quad \blacksquare$$

Remark 1. : The proof does not use the lipschitzeanity of $\bar{v}$. So, if it falls we can always assure that $\bar{v}(x)$ is an upper bound for the elements of $\overline{W}$, but if $\bar{v} \in W$ we can add that $\bar{v}$ is the maximum element for the partial order

$$w_1 \leq w_2 \quad \text{iff} \quad w_1(x) \leq w_2(x) \qquad \forall x \in \Omega\ .$$

§3. STATEMENT OF THE AUXILIARY PROBLEM.

Let us introduce the auxiliary problem

(3.1) Find $\displaystyle\max_{w \in W} F(w)$, where

$$(3.2)\qquad F(w) = \int_\Omega w(x)a(x)dx$$

$$(3.3)\qquad a(x) > 0\ /\ \int_\Omega a(x)dx < \infty\ .$$

Theorem 2 : If $\bar{v} \in W$, then $\bar{v}(x)$ is the unique solution of problem (3.1).

Proof : It is a consequence of the continuity of the elements belonging to W.

Remark 2. : In the following, we shall suppose that $\bar{v} \in W$; so, in order to find (1.7) we shall solve (after Theorem 2) the problem (3.1).

§4. A METHOD FOR SOLVING PROBLEM (3.1).

To solve Prob. (3.1), we consider a succession of approximation problems (P_h) (using internal approximations of a space S containing W) in such a form that each approximated problem is of linear programming type. We show now the explicit form of problems P_h :

(4.1) The set Ω is approximated with sets $\Omega_h \subset \Omega$.
Ω_h is union of simplexes (triangles in the bidimensional case). If $\|h\|$ is the maximum lengh of the edges of the simplexes, Ω_h will satisfy the following properties :

(4.1.1) $m(\Omega - \Omega_h) \to o$ as $\|h\| \to o$

(4.1.2) $\forall D$ compact set $\subset \Omega$, $\exists \delta > o \;/\; \forall \|h\| < \delta$, $\Omega_h \supset D$.

(4.1.3) $\forall h \;\exists$ a transformation $Q_h : \Omega_h \to R^n \;/\; Q_h \Omega_h \supset \Omega$; for $\|h\| \to o$:

$$Q_h x \to x \quad \text{uniformly in } \Omega_h$$

$$Q_h^{-1} x \to x \quad \text{uniformly in } \Omega$$

$$\frac{\partial}{\partial x} Q_h\, x = Z_h(x) \text{ is a continuous matrix and}$$

$$Z_h(x) \to I \quad \text{uniformly in } \Omega_h.$$

(4.1.4) $\exists \tau > o$ (fixed constant independent of h) such that there is a sphere of diameter greater or equal to $\tau\lambda$ inside the simplex (λ : diameter of a simplex).

We define :

$$(4.2) \qquad K_h = \{w_h : \Omega_h \to R \;/\; (4.2.1),(4.2.2),(4.2.3),(4.2.4)\}$$

(4.2.1) w_h is continuous, a.e. differentiable with $\frac{\partial w_h}{\partial x}$ constant in the interior of each simplex

(4.2.2) $\|\frac{\partial w_h}{\partial x}\|_{L^\infty(\Omega_h)} \leq L$, with $L \geq L(\bar{v})$

(4.2.3) $w_h(x_i) \leq g(x_i)$ $\qquad \forall x_i$ vertex of $\partial\Omega_h^+$

(4.2.4) $\frac{\partial wh}{\partial x}(x_{c_i})\, f\,(x_{c_i},u) + \ell(x_{c_i},u) + k(x_{c_i},u)\, w_h(x_{c_i}) \geq o$

$\forall u \in \mathcal{U}_h$, $\forall x_{c_i}$ baricenter of an elementary simplex of Ω_h.

We define the prolongation operator :

(4.3) $$p_h(w_h(x)) = w_h(Q_h^{-1}(x)) \qquad \forall x \in \Omega$$

We define :

(4.4) $$F_h(w_h) = \int_{\Omega h} a_h(x) w_h(x) dx$$

(4.4.1) with $a_h(x)$ constant in each simplex and $\int_{\Omega h} |a_h(x)-a(x)| dx \to o$ as $||h|| \to 0$.

Remark 3 : The functions w_h are determinated by the values $w_h(x_{hi})$; x_{hi} vertex of a simplex of Ω_h. So, (4.2.2),(4.2.3) and (4.2.4) are linear restrictions and the number of restrictions is finite if $\mathcal{U}_h$ is finite. Furthermore, from (4.4.1) $F_h(w_h)$ is also a linear function.

After (4.1), (4.2), (4.3) and (4.4) we put the

(4.5) problem P_h :Find $\bar{w}_h$ / $\max_{w_h \in K_h} F_h(w_h) = F_h(\bar{w}_h)$

and we shall give here the main ideas used to show the existence of solution of (4.5) and the sense in which we have convergence of $\bar{w}_h$ to the solution $\bar{v}(x)$ of (3.1) (and (1.7)!). For it we can prove (we shall omit the proofs) the following 5 lemmas :

Lemma 1. :

If i) f, ℓ, k are continuous functions of $u \in \mathcal{U}_{ad}$ compact

ii) $\lim_{||h|| \to o} \partial\Omega_h^+ = \partial\Omega^+$

iii) $\forall h$, $\mathcal{U}_h$ is finite ; $||h_1|| \geq ||h_2|| \Rightarrow \mathcal{U}_{h_1} \subset \mathcal{U}_{h_2}$; $\bigcup_h \mathcal{U}_h = \mathcal{U}_{ad}$

then :

(4.6) $$\overline{\lim_{h \to o}}\, p_h K_h \subset K_L$$

with

$$K_L = \{w \in W \;/\; |w(x_1) - w(x_2)| \leq L||x_1 - x_2|| \quad \forall x_1, x_2 \in \Omega\}$$

Lemma 2. :

If $\lim_{h \to o} d(\partial\Omega_h^+, \partial\Omega) = o$ then

(4.7) $$K_L \subset \underline{\lim}_{h \to o}\, p_h K_h$$

Lemma 3. :

If i) w_h / $\lim_{h \to o} p_h w_h = w_h$

ii) (4.1.1) and (4.4.1) are satisfied, then ,

(4.8) $$\lim F_h(w_h) = F(w)$$

Lemma 4. :

$\forall h$ the problem P_n has at least a solution.

In other words : if $Y_h = \{\bar{w}_h \,/\, F_h(\bar{w}_h) \geq F_h(w_h),\ \forall w_h \in K_h\}$, we prove :

(4.9) $$Y_h \neq \Phi$$

Lemma 5. :

(4.10) The set $\bigcup_h p_h Y_h$ is totally bounded set of $C(\Omega)$ $\left(\|w\|_{C(\Omega)} = \sup_{x\in\Omega} |w(x)|\right)$

Remark : (4.9) gives the answer to the existence of solutions of (4.5). For the convergence of the device we have the following

Theorem 3. :

If (4.6), (4.7), (4.8), (4.9) and (4.10) hold

then :

(4.11) $$\bar{v} = \overline{\lim_{h\to o}}\, p_h Y_h = \underline{\lim_{h\to o}}\, p_h Y_h .$$

Proof : By virtue of (4.9) and (4.10) we obtain

(4.12) $$\overline{\lim_{h\to o}}\, p_h Y_h \neq \Phi .$$

Let be $\tilde{w} \in \overline{\lim_{h\to o}}\, p_h Y_h$; so there exists a succession $\{w_{h_\nu}\}/$

(4.13) $$\tilde{w} = \lim_{\nu\to\infty} p_{h_\nu} w_{h_\nu}$$

and for (4.6) it follows

(4.14) $$\tilde{w} \in K_L \quad ;$$

then, by remark 1 and (3.2)

(4.15) $$\tilde{w} \leq \bar{v} \Rightarrow F(\tilde{w}) \leq F(\bar{v})$$

and, by lemma 3 :

(4.16) $$F(\tilde{w}) = \lim_{\nu\to\infty} F_{h_\nu}(w_{h_\nu}) .$$

Now, using (4.7), $\forall h\ \exists\ \approx{w}_h$ /

(4.17) $$\bar{v} = \lim_{h\to o} p_h \approx{w}_h \quad ,$$

and, as $w_{h_\nu} \in Y_{h_\nu}$ it follows $F_{h_\nu}(w_{h_\nu}) \geq F_{h_\nu}(\tilde{\tilde{w}}_{h_\nu})$.

So, taking limit $\nu \to \infty$ in the last inequality we obtain using (4.16), (4.17) and lemma 3 :

(4.18) $F(\tilde{w}) \geq F(\bar{v})$ that implies

(4.19) $F(\tilde{w}) = F(\bar{v})$ that is to say, after theorem 2

(4.20) $\tilde{w} = \bar{v}$.

Then

(4.21) $\overline{\lim_{h\to o}}\, p_h Y_h = \bar{v}$, while the second part of (4.11) follows from (4.21) and (4.10). In this form Theorem 3 is proved.

BIBLIOGRAPHY

[1] J. CEA : "Approximation variationnelle des problèmes aux limites", Ann. Inst. Fourier 14 (1964), 345-444.

[2] A. FRIEDMAN : "Differential Games", Wiley Interscience (1971).

[3] R. GONZALEZ : "Sur l'existence d'une solution maximale de l'équation de Hamilton-Jacobi", C.R. Acad. Sc. Paris, t. 282 (1976).

[4] R. GONZALEZ : "Solution globale de quelques problèmes de contrôle", to appear.

[5] R. GONZALEZ ; E. ROFMAN : "Coût optimal et solution maximale dans des problèmes de contrôle stationnaire", Cahier de Mathématiques de la Décision, Univ. Paris IX-Dauphine (1978) to appear.

[6] U. MOSCO : "An introduction to the approximate solution of variational inequalities", Constructive Aspects of Functional Analysis. Edizioni Cremonese , Roma 1973, pp.497-685.

[7] G. STRANG ; G. FIX : "An analysis of the finite element method", Prentice Hall (1973).

[8] K. YOSIDA : "Functional Analysis", Springer Verlag, (1974).

TIME OPTIMAL CONTROL OF STATE CONSTRAINED LINEAR DISCRETE SYSTEMS

M.P.J. SCOTT and A.A. DICKIE
Department of Electrical Engineering and Electronics
The University, Dundee, Scotland

ABSTRACT:- The direct Linear Programming method of solving the time optimal control problem for state constrained linear discrete systems is time consuming and requires a very large amount of storage, for even a modest problem. The algorithm presented here utilises a cutting-plane technique to efficiently deal with the state constraints, and exploits their special nature to further reduce the storage requirements. The computational effort depends on the severity of the state constraints. A sixth order example is used to illustrate the proposed method.

1. INTRODUCTION

It has been known for some time that the linear time optimal control problem can be solved by discretising the state equations and then using the Linear Programming (LP) or Simplex method. This was first pointed out by Zadeh and Whalen[1], and expanded upon by Torng[2]. Lesser and Lapidus[3] showed how the method could be directly extended to deal with state variable constraints. Fath[4] demonstrated the versatility of the approach by solving an example with complicated time-varying bounds on both control and state variables. However, whereas control variable constraints can be handled conveniently by using the bounded variable technique, the inclusion of state variable constraints vastly increases the size of the simplex tableau, and hence the overall computational requirements. Nevertheless the practical usefulness of this kind of control has been demonstrated by Nieman and Fisher[5].

Early workers formulated the problem in such a way that it could be solved by standard Linear Programming packages. Bashein[6] has demonstrated that a much more elegant solution is possible using a special purpose LP algorithm which takes into account the special nature of the time optimal control problem. This algorithm forms the basis of the present work. It is proposed to deal with the state variable constraints by using a cutting-plane technique.

2. MATHEMATICAL FORMULATION

The behaviour of an n-th order linear system can be described by

$$\dot{\underline{x}}(t) = A\underline{x}(t) + B\underline{u}(t) \qquad (1)$$

where $\underline{x}$ is the n-dimensional state vector, and $\underline{u}$ the r-dimensional input vector. For notational convenience the development will be made for the scalar input case, i.e. for r = 1. In sampled data mode the input is constrained to be constant between sampling intervals, that is $u(t) = u_i$ for $(i - 1)T < t \leqslant iT$ where i is a positive integer and T is the sampling interval.

Integrating (1) over one such sampling interval gives

$$\underline{x}(iT) = \phi(T)\underline{x}[(i-1)T] + H(T)u_i \tag{2}$$

where $\phi(T)$ is the state transition matrix exp (AT) and H(T) is the control transfer matrix given by

$$H(T) = \int_0^T \phi(T-t)B\,dt$$

Starting from an initial state $\underline{x}_o$, and using (2) repetitively over k sampling intervals yields

$$\underline{x}(kT) = \phi(kT)\underline{x}_o + \sum_{i=1}^{k} \phi[(k-i)T]H(T)u_i \tag{3}$$

This can be rearranged to give

$$\sum_{i=1}^{k} \phi(-iT)H(T)u_i = \phi(-kT)\underline{x}(kT) - \underline{x}_o \tag{4}$$

The time optimal control problem can be formulated in the following way. Find a feasible input sequence $u_1, u_2, \ldots u_k$ which will drive the system from its initial state $\underline{x}_o$ to a desired final state $\underline{x}_f$, in the minimum number of sampling periods. Here a feasible input sequence is one which fulfils the input constraints, and which does not at any time drive the system beyond its state constraints. The final state $\underline{x}_f$ will be assumed to be the origin of the state space. For an extension of the method to a more general set of final states, see Reference 12.

If the system is to achieve the desired transition from $\underline{x}_o$ to the origin in k sampling intervals, then the terminal constraints must be met. Therefore from (4)

$$\sum_{i=1}^{k} \phi(-iT)H(T)u_i = -\underline{x}_o \tag{5}$$

The input saturation constraints are of the form

$$u_L \leqslant u_i \leqslant u_B, \quad i = 1, 2, \ldots, k \tag{6}$$

The state variable constraints considered here are assumed to be of the form

$$D\underline{x}(t) \leqslant \underline{e}, \quad \text{for all } t$$

where D is an nxp matrix, p is the number of constraints, and $\underline{e}$ is a p-vector with non-negative elements. Having discretised the state equations these constraints can only be applied at the end of each sampling period. Substituting from Equation (3) gives the p.k inequalities

$$D\sum_{i=1}^{j} \phi[(j-i)T]H(T)u_i \leqslant \underline{e} - D\phi(jT)\underline{x}_o, \quad j = 1, 2, \ldots, k \tag{7}$$

In Linear Programming terms the problem is to find the minimum value of k which permits phase I feasibility of the problem defined by the linear constraints (5) - (7). Note that an objective function as such does not exist, as for the minimum time problem only a feasible solution is required. Torng[2] has shown how to select from the set of feasible solutions the on which requires minimum fuel consumption.

Attempting to solve directly by introducing slack and artificial variables as necessary into (5) and (7) leads to a Simplex tableau with a total of n + p.k equations in k + n + p.k unknowns. The constraints (6), due to their special simple form do not effect the dimensions of the problem. However, for most cases of interest $k \gg n$ and so the inclusion of state variable constraints rapidly increases the size of the LP problem.

The approach here is to first solve the much simpler problem defined by (5) and (6), and to deal with the state constraint violations only when and if they occur.

3. SOLUTION WITHOUT STATE CONSTRAINTS

Changing (5) into a form more suitable for the application of Linear Programming gives

$$S\underline{U} = \underline{b}$$

where $\underline{U}$ is the control sequence vector containing as elements $u_1, u_2, \ldots, u_k$, $\underline{b}$ is the n-dimensional requirements vector, initially containing $-\underline{x}_o$, and S is given by

$$S = [\phi(-T)H(T)\phi(-2T)H(T) \ \ldots \ \phi(-kT)H(T)]$$

S can be precalculated and stored for some value k = K, where K is a number greater than the number of samples likely to be needed.

At the start all the u_i, i = 1, 2, ..., K are set to zero and an initial basis is formed by introducing an n-vector of artificial variables, giving

$$S\underline{U} + I\underline{v} = \underline{b}$$

The object now is to find the minimum value of k, k_{min}, needed to permit all the artificial variables to be driven from the basis, subject to the constraints (6).

In standard LP codes all variables are automatically constrained to be non-negative, which is inconvenient for the present application. Bashein[6] has described a modified Simplex algorithm more suited to this particular problem. Using a modified version of Dantzig's[7] upper bounding technique to allow the u_i to be non-basic at three levels, u_L, zero and u_B, he has shown how the problem as posed can be solved directly, without further manipulation. To begin k is set to some integer known to be less than the number of samples actually needed, e.g. k = 1. The time optimal problem is solved by using the following rule:- A variable, u_{k+1}, is not considered for entry into the basis until it is determined that a basic feasible solution cannot be obtained using only u_i, $i = 1, 2, \ldots, k$.

For what follows it is important that the so-called revised Simplex method be used. This has the advantage that the matrix S will not be corrupted by the Simplex process. Instead the inverse basis B^{-1} is stored and continuously updated. The modifications to Basheins algorithm, which uses the ordinary Simplex method, are quite straightforward (see Hadley[8] Chapter 7). Also provision must be made for dealing with the slack variables which will arise when the state variable constraints are introduced.

4. SOLUTION WITH STATE VARIABLE CONSTRAINTS

Cutting-plane algorithms were first introduced by Kelley[9] for the solution of constrained convex programming problems. The application to optimal control was pointed out by Levitin and Polyak[10] in their well known treatise on constrained minimization techniques. Simply stated, the problem is first solved in the absence of state constraints, and then, if the resultant trajectory violates the state constraints, this information is used to form a new constraint which cuts off that portion of the solution space, thus rendering the original result infeasible. A new feasible solution is then sought, and this procedure is repeated until a solution is obtained which satisfies all the constraints. There are various ways of making the cuts. One method suggested is to form the new constraint by summing over all the violated constraints. This was proposed by Kapur and Van Slyke[11]. Levitin and Polyak[10] propose simply introcing that constraint which has suffered the worst violation, and this is the method that will be used here.

Assume that after solving without state constraints and then having run a simulation it is found that the q-th state variable constraint has been violated to an extent ϵ at the J-th sampling instant. Assume also that this is the worst (maximum) violation. From (7) the new cut-off constraint to be introduced is then given by

$$\underline{d}_q \sum_{i=1}^{J} \phi[(J - i)T]H(T)u_i \leqslant e_q - \underline{d}_q\phi(JT)\underline{x}_0 \tag{8}$$

where $\underline{d}_q$ is the q-th row of matrix D, and e_q is the q-th component of vector $\underline{e}$. If this will be the m-th constraint of the Simplex tableau, then it can be rewritten

$$[\underline{d}_q\phi[(J - 1)T]H(T) \ldots \underline{d}_qH(T), 0, 0, \ldots 0]\underline{U} + s_m - v_m = e_q - \underline{d}_q\phi(JT)\underline{x}_0 \tag{9}$$

where s_m is the slack variable associated with (8). As the solution obtained is infeasible this slack variable would be negative, and so an artificial variable v_m has been introduced into the basis. At the time when this new equation is added to the Simplex tableau, the value of the artificial variable is simply ϵ, which is known from the simulation, and so the right-hand side of (9) need not be evaluated explicitly. If the original basis was B, the new augmented basis is given by

$$B_n = \begin{vmatrix} B & 0 \\ \underline{\gamma} & -1 \end{vmatrix} \quad \text{and} \quad B_n^{-1} = \begin{vmatrix} B^{-1} & 0 \\ \underline{\gamma}B^{-1} & -1 \end{vmatrix} \tag{10}$$

The last column corresponds to the new artificial variable, and $\underline{\gamma}$ is the row vector containing as elements the coefficients in the new constraint of the variables in the original basis B. See Hadley[8], Chapter 11. As B^{-1} is already known, B_n^{-1} can be quickly calculated. It should be noted that introducing a new constraint in this manner into an already active problem is an established Linear Programming technique[7,8].

Now the Simplex method is used as before to drive out this new artificial variable, a procedure which may well require an increase in k. Then the process is repeated. Convergence for the method can be proved by noting that in a finite number of iterations either a feasible solution is obtained, or the full set of constraints given by (7) will have been introduced.

During the course of the solution a slack variable associated with a state variable constraint may well enter the basis. This implies that this constraint is no longer binding. The size of the problem can sometimes be limited by deleting these redundant constraints. However experience has shown that many constraints which are thus deleted have later to be re-introduced, leading to slower convergence. Therefore the following heuristic procedure has been used. A constraint introduced when $k = k_c$ is deleted if it becomes non-binding for $k > k_c$. This rule is justified by the fact that as k increases, the state trajectory can change more and more. Therefore constraints introduced early on, and which become redundant at a later stage, tend to remain redundant. The deletion is achieved by simply rubbing out the relevant row in the Simplex tableau, and deleting the relevant row and column from the inverse basis B^{-1}, in effect the reverse of the procedure for including a new constraint. The size of the basis will thus fluctuate as constraints are added and deleted. It can be easily seen that this process does not upset the convergence proof.

5. THE ALGORITHM

The algorithm can now be summarised as follows

1. Solve using Simplex method, without state variable constraints.
2. Run a simulation, using equation (2) repeatedly. Check for violation of the state constraints. If none go to step 6.
3. Create a new constraint from the worst violation and augment the inverse basis, using equations (9) and (10).
4. Obtain a new basic feasible solution using the Simplex method.

5. Delete the redundant constraints according to the above rule. Go to step 2.
6. Finish.

From (9) it can be seen that the state variable constraints are of a particularly simple form. If the values $D\phi[(K - i)T]H(T)$, $i = 1, \ldots, K$ are precalculated and stored, the coefficients for any particular constraint can be quickly accessed. Thus the new constraints do not need to be explicitly added to the Simplex tableau. In this way the amount of storage required can be reduced.

The algorithm has been implemented in Fortran IV, and is normally run on a 24k 16-bit Data General Nova 2/10 mini-computer. A program listing with documentation is given in Reference 12. The amount of dimensioned floating point data storage required is approximately $(n + r.K)(n + p + 2) + M^2 + 6M$, where M is the maximum size of the basis. The magnitude of M depends on the anticipated severity of the state constraints, i.e. on how many state constraints will need to be introduced to obtain a feasible solution. In the worst case situation, when the optimal trajectory lies on a state constraint boundary at nearly every sampling instant, M would be required to be of the order of k_{min}.

The amount of computer running time needed also depends on the severity of the state constraints. The number of floating point multiplications required to perform one Simplex pivot operation is approximately $m(r.k + 2) + m^2$, where m is the current size of the basis. This can be reduced however by exploiting the known sparseness of the state constraint equations (9). In addition each simulation requires $k(n^2 + n.r + n.p)$ floating point multiplications.

6. AN EXAMPLE

The performance of the algorithm is best illustrated by an example. Consider the sixth order gas absorber which is described by the equation (1), where

$$A = \begin{vmatrix} x & y & 0 & 0 & 0 & 0 \\ w & x & y & 0 & 0 & 0 \\ 0 & w & x & y & 0 & 0 \\ 0 & 0 & w & x & y & 0 \\ 0 & 0 & 0 & w & x & y \\ 0 & 0 & 0 & 0 & w & x \end{vmatrix} \qquad B = \begin{vmatrix} w & 0 \\ 0 & 0 \\ 0 & 0 \\ 0 & 0 \\ 0 & 0 \\ 0 & y \end{vmatrix}$$

$w = 0.539, \quad x = 1.1731, \; y = 0.6341$

This system was also used as an example by Lesser and Lapidus[3] and by Bashein[6]. The initial conditions vector is

$$\underline{x}_o = [-0.0306, -0.0568, -0.0788, -0.0977, -0.1138, -0.1273]^T$$

and the constraints are

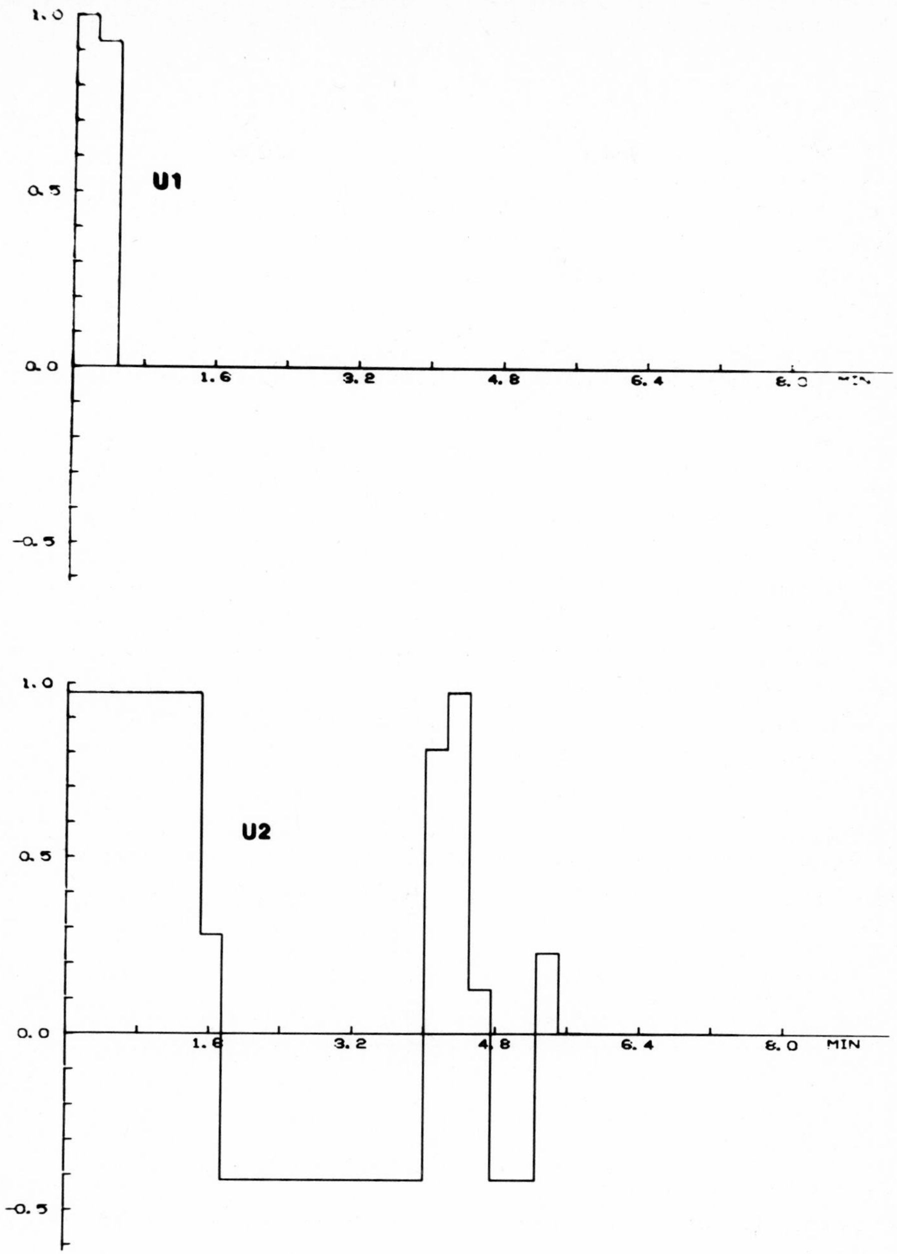

FIG. 1 Time-optimal inputs for h = 0.5

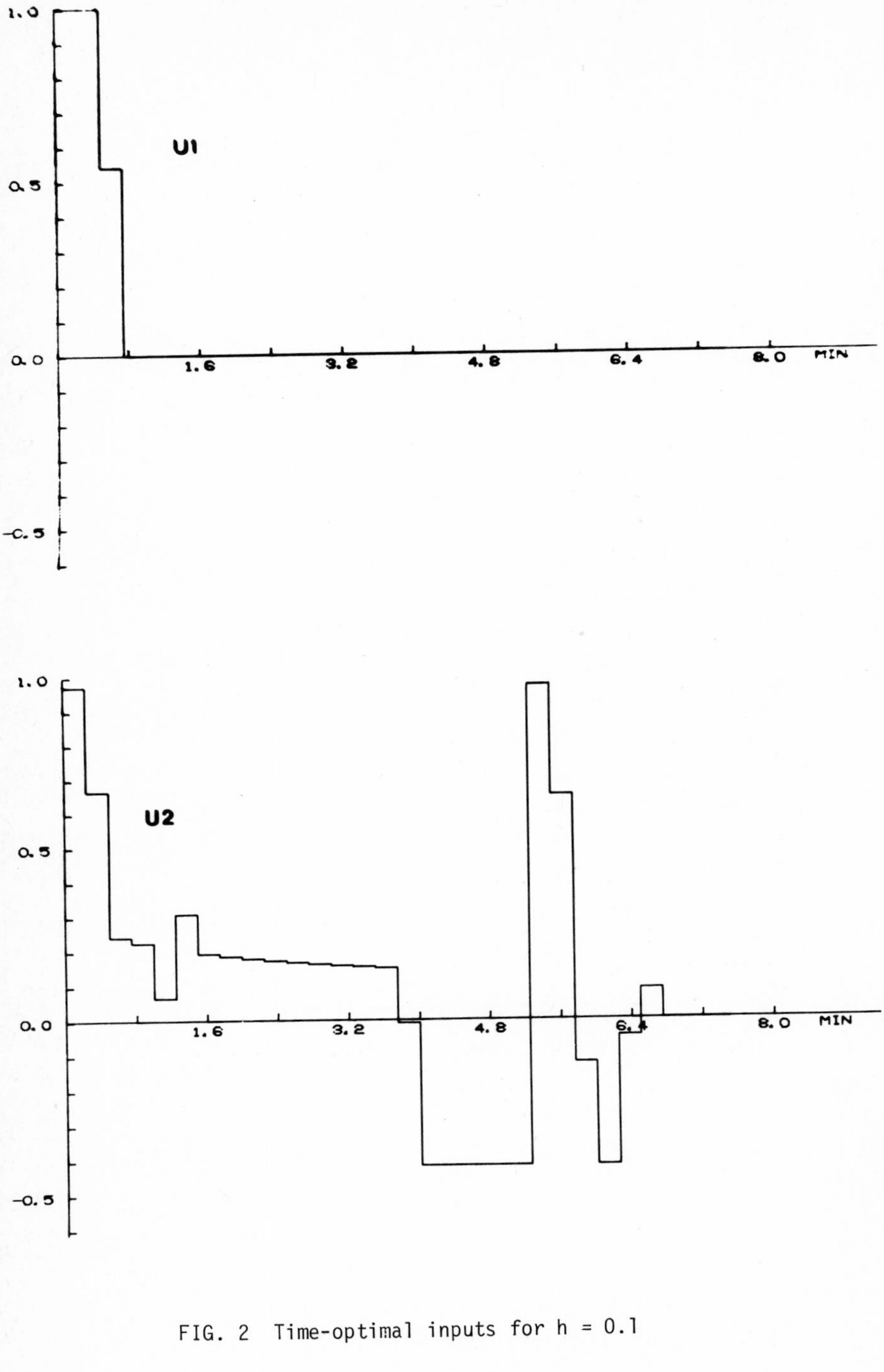

FIG. 2 Time-optimal inputs for h = 0.1

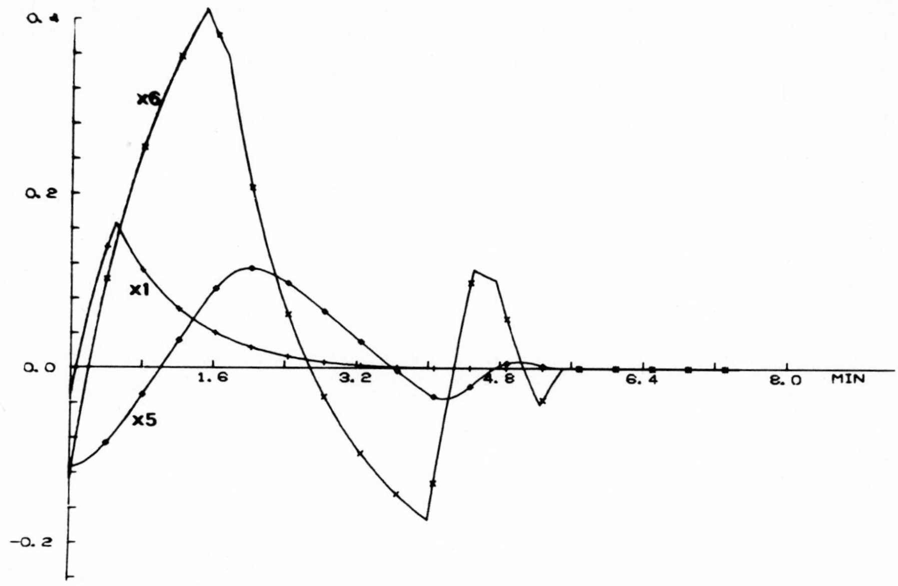

FIG. 3 State trajectories for h = 0.5

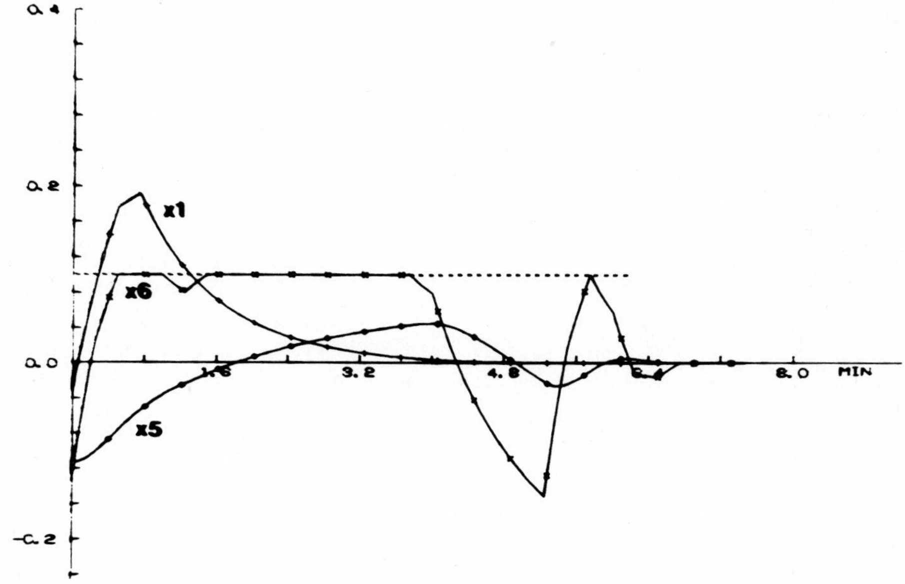

FIG. 4 State trajectories for h = 0.1

$0.0 \leq u_1(t) \leq 1.0$
$-0.4167 \leq u_2(t) \leq 0.972$
$x_1(t) \leq 0.2$
$x_6(t) \leq h$

A sampling period of 0.25 minutes was used, and K was set to 30. Results for various values of h are given in Table 1. The last column indicates t_r, the amount of computer running time required relative to that needed when the state constraints are not violated. It should be pointed out that the first state constraint, $x_1 \leq 0.2$, is never violated, and thus is never considered by the algorithm.

TABLE 1

h	k_{min}	M	t_r
> 0.411	23	6	1.00
0.4	23	7	1.10
0.3	23	11	1.66
0.2	24	12	2.17
0.1	27	21	6.02

The optimal controls and trajectories for the cases when h = 0.5 and h = 0.1 are shown in Figures 1-4.

7. CONCLUSIONS

A simple cutting-plane technique has been used to develop an efficient algorithm for the solution of the state constrained time optimal control problem. The computational requirements depend on the severity of the state constraints for the particular time optimal transition envisaged, rather than on the number of such constraints.

REFERENCES

1. ZADEH, L.A. and WHALEN, B.H.: "On optimal control and linear programming", IRE Trans. Autom. Contr. (Corres.), 1962, Vol. AC-7, pp 45-46.
2. TORNG, H.C.: "Optimization of discrete control systems through linear programming", J. Franklin Inst., 1964, Vol. 278, pp 28-44.
3. LESSER, H.A. and LAPIDUS, L.: "The time optimal control of discrete time linear systems with bounded controls", A.I.Ch.E.J., 1966, Vol. 12, pp 143-152.
4. FATH, A.F.: "Approximation to the time optimal control of linear state constrained systems", J.A.C.C., 1968, pp 962-969.
5. NIEMAN, R.E. and FISHER, D.G.: "Experimental evaluation of time optimal, open-loop control", Trans. Inst. Chem. Engrs., 1973, Vol. 51, pp 132-140.
6. BASHEIN, G.: "A simplex algorithm for on-line computation of time optimal controls", IEEE Trans. Autom. Contr., 1971, Vol. AC-16, pp 479-482.

7. DANTZIG, G.B.: "Upper bounds, secondary constraints, and block triangularity", Econometrica, 1955, Vol. 23, pp 174-183.

8. HADLEY, G.: "Linear programming", (Addison-Wesley, 1962).

9. KELLEY, J.E.: "The cutting-plane method for solving convex programs", J. SIAM, 1960, Vol. 8, pp 703-712.

10. LEVITIN, E.S. and POLYAK, B.T.: "Constrained minimization methods", USSR Comp. Math. & Math. Phys., 1968, Vol. 6, pp 1-50.

11. KAPUR, K.C. and VAN SLYKE, R.M.: "Cutting-plane algorithms and state constrained linear optimal control problems", J. Comp. Sys. Sci., 1970, Vol. 4, pp 570-605.

12. SCOTT, M.P.J.: "Computational Algorithms for time-optimal control', Ph.D. Thesis, University of Dundee, 1977.

A ROBUST ADAPTIVE PROCEDURE FOR SOLVING A NON GAUSSIAN IDENTIFICATION PROBLEM

A. Benveniste*, M. Goursat**, G. Ruget*

*IRISA - Université de Rennes
Avenue du G-al Leclerc BP 25A
35031 Rennes Cedex
FRANCE

**IRIA - LABORIA
Domaine de Voluceau BP 105
78150 Le Chesnay
FRANCE

ABSTRACT.

Consider an unknown linear time-invariant system S without control, drived by random variables with known law. We are interested in the identification of S from the output. The usual results work only under the major assumption : S is minimum phase. In our case the system S is non minimum phase and the litterature gives only a negative result : the identification is impossible for a gaussian driving noise. For a large class of other input laws we give here a solution to this problem and present some numerical results for a concrete case, origin of our study: the blind settling phase of an equalizer in data communication.

1. SETTING THE PROBLEM - DESCRIBING THE SOLUTION.

1.1. The physical problem.

The problem we describe here is related to the data transmission with a telephone line. The transmisssion system may be considered as an unknown time-invariant linear system S without control. The input is a sequence of independent identically distributed (i.i.d.) random variables (r.v.) which are the data to be transmitted. The duration of the impulse response of S being large with respect to each data, we have intersymbol interferences in the output. The problem is to restore the unknown transmitted sequence or, equivalently, identify the inverse channel S^{-1} from the observations. The solution consists in placing a serial adaptive system called equalizer which is a linear filter ; its tap weights are adjusted in order to minimize the mean square error for its output. The adjustment is done by a stochastic gradient algorithm. This technique requires the knowledge of the transmitted data: this difficulty is avoided in practice by a settling phase during which the emitter transmits an a priori known sequence with which we get an equalizer close to the inverse.The obtained tap weights give good estimates of the input ; the algorithm continues by using the estimates and the equalizer follows the evolution of the inverse (in practice, obviously, the channel is not time-invariant but the adaptivity of the algorithm is quick enough for the non-stationarity of the channel). In some cases (for example : break in a multipoint communication) the transmission of a known sequence is impossible and the receiver is obliged to inverse the system S only with the receiver output flow : this is our blind adjustment problem.

Remark : we do here a little mistake because we neglect the additive noise on the output. We do not exactly look for S^{-1}, we also have to filter the output. The previous procedure (minimizing the mean square error) realizes simultaneously this work. We will see(in 1.2) that neglecting the noise is not a worry in our case.

1.2. Mathematical formulation.

The input is $(a_n)_{n\in Z}$: sequence of i.i.d. r.v. with a known distribution ν (input law). In data communication (a_n) are equally distributed on a finite set (for example on $\{\pm 7, \pm 5, \pm 3, \pm 1\}$), but we give here a solution for a large class of laws ν.

The unknown channel is $S = (s_k)_{k\in Z}$: in practice (s_k) corresponds to the sampling of the impulse response with a given time step Δ.

The observation is a stationary sequence $(x_n)_{n\in Z}$ given by :

(1.1) $$x_n = \sum_{k\in Z} s_k \, a_{n-k}$$

The equalizer is $H = (h_k)_{k\in Z}$ and its output is (c_n), $c_n = \sum_{k\in Z} h_k \, x_{n-k}$

$$(a_n) \xrightarrow[\text{unknown}]{S = (s_k)} x_n \xrightarrow[\text{adjustable}]{H} c_n$$

In data communication we use the fact that (a_n) belongs to a finite set : c_n is quantized to the nearest data level (if $|c_n - a_n| < 1$ the estimate is exact). This point is one of the reasons (with a small level of the noise and also with the fact we are only looking for a settling phase) for which we neglect here the additive noise on the output (x_n).

Our problem is : adjust H in order to have $c_n \simeq a_n$.

Notations. :

$X = (x_n)_{n\in Z}$; $x_n = (S*A)(n)$; $x_n = X(n)$;

$\check{H} = (h_{-k})_{k\in Z}$ if $H = (h_k)_{k\in Z}$

ℓ^2 = Hilbert space of square convergent series ; $\langle .,. \rangle$ and $\|\cdot\|$: inner product and associated norm in ℓ^2.

1.3. A classical solution ?

The identification problem we have described seems classical. If S is stable and with a stable inverse (minimum phase system) the solution is given by the well-known least squares identification method and a_n is simply the innovation given by :

(1.2) $$a_n = \lambda(x_n - \mathbb{E}(x_n \,/\, x_{n-1}, x_{n-2}, \ldots))$$

where $\mathbb{E}(.|.)$ is the least squares estimation operator and λ chosen in order to adjust the variance.

But in our case, we have a non minimum phase system. The factorization of the spectrum of (x_n) is not unique and with the second order statistics we can only identify the amplitude spectrum, we cannot identify the phase. If the entry law ν is gaussian the 2nd order statistics are exhaustive and the problem is impossible.

(1.3) The two basic assumptions are :

A1 ν is symmetric with finite variance

A2 The energy of S is finite ; S is invertible i.e.

$\Lambda = (\Lambda_{ij})_{i,j\in Z}$ with $\Lambda_{ij} = \sum_k s_k s_{k+i-j}$ (covariance matrix of (x_n) up to a multiplying constant) is bounded and positive definite in ℓ^2.

Remark 1 : S non-minimum phase => S^{-1} is not a one-sided function of time. We cannot restore (a_n) on-line but only the global sequence. In practice, the lines are truncated and (a_n) is restored with a constant delay and S^{-1} is defined up to a time shift.

Remark 2 : ν being symmetric , $(-a_n)$ has the same law => S^{-1} and $-S^{-1}$ are two possible solutions.

Therefore our goal is :

(G) find H so that the global line H.S is ±I up to a time shift (I : only one non zero coefficient, equal to 1).

1.4. Describing the method.

The previous section shows that the major point is that ν is non gaussian. Using this fact a characterization of a solution H is given by the following results :

Lemma 1.1.

(a_n) is a sequence of i.i.d. r.v. with law ν satisfying to A1 (1.3). Suppose there exists $T = (t_k)_{k\in Z}$ with at least two non zero coefficients such that $\sum_k t_k^2 = 1$ and for which the law of $c = \sum_{k\in Z} t_k a_{-k}$ is ν.

Then, ν is necessarily gaussian ,

and we immediately get a characterization of H :

Corollary 1.2. : ν non gaussian ; H is such that the law of c is ν ; then we necessarily have H.S = ±I (up to a time shift).

Proof of Lemma 1.1 and Corollary 1.2. : see [1]

We have here the basic explanation :

- ν gaussian <=> no distortion by a rotation ($\neq$ identity) : problem impossible if ν gaussian
- ν non gaussian : we will get our solution if we adjust H so that the instantaneous law of c converges to the entry law ν.

We have now a precise view on the problem ; the method to get the solution is a classical approach :

- define a potential $\mathcal{J}$: $\mathcal{J}(H) = \mathbb{E}[\varphi(c_n)]$

(1.4) $\begin{cases} \text{where } \varphi \text{ (depending on } \nu) \text{ is an even function } \mathbb{R} \to \mathbb{R} \text{ to be chosen} \\ \text{such that the minima of } \mathcal{J} \text{ are } \pm S^{-1} \end{cases}$

- Realize by a stochastic gradient algorithm :

(1.5) $$\min_{H \in \ell^2} \mathcal{J}(H)$$

If the derivative of φ is Ψ we can formally write the algorithm :

(1.6) $$\begin{cases} H^{n+1} = H^n - \tau X_n \Psi(c_n) \qquad c_n = \sum_{k=-N}^{+N} h_k^n x_{n+k} \ , \ H^n = (h_k^n)_{-N \leq k \leq N} \\ X_n = (x_{n-N}, \ldots, x_n, \ldots, x_{n+N}) \\ \tau > 0 \text{ is the step of the gradient} \end{cases}$$

Remark 1 : With (1.5) we formally get (1.6) with $N = +\infty$. The first step to be justified is the differentiation. After that we have another difficulty : in practice we use (1.6) with N finite. We have to justify the truncating : in the following we omit this point (see [1]) which is intuitively easy to admit.

Remark 2 : We will see in §3 how to converge to S^{-1} rather than to $-S^{-1}$. This point is depending on the initial value for (1.6) and may be solved with a rough information about S (available in practice).

2. DEFINITION AND STUDY OF THE POTENTIAL.

We present here only some results with the related sets of assumptions. The complete results and the proofs are given in [1].

2.1. Using the global line.

We consider the following scheme :

(2.1)

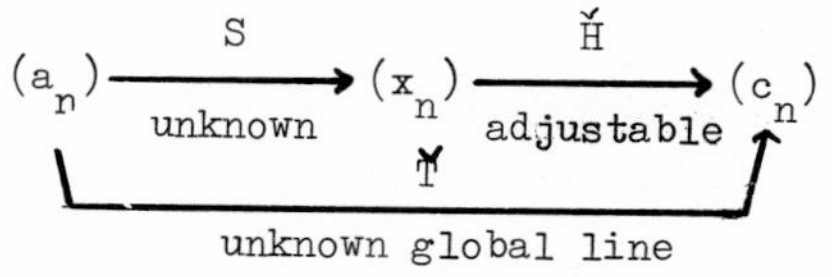

$T = H * \check{S}$

An essential point is the use of the global line thanks to :

(2.2) $$H * \check{S} = T \in \ell^2 \Leftrightarrow H \in \ell^2 \qquad \text{(thanks to (1.3)A2)}$$

So, if we define the potential

(2.3) $$\mathcal{V}(T) = \mathbb{E}(\varphi(c)) \quad \text{where} \quad c = \langle T, A \rangle$$

we have

(2.4) $$\mathcal{V}(T) = \mathcal{J}(H) \quad \text{for} \quad T = H * S$$

and the problem (1.4), impossible because S^{-1} is unknown, becomes :

(2.5) choose φ such that the minima of $\mathcal{V}$ are $\pm I$

We can solve (2.5) because we know the entry law for T ; our work is now :

- solve (2.5)
- study the steepest descent lines (s.d.l.) of $\mathcal{V}$
- study of the s.d.l. of $\mathcal{J}$.

Remark : we can give a formal explanation on the effect of T. Consider the simple case with ν = uniform law on $[-a,+a]$ and $T=(t_1,t_2)$ $t_1=\cos\theta, t_2=\sin\theta$. The law of the output c is a marginal distribution of the law of the transformed by T of $\nu\times\nu$: starting with $\theta=0$, we have a displacement of the probability mass growing until $\theta=\frac{\pi}{4}$, decreasing on $[\frac{\pi}{4},\frac{\pi}{2}]$ and so on.

2.2. Definition of a regular potential.

Denoting the derivative of φ by Ψ we introduce the vectors fields :

(2.6)
$$\begin{cases} V_T = -\mathbb{E}(A\Psi(c)) & T\in\ell^2 \\ V_T^c = -\{\mathbb{E}(A\Psi(c)) - T\,\mathbb{E}(c\Psi(c))\} & T\in s^2 \quad s^2 \text{ unit sphere of } \ell^2. \end{cases}$$

$\mathbb{E}(A\Psi(c))$ is formally the gradient of $\mathcal{V}$: this point is justified by the following theorem. The situation is the same for V_T^c with respect to $\mathcal{V}/_{s^2}$ (restriction of $\mathcal{V}$ on s^2). V_T^c is introduced because we use the spherical coordinates and the study of V_T^c will give the major result for V_T : the behaviour when θ changes.

(2.7) Assumptions :

We take $\Psi = -\alpha\,\mathrm{sgn} + \tilde{\Psi}$ with

B1 : $\tilde{\Psi}$ is odd, C^1 and satisfies to

(2.8) $$\sup_{\|T\|\le K} \mathbb{E}((\Psi'(\langle T,A\rangle))^2) < +\infty \qquad \forall K<+\infty$$

(2.9) ν has a fourth moment, is absolutely continuous with respect to the Lebesgue measure and with a bounded density f.

B2 : $\tilde{\Psi}$ is odd, C^3, satisfies to (2.8) and to :

(2.10) $$\sup_{\|T\|\le K} E((\Psi^{(3)}(\langle T,A\rangle))^2) < +\infty \qquad \forall K<+\infty$$

ν satisfies to (2.9) with f of C^2-class, bounded and with a bounded 2nd derivative.

Theorem 2.1. :

(i) Under the assumption B1 (2.7) :

$(V_T)_{T\in\ell^2}$ (resp. $(V_T^c)_{T\in s^2}$) is bounded in $\ell^2-\{0\}$ (resp. s^2) and is the opposite of the gradient field of $\mathcal{V}$ (resp. $\mathcal{V}/_{s^2}$).

(ii) under the assumption B2 (2.7)

(V_T) (resp.(V_T^c)) is locally Lipschitz in $\ell^2-\{0\}$ (resp s^2)

Our situation is now : the gradient field of $\mathcal{V}$ is well defined and regular enough to have the flow of its integral curves. The same result is available for (V_T^c).

2.3. The potential giving the solution.

The first point is : in accord with the previous assumptions, choose Ψ such that $\mathcal{V}/_{s^2}$ is a measurement of the distortion by a rotation i.e. $\mathcal{V}/_{s^2}$ is increasing with θ on $[0,\pi/4]$. (see Remark of §2.1).

Consider $T_\theta \in s^2$ and (i,j) fixed, with $i\neq j$; $T_\theta = (t_k)$ with

$$t_i = R\cos\theta \quad , \quad t_j = R\sin\theta \quad , \quad R^2 = 1 - \sum_{k\neq i,j} t_k^2 > 0$$

Let $\frac{\partial}{\partial\theta_{ij}}\mathcal{V}(T)$ be the derivative at $T = T_\theta$ of the function $\theta \to \mathcal{V}(T_\theta)$. The above property ($\mathcal{V}/_{s^2}$ increasing on $[0,\pi/4]$) is obtained by $\frac{\partial}{\partial\theta_{ij}}\mathcal{V}(T) > 0$; this result is given by the following result but we obviously need some additional assumptions on ν :

(2.11) Assumptions :

C1 : ν is uniform on $[-a, +a]$

C2 : $\nu(dx) = k\, e^{-g(x)}dx$ g even, $g(x)$ and $g'(x)/x$ strictly increasing on $\mathbb{R}_+$

Lemma 2.2.

Under the assumption C1 or C2 and Ψ such that :

$$(2.12)\qquad \begin{cases} \Psi \text{ odd, } C^2 \text{ except at } 0, \\ \Psi(0_+) \leq 0, \quad \Psi''(x) \geq 0 \text{ for } x \geq 0 \text{ (with at least one inequality being strict)} \end{cases}$$

we have

$$\frac{\partial}{\partial\theta_{ij}}\mathcal{V}\Big|_{T=T_\theta} > 0 \quad \text{for } 0 < \theta < \pi/4$$

With this lemma we have a good behaviour with respect to θ. We consider now $T \in \ell^2$: we have $T = \rho T^c$, $T^c \in s^2$. The study will be complete if we have for ρ a result analog to Lemma 2.2 for θ.

Writing $v(\rho) = \mathcal{V}(\rho T)$ $\quad T \in s^2$, $\rho \geq 0$ $\quad$ we have

$v'(\rho) = \mathbb{E}(c\Psi(\rho c)) = \int x\,\Psi(\rho x)\nu_c(dx)$ where ν_c is the law of c.

We want $v' = 0$ for the solution i.e. for $\rho = 1$ and $\nu_c = \nu$; if we take $\Psi = -\alpha$ sgn $+ \tilde{\Psi}$ we obtain the condition :

$$(2.13)\qquad \int x\,\tilde{\Psi}(x)\gamma(dx) = \alpha\int |x|\,\nu(dx)$$

with such a choice $\pm I$ may be minima of $\mathcal{V}$; if we obtain a good property of convexity $\pm I$ may be the unique minima : this is the following result :

Theorem 2.3. :

Under the assumption C1 or C2 of (2.11) and with $\Psi = -\alpha$ sign $+ \tilde{\Psi}$ satisfying (2.13) we have :

the unique local minima of $\mathcal{V}$ in ℓ^2 are $\pm I$ (up to a time shift) and these lines are the unique stable attractive points for the flow of the s.d.l. of $\mathcal{V}$ in $\ell^2-\{0\}$.

The situation is represented (for 2 coordinates) on the figure 0

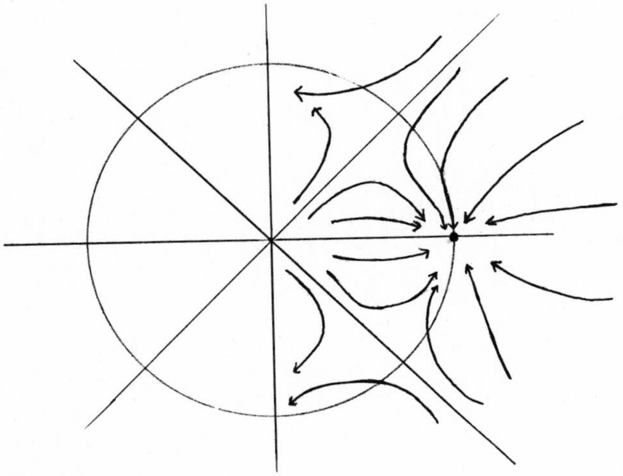

Figure 0 s.d.l of $\mathcal{V}$

Example : We take again (1.6). A simple example (used in the applications) of algorithm is :

$$(2.14) \text{ and } \begin{cases} \Psi(x) = x - \alpha \quad \text{for } x > 0 \ ; \ \alpha \text{ defined by } \alpha = \int x^2 \nu(dx) \,/ \int |x| \nu(dx) \\ H^{n+1} = H^n - \tau X_n (c_n - \alpha \operatorname{sgn}(c_n)) \end{cases}$$

3. GENERAL REMARKS - CONVERGENCE OF THE ALGORITHM.

a) Study of $\mathcal{J}$.

We omit here this point. We can simply remark that : $\mathcal{J} \xrightarrow{\Lambda} \mathcal{V}$ (Λ defined in (1.3)A2) or $\mathcal{V} \xrightarrow{\Lambda^{-1}} \mathcal{J}$: it is intuitively evident that we will get the same results for $\mathcal{J}$;

b) Convergence of the gradient algorithm.

The previous results show that we have a good behaviour of the s.d.l.. The solution of our problem will be complete only if we prove that the random trajectories of the stochastic gradient algorithm have their evolution close to the s.d.l.

This point is a consequence of a general result given in [1] about the convergence of stochastic approximation.

c) Convergence to S^{-1} rather than to $-S^{-1}$.

On the figure 0 we see that a convergence point is associated with a domain D_I (quarter of R^2)and an initial value in D_I gives the convergence to I. The representation is of course more difficult for $\mathcal{J}$. In practice we have some additional informations about S : for example the maximum module coefficient of S is positive ; this fact is sufficient to have a good initialization.

d) Treatment of other entry laws ν.

Considering the assumptions (2.11) C1-C2 we see that we have the solution for the subgaussian laws and the limit case of the uniform law. We can also consider the supergaussian laws : g increasing, $g'(x)/x$ strictly decreasing on R_+. We cannot apply directly the previous method because we have a compression to 0. Nevertheless we can get the solution with a little modification : between S and H we introduce a whitening filter and after that we keep the norm of H equal to 1. This technique is available for (2.11)C1-C2.

4. NUMERICAL RESULTS.

- Example 1 :

We have taken a usual telephone line : the figure 1 gives the impulse response. The figure 2 is the sampled response with $\Delta t = \frac{1}{3600}s$. The data to be transmitted are with 8 levels : $\{\pm 7, \pm 5, \pm 3, \pm 1\}$; the binary output is then 9600 bits/s.

The algorithm is given by (2.14) with $\alpha = E(a^2) / E(a) = 5.25$
The simulations have been done with an additive noise on the output given by the simulation of a zero-mean gaussian law with standard deviation $\sigma = 0.1\sqrt{2}$.
The lengh of the window is N = 10 i.e. 21 points for H.

Signification of the figures :

- figure 3 ; exact inverse (computed with known input)
- figure 4 : inverse computed with (2.14)
- figure 5 : evolution of errored data
  - ① directly on the line output $\cong$ 80 %
  - ② during the evolution of H^n
- figure 6 : evolution of the mean square error.

Remark : In practice we have some changes with respect to this example : use of the QAM (quadrature amplitude modulation) and double sampling. The presentation is a little more complicated but the algorithm is exactly the same and the numerical results are better (see [1]).

- Example 2 :

We take here a test example analog to [2] p.150. The entry law is now the uniform law on [-2, +2].

The figures are :

- figure 7 : the impulse response
- figure 8 : the obtained inverse
- figure 9 : evolution of the mean square error.

We give in [1] the effect of the use of a whitening filter for such a line.

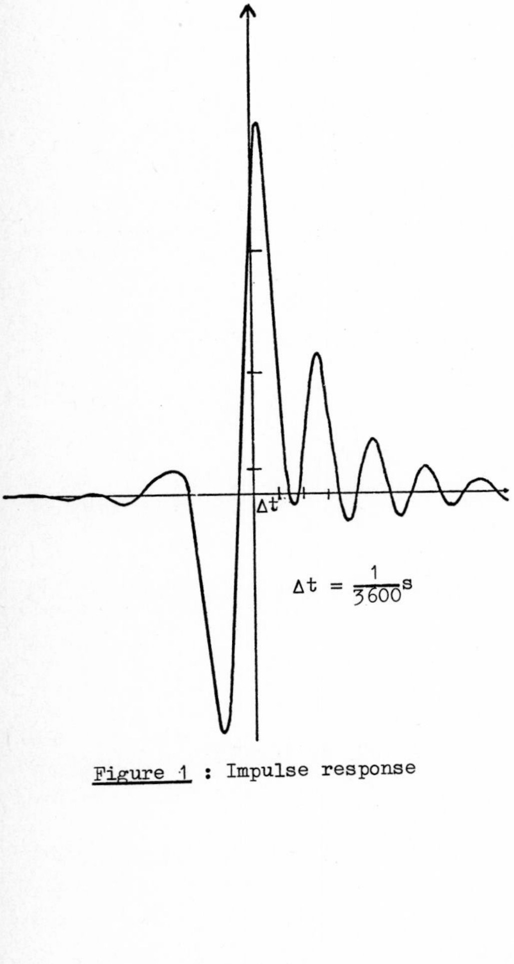

Figure 1 : Impulse response

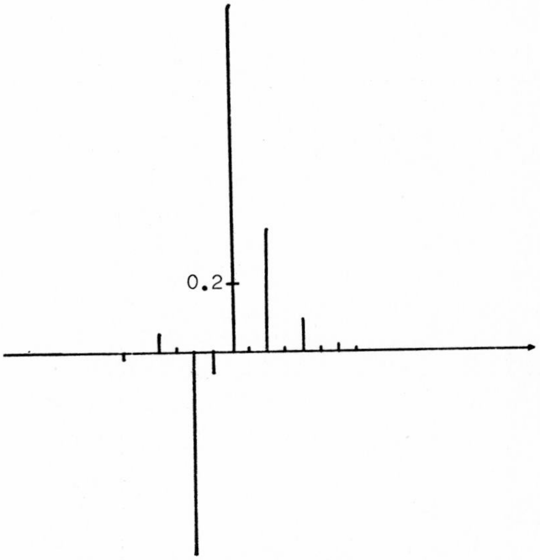

Figure 2 : Sampled response

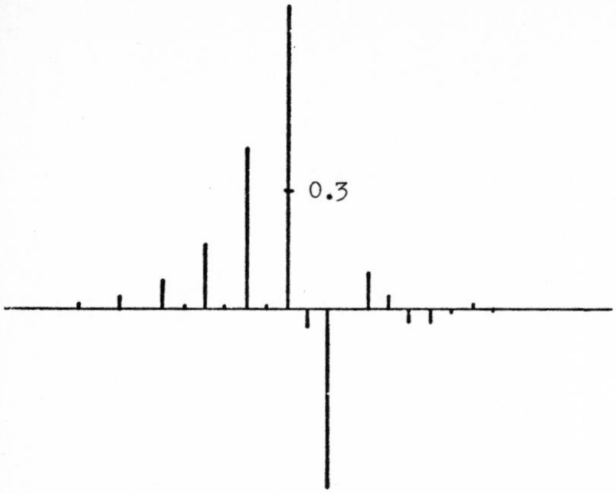

Figure 3 : Exact inverse

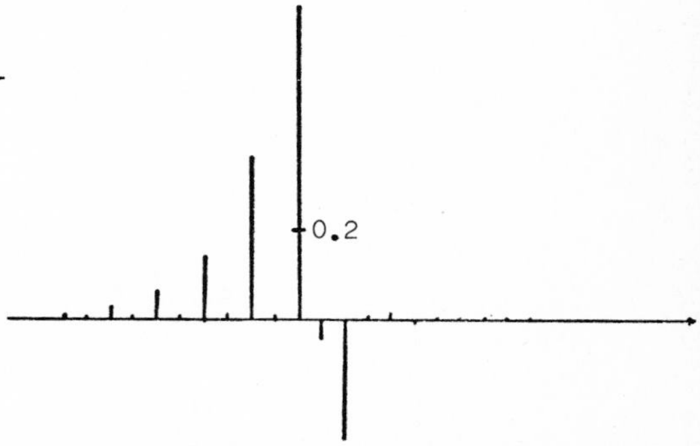

Figure 4 : Computed inverse

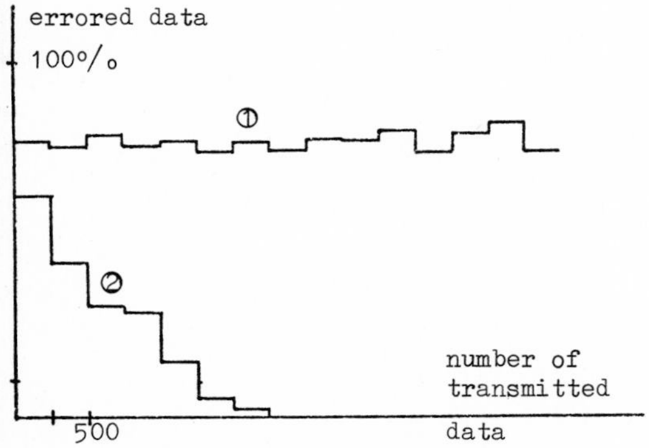

Figure 5 : Errored data

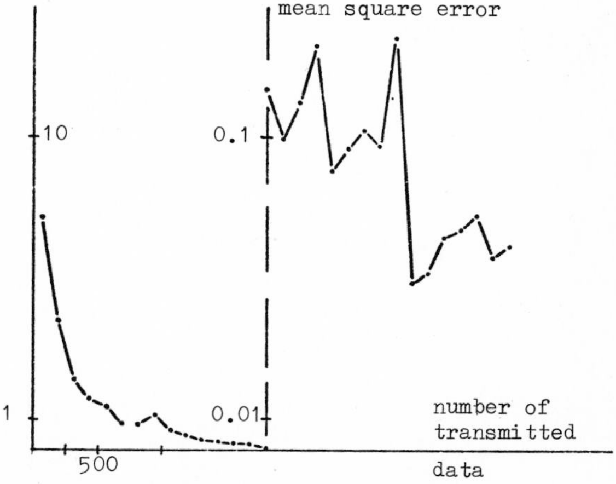

Figure 6 : Mean square error

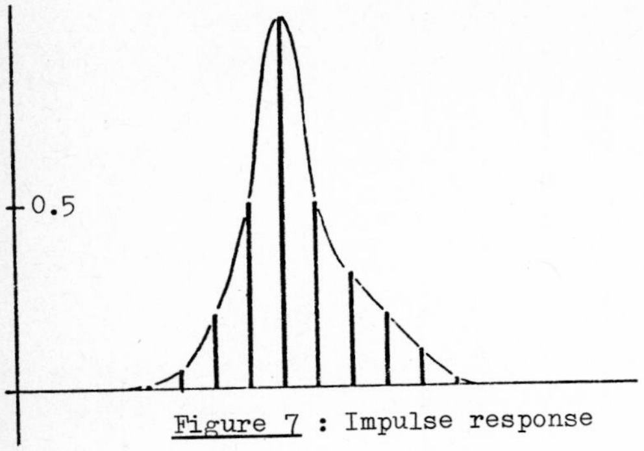

Figure 7 : Impulse response

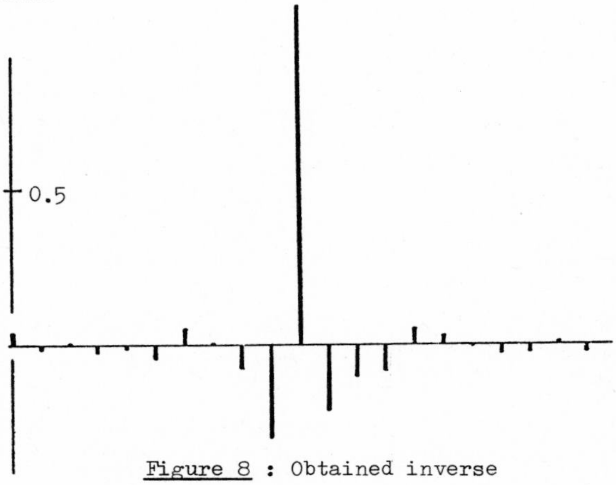

Figure 8 : Obtained inverse

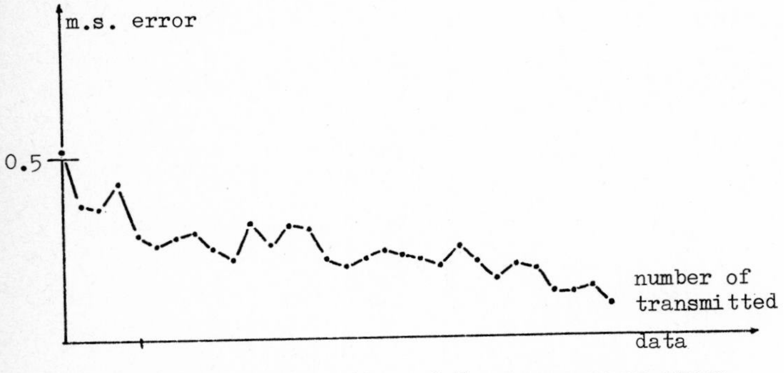

Figure 9 : Evolution of the mean square error

BIBLIOGRAPHY

[1] A. BENVENISTE ; M. BONNET ; M. GOURSAT ; C. MACCHI ; G. RUGET : Laboria Report to appear.

[2] J.G. PROAKIS : Advances in Equalization for Intersymbol Interference. Advances in Communication Systems pp 124-194. Vol. 4, Academic Press (1975).

OPTIMIZATION AND UNCERTAINTY

R.F. Drenick
Polytechnic Institute of New York
333 Jay Street, Brooklyn, N.Y. 11201/USA

1. Introduction

In many situations in practice, optimizations are desired when the data of the problem are not completely known. It has been customary in such cases to assume that the uncertainties can be represented by probability measures, or, more recently, by the membership functions of the theory of fuzzy sets. However, it is not always clear that such representations are actually appropriate. This paper deals with questions of what can be done when they are in fact considered inappropriate. Evidence is reviewed which shows that an optimization, in the usual sense of the term, may not be possible at all in such cases, and that some substantial recasting of the problem may be necessary even when it is. The ideas are illustrated with an example drawn from a recent attempt at developing a mathematical approach to organization theory.

Optimization, in the usual sense of the term, is the problem of finding the maximum of a "performance" function $\psi(\cdot)$ with respect to a "control" or "optimization" parameter u; ψ is scalar but u is typically a vector of finite or infinite dimension. The problem can nevertheless be visualized as that of determining the value of u* of u in the left half of Figure 1.

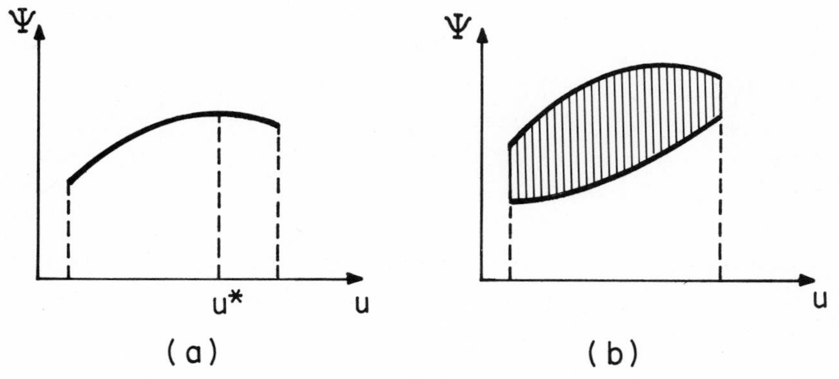

FIGURE 1: Performance, Without and With Uncertainty

The effect of the uncertainty on this problem is that of introducing a second parameter, ω say, which is sometimes called the "uncertainty parameter." It, too, is typically a vector variable. Its presence, however, entails the situation shown on the right of Fig. 1: for every value of u, ψ assumes a set of values. The performance in other words is no longer a point-valued, but a set-valued, function of the control parameter.

In such a predicament the notion of preference among control parameter values and, with it, that of optimality is often lost. In order to re-establish it, two steps must be possible. First a preference relation $u_1 \prec u_2$ must be redefined among the control parameter values; and second a new performance function must be defined which is again point-valued and which has the property that

(1.1) $$\Psi(u_1) < \Psi(u_2) \quad \text{iff} \quad u_1 \prec u_2.$$

Conditions under which these two steps actually can be taken have been under study for some time among mathematical economists (see, e.g., [3]), who use for Ψ the term "utility function" (or more precisely, "single-attribute utility function"). It will also be used below.

The conditions for the feasibility of the step may be easily fulfilled on occasion. In some situations for instance one can justify the relation

$$u_1 \prec u_2 \quad \text{iff} \quad \min_\omega \psi(u_1,\omega) < \min_\omega \psi(u_2,\omega),$$

i.e., u_2 is preferred to u_1 if, and only if the worst performance under u_2 is better than under u_1. Then $\Psi(u) = \min_\omega \psi(u,\omega)$ is point-valued, and an optimum can be defined relative to it. It is in fact the minimax value of u (i.e., the one on the right end of the u-interval in Figure 1).

The introduction into the problem of probability measures or membership functions can be interpreted as a similar strategy for replacing the interval-valued performance function ψ with a point-valued utility Ψ. When one can justify using it in one form or the other, it will restore the possibility of optimization. However, in some and perhaps even many cases in practice, it is not clear that this strategy does not merely compound the uncertainties of the problem, by injecting into it quantities which are themselves quite uncertain. One may then wish to avoid it, even at the risk of potential trouble with the optimization process.

It is conceivable that this is in fact what is being done by decision-makers in practice and what lies at the root of some of their decision-making troubles: preference relations, let alone utilities such as those in (1.1), are probably established only with difficulty. It should perhaps be mentioned that Simon suggested some time ago the consideration of a concept, which he called "satisficing," as a substitute for optimization in such cases (see, e.g., [1] for a more recent attempt at mathematical formulation of it, and [2] for another).

This paper will however remain in the spirit of this Conference and treat only those situations in which optimization can be defined. Section 2 reviews conditions under which this is possible, and Sections 3 to 5 develop the organizational illustration mentioned above.

2. Conditions for Compatibility Between Optimality and Uncertainty

The main complication of optimization in the presence of uncertainty as has just been explained lies with the fact that performance function may not be point-valued but interval-valued. It is then of interest to inquire under what circumstances optimization in the conventional sense is even possible.

It has also been explained that the idea of optimization presumes that a preference relation exists among the intervals or, equivalently, among the control parameters u. The question of how to best define such a relation has received considerable attention among mathematical economists, and a number of acceptable definitions have been advanced. One called a "weak order" seems sufficiently general, yet allows useful results to be derived from it. According to it, a control parameter value u_2 is preferred to u_1 ($u_1 \prec u_2$), if the relation obeys two axioms, namely ([3], p.11)

$$u_1 \prec u_2 \rightarrow u_2 \not\prec u_1,$$

(2.1)

$$u_1 \not\prec u_2 \;\&\; u_2 \not\prec u_3 \rightarrow u_1 \not\prec u_3.$$

The existence of a weak order among control parameter values or among the intervals they index, is not sufficient for that of a utility function, i.e., a function Ψ with the property (1.1). Three more conditions are needed to insure that such a Ψ exists. Thus, if u_1 is the index of the interval $[x_1, x_2]$ and u_2 that of $[y_1, y_2]$, one must satisfy the following ([3], p. 32).

(a) the range of x_1, x_2 is a rectangle;

(b) $x_1 < x_2 \;\&\; y_1 < y_2 \rightarrow u_1 \prec u_2$;

(2.2)

(c) $u_1 \prec u_2$, $u_2 \prec u_3 \rightarrow$ there exist λ, μ in (0,1) such that

$$\lambda u_1 + (1-\lambda)u_3 \prec u_2 \;\&\; u_2 \prec \mu u_1 + (1-\mu)u_3.$$

As already discussed, even the rather weak conditions (2.1) often are difficult to meet in practice unless a point-valued Ψ is readily discernible which represents the ordering. Among the remaining three conditions (2.2), the first seems to be the stumbling block in many cases. The lexicograpic order on the other hand (such as the one which makes u_2 preferable to u_1 if $x_2 > x_1$ and, when $x_2 = x_1$, when $y_2 > y_1$) violate the third.

In some applications, for instance in the one to be discussed below, the control parameter is a matrix Q of joint probabilities $q(i,k)$, $i = 1,2,\ldots, n$ and $k = 1,2,\ldots, m$. In such cases, the utility function Ψ is apparently best defined as a matrix with elements ψ_{ik} and, in place of (1.1), with the property that

$$E\{\Psi | Q_1\} < E\{\Psi | Q_2\} \quad \text{iff} \quad Q_1 \prec Q_2 \tag{2.3}$$

where

$$E\{\Psi | Q\} = \Sigma_{ik} \psi_{ik} q(i,k)$$

Again, conditions are of interest under which such a Ψ exists for all Q in some domain D of such matrices. A very weak set of such conditions has been advanced by Aumann ([3], p. 121). They are (in slightly paraphrased version):

(2.4)

(a) the domain D of matrices Q is closed and convex;

(b) $Q_1 \prec Q_2$ & $Q_2 \prec Q_3 \rightarrow Q_1 \prec Q_3$;

(c) $Q_1 \prec Q_2$ iff $\lambda Q_1 + (1-\lambda)Q \prec \lambda Q_2 + (1-\lambda)Q$ for all $Q \in D$ and all $\lambda \in (0,1)$;

(d) If $\lambda Q_1 + (1-\lambda)Q_3 \prec \lambda Q_2 + (1-\lambda)Q_4$ for all $\lambda \in (0,1]$, then $Q_4 \not\prec Q_3$.

3. Illustration: Optimal Staff Protocol in an Organization

As an illustration of the ideas presented in the preceding section, a problem will be discussed that has been encountered in the course of the development of a mathematical approach to organization theory [4]. It is the design of an optimal staff protocol. In order to clarify this piece of terminology, a brief discussion is needed of its background. This will be given here. One of the sources of uncertainty which injects uncertainty into the problem, and which seems rather unusual, is described in the next section. A procedure for its solution is presented in the last.

The mathematical approach to organization theory is based on the assumption that organizations are systems made up of components, namely their members, which have certain common characteristic properties. The problem of organizational design is to determine the interconnections and communications among these members which enable the organization to excute its task optimally.

This task is assumed to be of a simple decision-making kind. The organization receives, on the average once every Δ sec, one from a finite set of n input symbols x_i, and does so with a probability $p(x_i)$. It responds at the same average rate with one of m output symbols y_k, and does so with the probability $p(y_k | x_i)$. A utility ψ_{ik} is attached to every (x_i,y_k)-pair, and the expected utility

(3.1) $$E\{\Psi\} = \Sigma_{ik}\psi_{ik}\ p(x_i)\ p(y_k|x_i)$$

is its performance. The maximum, with respect to $p(y_k|x_i)$, of $E\{\Psi\}$ is the optimum This formulation is similar to that of team theory [5].

Organization members function in essentially the same way. They receive inputs u_i from, and transmit outputs v_k to, points within or without the organization. However, they require a certain "processing time" t_{ik} for the conversion of u_i into v_k. In order that none fall indefinitely behind schedule it is necessary that their mean processing times obey

(3.2) $$E\{T\} = \Sigma_{ik}\ t_{ik}\ p(u_i)\ p(v_k|u_i) \le \Delta,$$

where T stands for the matrix formed from the t_{ik}. If this inequality is violated for a member, he (or she or it) is said to be "overloaded".

The objective of organizational design in this view is the determination of a set of transition probabilities $p(v_k|u_i)$, if any, for all organization members which satisfy (3.2) and which, combined into an overall set $p(y_k|x_i)$, maximizes $E\{\Psi\}$ in (3.1). A set of probabilities $p(v_k|u_i)$, optimal or not, which characterizes the operational procedures followed by a member has been called his "protocol" (using a term which is used in roughly the same sense in computer network design). The collections of member protocols specifies also what is often called the "structure" of an organization, namely those transmission paths among members over which some $p(v_k|u_i)$ differ from zero.

The determination of an optimal organizational design would, in principle, be a rather straightforward linear programming problem, though typically one of quite appalling dimension. The problem is made more difficult by two kinds of complication. One is that of uncertainty. It arises from the fact that in practice none of the data of the problem, i.e., the input probabilities $p(x_i)$, the utility matrix Ψ and the processing time matrix T are precisely known. In fact, one is led to rather unrealistic structures and protocols if one assumes they are. One is however also led to unrealistic results if one assumes that these uncertainties can always be represented by prior probabilities. The ones in the $p(x_i)$ or in the ψ_{ik} would then merely induce changes in the $p(x_i)$-values that enter (3.1) and (3.2), and hence would merely alter some of the data but not the basic nature of the problem.

The second complication is a group of phenomena which are collectively called "task-dependence". The term refers to the fact that, for various reasons, the processing times t_{ik} often are not constants but vary with the parameters of the member's task, namely the number of sources from which the member acquires his inputs u_i, the probabilities $p(u_i)$ with which he does so, the probabilities $p(v_k|u_i)$ which make up his protocol, and the number of destinations to which he delivers his outputs. (Of the many other factors that influence a person's productivity, such as fatigue, most seem to be largely beyond effective control by organizational design and hence not pertinent to this discussion.) As will be explained in the next section, uncertainty and task-dependence are often interrelated in a rather curious way.

In view of these various complications, one can question how it has been possible to draw up designs for many large, yet rather well-functioning organizations in practice. There are no doubt many reasons for this but one may be that they are relatively mild perturbations on an arch-type which may be called "centralized". The latter is characterized by the fact that all inputs to the organization, and all feedback from its outputs are required to be processed, directly or indirectly, by an especially distinguished member, the "Executive". The coarse structure of such an organization is shown in Figure 2. The reasoning that leads to it is

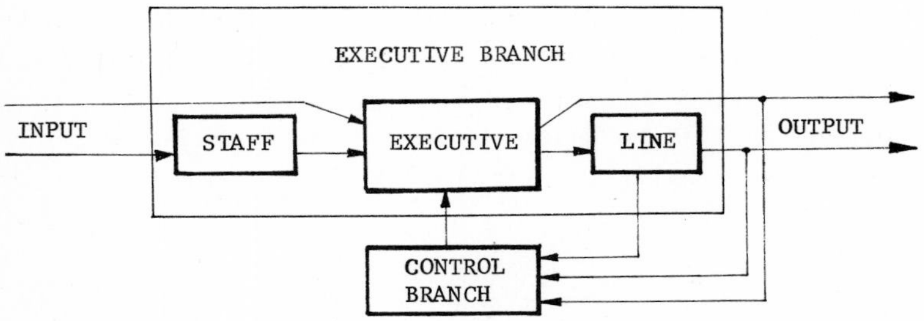

FIGURE 2: COARSE ORGANIZATIONAL STRUCTURE

straightforward and essentially non-mathematical, namely this. If indeed all input/output conversions must be acted upon by the Executive himself he is liable to be overloaded. It may then be possible to lighten his load by supplementing him with some pre-processors, and post-processors, called the "staff" and the "line" in Figure 2. The control branch then informs him of any departures from the desired performance.

The scope of this paper would be exceeded by a discussion here of the reasons for the choice of this terminology or, for that matter, for the obvious discrepancies between the block diagram of Figure 2 and innumerable examples of hierarchical organization charts that supposedly characterize organizational functioning. The discussion below will in fact deal with only one special topic, namely that of how to determine the staff protocol of a centalized organization. The point of interest here is the effect of uncertainties on this determination. One source of such uncertainty is discussed in the next section.

4. Uncertainty Due to Autonomy Among Members

A coupling exists between uncertainty and task-dependence in organizations involving human members which may be of interest because it probably has no counterpart in purely technological systems. It is due to the desire for autonomy among human members.

Social psychologists present assuring observations (e.g., in [6]), if they are needed, that the productivity of human organization members tends to increase if they are given a greater "degree of autonomy". One can presumably interpret this as meaning that there exists a quantity α^X, associated with member X and defining the degree of autonomy under which he works, such that his processing times t_{ik}^X are monotone non-increasing in α^X. The question is how α^X is to be defined. One way which has the advantage of the clarity of the axiomatic approach is the following.

One starts with the observation that autonomy presumably has something to do with the protocol $\bar{p}^X(v_k|u_i)$ which a member is expected to follow. If it is deterministic and of the form

$$\bar{p}^X(v_k|u_i) = \delta_{ik},$$

i.e., if X's response v_k is uniquely prescribed for every input u_i he receives, one would no doubt say that he has no autonomy. One might then say the same also if a stochastic protocol $\bar{p}^X(v_k|u_i)$ is prescribed for him since his response could be dictated to him by a suitable coin-tossing machine. If this argument is accepted it follows that X has autonomy only if his protocol is not fully prescribed for him but if he can choose any $p(v_k|u_i)$ from some domain D^X. This domain may be a set of transition probability matrices with elements $p^X(v_k|u_i)$ or, equivalently, of joint probability matrices Q with elements $q(i,k) = p(u_i)p^X(v_k|u_i)$.

The idea of choice however implies the idea of preference, and that in turn may be governed by a set of axioms such as the four in equation (2.4). Suppose that it is. Suppose also that in addition to these, the following fifth holds:

The degree of autonomy α^X is increased if the domain D^X is increased to the convex hull of D^X and

(a) any $Q' \notin D^X$ such that $Q \prec Q'$ for all $Q \in D^X$, or

(b) any $Q'' \notin D^X$ such that $Q'' \prec Q$ for all $Q \in D^X$.

One can then show easily that the member's choice is governed by a personal utility matrix Ψ^X which satisfies (2.3) and that

$$\alpha^X = f(\max_Q E\{\Psi^X|Q\}, \min_Q E\{\Psi^X|Q\})$$

where f(.,.) is monotononely increasing in its first argument, and monotonely decreasing in its second.

The upshot of this result is that the degree of autonomy which an organization member X attaches to his position is the range of his expected personal utility Ψ^X. This utility may or may not coincide with Ψ in (3.1) that governs the performance of the organization as a whole. If it does, the member will be happier (and correspondingly more productive) if his autonomy enables him to contribute more to the overall performance, perhaps because of the confidence in his judgment that is expressed by such an increase. He is what might be called a "trust seeker". His opposite would then be the obstructionist whose personal utility matrix is $\Psi^X = -\Psi$.

The matrix Ψ^X is presumably among the uncertainties that beset the person who wishes to optimize the performance (3.1) of the organization. The main point to be made here however is that, even if it were known, this performance could not be optimized in many cases. For such an optimization would imply the prescription of protocols throughout the organization and consequently the abolition of autonomy among all of its members. Their individual performances would suffer and so would in general that of the organization.

The conclusion here is that the performance of a well-functioning organization is inevitably uncertain.

5. Determination of the Staff Protocol

Suppose that the member X introduced in Section 4 above is the Executive. Disregard the line and the control branch of the organization, as well as the Executive's need for autonomy. The function of the staff is then the acquisition and interpretation of th organizational inputs x_i and their conversion to some others, u_k say, which avoid Executive overload. The conversion may be deterministic and one-to-one but, more generally, it will be governed by a protcol $p(u_k|x_i)$ which is so designed that, as in (3.1),

$$E\{\Psi\} = \Sigma_{ijk}\ \psi_{ij}\ p(x_i)\ p(u_k|x_i) p^X(y_j|u_k)$$

is maximized, subject to

$$E\{T^X\} = \Sigma_{jk}\ t^X_{jk} p(x_i)\ p(u_k|x_i) p^X(y_j|u_k) \leq \Delta.$$

This formulation assumes, for the sake of simplicity, that all inputs to the Executive are monitored by the staff and that none reach him directly (in the way indicated by Figure 2). It is then no restriction to simplify the problem by setting $p^X(y_k | u_j) = \delta_{jk}$. One can then consider first the case in which there is no uncertainty. In such a case there exists an optimal staff protocol $p^*(u_k | x_i)$ and it can be obtained by linear programming. Organizations in practice however do not find it by the simplex or a similar method. For at least one reason, namely the presence of uncertainty, they solve the problem by a process of successive approximation, i.e., by feedback control. It is thus perhaps of some interest that control schemes exist which converge on the optimal $p^*(u_k | x_i)$. One such scheme is given by the prescription that if $p^{(\nu)}(u_k | x_i)$ is the staff protocol used at the time $\nu\Delta$, then at $(\nu+1)$ one should use

$$p^{(\nu+1)}(u_k|x_i) = d^{-1}[p^{(\nu)}(u_k| x_i) + \rho r(- \lambda_i^{(\nu)} + p(x_i)t_{ik}^X\mu^{(\nu)} - p(x_i)\psi_{ik})]$$

$$\text{(5.1)} \qquad \lambda_i^{(\nu+1)} = d^{-1}[\lambda_i^{(\nu)} + \rho r(\Sigma_k p(y_k| x_i) -1)]$$

$$\mu^{(\nu+1)} = d^{-1} [\mu^{(\nu)} + \rho r(\Sigma_{ik} t_{ik} p(y_k| x_i)p(x_i) -\Delta)]$$

where $0 < \rho < 1$ and r is the ramp function, i.e.,

$$r(\xi) = \begin{cases} 0 & \text{for } \xi < 0 \\ \xi & \text{for } \xi \geq 0 \end{cases}.$$

and

$$d = 1 + \rho r[\mu^{(\nu)}\Delta + \Sigma_i \lambda_i^{(\nu)} - \Sigma_{ik}\psi_{ik}p(y_k| x_i)\, p(x_i)]$$

It will be noted that this is indeed a feedback control law: the arguments of r in it represent the error signals. It is nonlinear and, by all indications, unavoidably so. The optimal performance of the organization under this scheme is the limit, as ν goes to infinity, of $\mu^{(\nu)}\Delta$.

The proof of this assertion utilizes the equivalence beween the solutions of linear programming and game theory problems, (e.g., [7], p. 419) as well as the fact that the latter can be solved by feedback control schemes (ibid., p. 438). This scheme applies also in the presence of uncertainty (and, under certain restrictions on the nature of t_{kk}^X, even in the presence of task-dependence), provided a preference ordering is possible among the staff protocols as in (2.4). The matrix Ψ of elements ψ_{ik} which enter (5.1) must however be replaced with one whose existence is guaranteed under the conditions.

Of greater interest however may be the case in which no such preference ordering is possible. Some recent work [2] encourages the conjecture that feedback control schemes can also be devised for this case and, under certain circumstances, when it is combined with task-dependence. These schemes exploit an opportunity that is missed in (5.1), namely that information becomes available in the course of the control process which often enables one to reduce the uncertainty. The control, in other words, assumes adaptive charcter but the concept of optimality is lost.

References

1. R. Radner, Satisficing, Jour. Math. Economics, Vol. 2 (1975), pp. 253-262.

2. R.F. Drenick, Feedback Control of Partly Unknown Systems, SIAM Jour. Control and Optimization, Vol. 15 (1977), pp.506

3. P. Fishburn, Utility Theory in Decision Making, Wiley, New York, 1970.

4. R.F. Drenick, A Mathematical Approach to Organization Theory, to be published in these Lecture Notes, 1978.

5. J. Marachak and R. Radner, Economic Theory of Teams, Yale University Press, New Haven, 1969.

6. L.W. Porter, et al., Behavior in Organizations, McGraw-Hill, New York, 1975.

7. R.D. Luce and H. Raiffa, Games and Decisions, Wiley, New York, 1957.

THE SEPARATION PRINCIPLE FOR THE CONTROL OF LINEAR STOCHASTIC SYSTEMS WITH ARBITRARY INFORMATION STRUCTURE

Arunabha Bagchi and Huibert Kwakernaak
Department of Applied Mathematics
Twente University of Technology
P.O. Box 217, Enschede
THE NETHERLANDS

1. Introduction

We consider a stochastic control problem with linear dynamics and quadratic criterion, where the system state is disturbed by noise which is not necessarily Gaussian and where the observation is quite arbitrary. We look for the structural property that ensures the so-called separation principle to hold in such a general setting. This property is called neutrality.

The concept of separation appeared in control literature with the work of Joseph and Tou [1]. The idea was around in economics even earlier with the works of Simon [2] and Theil [3]. In a stochastic control problem with noisy observation, we say that the separation principle holds if the control law can be obtained in two steps: first, the system state at any instant conditioned on all the information available at that instant is calculated and then, the control can be computed only as a function of this estimated state. In [1], the standard LQG problem was considered. Wonham [4] developed a rigorous theory in this direction and proved a separation theorem with nonquadratic criterion. Although the conditions imposed were somewhat unrealistic in the general case, the theory provided a correct dynamic programming formalism, at least in the linear case. Alternate direct methods soon appeared [5,6] for proving separation theorems in LQG problems. Various extensions then followed. A separation theorem in distributed systems has been proved in various ways in [7,8,9]. Recently, the extension to point process type measurements was considered [10]. Our present work is an attempt to understand the structural property that makes separation possible. In a sense, our result is an extension of an earlier work [11], where it has been shown that if the state of the system may be accurately and directly observed, the optimal control law is the same for a broad class of state disturbances.

We should mention that considerable advances were made in separation type results in discrete-time systems, even with unknown parameters and general noise models, both in the control literature [12] and in economics. A good source of reference for works in this field is [13].

2. Problem Formulation

Consider the linear dynamical system

(1) $$dX_t = A(t)X_t dt + B(t)U_t dt + dM_t, \qquad t \geq 0,$$

where $A(t)$ and $B(t)$ are appropriate dimensional matrices, white M_t, $t \geq 0$, is an n-dimensional local martingale (see [14] for a definition) with respect to a given growing family of σ-algebras $\mathcal{F}_t$, $t \geq 0$. U_t, a p-vector, is the input to the system. We do not put any constraint on the magnitude of the input. For LQG problems with bounded input, a separation theorem was recently proved in [15]. For each t, we require U_t to be measurable with respect to a given σ-algebra $\mathcal{Y}_t \subset \mathcal{F}_t$, such that (1) has a unique solution. $\mathcal{Y}_t$, $t \geq 0$, is a growing family of σ-algebras where, for each t, $\mathcal{Y}_t$ represents the information about the system obtained by observing it. For example, $\mathcal{Y}_t$ could be the σ-algebra generated by $\mathcal{Y}_0$ and Y_s, $0 < s \leq t$, where Y_t, $t \geq 0$, is the observation process, specified by

(2) $$dY_t = C(t)X_t dt + G(t)dW_t,$$

and $\mathcal{Y}_0 \subset \mathcal{F}_0$, where W_t, $t \geq 0$, is a Brownian motion and Y_0 is given. Another possibility is that $\mathcal{Y}_t$ is generated by $\mathcal{Y}_0$ and $Y_{t_1}, \ldots, Y_{t_i}$, $0 < t_1 < \ldots < t_i \leq t$, with $\mathcal{Y}_0$ as before and

(3) $$Y_{t_i} = C_i X_{t_i} + G_i W_i, \qquad i = 1, 2, \ldots.$$

with W_i, $i = 1,2,\ldots$, being independent random vectors. There are many admissible ways in which the family $\mathcal{Y}_t$, $t \geq 0$, can be generated, as long as one important requirement for separation is satisfied, which is that the system plus the observation mechanism is <u>neutral</u>. Let us denote $\hat{X}_t = E[X_t|\mathcal{Y}_t]$, $t \geq 0$. Then the system plus observation mechanism is said to be neutral if $E[(X_t-\hat{X}_t)(X_t-\hat{X}_t)^*]$, $t \geq 0$, is invariant under different choices for the process U_t, $t \geq 0$, where '*' denotes transpose and we require U_t to be $\mathcal{Y}_t$-measurable, for each t.

The optimal control problem is now to determine the input process U such that

(4) $$J(U) = E\{\int_0^T ([Q(t)X_t,X_t] + \lambda[U_t,U_t])dt + [Q_f X_T,X_T]\}$$

$\lambda > 0$, is minimized, where $Q(t), Q_f \geq 0$, T fixed and $[.,.]$ denotes the inner product in appropriate Euclidean spaces.

3. An Auxiliary Result

Let X_t, $t \geq 0$, be an n-dimensional vector semi-martingale

$$dX_t = P_t dt + dM_t,$$

where P is an $\mathcal{F}$-adapted process and M is an $\mathcal{M}^2$-martingale [14]; that is, an $\mathcal{F}$-martingale such that $\sup_{t\geq 0} E\|M_t\|^2 < \infty$ ($\|.\|$ denotes Euclidean norm), with quadratic variation $\langle M,M\rangle_t$. Writing

$$X_t = Q_t + M_t, \qquad Q_t = X_0 + \int_0^t P_s ds,$$

and using the fact that Q is absolutely continuous, we find

$$\begin{aligned} X_t X_t^* &= Q_t Q_t^* + Q_t M_t^* + M_t Q_t^* + M_t M_t^* \\ &= Q_t Q_t^* + \int_0^t P_s M_s^* ds + \int_0^t Q_s dM_s^* + \int_0^t (dM_s) Q_s^* \\ &\quad + \int_0^t M_s P_s^* ds + \langle M,M\rangle_t + M_t', \end{aligned}$$

where we assumed $M_0 = 0$, and where M' is the (matrix-valued) $\mathcal{F}$-martingale obtained in the decomposition $M_t M_t^* = \langle M,M\rangle_t + M_t'$. Rearranging, we find that

$$X_t X_t^* = Q_t Q_t^* + \int_0^t P_s M_s^* ds + \int_0^t M_s P_s^* ds + \langle M,M\rangle_t + M_t'',$$

where M_t'' is another (matrix-valued) $\mathcal{F}$-martingale. In differential form, we may write equivalently

$$d(X_t X_t^*) = P_t Q_t^* dt + Q_t P_t^* dt + P_t M_t^* dt + M_t P_t^* dt + d\langle M,M\rangle_t + dM_t''. \tag{5}$$

Let $S(t)$, $t \geq 0$, be a differentiable $n \times n$ symmetric matrix-valued function. Then

$$S(t) X_t X_t^* = S(0) X_0 X_0^* + \int_0^t \dot{S}(s) X_s X_s^* ds + \int_0^t S(s) d(X_s X_s^*)$$

or, in differential form,

$$d(S(t) X_t X_t^*) = \dot{S}(t) X_t X_t^* dt + S(t) d(X_t X_t^*).$$

Substituting (5) and taking the trace of both sides, we get

$$\begin{aligned} d[S(t)X_t, X_t] &= [\dot{S}(t) X_t, X_t] dt + 2[S(t)X_t, P_t] dt \\ &\quad + \mathrm{Tr}.S(t) d\langle M,M\rangle_t + dM_t''' \end{aligned}$$

with M_t''' a (scalar) F-martingale. Assume now that $\langle M,M\rangle_t$ is absolutely continuous with $\langle M,M\rangle_t = \int_0^t \Lambda_s ds$. From the well-known elementary result in estimation theory (see e.g. [16]) that if X is a semi-martingale $dX_t = P_t dt + dM_t$, with P F-adapted and M an F-martingale, then $dE(X_t|\mathcal{Y}_t) = E(P_t|\mathcal{Y}_t)dt + dN_t$, with N a $\mathcal{Y}$-martingale, we get the corollary

$$dE(X_t^* S(t)X_t|\mathcal{Y}_t) = E(X_t^*\dot{S}(t)X_t|\mathcal{Y}_t)dt + 2E(X_t^*S(t)P_t|\mathcal{Y}_t)dt$$
$$+ \text{Tr.}S(t)E(\Lambda_t|\mathcal{Y}_t)dt + dN_t,$$

where N is a $\mathcal{Y}$-martingale.

4. Solution of the Stochastic Control Problem

Let S(t) be the unique nonnegative definite solution of the matrix Riccati equation [6]

$$\text{(6)}\quad \begin{aligned} &\dot{S}(t) + A^*(t)S(t) + S(t)A(t) + Q(t) - S(t)B(t)B(t)^*S(t)/_\lambda = 0 \\ &S(T) = Q_f. \end{aligned}$$

In the notation of section 3, $P_t = A(t)X_t + B(t)U_t$, which is clearly F-adapted. Without loss of generality, we take M_t to be an M^2-martingale by working with $M_{\min(t,T)}$ instead of a general local martingale M_t. Using the corollary above, with some rearrangement of terms and using (6), we get

$$\begin{aligned} dE(X_t^*S(t)X_t|\mathcal{Y}_t) &= \lambda \| U_t + B(t)^*S(t)\hat{X}_t/\lambda\|^2 dt \\ &-\lambda\| U_t \|^2 dt - E(X_t Q(t)X_t^*|\mathcal{Y}_t)dt \\ &+E(X_t^*S(t)B(t)B(t)^*S(t)X_t|\mathcal{Y}_t)dt/\lambda \\ &-\hat{X}_t^* S(t)B(t)B(t)^*S(t)\hat{X}_t dt/\lambda + \text{Tr.}S(t)\hat{\Lambda}_t dt + dN_t. \end{aligned}$$

Integrating over [0,T], rearranging terms and taking expectation, we get

$$J(U) = EX_0^*S(0)X_0 + E\int_0^T \text{Tr.}S(t)\Lambda_t dt + \lambda E\int_0^T \| U_t + B(t)^*S(t)\hat{X}_t/\lambda\|^2 dt$$
$$+ \int_0^T \text{Tr.}S(t)B(t)B(t)^*/\lambda E[(X_t-\hat{X}_t)(X_t-\hat{X}_t)^*]dt,$$

where we used the obvious result

$$E(X_t X_t^*|\mathcal{Y}_t) - \hat{X}_t\hat{X}_t^* = E((X_t-\hat{X}_t)(X_t-\hat{X}_t)^*|\mathcal{Y}_t).$$

From the neutrality assumption, the only term in the criterion depending on the control is the third term on the right of the above expression for J(U), and J(U) is minimum if and only if

$$U_t = -B(t)^* S(t)\hat{X}_t/\lambda,$$

provided (1) with this control has a solution. This, in general, involves the question of existence of stochastic functional-differential equations with martingale forcing terms. This is a recent field of investigation [16].

5. Examples of Neutrality

1. Standard LQG Problem:

We take the martingale disturbance in (1) as

$$M_t = \int_0^t F(s)dW_{1s} \tag{7}$$

where $F(t)$ is a deterministic matrix-valued function and W_{1t}, $t \geq 0$, a vector Brownian motion. We consider the observation mechanism (2). For a given process U_t, $t \geq 0$, measurable with respect to $\mathcal{Y}_t$, we let

$$X_t^u = \int_0^t A(s)X_s^u ds + \int_0^t B(s)U_s ds$$

$$\tilde{Y}_t = Y_t - Y_t^u, \quad Y_t^u = \int_0^t C(s)X_s^u ds,$$

and let $\tilde{X}_t = X_t - X_t^u$, $\hat{\tilde{X}}_t = E(\tilde{X}_t|\tilde{\mathcal{Y}}_t)$ where $\tilde{\mathcal{Y}}_t$ is the smallest σ-algebra generated by Y_s, $0 \leq s \leq t$. Then

$$E(X_t|\tilde{\mathcal{Y}}_t) = \hat{\tilde{X}}_t + X_t^u. \tag{8}$$

Defining the "innovation"

$$Z_{0t} = \tilde{Y}_t - \int_0^t C(s)\hat{\tilde{X}}_s ds$$

we know $\tilde{\mathcal{Y}}_t \equiv \mathcal{Z}_{0t}$. Consider controls U_t such that

$$\mathcal{Y}_t \equiv \mathcal{Z}_{0t}.$$

This class of admissible controls includes those which are linear functionals of the observation. In this class,

$$E(X_t|\tilde{\mathcal{Y}}_t) = E(X_t|\mathcal{Y}_t) = \hat{X}_t$$

and using (8), we get $X_t - \hat{X}_t = \tilde{X}_t - \hat{\tilde{X}}_t$, so that

$$E[(X_t-\hat{X}_t)(X_t-\hat{X}_t)^*]$$

is independent of the control term U_t, implying that the system plus observation mechanism is neutral.

2. Nonlinear Measurement:

Consider a measurement mechanism which allows feedback

$$dY_t = [C(t)X_t + f(t,Y.)]dt + dW_t \tag{9}$$

$$Y_0 = 0$$

where C(t) is the output matrix, and f(t,Y.) is the feedback term. Observations of this type have been considered in [17]. We impose conditions on f that guarantee unique solution of (9). Let us consider the same state equation as in the previous example and let

$$dX_t^0 = A(t)X_t^0 dt + F(t)dW_{1t}.$$

Define

$$dY_t^0 = C(t)X_t^0 dt + dW_t. \tag{10}$$

Then (8) can be written as

$$\begin{aligned} dY_t &= dY_t^0 + [f(t,Y.) + C(t)\int_0^t \Phi(t,s)B(s)U_s ds]dt \\ &= dY_t^0 + F(t,Y.)dt \end{aligned} \tag{11}$$

where $\Phi(t,s)$ is the state transition matrix and we restrict the class of control strategies to the form $U_t = \gamma(t,Y.)$ with γ a non-anticipative functional of the observation. Existence of solution of (11) was considered in [18].

Sufficient Condition. The integral equation (11) has a unique solution Y and Y_t is $\mathcal{Y}_t^0$-measurable for each t.

Lemma. Under the above sufficient condition, neutrality holds.

Proof. $\hat{X}_t = E(X_t|\mathcal{Y}_t) = E(X_t^0 + \int_0^t \Phi(t,s)B(s)U_s ds|\mathcal{Y}_t)$

$$= \int_0^t \Phi(t,s)B(s)U_s ds + E(X_t^0|\mathcal{Y}_t)$$

$$= \int_0^t \Phi(t,s)B(s)U_s ds + E(X_t^0|\mathcal{Y}_t^0)$$

so that $X_t - \hat{X}_t = X_t^0 - E(X_t^0|\mathcal{Y}_t^0)$ and neutrality holds.

6. Systems with Time Delay

Consider the linear time delayed system

$$\text{(12)} \qquad dX_t = \sum_{i=0}^{k} A_i(t)X_{t-h_i}dt + B(t)U_t dt + dM_t$$

$$X_t = 0 \quad \text{for } t \leqslant 0$$

the symbols having the same interpretation as in (1) with scalars h_i, $0 = h_0 < h_1 < \dots < h_k$ denoting time delays. The observation mechanism is general and we minimize the same criterion as in section 2.

This is an extension of an earlier work [18] where separation was proved when M_t was of the form (7) and observation mechanism was of the form (2). As in [19], let $P_t(\theta,s)$ be the symmetric matrix solution of the partial differential equation

$$\left(-\frac{\partial}{\partial t} + \frac{\partial}{\partial \theta} + \frac{\partial}{\partial s}\right) P_t(\theta,s) = -P_t(\theta,0)\ B(t)B(t)^* P_t(0,s)/_\lambda$$

with the boundary conditions

$$\left(-\frac{\partial}{\partial t} + \frac{\partial}{\partial s}\right) P_t(0,s) = [A_0^*(t) - P_t(0,0)B(t)B(t)^*/\lambda]P_t(0,s) + \sum_{i=1}^{k} A_i^*(t+h_i)P_t(h_i,s)$$

$$-\frac{\partial}{\partial t} P_t(0,0) = A_0^*(t)P_t(0,0) + P_t(0,0)A_0(t) + Q(t)$$

$$- P_t(0,0)B(t)B(t)^*P_t(0,0)/\lambda + \sum_{i=1}^{k} A_i^*(t+h_i)P_t(h_i,0)$$

$$+ \sum_{i=1}^{k} P_t(0,h_i)A_i(t+h_i)$$

with the final conditions

$$P_T(0,0) = Q_f, \quad P_T(\theta,s) = 0 \text{ for } \theta > 0 \text{ or } s > 0.$$

Denote $E(X_\sigma|\mathcal{Y}_t)$ by $\hat{X}_{\sigma|t}$ with $\hat{X}_{t|t} \equiv \hat{X}_t$. In this case, we proceed just as in section 4 except that we apply the corollary of section 3 to an expression of the form

$$dE([X_t^*P_t(0,0)X_t + 2\sum_{i=1}^{k}\int_{t-h_i}^{t} X_t^*P_t(0,\sigma+h_i-t)A_i(\sigma+h_i)X_\sigma d\sigma$$

$$+ \sum_{i=1}^{k}\sum_{j=1}^{k}\int_{t-h_i}^{t}\int_{t-h_j}^{t} X_\tau^* A_j^*(\tau+h_j)P_t(\tau+h_j-t,\sigma+h_i-t)A_i(\sigma+h_i)X_\sigma d\tau d\sigma]|\mathcal{Y}_t).$$

The optimal control law in this case, is given by

$$U_t = -\frac{1}{\lambda} B(t)^* \{P_t(0,0)\hat{X}_t + \sum_{i=1}^{k} \int_{t-h_i}^{t} P_t(0,\sigma+h_i-t)A_i(\sigma+h_i)\hat{X}_{\sigma|t} d\sigma\}$$

provided (12) with this control has a solution.

7. Conclusion

We have shown that the optimal control law for a linear system disturbed by general noise process (not necessarily Gaussian) and with a quadratic performance criterion is independent of the noise and measurement characteristic under the neutrality assumption. The optimal control law depends on $\hat{X}_t$ (and $\hat{X}_{\sigma|t}$ for time delayed systems).

The problem of determining $\hat{X}_t$ is, however, nontrivial. In the LQG case, it can be effected by using the Kalman filter. When M_t is of the form (7) and the observation mechanism (3) is considered, a modification of Kalman filter works [20]. The case, when M_t is a centralized Poisson process and the observation mechanism is given by (2), was also studied [21]. The last problem, however, does not admit a finite-dimensional filter as a solution.

REFERENCES

[1] P.D. Joseph and J.T. Tou, "On linear control theory", AIEE Transactions, Applications and Industry, 30, 193-196, 1961.

[2] H.A. Simon, "Dynamic programming under uncertainty with a quadratic criterion function", Econometrica, 24, 74-81, 1956.

[3] H. Theil, "A note on certainty equivalence in dynamic planning", Econometrica, 25, 346-349, 1957.

[4] W.M. Wonham, "Random differential equations in control theory", in Probabilistic Methods in Applied Math., 2, Academic Press, 1970.

[5] K.J. Åström, Introduction to Stochastic Control Theory, Academic Press, 1970.

[6] A.V. Balakrishnan, Stochastic Differential Systems, Lecture Notes in Economics and Math. Systems, Springer-Verlag, 1973.

[7] A.V. Balakrishnan, "Stochastic optimization theory in Hilbert spaces-1", Applied Math. and Optimization, 1, 97-120, 1974.

[8] R.F. Curtain and A. Ichikawa, "The separation principle for stochastic evolution equations", SIAM J. Contr. and Opt., 15, 367-382, 1977.

[9] A. Bensoussan, "On the separation principle for distributed parameter systems", IFAC conference on Distributed Parameter Systems, Banff, 1971.

[10] D.L. Snyder, I.B. Rhodes and E.V. Hoversten, "A separation theorem for stochastic control problems with point-process observation", Automatica, 13, 85-87, 1977.

[11] H. Kwakernaak, "An extension of the stochastic linear regulator problem", IEEE Trans Automat. Contr., AC-19, 121-123, 1974.

[12] E. Tse, "Sequential decision and stochastic control", in Mathematical Programming Study, 5, 227-243, North Holland, 1976.

[13] C.Aldrich and D.W. Peterson, "Quadraticity and neutrality in discrete time stochastic linear quadratic control", Automatica, 13, 307-312, 1977.

[14] C. Doléans-Dade and P.A. Meyer, "Intégrales stochastiques par rapport aux martingales locales", in Séminaire de Probabilités IV, Lecture Notes in Mathematics, Springer-Verlag, 1970.

[15] J. Ruzicka, "On the separation principle with bounded controls", Applied Math. and Optimization, 3, 243-262, 1977.

[16] C. Doléans-Dade, "On the existence and unicity of solutions of stochastic integral equations", Z. Wahrscheinlichkeitstheorie, 36, 93-101, 1976.

[17] K. Uchida and E. Shimemura, "On certainty equivalence in linear-quadratic control problems with nonlinear measurements", Conference of Decision and Control of Socio-economic Systems, Nagoya, 1976.

[18] W.H. Fleming and M. Nisio, "On the existence of optimal stochastic controls", J. Math. Mech., 15, 777-794, 1966.

[19] A. Bagchi, "Control of linear stochastic time delayed systems", J. Math. Analysis and Applications (to appear).

[20] A.H. Jazwinski, Stochastic Processes and Filtering Theory, Academic Press, 1970.

[21] H. Kwakernaak, "Filtering for systems excited by Poisson white noise", in Control Theory, Numerical Methods and Computer Systems Modelling, Lecture Notes in Economics and Mathematical Systems, 107, Springer-Verlag, 1975.

A DECOMPOSITION SCHEME FOR THE HAMILTON-JACOBI EQUATION

Sady MAURIN
IRIA - LABORIA
Domaine de Voluceau
Rocquencourt
78150 Le Chesnay, FRANCE

ABSTRACT.

One way to compute optimal control in completely observable stochastic optimal control problems is to solve the Hamilton-Jacobi Equation. We propose to replace in the quasi linearization technique of Bellman [1] the step of inversion of a linear system of dimension n by several steps of n linear equations of dimension 1 in parallel or in sequence. We are then led to apply for instance splitting up methods-studied in particular in [3], [6], [8], [9]. We obtain what we call a decomposition scheme to approximate evolutive Hamilton-Jacobi Equation. We study convergence results as time steps converge to zero with compacity methods adapted from [4] and give some numerical results.

1. INTRODUCTION.

Let Ω be an open set of R^n and Γ his boundary.

We consider in the domain Ω the following Hamilton-Jacobi Equation with Dirichlet or Neuman boundary conditions.

$$(1.1) \qquad \alpha U - \frac{1}{2}\sum_{i=1}^{n} \frac{\partial}{\partial x_i}[\sigma_i^2(x)\frac{\partial U}{\partial x_i}] - \inf_{v\in\mathcal{U}(x)} [\mu(v,x).\frac{\partial U}{\partial x} + \varphi(v,x)] = f$$

With the following definitions :

- $\mathcal{U}(x)$ is a multiapplication from R^n in a fixed compact K of R^m
- v is for each x an element of $\mathcal{U}(x)$
- $\sigma(x)$ is an application from R^n into R^n with components $\sigma_i(x)$ $i=1,2,\ldots,n$
- $\mu(v,x)$ is an application from $R^m \times R^n$ into R^n, with components $\mu_i(v,x)$ $i=1,2,..,n$
- φ is an application from $R^m \times R^n$ into R
- f is an application from R^n in R
- α is a real number

We note $\qquad \frac{\partial U}{\partial x} = \frac{\partial U}{\partial x_1},\ldots,\frac{\partial U}{\partial x_n}$

and $$\mu(v,x) \cdot \frac{\partial U}{\partial x} = \sum_{i=1}^{n} \mu_i(v,x) \frac{\partial U}{\partial x_i}$$

For conditions of existence and unicity of a solution to (1.1), we refer to A. BENSOUSSAN - J.L. LIONS [2]. These conditions are supposed to be fulfilled here.

To compute U one can use a general method of quasilinearization.

Formally it is defined as follows :

If we put :

$$F(x,v,p) = \mu(v,x).p + \varphi(v,x)$$
$$\text{for } p = p_1,\dots,p_n \in R^n$$

Given any $v_o(x)$.

Define :

Step 1 : Find U_1 solution of

$$\alpha U_1 - \tfrac{1}{2}\sum_{i=1}^{n} \frac{\partial}{\partial x_i}\left[\sigma_i^2(x) \frac{\partial U_1}{\partial x_i}\right] + F\left(x, v_o(x), \frac{\partial U_1}{\partial x}\right) = f \tag{1.2}$$

with boundary conditions.

Define then $v_1(x)$ such that :

$$F\left(x, v_1(x), \frac{\partial U_1}{\partial x}\right) = \inf_{v \in \mathcal{U}(x)} F\left(x, v, \frac{\partial U_1}{\partial x}\right)$$

Then, given U_p and $v_p(x)$ such that :

$$F\left(x, v_p(x), \frac{\partial U_p}{\partial x}\right) = \inf_{v \in \mathcal{U}(x)} F\left(x, v, \frac{\partial U_p}{\partial x}\right) \tag{1.3}$$

Define :

Step p+1 : Find U_{p+1} solution of

$$\alpha U_{p+1} - \tfrac{1}{2}\sum_{i=1}^{n} \frac{\partial}{\partial x_i}\left[\sigma_i^2(x) \frac{\partial U_{p+1}}{\partial x_i}\right] + F\left(x, v_p(x), \frac{\partial U_{p+1}}{\partial x}\right) = f \tag{1.4}$$

with boundary conditions.

Define $v_{p+1}(x)$ such that

$$F\left(x, v_{p+1}(x), \frac{\partial U_{p+1}}{\partial x}\right) = \inf_{v} F\left(x, v, \frac{\partial U_{p+1}}{\partial x}\right) \tag{1.5}$$

Then, one can expect see R. BELLMAN [1] that U_p converges to U in adapted topology.

2. DECOMPOSITION SCHEME.

In p+1 ^{e} step of precedent scheme, we have to solve a partial differential equation of elliptic type.

(2.1) $$\alpha U_{p+1} - \sum_{i=1}^{n} \frac{1}{2} \frac{\partial}{\partial x_i}[\sigma_i^2(x) \frac{\partial U_{p+1}}{\partial x_i}] - F(x, v_p(x), \frac{\partial U_{p+1}}{\partial x}) = f$$

with boundary conditions.

But (2.1) can be written

(2.2) $$\sum_{i=1}^{n} [\alpha_i U_{p+1} - \frac{1}{2} \frac{\partial}{\partial x_i}[\sigma_i^2(x) \frac{\partial U_{p+1}}{\partial x_i}] - F_i(x, v_p(x), \frac{\partial U_{p+1}}{\partial x_i})] = \sum_{i=1}^{n} f_i$$

with $$\alpha = \sum_{i=1}^{n} \alpha_i$$

$$f = \sum_{i=1}^{n} f_i$$

$$F_i(x, v_p(x), \frac{\partial U_{p+1}}{\partial x_i}) = \mu_i(v_p(x), x) \frac{\partial U_{p+1}}{\partial x_i} + \varphi_i(v_p(x), x)$$

$$\varphi = \sum_{i=1}^{n} \varphi_i$$

If we put formally :

$$A_i = \alpha_i I - \frac{1}{2} \frac{\partial}{\partial x_i} \sigma_i^2(x) \frac{\partial}{\partial x_i} - \mu_i(v_p(x), x) \frac{\partial}{\partial x_i}$$

where I represents the identity operator.

We can write (2.1) in the form :

(2.3) $$\sum_{i=1}^{n} A_i U_{p+1} = f - \varphi(v_p(x), x)$$

To solve (2.1) we can then use decomposition techniques (see A. BENSOUSSAN - J.L. LIONS - R. TEMAM [3].

But it would be very heavy and long to use complete Decomposition scheme ; so a question arises naturally : Can we just make a step of a decomposition scheme and go on with the quasilinearization method ?

In the case of splitting up scheme (M.M. YANENKO [9]) we show that the answer is positive. It is probably true for other decomposition methods.

Let us define formally the decomposition scheme for equation (1.1) for the case n=m=2 the method is general.

Given $v_o(x)$ and U_o $\Delta t > 0$

Define :

Step 1 : $U^{\frac{1}{2}}$ solution of

$$\frac{U^{\frac{1}{2}}-U_o}{\Delta t} + \alpha_1 U^{\frac{1}{2}} - \frac{1}{2}\frac{\partial}{\partial x_1}[\sigma_1^2(x)\frac{\partial U^{\frac{1}{2}}}{\partial x_1}] - F_1(x,v_o(x),\frac{\partial U^{\frac{1}{2}}}{\partial x_1}) = f_1 \tag{2.4}$$

with boundary conditions

Step 2 : U^1 solution of

$$\frac{U^1-U^{\frac{1}{2}}}{\Delta t} + \alpha_2 U^1 - \frac{1}{2}\frac{\partial}{\partial x_2}[\sigma_2^2(x)\frac{\partial U^1}{\partial x_2}] - F_2(x,v_o(x),\frac{\partial U^1}{\partial x_2}) = f_2 \tag{2.5}$$

with boundary conditions

Define $v_1(x)$ such that

$$F(x,v_p(x),\frac{\partial U^{\frac{1}{2}}}{\partial x_1},\frac{\partial U^1}{\partial x_2}) = \inf_{v\in\mathcal{U}(x)} F(x,v,\frac{\partial U^{\frac{1}{2}}}{\partial x_1},\frac{\partial U^1}{\partial x_2}) \tag{2.6}$$

Then :

Given U_p and $v_p(x)$ such that

$$F(x,v_p(x),\frac{\partial U^{p-\frac{1}{2}}}{\partial x_1},\frac{\partial U^p}{\partial x_2}) = \inf_{v\in\mathcal{U}(x)} F(x,v,\frac{\partial U^{p-\frac{1}{2}}}{\partial x_1},\frac{\partial U^p}{\partial x_2}) \tag{2.7}$$

Define :

Step 2p+1 : $U^{p+\frac{1}{2}}$ solution of

$$\frac{U^{p+\frac{1}{2}}-U^p}{\Delta t} + \alpha_1 U^{p+\frac{1}{2}} - \frac{1}{2}\frac{\partial}{\partial x_1}[\sigma_1^2(x)\frac{\partial U^{p+\frac{1}{2}}}{\partial x_1}] - F_1(x,v_p(x),\frac{\partial U^{p+1}}{\partial x} = f_1 \tag{2.8}$$

with boundary conditions

Step 2(p+1) : U^{p+1} solution of

$$\frac{U^{p+1}-U^{p+\frac{1}{2}}}{\Delta t} + \alpha_2 U^{p+1} - \frac{1}{2}\frac{\partial}{\partial x_2}[\sigma_2^2(x)\frac{\partial U^{p+1}}{\partial x_2}] - F_2(x,v_p(x),\frac{\partial U^{p+1}}{\partial x}) = f_2 \tag{2.9}$$

with boundary conditions.

Then, define $v_{p+1}(x)$ such that :

$$F(x,v_{p+1}(x),\frac{\partial U^{p+\frac{1}{2}}}{\partial x_1},\frac{\partial U^{p+1}}{\partial x_2}) = \inf_{v\in\mathcal{U}(x)} F(x,v,\frac{\partial U^{p+\frac{1}{2}}}{\partial x_1},\frac{\partial U^{p+1}}{\partial x_2}) \tag{2.10}$$

and so on :

We are going now to give conditions for convergence of the sequence U_p to U solution of (1.1).

3. CONVERGENCE RESULTS.

Let us write steps 2p+1 and 2(p+1) in general form for i = 1, 2

$$(3.1)\qquad \frac{U^{p+i/2}-U^{p+\frac{i-1}{2}}}{\Delta t} + \alpha_i U^{p+i/2} - \frac{1}{2}\frac{\partial}{\partial x_i}\Big[\sigma_i^2(x)\frac{\partial U^{p+i/2}}{\partial x_i}\Big] - F_i\Big(x, v_p(x), \frac{\partial U^{p+i/2}}{\partial x_i}\Big) = f_i$$

Resolution of (3.1).

Let us define $\mathcal{U}_i = \{u \in \mathcal{L}^2(\Omega)\ \frac{\partial u}{\partial x_i} \in \mathcal{L}^2(\Omega)\}$

If we solve (1.1) with Neuman boundary conditions we define $V_i = \mathcal{U}_i$ with the norm on $\mathcal{U}_i$:

$$\|u\|^2_{\mathcal{U}_i} = \|u\|^2_{\mathcal{L}^2(\Omega)} + \Big\|\frac{\partial u}{\partial x_i}\Big\|^2_{\mathcal{L}^2(\Omega)}$$

and we define V_i the closure in the precedent space $\mathcal{U}_i$ of $\mathcal{D}(\Omega)$ with the precedent norm if we resolve (1.1) with Dirichlet boundary conditions (see J.L.LIONS - E. MAGENES [5].

If we introduce the bilinear form on V_i

$$(3.2)\qquad a_i(v_p,u,v) = \int_\Omega \frac{\sigma_i^2(x)}{2}\frac{\partial U}{\partial x_i}\frac{\partial V}{\partial x_i} - \int_\Omega \mu_i(v_p,x)\frac{\partial U}{\partial x_i}v + \alpha_i\int uv$$

$$u \in V_i,\ v \in V_i$$

We suppose

$$(3.3)\qquad a_i(v_p,u,u) \geq \beta_i\|u\|^2_{V_i} \qquad \forall v_p \in K$$

Condition (3.3) is satisfied if :

$$|\mu_i(u,x)| \leq C \qquad \forall u \in K \quad \forall x \in \Omega$$

We suppose that

$$|\varphi_i(u,x)| \leq C \qquad \forall u \in K \quad \forall x \in \Omega$$

$$f_i \in \mathcal{L}^2(\Omega)$$

We can then speak of a solution in V_i of equation (3.1) in variational form.

$$(3.4)\qquad \Big\langle\Big(\frac{1}{\Delta t}+\alpha_i\Big)U^{p+i/2}.\ v\Big\rangle_{\mathcal{L}^2(\Omega)} + \int_\Omega \frac{\sigma_i^2(x)}{2}\frac{\partial U^{p+i/2}}{\partial x_i}\frac{\partial V}{\partial x_i} - \int_\Omega F_i\Big(x,v_p(x),\frac{\partial U^{p+i/2}}{\partial x_i}\Big)v = \int_\Omega f_i v + \int_\Omega \frac{1}{\Delta t}U^{p+\frac{i-1}{2}}v$$

$$\forall v \in V_i$$

Convergence of scheme.

If we define :

$$u^i_{\Delta t}(t) = u^{p+\frac{i-1}{2}}\left(1-\frac{(t-p\Delta t)}{\Delta t}\right) + u^{p+i/2}\,\frac{(t-p\Delta t)}{\Delta t} \tag{3.5}$$

for $p\Delta t \leq t \leq (p+1)\Delta t \quad p \in N \quad i=1,2$

$$\mathcal{L}^2(OT,V_i) = \{u:[0,T] \text{ in } V_i \; \int_0^T \|u(\tau)\|^2_{V_i}\, d\tau < +\infty\} \tag{3.6}$$

$$\mathcal{L}^2(OT,V) = \{u\ [0,T] \text{ in } V \; \int_0^T \|u(\tau)\|^2_V d\tau < +\infty\}$$

with $V = H^1(\Omega)$ or $H^1_o(\Omega)$ according to Neuman or Dirichlet conditions. (see J.L.LIONS-E. MAGENES [5].)

$$\Pi(x,\frac{\partial U}{\partial x}) = \inf_{v\in\mathcal{U}(x)} F(x,v,\frac{\partial U}{\partial x}) \tag{3.7}$$

$$b(\varphi,\Psi) = \sum_{i=1}^{n} \int_\Omega \frac{\sigma_i^2(x)}{2}\,\frac{\partial\varphi}{\partial x_i}\,\frac{\partial\Psi}{\partial x_i} + \int_\Omega -\Pi(x,\frac{\partial\varphi}{\partial x})\Psi + \alpha\int\varphi\Psi \tag{3.8}$$

$\forall\varphi,\ \Psi \in V$

and $u(t) \in \mathcal{L}^2(OT,V)$ a solution of

$$\int_0^T \langle\frac{\partial u}{\partial t} \,.\, v\rangle_{V'\times V} + \int_0^T b(u,v) = \int_0^T \langle f.v\rangle \qquad \forall v \in V \tag{3.9}$$

We have the following result :

<u>Theorem</u>. :

$u^i_{\Delta t}(t)$ converges to $u(t)$ solution of (3.9) in $\mathcal{L}^2(OT,V_i)$.

Moreover, if b is supposed to be a monotone operator :

$\forall\varepsilon \qquad \|u^i_{\Delta t}(T) - u\|^2_{\mathcal{L}^2(\Omega)} \leq \varepsilon$. For T large enough and Δt small enough,

where u is a solution of (1.1). For this result see S.MAURIN [7]

<u>Remarks</u> : 1. Instead of a recurrent scheme, we can define a parallel scheme.

2. The scheme can be applied to more general non linear parabolic equations, if we can approximate the non linear term by a linearized term, for instance (convex case)

4. NUMERICAL EXAMPLE.

We take :

$$\mu_1(x,v_1,v_2) = v_1+v_2$$
$$\mu_2(x,v_1,v_2) = v_2$$
$$\varphi(v_1,v_2) = v_1^2 + v_2^2 + v_1v_2$$
$$\sigma_1^2(x) = 2$$
$$\sigma_2^2(x) = 2$$
$$\alpha_1 = \alpha_2 = \tfrac{1}{2}$$

We compute f as to make $\cos \pi x_1 \cos \pi x_2$ solution of (1.1) with Neuman boundary conditions on $\Omega =]0,1[\times]0,1[$.

We note that this example is not included in the case of our study since μ_1, μ_2, and φ are not bounded but : if $p = p_1, p_2$

$$\mu_1(v_1,v_2)p_1+\mu_2(v_1,v_2)p_2+\varphi(v_1,v_2) =$$

$$= (v_1+v_2)p_1 + v_2p_2 + v_1^2 + v_2^2 + v_1v_2 = F(x,v,p)$$

then F is Gâteaux differentiable in v and :

$$\frac{\partial F}{\partial v_1} = p_1 + 2v_1 + v_2$$

$$\frac{\partial F}{\partial v_2} = p_1 + p_2 + 2v_2 + v_1$$

Then, $\inf_v F(x,v,p) = F(x,\hat{v},p) \iff \hat{v} = \hat{v}_1, \hat{v}_2$ with

$$\hat{v}_1 = \frac{p_2 - p_1}{3} \qquad \hat{v}_2 = -\left(\frac{p_1 + 2p_2}{3}\right)$$

Hence :

$$F(x,\hat{v},p) = -\frac{1}{3}[p_1^2 + p_2^2 + p_1p_2]$$

So the non linear part of Hamilton-Jacobi Equation is convex, we are in the case of remark 2.

We have discretized equation (3.1) by finite differences and use a method described in YANENKO [9] to solve the tridiagonal matrices involved.

Results are summarized in Tables 1 and 2.

-TABLE 1-

Δt	NI	T.E	CEPS
1/12	8	0.29	0.008
1/25	13	0.34	0.007
1/50	21	0.40	0.006
1/100	37	0.58	0.006

Convergence of optimal control

NI : number of iterations
TE : time of execution in seconds on IBM 370-168
CEPS : average error on computed controls.

-TABLE 2-

Δt	T.E	EPS
1/12	2	0.081
1/25	5	0.042
1/50	10	0.026

Convergence of cost u

EPS : average error on computed cost.

REFERENCES

[1] R. BELLMAN : Adaptive control processes A guided tour. Princeton N.J. , Princeton University Press 1961.

[2] A. BENSOUSSAN ; J.L. LIONS : Book to appear.

[3] A. BENSOUSSAN ; J.L. LIONS ; R. TEMAM : Méthode de décomposition. Cahier IRIA n°11, Juin 1972.

[4] J.L. LIONS : Quelques méthodes de résolution des problèmes aux limites non-linéaires, Dunod, Paris 1969.

[5] J.L. LIONS ; E. MAGENES ; Problèmes aux limites non homogènes, Dunod, 1968.

[6] G.I. MARCHUK : Methods of numerical mathematics. Springer Verlag, New York, Heidelberg Berlin, 1975.

[7] S. MAURIN : Schéma de décomposition pour l'équation de Hamilton-Jacobi. Rapport LABORIA (IRIA), May 1977.

[8] R. TEMAM : Thesis, Paris 1967.

[9] N.N. YANENKO : Méthode à pas fractionnaire, Armand Colin, 1968.

CALCULATION OF OPTIMAL MEASUREMENT POLICIES FOR FEEDBACK CONTROL OF LINEAR STOCHASTIC SYSTEMS

by D.J. Mellefont and R.W.H. Sargent

Introduction

In the control of processes, measurements are required to provide information about plant states so that corrective action can be taken to counter the effects of random disturbances on the system. Measurements themselves are subject to random errors and are costly to make. In modern computer control applications, the computing time necessary to process raw measurement data must be considered a cost when applying on-line control. For control system design, it is necessary to choose between alternative sources of information to achieve some compromise between cost and accuracy of measurement. Similarly, in multi-stage processes common in the chemical industry, a large number of possible measurement locations exist. The problem, therefore, is to show how some subset of the available measurements can be selected and to calculate an optimal measurement policy for the use of those measurements.

This paper will consider only linear stochastic systems in the form of the wellknown LQG problem. For measurement control in non-linear systems see Mellefont (1977). A number of workers have endeavoured to find optimal measurement policies for the LQG problem. From the separation principle of Wonham (1968), control is a linear feedback of the state estimate so that the measurement optimization separates as a deterministic control problem. Herring and Melsa (1974) apply Athans' (1968) matrix minimum principle to obtain a two point boundary value problem (TPBVP). With suitable restrictions on the problem, they derive a switching function which is used to determine optimal subsets of measurements but do not allow for the use of those measurements for control purposes. Very few algorithms are presented for numerical solution to the problem. Athans (1972) iteratively minimizes the Hamiltonian but has the constraint that only one measurement at a time can be used. This approach is shown to be inefficient for general subset selection by Herring (1972) who presents an heuristic algorithm, although he reports this to have convergence problems. Selection of measurement subsets is inherently a combinatorial problem and various assumptions or restrictions have been necessary to facilitate solution. Mellefont and Sargent (1977a) show that these restrictions are unnecessary and the problem can be solved by standard non-linear programming techniques. This approach will be extended here to problems subject to combinatorial constraints. When capital costs are included, the combinatorial problem cannot be satisfactorily expressed in non-linear programming form and an implicit enumeration algorithm is proposed for such design problems.

Problem Formulation

Consider a linear stochastic system

$$dx(t) = [A(t)x(t) + B(t)u(t)]dt + dw_1(t) \tag{1}$$

where $dw_1(t)$ is a Wiener process and

$$E\{x(t_o)\}=x_o \ , \ cov\{x(t_o)\} = Q_o$$
$$E\{dw_1(t)\}=0 \ , \ cov\{dw_1(t)\}= V_1(t)dt$$

Since state $x(t)$ is not known precisely, it is necessary to take measurements. These measurements are costly so a selection vector $s(t)$ is introduced such that

$s_i(t) = 1$ if the ith measurement is to be used

$s_i(t) = 0$ if not.

Measurements satisfy

$$dz(t) = C(s,t)x(t)dt + dw_2(s,t) \tag{2}$$

where $dw_2(s,t)$ is a Wiener process

$$z(t_o) = 0, \ E\{dw_2(s,t)\}=0, \ cov\{dw_2(s,t)\}=V_2(s,t)dt$$

Note that equation (2) is of variable dimension depending on the active set of measurements at each time t.

For optimal control, the objective to be minimized is

$$J = E\{\int_{t_o}^{t_f} x'R_1x + u'R_2u \, dt + x'(t_f)P_fx(t_f) \tag{3}$$

Cost of measurement can be represented as a quadratic term $s'R_3s$ for inclusion in (3). Since $s(t)$ is a zero-one vector, off diagonal terms in $R_3(t)$ will allow for additional costs or discounts associated with using any given combination of measurements.

The optimal estimator for the linear system is given by the Kalman filter

$$d\hat{x}(t) = [A(t)\hat{x}(t)+B(t)u(t)]dt+Q(t)C'(s,t)V_2^{-1}(s,t)dv(t) \qquad \hat{x}(t_o)= x_o \tag{4}$$

$$\dot{Q}(t) = A(t)Q(t)+Q(t)A'(t)+V_1(t) -Q(t)C'(s,t)V_2^{-1}(s,t)C(s,t)Q(t) \qquad Q(t_o) = Q_o \tag{5}$$

where innovation $dv(t) = dz(t) - C(s,t)\hat{x}(t)dt$

Expanding the objective function in a Taylor Series about $\hat{x}(t)$ gives

$$J = E\{\int_{t_o}^{t_f} \hat{x}'R_1\hat{x}+u'R_2udt+\hat{x}'(t_f)P_f\hat{x}(t_f)\}$$
$$+Tr\{\int_{t_o}^{t_f} R_1Q + s'R_3sdt + P_fQ(t_f)\} \tag{6}$$

From the separation principle, optimal control $u(t)$ is given by

$$u(t) =-R_2^{-1}(t)B'(t)P(t)\hat{x}(t) \tag{7}$$

where

$$-\dot{P}(t)=A(t)'P(t)+P(t)A(t)+R_1(t)-P(t)B(t)R_2^{-1}(t)B'(t)P(t) \qquad P(t_f) = P_f \tag{8}$$

Incorporating into objective (6) leaves

$$J=Tr\{\int_{t_o}^{t_f} R_1Q + PQC'V_2^{-1}CQ+s'R_3s \, dt +P_fQ(t_f)\} \tag{9}$$

as the objective for measurement control subject to matrix state equation (5) Note tł term $PQC'V_2^{-1}CQ$ which acts as an additional cost in the objective function and reflects the use of measurement information for control. This problem is a deterministic optimal control problem with a bang-bang control s(t).

Solution as a Non-Linear Program

Non-linear programming is a well established means of solving an optimal control problem (Sargent and Sullivan, 1977). Usually a piecewise constant approximation to the control is made to give a reasonable number of optimization variables. For measurement subset selection, the control s(t) is piecewise constant by definition and it is only necessary to optimize the switching times when measurements become alternately active and inactive. The advantage of this approach is that any standard optimization algorithm can be used. In the examples presented here, the variable metric projection (VMP) method of Sargent and Murtagh (1973) was used.

When there are no constraints on the subsets of measurements used, the combinatorial problem is completely eliminated by defining the switching times for each measurement as optimization variables, since the switching times can be adjusted independently for each measurement. Gradients of the objective function with respect to switching times are given by

$$\frac{\partial J}{\partial T_{ij}} = \Delta\psi_{ij}(Q,s,\Lambda,t) \tag{10}$$

where $\psi = s'R_3 s + \mathrm{Tr}QC'V_2^{-1}CQ(P-\Lambda)$ (11)

$\Delta\psi_{ij}$ is the change in ψ across switch T_{ij}

T_{ij} is the jth switch of the ith measurement

Λ is the matrix adjoint

$$-\dot{\Lambda} = R_1 + A'\Lambda + \Lambda A + (P-\Lambda)QC'V_2^{-1}C + C'V_2^{-1}CQ(P-\Lambda) \qquad \Lambda(t_f) = P_f \tag{12}$$

To solve the non-linear program it is necessary to prespecify the number of switches allowed for each measurement. There is no way to predict this number a priori but a simple test can be applied to the converged solution to see if more switches would give a significantly better result.

Test for more switches

Switches determine when measurements become alternately active and inactive. Thus if two consecutive switches coincide, their effect is cancelled. Alternatively, it is possible to consider the converged solution to the non-linear program as consisting of an infinite number of switches of which only a finite number are not cancelled. Equation (10) can be used to calculate gradients of these hypothetical switches and if these gradients indicate that the paired switches should move apart then an increase in the number of optimization variables is indicated. The magnitude of resulting gradients should be related to the convergence criteria for the particular problem to assess the significance of the local improvement that is indicated. This test is simple to apply and involves at most one extra set of gradient calculations.

If more switches are indicated, this test also reveals the best locations to introduce additional switches for the continued optimization.

Measurement Policies for a Chemical Reactor

The above approach was used to determine optimal measurement policies for control of a constant flow reactor under steady operating conditions. A linearized model was derived by Mellefont (1977) with coefficient matrices:

$$A = \begin{bmatrix} -11 & -23.26 \\ -5.46 & -16.10 \end{bmatrix} \qquad B = \begin{bmatrix} 0 \\ 5.46 \end{bmatrix} \tag{13}$$

where x_1 = concentration, x_2 = temperature, u = heat input.

The system was assumed to be subject to a random heat input with standard deviation of 10% of the steady state value. Measurements of concentration and temperature are considered.

FIGURE 1

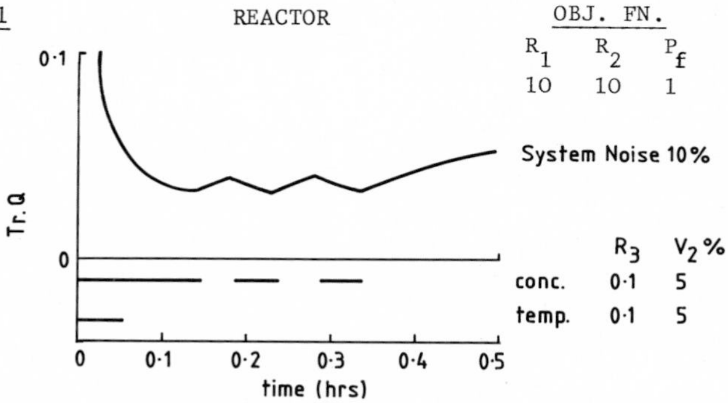

Fig.1 shows the optimal policy when both measurements are of equal cost and accuracy. Concentration is found to be the dominant measurement and enters a cylic pattern during the central region of the control. To reduce a high initial variance, temperature measurement is introduced to improve estimation. Towards the end of the control interval, measurement ceases since there is not time to utilize the information gained. If the temperature measurement error is reduced to 3% standard deviation, fig.2 shows essentially a reversal in the measurement policy. However, if both measurements are of equal accuracy, but temperature measurement is half as costly, fig.3 shows the use of temperature measurement as a background with short periods of concentration measurement to improve the estimation.

Constrained Measurement Subset Policies

Physical constraints often exist on the control system. Athans (1972) considers an aerospace situation where only one measurement at a time can be accessed. This can be generalized to an imposed minimum and maximum number of measurements giving the constraint

$$\text{NMIN} \leq \sum_{i=1}^{m} s_i(t) \leq \text{NMAX} \tag{14}$$

FIGURE 2

REACTOR

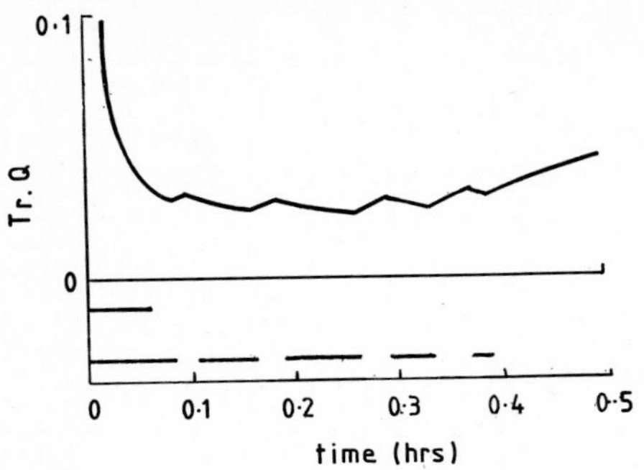

OBJ. FN.

R_1 P_2 P_f
10. 10. 1.

System Noise 10%

	R_3	V_2 %
conc.	0·1	5
temp.	0·1	3

FIGURE 3

REACTOR

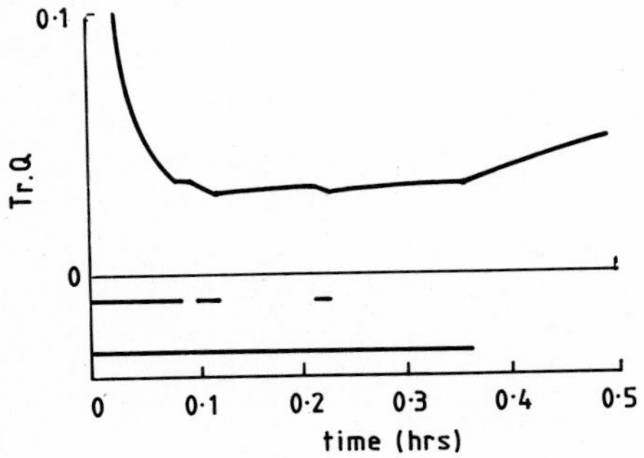

OBJ. FN.

R_1 P_2 P_f
10. 10. 1.

System Noise 10%

	R_3	V_2 %
conc.	0·1	5
temp.	0·05	5

The number of input channels to the controller determines NMAX and a minimum may be required for safety to ensure that the system is always under observation. Centralized control systems use information from remote sensors and transmission lines are expensive, so that local multiplexors must select one measurement at a time for transmission to the central controller. Such situations of hardware sharing lead to constraints of the type $s_k + s_j \leqslant 1$ (15)
if measurements k and j cannot be used simultaneously.
Switching times can no longer be moved independently for each measurement, and we have a mixed integer programming problem to select zero one vectors s_k satisfying the constraints for each of k control intervals of variable size. Consider a single control interval of size δt shown in fig.4 subject to the constraints **that no** than two measurements can be used at any one time.

FIGURE 4

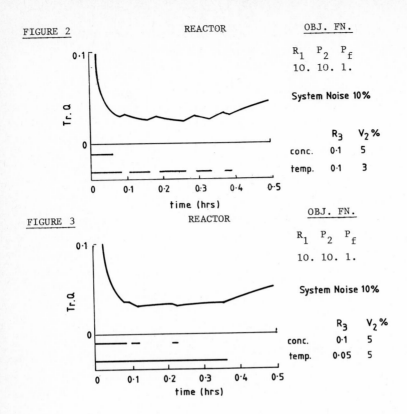

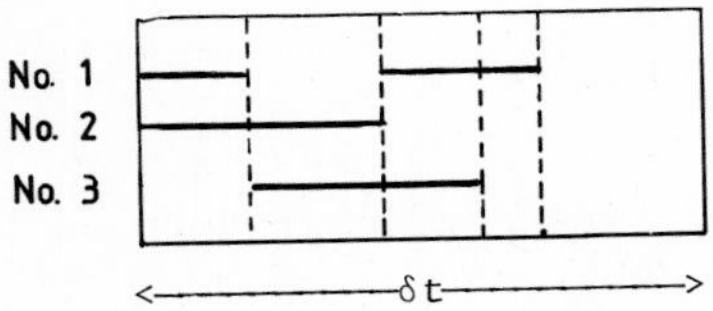

— represents active measurement

Define r_i as the fraction of δt over which the ith feasible combination is active. Then

$$\sum_{i=1}^{NC} r_i = 1 \qquad (16)$$

for NC combinations. In the limit as δt approaches zero, the covariance Q satisfies

$$\dot{Q} = AQ+QA'+V_1 - \sum_{i=1}^{NC} r_i \, QC_i' V_{2_i}^{-1} C_i Q \qquad (17)$$

where $C_i' V_{2_i}^{-1} C_i$ corresponds to the ith combination. The limiting case involves an infinitesimal δt, and then (17) is independent of the order of the subsets.

Although $r_i(t)$ can be treated as a continuous control variable and a non-linear programming problem derived, the dimension of r rapidly increases as the number of available measurements increases. In fig.4 there are seven allowable combinations (of which 5 are in use) from three measurements. It would be preferable to define the problem in terms of the fractional usage of each measurement, which is related to r by

$$s^o = \sum_{i=1}^{NC} r_i \, s_i \qquad (18)$$

where s_i is the zero one vector corresponding to the ith combination. Since as $\delta t \to 0$ the order of appearance of measurement combinations is arbitrary, we can prove the following simplifying result:

<u>Theorem</u>

Let s^o be an m vector whose jth component value corresponds to the fraction of time that measurement j is used in a given interval . Suppose that measurement combinations must satisfy

$$N + Gs_i \leq 0 \qquad i = 1,2 \ldots\ldots \qquad (19)$$

where N is an n vector of integer values

G is an nxm matrix with coefficients from $\{-1,0,1\}$

s_i is an m vector of zero-one variables representing the ith combination.

Then a necessary and sufficient condition that there exist a set of cominbations s_i which satisfy (19) is that s^o satisfies

$$N + Gs^o \leq 0 \qquad (20)$$

$$0 \leq s^o \leq 1$$

For a proof of this theorem, see Mellefont (1977).

For the limiting case, the combinatorial problem can therefore be completely eliminated and a non-linear programming problem derived for constrained measurement subset policies. However, it is not always acceptable to allow a high rate of switching between subsets, and when δt is chosen to reflect achievable switching rates it becomes necessary to specify the order of appearance of combinations in each control interval. The approach taken here is to specify a priority for each measurement or combination (which may change over the control interval).

Then for the given measurement structure, a non-linear program can be solved to obtain an optimal solution. It should be emphasised that this approach is not restrictive since the continuous s_k^o variables can be chosen by the optimization routine so that any combination order occurs provided that sufficient control intervals are allowed. This is illustrated in Fig.5, where there are five measurements subject to the constraints:

1. No more than two measurements at the same time
2. Measurements Nos 3 and 4 must be made together.

Figure 5a shows the optimum order, achieved in one interval with a good priority order, while Figure 5b, using a poor priority order, achieves the same policy over two intervals.

FIGURE 5 Arrangement of Measurements to Satisfy Constraints

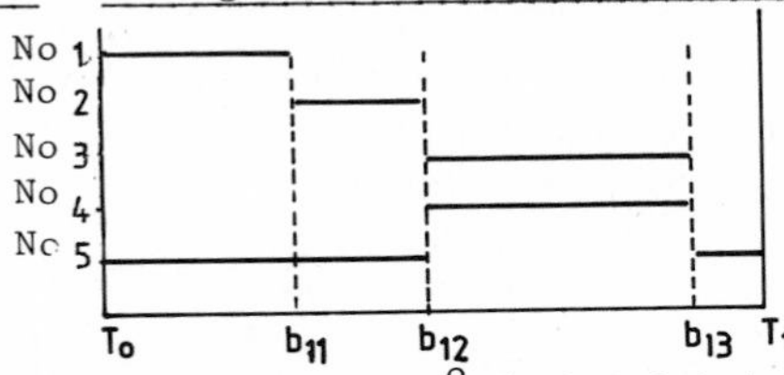

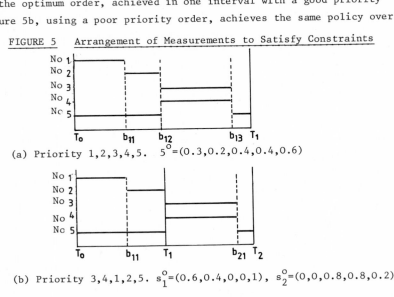

(a) Priority 1,2,3,4,5. $5^{o}=(0.3,0.2,0.4,0.4,0.6)$

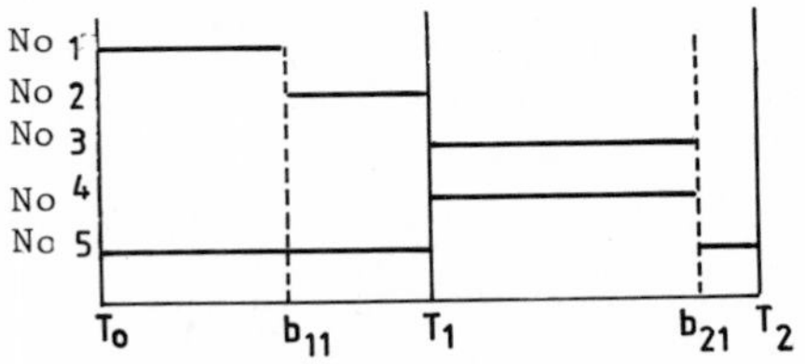

(b) Priority 3,4,1,2,5. $s_1^o=(0.6,0.4,0,0,1)$, $s_2^o=(0,0,0.8,0.8,0.2)$

The simplicity of testing for the need to increase the number of control intervals makes this approach effective for an arbitrarily specified priority order, but clearly much computation will be saved if a good priority order can be specified. It can be shown that the objective function (9) is equivalently written as

$$J = \int_{t_0}^{t_f} Tr.PBR_2^{-1}B'PQ + s'R_3 s \, dt + \text{const. terms} \tag{21}$$

On the basis that a decrease (or minimum increase) in Q is required over each control interval, measurement priority can be assigned such that the rate of decrease of Q is maximized. Neglecting off diagonal terms in measurement noise covariance matrix V_2, Mellefont(1977) shows that measurement priority for the ith measurement is obtained from

$$p_i = Tr.(PBR_2^{-1}B'P)(QC_i'V_{2i}^{-1}C_iQ) \tag{22}$$

with maximum priority given to the measurement with largest p_i value. Measurement priorities are assigned at appropriate times during the control period, for example at the beginning of each control interval. It was found that the priority order is often constant over successive iterations, and hence the initial priority order can be used throughout.

Whatever method is used to fix the measurement structure in each control interval, an allocation routine is needed to convert optimization variables s_k^o, T_k into a set of bang-bang controls s_i so that objective (9) and gradients can be calculated. Objective function calculation is straightforward but from fig.5 it can be seen that changing s_k^o and T_i will move the boundaries between combinations according to the allocation routine. Using the priority method described above, the boundaries are linearly related to s_k^o, T_k, and the linear coefficients are most easily obtained by finite differences. Gradients with respect to s_k^o, T_k are then linear combinations of the gradients for individual boundaries calculated from (10), with T_{ij} replaced by b_{ki}, the boundary dividing combinations i and i+1 in the kth control interval.

This method was used to find the constrained measurement policy for a chemical reactor with linearized coefficient matrices (Mellefont,1977)

$$A = \begin{bmatrix} -1. & 0 & 0. \\ -9. & -11 & -23.26 \\ -5.46 & -5.46 & -16.10 \end{bmatrix} \qquad B = \begin{bmatrix} 2. & 0. \\ 9. & 0. \\ 0. & 5.46 \end{bmatrix} \qquad (23)$$

where x_1 = holdup x_2 = concentration x_3 = temperature

u_1 = inlet flowrate u_2 = heat input.

The system was assumed to be subject to fluctuations in flowrate and heat input of 10% standard deviation. Measurement of each state is possible but only one measurement at a time was allowed. The priority order was found to be concentration, temperature, holdup and fig.6 shows the optimal measurement policy obtained after 11 iterations of the optimization routine. Reversing this priority, fig.7 shows that the optimization routine was able to force concentration measurement to be made first during the first control interval but at the expense of 21 iterations. Since the high initial covariance has been largely reduced by this stage, it was not found necessary to alter the assumed priority. The insensitivity of measurement policies to the order of

FIGURE 6.

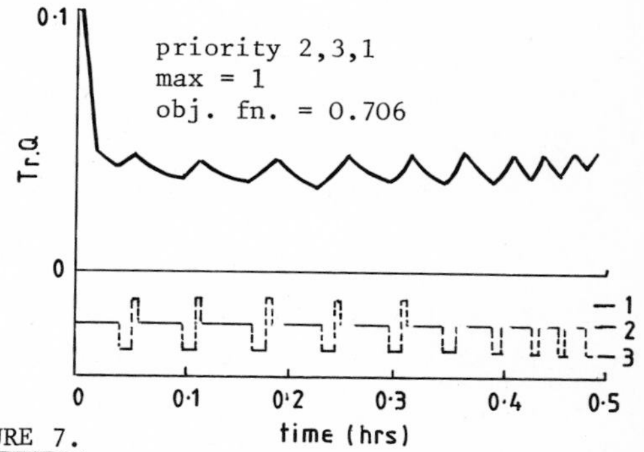

FIGURE 7.

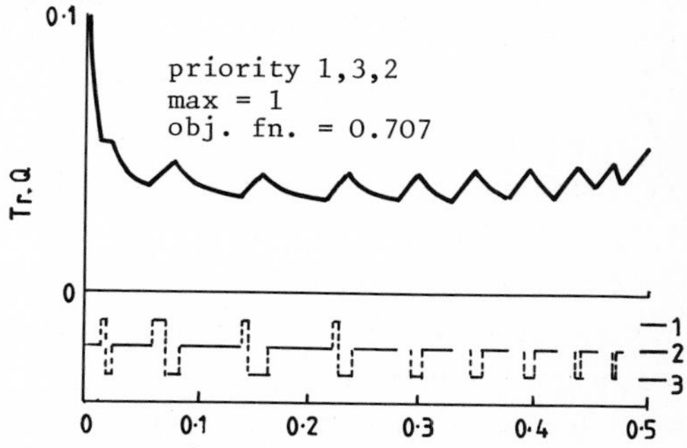

1. holdup, 2. conc., 3. temp.

priority during cyclic operation is discussed below. This example demonstrates the ability of the above method to determine constrained measurement policies even when a poor priority order is specified.

Cyclic Measurement Policies for Processes at Steady State

Results indicate that the optimal measurement policy adopts a cyclic pattern when the system is under steady operating conditions. Considering equation (17) at steady state, the optimal fraction r can be found by non-linear programming. However, hardware limitations impose a minimum practicable cycle time, and fig.8 shows the effect of applying the optimal fraction r for the chemical reactor example over cycles of increasing size. For cycles up to 3.6 minutes the degradation in performance was less than 1% so that a good approach is to apply the optimal fraction to the minimum cycle that can be achieved.

As cycle time increases, the effect of the order of measurement becomes more significant. Figs 8 and 9 show that it is best to spread measurement effort over the cycle. The effect of measurement order is seen to be small over realistic cycle times and hence, for constrained problems, the order of measurement is only important when the covariance is high.

FIG. 8 STEADY STATE CYCLES

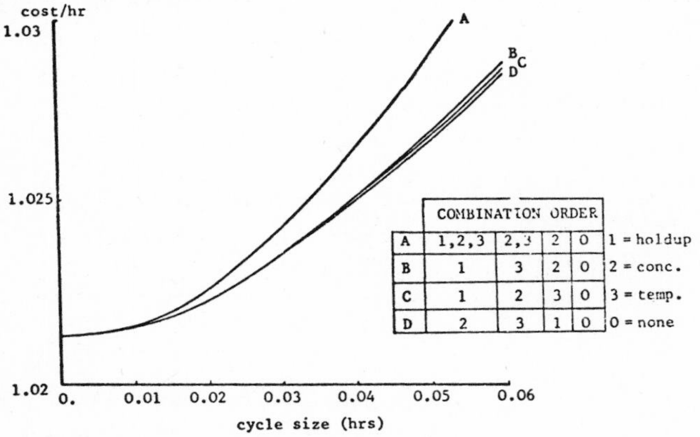

FIG. 9 EFFECT OF MEASUREMENT POLICY ON COVARIANCE

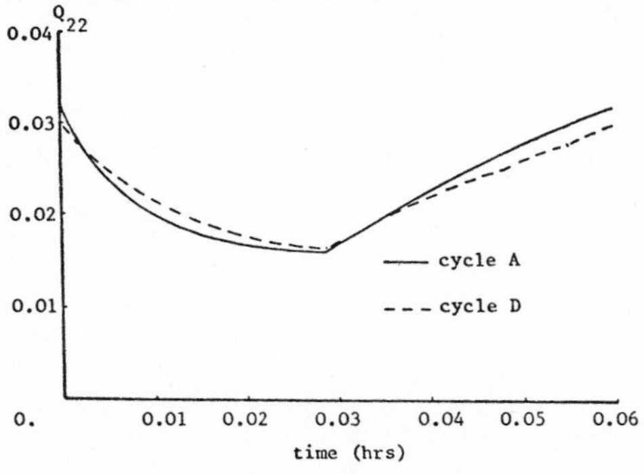

A Design Problem

Often there are alternative means of providing information about system states. For example, concentration can be either measured directly or inferred from related physical properties such as pH, refractive index or temperature. In multistage processes such as distillation, these measurements can be made on any stage thus creating a potentially large set of measurements from which to choose. In designing the control system, it is necessary to select a minimum number of measurements to provide the required information. Capital costs are of particular importance but introduce step changes in the objective function which make non-linear programming unsuitable. Similarly if constraints are placed on the total number of measurements that can be used, the combinatorial nature of the problem cannot be eliminated.

Since the measurement selection vector s is binary, the mixed integer non-linear programming problem can be solved by implicit enumeration. Mellefont and Sargent (1977) describe an implicit enumeration algorithm for the selection of measurements to be used over the entire control interval, and this algorithm is easily extended to deal with combined capital and operating costs by including a non-linear programming stage at the termination of appropriate branches. However, experience with this algorithm has shown that a good suboptimal approach is to determine an optimal set of measurements based on full use of each measurement and to perform a single non-linear optimization to find the optimal policy for that set. In this way considerable computation savings can be realized.

Conclusions

The algorithms outlined above enable the design of the measurement subsystem to be made so that combinatorial constraints are met and costs are minimized. The use of a standard non-linear programming algorithm ensures that the optimization proceeds efficiently.

The problem is presented at several levels of complexity and different approximations are suggested to minimize computation costs. Results show that measurements should be located so that the most sensitive state variables are under observation. For steady state operation, cyclic policies develop once a reasonable level of estimation accuracy has been achieved, and these can be approximated with minimum computational effort.

References

Athans, M. (1968); Information and Contr. 11 pp592-606

Athans, M. (1972); Automatica, 8 pp397-412

Herring, K.D. (1972); PhD thesis, Southern Methodist University,

Herring, K.D. and Melsa, J.L. (1974); I.E.E.E. Trans Auto Contr., AC19(6), pp264-266

Mellefont, D.J. (1977); PhD thesis, University of London

Mellefont, D.J. and Sargent, R.W.H. (1977a); Int.J.Contr. 26(4) 595-602pp

Mellefont, D.J. and Sargent, R.W.H. (1977b); to appear

Sargent, R.W.H. and Murtagh, B.A. (1973); Math.Program.,4,pp245-268

Sargent, R.W.H. and Sullivan, G.R. (1977); 85h IFIP Conference on Optimization Techniques, Würzburg F.R.G.

Wonham, W.M. (1968); SIAM J.Contr., 6(2), pp312-326.

ON THE OPTIMAL SEARCH FOR A MOVING TARGET IN DISCRETE SPACE

Ulla Pursiheimo
Department of Mathematical Sciences
University of Turku
SF-20500 Turku 50, Finland

ABSTRACT

A target is located at each moment in one of n boxes (or cells, regions) and it can move during the search time from boxes to other boxes. The motion is uniquely determined by a route function depending on a parameter and the initial box. A necessary and sufficient condition for the allocation of the search effort to be optimal will be determined in the following cases: a) when the probability of detection during a given time is to be maximized and b) when the expected search time is to be minimized. Finally, it is shown by an example that a uniformly optimal allocation does not always exist.

INTRODUCTION

A moving target in a discrete space, i.e., when target moves from one box to another, has been treated by Pollock [6], Iida [5] and Ciervo [1] in the discrete time case. The search then consists of successive looks at boxes. The motion in [6] is a two-cell Markov model with known constant transition probabilities. Dobbie [2] formulates a two-cell model of search for a moving target as a continuous time Markov-process with known constant transition rates and detection rates.

The present paper treats a somewhat different continuous time model. The motion of the target is deterministic after the initial box k and the parameter y are fixed, but the searcher does not know them. He only knows the joint a priori probability law of k and y. The respective model in continuous space is presented in [7].

MOTION AND DETECTION

Let a function $\beta(k,y,t)$ (route-function), piecewise constant with respect to t, having values $1,2,\ldots,n$, describe the motion of the target given that the target at the moment $t = 0$ is at box k ($1 \leq k \leq n$) and that the parameter is $y \in Y$. Further we suppose that for all values y and for almost all $t \in [0,T]$ function $\beta(k,y,t)$ is a one-to-one mapping from $\{1,2,\ldots,n\}$ onto itself. If we have $l = \beta(k,y,t)$ then the inverse function is written $k = \beta^{-1}(l,y,t)$. If k is the initial box we have to assume further that $\beta(k,y,0) = k$ for all routes that come into question.

The initial box k has a conditional distribution with respect to parameter y such that $p(k|y)$ is the probability of k subject to known value y. Parameter y has a priori probability distribution function $F(y)$ over space Y. Hence, the route $\beta(k,y,t)$ has a probability equal to $p(k|y)dF(y)$ and in general we have

$$(1) \qquad \int_Y \sum_{k=1}^{n} p(k|y)dF(y) = 1.$$

Let $\alpha(k,t)$, $k=1,2,\ldots,n$, $t \in [0,T]$, denote the amount of search effort at box k at the moment t. Allocation α must satisfy the following conditions

$$(2) \qquad \alpha(k,t) \geq 0, \quad k=1,2,\ldots,n, \quad t \in [0,T] \text{ a.e.}$$

$$(3) \qquad \sum_{k=1}^{n} \alpha(k,t) = 1, \quad t \in [0,T] \text{ a.e.}$$

Let $\omega(k)$ be a coefficient of efficiency of the search effort at box k. With these coefficients we make the different amounts of search effort in different boxes comparable with each other. When the target moves during the search time from boxes to other boxes according to route function $\beta(k,y,t)$ with k and y fixed, there accumulates amount

$$u(k,T,\alpha) = \int_0^T \omega(\beta(k,y,t))\alpha(\beta(k,y,t),t)dt$$

of this modified search effort upon the target. Given this amount u, the target is not detected with probability $\varphi(u)$. Function $\varphi(u)$ is the conditional non-detection function and is supposed to be a continuous, convex and monotonically non-increasing function of u, with $\varphi(0) = 1$. The probability of non-detection, after time T, will be

$$(4) \qquad P(T,\alpha) = \int_Y \sum_{k=1}^{n} p(k|y)\varphi(u(k,T,\alpha))dF(y).$$

MINIMIZATION OF PROBABILITY OF NON-DETECTION

As in the case of continuous space [7] we use the theorem of Dubovitskii and Milyutin [3], presented in a more applicable form in the book by Girsanov [4]. In the proof of the following theorem the results follow in a straightforward manner the corresponding theorem of [7]. Therefore the detailed proofs will be omitted.

Theorem 1 . A necessary and sufficient condition for α^* to be a solution of the problem

$$(5) \qquad P(T,\alpha^*) = \min\{P(T,\alpha) \mid \alpha \in Q_1 \cap Q_2\},$$

where Q_1 and Q_2 are the sets of functions α satisfying the conditions (2) and (3), respectively, is that there exists a function $\lambda(t) \leq 0$ such that

$$(6) \qquad \omega(k)D(k,t,\alpha^*) \begin{cases} = \lambda(t), & \text{if } \alpha^*(k,t) > 0, \\ \geq \lambda(t), & \text{if } \alpha^*(k,t) = 0, \end{cases}$$

where

$$(7) \qquad D(k,t,\alpha) = \int_Y p(\beta^{-1}(k,y,t)|y)\varphi'(u(\beta^{-1}(k,y,t),T,\alpha))dF(y).$$

<u>Proof</u>. We consider space $L^{(n)}$ of all vector-valued functions $\alpha(t) = (\alpha(1,t),\ldots,\alpha(n,t))$ such that

$$(8) \qquad \sum_{k=1}^{n} \operatorname*{ess\,sup}_{t\in K} |\alpha(k,t)| < \infty$$

where K is any compact subset of the interval [0,T]. Expressions (8) are seminorms and, hence, the space $L^{(n)}$ is a locally convex topological space. It further has the property that for all $\alpha \in L^{(n)}$ the search effort accumulated at any route is finite, i.e., $u(k,T,\alpha) < \infty$. In this space $P(T,\alpha)$ is a continuous and convex functional of α. Hence, the cone of directions of decrease at the optimal point α^* is

$$K_o = \{h \in L^{(n)} \mid P'(T,\alpha^*,h) < 0\}$$

$$= \{h \in L^{(n)} \mid \int_0^T \sum_{k=1}^{n} \Big\{\int_Y p(\beta^{-1}(k,y,t)|y) \times$$

$$\times\varphi'(u(\beta^{-1}(k,y,t),T,\alpha^*))dF(y)\Big\}\omega(k)h(k,t)dt < 0\}.$$

Above $P'(T,\alpha^*,h)$ stands for the directional derivative of $P(T,\alpha^*)$. Using the notation (7) we get further

$$K_o = \{h \in L^{(n)} \mid \int_0^T \sum_{k=1}^{n} \omega(k)D(k,t,\alpha^*)h(k,t)dt < 0\}.$$

The dual cone K_o^* of K_o is obtained directly as

$$K_o^* = \{g_o \in L^* \mid g_o(h) = \lambda\int_0^T \sum_{k=1}^{n} \omega(k)D(k,t,\alpha^*)h(k,t)dt,\ \lambda < 0\},$$

where L^* is the dual space of $L^{(n)}$.

Instead of considering the conditions (2) and (3) separately, we join them together as the following condition

$$(9) \qquad \operatorname*{ess\,sup}_{t\in[0,T]} \sum_{k=1}^{n} \alpha^+(k,t) \leq 1,$$

where

$$\alpha^+(k,t) = \begin{cases} \alpha(k,t), & \text{if } \alpha(k,t) > 0 \\ 0, & \text{if } \alpha(k,t) \leq 0. \end{cases}$$

Obviously this condition is satisfied if (2) and (3) are valid. On the contrary, if we have minimized $P(T,\alpha)$ by using the above condition (9), it is obvious by the nature of the problem that the same α gives also the minimum of $P(T,\alpha)$ with conditions (2) and (3). The set Q of functions α such that condition (9) is satisfied is a convex set with non-void interior Q^o. Hence, the cone of feasible directions at α^* is according to theorem 8.2. of [4]

$$K_1 = \{h \in L^{(n)} \mid h = \lambda(\alpha-\alpha^*),\ \alpha \in Q^o,\ \lambda > 0\}.$$

The dual cone K_1 consists of such continuous linear functionals which are non-negative at K_1, i.e., for all $h \in K_1$, $g_1 \in K_1^*$ there must be

$$g_1(h) = g_1(\lambda(\alpha-\alpha^*)) = \lambda g_1(\alpha) - \lambda g_1(\alpha^*) \geq 0,\ \alpha \in Q^o.$$

Hence, we see that every such functional g_1 is a support functional of Q at α^*. From theorem 6 in [4] it follows that the optimal α^* must satisfy for all $g_o \in K_o$, $g_1 \in K_1^*$ the equation $g_o + g_1 = 0$. This means as a consequence of the definition of K_o^* that

$$(10) \qquad g_1(h) = -\lambda\int_0^T \sum_{k=1}^{n} \omega(k)D(k,t,\alpha^*)h(k,t)dt.$$

It is now easily verified that $D(k,t,\alpha^*)$ is integrable, so that

$$\int_0^T \sum_{k=1}^{n} \omega(k)|D(k,t,\alpha^*)|dt < \infty.$$

Therefore we can make use of the example 10.5. of [4] and conclude that condition (10) leads to the condition

$$(11) \qquad \lambda \sum_{k=1}^{n} \omega(k)D(k,t,\alpha^*)\alpha(k,t) \leq \lambda \sum_{k=1}^{n} \omega(k)D(k,t,\alpha^*)\alpha^*(k,t)$$

for almost all $t \in [0,T]$ and for all $\alpha(\cdot,t) \in M \subset R^n$ where

$$M = \{x = (x_1,\ldots,x_n) \in R^n \mid \sum_{k=1}^{n} x_k^{+} \leq 1\}.$$

Taking into account that $\lambda D(k,t,\alpha^*)\omega(k) \geq 0$ for all $k = 1,2,\ldots,n$, $t \in [0,T]$ the inequality (11) is possible only if $\alpha^*(k,t) > 0$ for such k that

$$\lambda\omega(k)D(k,t,\alpha^*) = \max\{\lambda\omega(k)D(k,t,\alpha^*) \mid k=1,2,\ldots,n\} \equiv \lambda(t)$$

and $\alpha^*(k,t) = 0$ if

$$\lambda\omega(k)D(k,t,\alpha^*) < \lambda(t).$$

This result can be written in the form (6).

According to theorem 15.2. of ref. [4] these conditions are also sufficient for the optimality because $P(T,\alpha)$ is a convex functional and Q is such a convex set which has a non-void interior.

MINIMIZATION OF THE EXPECTED SEARCH TIME

In the previous chapter the only thing of interest was to get an as good probability of detection as possible. Now we consider this search process by taking into account the cost of the search. Let us start with the assumption that the search process will continue until the target is detected, not however, longer than up to the moment T. The cost of the search is assumed to be proportional to the duration of the search. so if the allocation α is applied, then the target will be found during time $[0,t]$, where $t \leq T$, with probability $1-P(t,\alpha)$. The search will be unsuccesful after time T with probability $P(T,\alpha)$. Therefore, the expected search time will be

$$(12)\qquad E(T,\alpha) = -\int_0^T tP'(t,\alpha)dt + TP(T,\alpha) = \int_0^T P(t,\alpha)dt.$$

For the optimality of α in this case we get the following theorem.

Theorem 2. A necessary and sufficient condition for $\tilde{\alpha}$ to be a solution of the problem

$$(13)\qquad E(T,\tilde{\alpha}) = \min\{E(T,\alpha) \mid \alpha \in Q_1 \cap Q_2\},$$

is that there exists a function $\mu(t) \leq 0$ such that

$$(14)\qquad \omega(k)G(k,t,\tilde{\alpha}) \begin{cases} = \mu(t), & \text{if } \tilde{\alpha}(k,t) > 0 \\ \geq \mu(t), & \text{if } \tilde{\alpha}(k,t) = 0, \end{cases}$$

where

$$(15)\qquad G(k,t,\alpha) = \int_Y p(\beta^{-1}(k,y,t)\mid y)\times$$

$$\times\left\{\int_t^T \varphi'(u(\beta^{-1}(k,y,t),s,\alpha))ds\right\}dF(y)$$

Proof. The constraints for α are the same as in the theorem 1. Therefore, we can use directly results obtained there. There remains then only the determination of cones K_o and K_o^* corresponding the functional $E(T,\alpha)$. First we notice that $E(T,\alpha)$ is a convex and continuous functional of α, and, hence, is regularly decreasing at $\tilde{\alpha}$. The directional derivative of $E(T,\alpha)$ at $\tilde{\alpha}$ is

$$E'(T,\tilde{\alpha},h) = \lim_{\varepsilon\downarrow 0} \frac{1}{\varepsilon}\left\{E(T,\tilde{\alpha}+\varepsilon h) - E(T,\tilde{\alpha})\right\}$$

$$= \lim_{\varepsilon\downarrow 0} \int_0^T \frac{P(t,\tilde{\alpha}+\varepsilon h) - P(t,\tilde{\alpha})}{\varepsilon}\,dt$$

$$= \int_0^T P'(t,\tilde{\alpha},h)dt$$

which we finally write in the form

$$(16) \qquad E'(T,\tilde{\alpha},h) = \int_0^T \left\{ \int_0^t \sum_{k=1}^n \omega(k) B(k,s,t,\tilde{\alpha}) h(k,s) \right\} dt.$$

Here we have applied the convexity of $E(T,\alpha)$ and of $P(T,\alpha)$ with the monotone convergence theorem and notation

$$B(k,s,t,\alpha) = \int_Y p(\beta^{-1}(k,y,s)\,|\,y)\varphi'(u(\beta^{-1}(k,y,s),t,\alpha))dF(y).$$

If $\varphi'(0)$ exists finite then the integral (16) is absolutely convergent and we can change the order of the integration and get

$$E'(T,\tilde{\alpha},h) = \int_0^T \sum_{k=1}^n \left\{ \int_s^T \omega(k) B(k,s,t,\tilde{\alpha}) dt \right\} h(k,s) ds,$$

which is a continuous linear functional of h. Hence, the cone of directions of decrease is

$$K_o = \{h \in L^{(n)} \mid E'(T,\tilde{\alpha},h) < 0\}.$$

Directly from this we get the dual cone as

$$K_o^* = \{g_o \in L^* \mid g_o(h) = \lambda \int_0^T \sum_{k=1}^n \left\{ \int_t^T \omega(k) B(k,t,s,\tilde{\alpha}) ds \right\} h(k,t) dt\}.$$

As in the previous chapter we get the equation $g_o + g_1 = 0$, where $g_i \in K_i^*$. This means that

$$g_1(h) = -\lambda \int_0^T \sum_{k=1}^n \left\{ \int_s^T \omega(k) B(k,s,t,\tilde{\alpha}) dt \right\} h(k,s) ds$$

for all $h \in L^{(n)}$. By proceeding quite in the similar way as in the previous theorem we get the proposition of the theorem.

UNIFORMLY OPTIMAL SOLUTION

Allocation α_u is called uniformly optimal, if it satisfies for all $T > 0$ the condition

$$P(T,\alpha_u) = \min_{\alpha} P(T,\alpha).$$

This means that the allocation is optimal even though we do not know the search time in advance. If there exists a uniformly optimal solution then it is easily demonstrated that the same allocation also minimizes the expected search time whatever the time T is. In the case of a stationary target with rather general assumptions the uniformly optimal solution exists. But when the target can move during the search we can show with an example that this is no more true. In general the optimal solution will depend essentially on the search time T.

EXAMPLES

We consider the following examples. With the first we show the nonexistence of the uniformly optimal allocation.

We suppose that there is only 2 boxes and the parameter space Y consists of pairs (y,i), such that i is an integer indicating if the target really moves or not and $y \in [0,1]$ is the moment when the target changes the box. Search time T = 1. We have following types of routes in question with two different values of probabilities respecting examples I and II.

probability		box	route	formula
case I	case II			
1/4	1/8	1 2		$\beta(1,(y,0),t)=1,\ 0\leq t\leq 1$
1/4	1/8	1 2		$\beta(2,(y,0),t)=2,\ 0\leq t\leq 1$
1/2	1/2	1 2		$\beta(1,(y,1),t)=\begin{cases}1, & 0\leq t<y\\ 2, & y\leq t\leq 1\end{cases}$
0	1/4	1 2	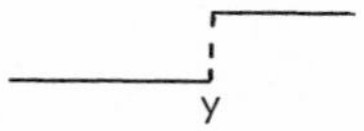	$\beta(2,(y,1),t)=\begin{cases}2, & 0\leq t<y\\ 1, & y\leq t\leq 1\end{cases}$

We further suppose that $\varphi(u) = \exp(-u)$, $\omega(1) = 1$, $\omega(2) = 2$ and that y is a random variable which is rectangularly distributed in the interval [0,1]. Integer i is 0 or 1. We can now easily demonstrate in the first case that the optimal solution is a bang-bang type allocation such that

$$(17) \qquad \alpha^*(1,t) = \begin{cases}1, & 0 \leq t < t_1\\ 0, & t_1 \leq t \leq 1\end{cases}$$

with $t_1 \approx 0.4314$ and $P(1,\alpha^*) \approx 0.409$. With this example it is easy to verify that a uniformly optimal allocation does not exist. If we calculate the probability of non-detection for the search time T = 0.5 by using the allocation α^* we get $P(0.5,\alpha^*) \approx 0.716$. But if we apply an other α similar to α^* but with $t_1 \approx 0.2638$, we get a smaller probability of non-detection $P(0.5,\alpha) \approx 0.685$. Hence, the allocation which was optimal for the search time T = 1, is not optimal for time T = 0.5. This just means that a uniformly optimal allocation cannot exist. Just for the curiousity , the optimal solution that minimizes the expected search time is of the same type (17) with $t_1 \approx 0.2941$.

In the case II we get by using theorem 1 that if the effort is positive at both boxes at some time interval $[t_1,t_2]$ then there must be satisfied

$$D(1,t,\alpha^*) = 2D(2,t,\alpha^*), \quad t \in [t_1,t_2] \text{ a.e.}$$

This gives the result that $\alpha^*(1,t) = 2/3$ and $\alpha^*(2,t) = 1/3$ on that interval. With theorem 1 we cannot conclude further the final solution. We now know that $\alpha^*(1,t)$ can have 3 different values: 0, 1 and 2/3. The final result was obtained by first defining a rough shape of $\alpha^*(1,t)$ with a direct search method. Then it seemed possible that $\alpha^*(1,t)$ would be of the following type

$$\alpha^*(1,t) = \begin{cases} 0, & 0 \le t < t_1,\ t_2 \le t \le 1 \\ 2/3, & t_1 \le t < t_2. \end{cases}$$

Among all this type of solutions the minimum is attained with values $t_1 \approx 0.1350$, $t_2 \approx 0.5184$ as $P(1,\alpha^*) \approx 0.4441$. With theorem 1 we can verify that it is really the optimal solution that was wanted.

Even in these simple examples there arose rather much computational difficulties, which will be a main trouble also in more complicated cases.

The functions $D(k,t,\alpha)$ in these examples were, if p_i, $i = 1,2,3,4$ are the probabilities in question, as

$$\begin{aligned} D(1,t,\alpha) &= p_1 \exp\left(-\int_0^1 \alpha(1,t)dt\right) + \\ &+ p_4 \int_0^t \exp\left(-2\int_0^y \alpha(2,t)dt - \int_y^1 \alpha(1,t)dt\right)dy + \\ &+ p_3 \int_t^1 \exp\left(-\int_0^y \alpha(1,t)dt - 2\int_y^1 \alpha(2,t)dt\right)dy \end{aligned}$$

$$\begin{aligned} D(2,t,\alpha) &= p_2 \exp\left(-2\int_0^1 \alpha(2,t)dt\right) + \\ &+ p_3 \int_0^t \exp\left(-\int_0^y \alpha(1,t)dt - 2\int_y^1 \alpha(2,t)dt\right)dy + \\ &+ p_4 \int_t^1 \exp\left(-2\int_0^y \alpha(2,t)dt - \int_y^1 \alpha(1,t)dt\right)dy. \end{aligned}$$

REFERENCES

[1] A.P.Ciervo, Search for Moving Targets, *Pacific-Sierra Research Corp. Report 619 B*, (1976) Santa Monica, California

[2] J.M.Dobbie, A Two-cell Model of Search for a Moving Target, *Operations Res.* 22 (1974) pp. 79-92

[3] A.Dubovitskii - A.Milyutin, Extremum Problems in the Presence of Constraints, *Z. Vycisl. Mat. i Mat. Fiz.*, 5 (1965) pp. 395-453 (Russian)

[4] I.Girsanov, *Lectures in the Mathematical Theory of Extremal Problems*, Springer-Verlag, Berlin (1972)

[5] K.Iida, Ido mokuhyobutsu no tansaku (The Optimal Distribution of Searching Effort for a Moving Target) (Japanese), *Keiei Kagaku* 16 (1972) pp. 204-215

[6] S.M.Pollock, A Simple Model of Search for a Moving Target, *Operations Res.* 18 (1970) pp. 883-903

[7] U.Pursiheimo, On the Optimal Search for a Target Whose Motion Is Conditionally Deterministic with Stochastic Initial Conditions on Location and Parameters, *SIAM J. Appl. Math.* 32, (1977) pp. 105-114

OPTIMAL MAINTENANCE AND INSPECTION : AN IMPULSIVE CONTROL APPROACH

Maurice Robin
IRIA-LABORIA
Domaine de Voluceau
Rocquencourt
78150 Le Chesnay
(France)

Abstract :

We studied here a particular case of stochastic control with imperfect information. Let us consider a typical example : we assume that we have a machine for which the degradation is modelled by a markov process. Two types of decision can be taken : replace the old machine by a new one or observe the degradation state of the operating machine. There is no information about the degradation except at decision times. Moreover costs are incurred for replacements or inspections. The problem is formulated as an impulsive control problem with partial information : under fairly general conditions, the existence of an optimal control is proved, and a characterization of the optimal cost function is given.

Introduction :

Let us consider a machine for which the degradation is modelled by a markov process x_t. At any time, one can observe the degradation or replace the machine by a new one. In the first case, the inspection gives the true state x_t (i.e. there is no inspection error), and an inspection cost is incurred. Moreover the machine must be stopped during h_o units of time for the inspection. In the case of a replacement, there is also a cost for the new machine, and also a delay h_r is needed; during this delay, the machine is stopped. The true state of the machine is known only at decision times.

This type of problem can be stated in the framework of impulsive control theory which was introduced and developped by A. Bensoussan - J.L. Lions, in particular for diffusion processes (cf. [2] , [3] , [4] and the bibliography of [2]).

Here the state of the machine is not known at any time, we have only a partial information. A particular case of partial information impulsive control problem was considered by R. Anderson and A. Friedman [1] , for a random evolution : their problem and methods are very different from ours. A problem similar to the one studied here, but with discrete time, was treated by Rosenfield [12].

The complete information case, (without a cost for inspection) was studied in [11].

In section 1, we give an heuristic formulation of the problem. In section 2 a precise statement is developped. In section 3 a characterization of the optimal cost function is given and we prove the existence of an optimal control. In section 4, we prove that the optimal cost function is the maximum solution of a quasi-variational inequality; stronger result will be obtained in the case of jump markov processes, in section 5.

1. Heuristic formulation

We introduce here some notations. Precise hypothesis will be stated in section 2. Let k_o be the observation cost which is incurred at the instant of decision and h_o the time delay needed for the inspection. If t is a decision time, the machine is stopped between t and $t+h_o$, and x_t is known at $t+h_o$. Let k_r the cost of replacement of the operating machine by a new one : we assume (for sake of simplicity) that the new machine is in a given state $\bar{x}$. Let h_r the time delay needed for replacement. There is also an operating cost $f(x)$ depending on the state of the machine, and a discount factor $\alpha < 0$.

We denote by A_x the infinitesimal generator associated with x_t, and we define y_t as the elapsed time, from the last decision-time, to time t ("decision-time") means inspection or replacement time). If we set

(1.1) $\tau_t = t - y_t$ (at t, we known x_{τ_t} and of course y_t).

Now (formally), a control is a sequence $v = (\tau^i, \xi^i, i \geq 1)$ of instants τ^i, and variables $\xi^i \in \{1,2\}$

$\xi^i = 1$ means that τ^i is an inspection time.

$\xi^i = 2$ means that τ^i is a replacement time.

The time τ^i must satisfy :

(1.2) $\tau^{i+1} \geq \tau^i + h_o$ if $\xi^i = 1$, $\tau^{i+1} \geq \tau^i + h_r$ if $\xi^i = 2$.

In a complete information problem the payoff would be

$$(1.3)\qquad J_x(v) = E\{ \int_0^{\tau^1} e^{-\alpha s} f(x_s)\, ds + \sum_{i\geq 1} \int_{\tau^i+h_i}^{\tau^{i+1}} e^{-\alpha s} f(x_s)\, ds + \sum_{i\geq 1} e^{-\alpha\tau^i} k_i \,|x_o = x\}$$

where $h_i = h_o\,\chi_{\xi^i=1} + h_r\,\chi_{\xi^i=2}$, $k_i = k_o\,\chi_{\xi^i=1} + k_r\,\chi_{\xi^i=2}$, $\chi_{\xi^i=m} = 1$ if $\xi^i = m$, 0 otherwise.

Here, what is known at time 0 (for any time t, the argument will be the same because we consider a stationnary problem), is y, the elapsed time from the last decision instant, and x, the true state at t = - y.

Therefore the payoff is

$$(1.4)\qquad J_{xy}(v) = E\{ \int_0^{\tau^1} e^{-\alpha s} f(x_s)\, ds + \sum_{i\geq 1} \int_{\tau^i+h_i}^{\tau^{i+1}} e^{-\alpha s} f(x_s)\, ds + \sum_{i\geq 1} e^{-\alpha\tau^i} k_i \,|x_{-y} = x\}$$

We set

$$(1.5)\qquad u(x,y) = \operatorname*{Inf}_v J_{xy}(v),$$

and we now use the formal argument of dynamic programming to obtain the evolution of u.

At a replacement time the "known state" goes from (x,y) to $(\bar{x},o)$ with a delay h_r : therefore, if the future decisions are optimal the cost for this situation is

$$(1.6) \qquad c_1 = k_r + e^{-\alpha h_r} u(\bar{x},o)$$

(we recall that no operating cost is incurred when the machine is stopped).

At an inspection time, considering the last decision time as the time origin, we observe x_y, that is, the future cost is given by

$$(1.7) \qquad c_2 = k_o + e^{-\alpha h_o} E_x\, u(x_y,o).$$

If no decision is made during $[0,\delta]$ we shall have a future cost(*).

$$(1.8) \qquad c_3 = E_x\{E_{x_y} \int_o^\delta e^{-\alpha s} f(x_s)\, ds + e^{-\alpha\delta} u(x,y+\delta)\}.$$

Now we set

$$(1.9) \qquad V(x,y) = E_x\, f(x_y)$$

which is the solution of the Cauchy problem :

$$(1.10) \qquad \frac{dV}{dy} = A_x V, \quad y > 0, \quad V(x,o) = f(x)$$

Then, the optimal cost, when we begin with the information (x,y) (i.e. the last decision time is passed since y units of time and the true state was x at this date), is given by

$$(1.11) \qquad u(x,y) = \min(c_1, c_2, c_3)$$

which can be written

$$(1.12) \qquad u(x,y) \le c_3,\ u(x,y) \le \min(c_1,c_2),\ (u(x,y) - c_3)(u(x,y) - \min(c_1,c_2)) = 0$$

Then, assuming u smooth enough, and going to the limit $\delta \downarrow 0$ we obtain the following system :

$$(1.13) \qquad \begin{cases} \dfrac{\partial u}{\partial y} - \alpha u + V(x,y) \ge 0 \\ u(x,y) \le Mu(x,y) \\ (\dfrac{\partial u}{\partial y} - \alpha u + V)(u - Mu) = 0 \end{cases}$$

Where we used the notation

$$(1.14) \qquad Mu(x,y) = \min \begin{cases} k_o + e^{-\alpha h_o} E_x\, u(x_y,o) \\ k_2 + e^{-\alpha h_r} u(\bar{x},o) \end{cases}$$

and V is given by (1.9).

In the terminology of A. Bensoussan - J.L. Lions [2], [3], the system (1.13) is a quasi-variational inequality (Q.V.I.).

(*) E_x denoting the probability measures associated to the uncontrolled markov process.

In the formal derivation of the system (1.13), it is clear that optimal decision times should be given by the time when the "known state" (x,y) is such that $u(x,y) = Mu(x,y)$ and the decision will be to observe if the min in (1.15) is realized by the first term, to replace if the min is realized by the second one.

2. Precise statement of the control problem

Let E be a locally compact space with countable base endowed with its borel σ-field. Let $\Omega = D(0, \infty; E)$ the space of right continuous, left limited functions from R^+ into E. Let $x_t(\omega) = \omega(t)$ for $\omega \in \Omega$ and F_t = σ-algebra generated by $\{x_s \;\; s \le t\}$. Let θ_t the translation operator i.e. $\theta_t(\omega) = \{s \to \omega(t+s) \quad s \ge 0\}$.

We assume that

(2.1) $\quad X = (\Omega, F_t, \theta_t, x_t, P_x)$ is a given markov process with values in E, and we assume

(2.2) $\quad$ X is a Feller process, that is, if C is the space of bounded continuous functions on E, with the sup norm (noted $||.||$),

$$(2.3)\qquad \begin{cases} \Psi(x) = E_x\, f(x_t) \in C, & \forall f \in C, \quad \forall t > 0 \\ \lim\limits_{t \downarrow 0} E_x f(x_t) = f(x) & \forall f \in C. \quad \forall x \in E. \end{cases}$$

We set also

$$(2.4)\qquad \mu_{xy}(\Gamma) = P_x(x_y \in \Gamma),\ x \in E,\ y \in R^+,\ \Gamma \text{ borel set of } E.$$

$$(2.5)\qquad P_\mu(A) = \int_E \mu_{xy}(dz)\, P_z(A).$$

X defines the uncontrolled process. We must precise now the information structure and the admissible controls.

We assume for sake of simplicity that

$$(2.6)\qquad h_o = h_r = h > 0 \quad \text{a given constant.}$$

Assume that the initial information is (x,y) (that is : the last decision point was $t = -y$, and x was the true state at this date). Denote by B the borel σ-field of $E \times R^+$.

An admissible control v will be a sequence

$$v = (\tau^i, \xi^i, i \ge 1) \text{ such that}$$

$$(2.7)\qquad \begin{aligned} \tau^1 &= \varphi(x,y) \text{ where } \varphi \text{ is a measurable function from } E \times R^+ \text{ into } R^+ \cup \{\infty\}, \\ \xi^1 &= \Psi(x,y) \text{ where } \Psi \text{ is a measurable function from } E \times R^+ \text{ into } \{1,2\}. \end{aligned}$$

As previously, $\xi^1 = 1$ means that τ^1 is an inspection time, $\xi^1 = 2$, a replacement time.

τ^2 will be a random variable on (Ω, F_∞), such that, if $z^1 = x_{\tau^1} \chi_{\xi^1=1} + \bar{x}\, \chi_{\xi^1=2}$, τ^2 is measurable with respect to $\sigma\{z^1\}$, (the σ-algebra generated by z^1), with values in R^+, and $\tau^2 \ge \tau^1 + h$,

ξ^2 will be $\sigma(z^1)$ measurable with values in $\{1,2\}$.

In general τ^n is a.r.v. $\sigma\{z^1, \dots, z^{n-1}\}$ measurable, with values in R^+, and

(2.8) $\quad \tau^n \geq \tau^{n-1} + h$

ξ^n is $\sigma(z^1, \dots, z^{n-1})$ measurable, with values in $\{1,2\}$.

One could verify that τ^i is an F_t stopping time.

The cost structure is the following

(2.9) $\quad f \geq 0,\ f \in C$ will be the operating cost, and

(2.10) $\quad k_o, k_r > 0$, will be the inspection cost and the replacement cost respectively.

We now define the process corresponding to a given admissible control as a probability measure on the canonical space (Ω, F_∞). We shall use the following lemma :

Lemma 2.1 : Let $x \in E$, τ an F_t stopping time, z a random variable with values in E, F_τ-measurable, then there exists a probability measure $\tilde{P}$ on (Ω, F_∞) such that

i) $\quad \begin{cases} \tilde{P} = P_x \text{ on } F_{(\tau+h)^-} \\ \tilde{E}\,[\varphi(x_{\tau}+h+s) \mid F_{(\tau+h)^-}] = E_z\, \varphi(x_s) \end{cases}$

a.s P_x, on $\{\tau < +\infty\}$, $\forall$ s ≥ 0, $\forall\ \varphi$ bounded, measurable on E. ($\tilde{E}$ meaning the expectation w.r.t. the measure $\tilde{P}$).

ii) $\quad \tilde{E}\,[\varphi(x_{\tau+h+s+t}) \mid F_{\tau+h+s}] = E_{x_{\tau+h+s}}(\varphi(x_t))$

a.s. $\tilde{P}$ on $\{\tau < +\infty\}$, $\forall$ s, t ≥ 0 .

Proof : The proof is given in [12] □

Now, if v is an admissible control, we define the following sequence of probability measures

$$P^1 = P_\mu \qquad \text{(x, y being given)}$$

$$\text{(2.11)} \quad \begin{cases} P^2 = P^1 \text{ on } F_{(\tau^1+h)^-} \\ E^2\,[\varphi(x_{\tau^1+h+s}) \mid F_{(\tau^1+h)^-}] = E_{z^1}(\varphi(x_s)) \end{cases}$$

a.s. P^1 on $\{\tau^1 < +\infty\}$, $\forall\ \varphi$ bounded, measurable.

Where

$$z^1 = x_{\tau^1}\ \chi_{\xi^1=1} + \bar{x}\ \chi_{\xi^1=2}$$

...

$$\text{(2.12)} \quad \begin{cases} P^n = P^{n-1} \text{ on } F_{(\tau^{n-1}+h)^-} \\ E^n\,[\varphi(x_{\tau^n+h+s}) \mid F_{(\tau^n+h)^-}] = E_{z^n}(\varphi(x_s)) \end{cases}$$

with

$$z^n = x_{\tau^n} \chi_{\xi^n=1} + \bar{x} \chi_{\xi^n=2}$$

Then it is possible to define

(2.13) $\quad P^v_{xy} = P^n$ on $F_{(\tau^n+h)^-}$

(Notice that $\tau^n(\omega) \nearrow +\infty \ \forall \omega \in \Omega$ because of (2.8)).

(2.13) is taken as the definition of the process corresponding to the control v. We now define the payoff

(2.14) $$J_{xy}(v) = E^v_{xy} \{\int_o^{\tau^1} e^{-\alpha s} f(x_s)\, ds + \sum_{i\geq 1} \int_{\tau^i+h}^{\tau^{i+1}} e^{-\alpha s} f(x_s)\, ds + \sum_{i\geq 1} e^{-\alpha\tau^i} k(\xi^i)\}$$

where $k(\xi^i) = k_o \chi_{\xi^i=1} + k_r \chi_{\xi^i=2}$.

Moreover we set

(2.15) $$u(x,y) = \operatorname*{Inf}_v J_{xy}(v) .$$

Remark 2.1.

a) It can be seen on the definition of z^i, that the state at τ^i is modified only when τ^i is a replacement time, if τ^i is an inspection time, the law of x_{τ^i} is the same for p^{i-1} or p^i.

b) Because of the properties (2.3) $w_x(t) = E_x\ \varphi(x_t)$, for fixed x, is a continuous function of t if $\varphi \in C$.

□

3. Characterization of the optimal cost

3.1. Some properties of optimal stopping time problems

We give here, without proof, some results about optimal stopping time problems which will be used below. The proofs can be obtained as in [3] for example.

Let V, $\Psi : R^+ \to R$ bounded, coutinuous

and let us define

(3.1) $$u(y) = \operatorname*{Inf}_\tau [\int_o^\tau e^{-\alpha s} V(y+s)\, ds + e^{-\alpha\tau} \Psi(y+\tau)]$$

where the infinum is taken on $\tau \in R^+$.(3.1) is the lower bound for an optimal stopping problem for the deterministic markov process $Z_y(t) = y + t$, $y \in R^+$ and the methods of [4] can be used to obtain the following results :

Lemma 3.1 : i) u is a bounded continuous function

ii) there exists an optimal stopping time $\hat{\tau}$, given by

$$\hat{\tau} = \text{Inf } (s \geq 0,\ u(y+s) = \Psi(y+s))$$

iii) u is the maximum element of the set of functions w verifying

$$w \in C_b^o(R^+)$$

$$w \leq \Psi$$

$$w(y) \leq e^{-\alpha t}\, w(y+t) + \int_o^t e^{-\alpha s}\, V(y+s)\, ds \qquad \forall\, t \geq 0$$

Moreover, it is solution of a variational inequality cf [3], [8].

We shall need of the supplementary result (cf. [3], [4] , (which is used in the proof of lemma 3.1).

If we define

$$(3.2) \qquad u_\varepsilon(y) = \underset{v}{\text{Inf}} \int_o^{+\infty} e^{-\alpha s}\, e^{-\frac{1}{\varepsilon}\int_o^s v(s)\, d\lambda} \left[V(y+s) + \frac{1}{\varepsilon}\, v(s)\, \Psi(y+s)\right] ds$$

where the infimum is taken for v measurable function in R^+, with values in [0,1], then u_ε is continuous and converges uniformly on every compact to u, and u_ε decreases to u when ε decreases to zero.

3.2. Main result

As in section 1, we use the following notation

$$V(x,y) = E_x\, f(x_y)\ , \qquad x \in E, \quad y \in R^+,$$

it must be noticed that

$$V(x,y+s) = E_x\, f(x_{y+s}) = E_x\, E_{x_y}\, f(x_s) = E_\mu\, f(x_s)$$

(where we used the markov property and where μ was defined in (2.4)). We define also Mu as in (1.15).

This section deals with the proof of the following result

Theorem 3.1. Under the assumptions (2.1), (2.3), (2.9), (2.10)
i) There exists an optimal control for the problem (2.15).
ii) u is the maximum element of the set of functions w such that

$$w(x,.) \in C_b^o(R^+)\ , \quad w(.,y) \in C_b^o(E)$$

$$w(x,y) \leq Mw(x,y)$$

$$w(x,y) \leq e^{-\alpha t}\, w(x,y+t) + \int_o^t e^{-\alpha s}\, V(x,y+s)\, ds \qquad \forall\, t \geq 0$$

the proof of theorem 3.1 is made in several steps that we indicate briefly. Details on these technics can be found in [12].

We begin by defining the sequence of approximate problems :

$$(3.3) \qquad u^o(x,y) = E_\mu \int_o^{+\infty} e^{-\alpha s}\, f(x_s)\, ds = \int_o^{+\infty} e^{-\alpha s}\, V(x+y+s)\, ds$$

(3.4) $$u^n(x,y) = \operatorname*{Inf}_{\tau\in R^+} \left[\int_o^\tau e^{-\alpha s} V(x,y+s)\, ds + e^{-\alpha\tau} Mu^{n-1}(x,y+\tau)\right]$$

It is clear here that x is only a parameter; (3.4) define a stopping problem similar to those considered in § 3.1.

Lemma 3.2. Under the hypothesis of theorem 3.1

i) $0 \le u^{n+1} \le u^n \le u^o \le \frac{\|f\|}{\alpha}$ $\quad\forall\ n \ge 0$

ii) $y \to u^n(x,y)$ is continuous $\forall\ x \in E$, $\quad\forall\ n \ge 0$

$x \to u^n(x,y)$ is continuous $\forall\ y \in R^+$, $\quad\forall\ n \ge 0$.

Lemma 3.3. u^n converges uniformly to a function $\tilde{u}$.

Proof. It is easy to show that :

$$\| u^n = u^{n-1} \| \le e^{-\alpha h} \| u^{n-1} - u^{n-2} \|$$

Therefore u^n converges in the space of bounded measurable functions on $E \propto R^+$ (with the sup norm $\| . \|$). We denote by $\tilde{u}$ its limit. From lemma 3.2 we have that $\tilde{u}(x,.) \in C_b^o(R^+)$, and $\tilde{u}(.,y) \in C_b^o(E)$.

Corollary 3.1. $\tilde{u}$ is the unique solution of the equation

(3.6) $$\begin{cases} \tilde{u}(x.y) = \operatorname*{Inf}_{\tau} \left(\int_o^\tau e^{-\alpha s} V(x,y+s)\, ds + e^{-\alpha\tau} M\tilde{u}(x,y+\tau)\right). \\ \tilde{u} \text{ bounded measurable on } E \times R^+. \end{cases}$$

Proof. The previous lemma shows that the application $w \to \Pi w$ defined by

$$\Pi w = \operatorname*{Inf}_{\tau} \left(\int_o^\tau e^{-\alpha s} V(x,y+s)\, ds + e^{-\alpha\tau} Mw(x,y+\tau)\right)$$

is a contraction, and therefore, Π has a unique fixed point, which is clearly $\tilde{u}$ by the definition of u^n.

Lemma 3.4. Under the assumptions of theorem 3.1 there exists an admissible control $\hat{v}$ such that

(3.7) $$\tilde{u}(x,y) = J_{xy}(\hat{v}).$$

Proof. We set $\tau^1 = \operatorname{Inf}(s \ge 0,\ \tilde{u}(x,y+s) = M\tilde{u}(x,y+s))$. From (3.6) and the section 3.1, we have that τ^1 is an optimal stopping time for the problem corresponding to $\tilde{u}$ (verifying (3.6)), i.e.

(3.8) $$\tilde{u}(x,y) = \int_o^{\tau^1} e^{-\alpha s} V(x,y+s)\, ds + e^{-\alpha\tau^1} M\tilde{u}(x,y+\tau^1)$$

Define

$$\xi^1 = 1 \text{ if } k_o + e^{-\alpha h} E_x\, \tilde{u}(x_{y+\tau^1}, 0) < k_r + e^{-\alpha h}\, \tilde{u}(\bar{x},o)$$

$$\xi^1 = 2 \text{ otherwise.}$$

This gives

$$\tilde{u}(x,y) = \int_o^{\tau^1} e^{-\alpha s} V(x,y+s)\, ds + e^{-\alpha\tau^1}\{\chi_{\xi^1=1}[k_o + e^{-\alpha h} E_x\, \tilde{u}(x_{y+\tau^1},0)] +$$

$$+\chi_{\xi^1=2}[k_r + e^{-\alpha h}\, \tilde{u}(\bar{x},o)]\}$$

But $E_x \tilde{u}(x_{y+\tau^1},0) = E_x E_{x_y} \tilde{u}(x_{\tau^1}, 0) = E^1 \tilde{u}(x_{\tau^1},0)$

implies

$$\tilde{u}(x,y) = E^1\{ \int_0^{\tau^1} e^{-\alpha s} f(x_s)ds + k(\xi^1)\, e^{-\alpha\tau^1} + \chi_{\xi^1=1}\, e^{-\alpha(\tau^1+h)}\tilde{u}(x_{\tau^1},0) + \chi_{\xi^1=2}\, e^{-\alpha(\tau^1+h)}\tilde{u}(\bar{x},o)\}$$

where $P^1 = P_\mu$ as in section 2.

If we define now

$$z^1 = x_{\tau^1}\, \chi_{\xi^1=1} + \bar{x}\, \chi_{\xi^1=2}$$

we have

$$u(x,y) = E^1 \{\int_0^{\tau^1} e^{-\alpha s} f(x_s)\, ds + k(\xi^1)\, e^{-\alpha\tau^1} + e^{-\alpha(\tau^1+h)}\, \tilde{u}(z^1,0)\}.$$

Now, assuming the formula

$$\tilde{u}(x,y) = E^n\{ \int_0^{\tau^1} e^{-\alpha s} f(x_s)\, ds + \sum_{i=1}^{n-1} \int_{\tau^i+h}^{\tau^{i+1}} e^{-\alpha s} f(x_s)\, ds\ \sum_{i=1}^{n} e^{-\alpha\tau^i}\, k(\xi^i) + e^{-\alpha\,(\tau^n+h)}\, \tilde{u}(z^n,o)\} \tag{3.10}$$

where z^n, τ^n, ξ^n are defined by induction

$$\tau^m = \text{Inf } (s \geq \tau^{m-1} + h,\ \tilde{u}(z^{m-1},\ s-\tau^{m-1} - h) = M\tilde{u}(z^{m-1},\ s-\tau^{m-1} - h))$$

$$\xi^m = 1 \text{ if } k_o + e^{-\alpha h}\, E_{z^{m-1}}\, \tilde{u}(x_{\tau^m-\tau^{m-1}-h},0) < k_r + e^{-\alpha h}\, \tilde{u}(x,o)$$

$$\xi^m = 2 \text{ otherwise}$$

$$z^m = x_{\tau^m}\, \chi_{\xi^m=1} + \bar{x}\, \chi_{\xi^m=2}$$

we show (3.10) for n+1 :

for any z,

$$\tilde{u}(z,o) = \int_0^{\tilde{\tau}} e^{-\alpha s}\, V(z,s)\, ds + e^{-\alpha\tilde{\tau}}\, M\tilde{u}(z,\tilde{\tau})$$

where

$$\tilde{\tau}(z) = \text{Inf}(s \geq 0,\ \tilde{u}(z,s) = M\tilde{u}(z,s)).$$

But $V(z,s) = E_z f(x_s)$ implies

$$\tilde{u}(z,o) = E_z \int_0^{\tilde{\tau}} e^{-\alpha s}\, f(x_s)\, ds + e^{-\alpha\tilde{\tau}}\, Mu(z,\tilde{\tau}) \tag{3.12}$$

Defining

$$P^{n+1} = P^n \text{ on } \mathcal{F}_{(\tau^n+h)^-}$$

$$E^{n+1}[\varphi\, (x_{\tau^n+h+s}) \mid \mathcal{F}_{(\tau^n+h)^-}] = E_{z^n}\, \varphi(x_s) \tag{3.13}$$

a.s. P^n, on $\{\tau^n < +\infty\}$, as in lemma 2.1, we have, using (3.12) (3.13) in (3.10), if we set $\tau^{n+1} = \tau^n+h+\tilde{\tau}$

$$\tilde{u}(x,y) = E^{n+1}\{\int_0^{\tau^1} e^{-\alpha s} f(x_s)ds + \sum_{i=1}^{n}\int_{\tau^i+h}^{\tau^{i+1}} e^{-\alpha s} f(x_s)ds + \sum_{i=1}^{n} e^{-\alpha\tau^i} k(\xi^i) + e^{-\alpha\tau^{n+1}} M\tilde{u}(z^n,\tilde{\tau})\}.$$

But, if we use

$$M\tilde{u}(z^n,\tilde{\tau}) = k(\xi^{n+1}) + E^{n+1}[e^{-\alpha h}\tilde{u}(z^{n+1},o) \,|\, F_{(\tau^n+h)^-}]\; P^n \text{ a.s., on } \{\tau^n < +\infty\},$$

we obtain (3.10) for n+1.

Now since $\tilde{u}$ is bounded and $\tau^n \uparrow +\infty$ ($\forall\,\omega$), we can go to the limit in (3.10) ($n \uparrow +\infty$) and we obtain

$$\tilde{u}(x,y) = J_{xy}(\hat{v})$$

where $\hat{v} = (\tau^m, \xi^m, m \geq 1)$ is defined by (3.11), and is admissible.

Lemma 3.5. $\tilde{u} = u$

Proof. Clearly, from the lemma 3.4, it is enough to show that $\lim_{n\uparrow\infty} u^n \leq u$.

Let $v' = (\tau^i, \xi^i, i \geq 1)$ be an admissible control. We have

$$u^n(x,y) \leq E^1 \int_0^{\tau^1} e^{-\alpha s} f(x_s)\, ds + e^{-\alpha\tau^1} Mu^{n-1}(x,y+\tau^1)$$

and also

$$Mu^{n-1}(x,y+\tau^1) \leq k(\xi^1) + \chi_{\xi^1=1}\, e^{-\alpha h} E_x\, u^{n-1}(x_{y+\tau^1},0) + \chi_{\xi^1=2}\, e^{-\alpha h}\, u^{n-1}(\bar{x},o).$$

But $E_x\, u^{n-1}(x_{y+\tau^1},0) = E^1\, u^{n-1}(x_{\tau^1},0)$ and is z^1 is defined as previously, we have

$$u^n(x,y) \leq E^1(\int_0^{\tau^1} e^{-\alpha s} f(x_s)ds + e^{-\alpha\tau^1} k(\xi^1) + e^{-\alpha(\tau^1+h)}\, u^{n-1}(z^1,0)).$$

Using the same arguments that those of lemma 3.4 (with inequality taking place of equality) the recurrence gives

$$u^n(x,y) \leq E^n\{\int_0^{\tau^1} e^{-\alpha s} f(x_s)\, ds + \sum_{i=1}^{n-1}\int_{\tau^i+h}^{\tau^{i+1}} e^{-\alpha s} f(x_s)\, ds + \sum_{i=1}^{n} e^{-\alpha\tau^i} k(\xi^i) + e^{-\alpha(\tau^n+h)}\, u^o(z^n,o)\}.$$

This implies ($f \geq 0$ and k_o, $k_n > 0$)

$$u^n(x,y) \leq J_{xy}(v) + E^n\, e^{-\alpha(\tau^n+h)}\, u^o(z^n,o).$$

But, u^o is bounded, and $\tau^n \uparrow +\infty$, therefore

$$\lim_{n\uparrow\infty} u^n(x,y) \leq J_{xy}(v)$$

for any admissible control v.

This completes the proof of theorem 3.1 (i). □

In order to complete the proof of theorem 3.1 (ii), we use the fact that $u = \tilde{u}$ verifies the equation (3.6) to obtain easily that u has the properties (ii). It is enough to show that u is the maximum element of the set of functions satisfying these properties and this is done like in [12].

4. Quasi variational inequality

Using the methods of [4], [12], one can show, for the problem (3.1).

Lemma 4.1 (under the assumptions of section 3.1) u_ε given by (3.2) is the unique solution of the equation

$$(4.1)\qquad -\frac{du_\varepsilon}{dy} + \alpha\, u_\varepsilon + \frac{1}{\varepsilon}(u_\varepsilon - \Psi)^+ = V,\ u_\varepsilon \in C_b^1(R). \quad \square$$

Theorem 4.1 : Under the assumptions of section 3.1. u is solution of the variational inequality

$$(4.2)\qquad -\left(\frac{dw}{dy}, w-u\right) + \alpha(u,w-u) + \gamma\,|w-u|^2 \geq (V, w-u)$$

$$u \leq \Psi,\ \forall\ w \leq \Psi,\ w, \frac{dw}{dy} \in L^{2,\gamma}(R^+)$$

Going back to the impulse control problem, that is u given by (2.15), we have

Theorem 4.2 : Under the assumptions of theorem 3.1 the optimal cost function u (cf. (2.15)) is the maximum solution of the quasi variational inequality

$$(4.3)\qquad \left(-\frac{dw}{dy}, w-u\right) + \alpha(u,w-u) + \gamma|w-u|^2 \geq (V, w-u)$$

$$(4.4)\qquad u \leq Mu$$

$$(4.5)\qquad 0 \leq u \leq \frac{\|f\|}{\alpha},\ u \text{ measurable on } E \times R^+,$$

$$(4.6)\qquad \forall\, w(y) \leq Mu,\ w, \frac{dw}{dy} \in L^{2,\gamma}(R^+).$$

(4.3) to (4.6) are written for fixed x in E (that is : x is only a parameter).

Proof. Using the lemma 4.1, we have that u^n is solution of the following variational inequality

$$(4.7)\qquad \begin{aligned} &-\left(\frac{dw}{dy}, w-u^n\right) + \alpha(u^n,w-u^n) + \gamma|w-u^n|^2 \geq (V,w-u^n) \\ &u^n \leq Mu^{n-1} \\ &\forall\, w \leq Mu^{n-1} \end{aligned}$$

with the regularity indicated in the theorem.
We have also $u^n(x,y) \searrow u(x,y)$ in each point (x,y), in particular $u^n \to u$ in $L^{2,\gamma}(R^+)$ strongly; these two properties are sufficient to go to the limit in (4.7) for any $w \leq Mu \leq Mu^n, \forall\, n$, (recalling that $u \leq u^n\ \forall\, n$), with the regularity (4.6), (4.4) and (4.5) was already known. We can prove that u is the maximum solution by adaptation of [9].

5. The case of markov jump processes

As it can be seen in the previous section, the formulation we obtain for the quasi variational inequality is a very "weak" one. In particular, we did not prove that, in general, the inequalities can be written as in section 1. We prove such a result in a particular case : the markov jump processes.

We assume (for sake of simplicity) that E is a countable set and we consider a markov jump process given by the following infinitesimal generator

$\Lambda \varphi(x) = \lambda(x) \left(\int_E q(x,dy) \varphi(y) - \varphi(x) \right)$, $\forall \varphi$, bounded measurable in E.

with the hypothesis

(5.1) $\quad 0 \le \lambda(x) \le M \quad \forall x \in E.$

As usual, E is endowed with the discrete topology, and the corresponding Borel σ-algebra.
Then, u, being the optimal cost function of the impulsive control problem, is a bounded measurable function. Therefore u(x,o) belongs to the domain of the infinitesimal generator A, denoted by D_A. This implies (cf. Dynkin [6]) that there exits exists a unique solution of the problem

(5.2) $\quad \frac{\partial \Phi}{\partial t} = A\Phi \quad , t > 0 \qquad \Phi(x,o) = u(x,o)$

and this solution is

(5.3) $\quad \Phi(x,t) = E_x\, u(x_t,o).$

Moreover since Φ is bounded on $E \times R^+$,

(5.4) $\quad \left\| \frac{\partial \Phi}{\partial t} \right\| \le$ constant

Theorem 5.1. Under the assumptions of the theorem 3.1 and 5.1 u is the unique solution of the quasi variational inequality ("strong" formulation) :

$$(5.5) \quad \begin{cases} -\frac{\partial u}{\partial y} + \alpha u \le V \\ u \le Mu \\ \left(-\frac{\partial u}{\partial y} + \alpha u - V \right) (u - Mu) = 0 \end{cases}$$

(5.6) $\quad u(x,.) \in C_b^o(R^+), \quad \frac{\partial u}{\partial y}(x,.) \in L^\infty(R^+), \quad$ u bounded on $E \times R^+$

Proof : Itwas shown that u is the unique solution of

$$(5.7) \quad u(x,y) = \operatorname*{Inf}_{\tau} \left(\int_o^\tau e^{-\alpha s} V(x,y+s)\, ds + e^{-\alpha\tau} Mu(x,y,\tau) \right)$$

That is, u is the optimal cost for a stopping time problem, with Mu as the final cost.
Moreover, (5.4) shown that Mu(x,y) is uniformly Lipchitz w.r.t. y and also V has the same property (It is easy to show on (5.7) that this implies the Lipschitz property for u).

Now we introduce the penalized problem associated to (5.7)

$$-\frac{\partial u_\varepsilon}{\partial y} + \alpha u_\varepsilon + \frac{1}{\varepsilon} (u_\varepsilon - Mu)^+ = V$$

for which, the unique bounded solution is

$$(5.8) \quad u_\varepsilon(y) = \operatorname*{Inf}_{v} \int_o^{+\infty} e^{-\alpha s}\, e^{-\frac{1}{\varepsilon}\int_o^s v(\lambda)d\lambda} \left[V + \frac{1}{\varepsilon} v\, Mu\right] (y+s)\, ds$$

where the infimum is taken for the measurable functions from R^+ into [0,1].

Using the Lipschitz property of V and Mu, we show directly in (5.8) that

(5.9) $\quad |u_\varepsilon(x,y) - u_\varepsilon(x,y')| \le c_1 |y-y'|$

with c_1 independant of ε, (and x).

This implies that u_ε has a derivative (a.e) which belongs to a bounded subset of $L^\infty(R^+)$ (this bounded set being independant of x).
Then we can go to the limit in the equation satisfied by u_ε, for every x, in $H^{1,\gamma}(R^+)$. Weakly, ($H^{1,\gamma}$ is the space of functions φ such that $\varphi \in L^{2,\gamma}(R^+)$ and $\frac{d\varphi}{dy} \in L^{2,\gamma}(R^+)$). Then the result is obtained. The uniqueness is proved by adaptation of Laetsch[7]. see also [12].

REFERENCES

[1] R. Anderson and A. Friedman : A quality control problem. To appear.

[2] A. Bensoussan and J.L. Lions : Nouvelles méthodes en contrôle impulsionnel. Applied Math and Optimization n° 4. 1974. pp. 289-312.

[3] A. Bensoussan and J.L. Lions : Temps d'arrêt et contrôle impulsionnel : inéquations variationnelles et quasi variationnelles d'évolution. Cahiers de mathématiques de la Décision, 1975. Paris IX - University n° 7523.

[4] A. Bensoussan and J.L. Lions : Contrôle impulsionnel et Applications. Book to appear. Dunod 1978.

[5] C. Dellacherie and P.A. Meyer : Probabilités et Potentiels, Hermann, Paris 1976

[6] E. Dynkin : Markov processes. Vol. 1. Springer Verlag. 1965.

[7] T. Laetsch : A unicity theorem for elliptic quasi variational inequalities. J. of Functional Analysis, 1975, Vol. 18, n° 3.

[8] J.L. Lions : Quelques méthodes de résolution des problèmes aux limites non linéaires. Dunod, Paris 1969.

[9] F. Mignot and J.P Puel : Solution maximum d'inéquations variationnelles et quasi variationnelles d'évolution. Comptes rendus Acad. Sciences, 1973, tome 280, pp. 259-262.

[10] M. Robin : Impulsive control with Time lag. Joint Automatic Control Conference 1976, La Fayette Indiana.

[11] M. Robin : Sur le contrôle impulsionnel des processus markoviens et semi markoviens, Comptes Rendus Acad. Sciences, tome 282, 1976, pp. 631-634.

[12] M. Robin : Thèse Paris 1977.

[13] D. Rosenfield : Markovian deterioration with uncertain information; to appear.

APPLICATION OF OPEN LOOP CONTROL TO THE DETERMINATION OF OPTIMAL TEMPERATURE PROFILE IN THE CHEMICAL REACTOR

Lesław Socha
Department of Mathematics and Physics,
Silesian Polytechnical Institute
Gliwice, POLAND

Jerzy Skrzypek
Research Centre of Chemical Engineering
and Equipment Design
Polish Academy of Sciences, Gliwice POLAND

Introduction

The determination of the optimal temperature profile in the chemical reactor is very important in the problems of optimization of chemical engineering.

The solutions of those problems in deterministic models were presented in [1],[2],[3],[4]. Since random disturbances act in real physics systems, the stochastic models were applied to studies in chemical engineering in the last years.

In this paper the stochastic open loop control for the determination of optimal temperature profile in the chemical reactor will be applied. It is assumed in the deterministic model that the constants of the speed of the reaction which determines the activity of catalyst in the catalytic processes are invariable in time (in tubular reactors the time of contact is considered).

We assume that in the stochastic model the constants of the speed of the reaction (i.e. of the catalyst activity) are influenced by some random disturbances which in turn are assumed to be the "white noise". We will consider the system of two parallel reactions which are described by stochastic differential equations.

$$d\Gamma_A = -(k_1+k_2)\Gamma_A dt - K(k_1+k_2)\Gamma_A\, d\xi \qquad (1)$$

$$d\Gamma_B = -k_1\Gamma_A\, dt + Kk_1\Gamma_A\, d\xi \qquad (2)$$

where

$$k_1 = k_{10}\exp\left(-\frac{E_1}{R\,T(t)}\right) \qquad k_2 = k_{20}\exp\left(-\frac{E_2}{R\,T(t)}\right) \qquad (3)$$

E_1, E_2, k_{10}, k_{20}, R some positive parameters of the system

K - parameter of noise

$d\xi$ - differential in "Stratonovich sense"

T(t) - temperature (control)

Γ_A, Γ_B, concentration of components A,B

We take the initial conditions for the above system as deterministics

$$\Gamma_A(o) = 1 \quad ; \quad \Gamma_B(o) = 0 \qquad (4)$$

We assume the temperature T(t) satisfies the following inequalities

$$T_{min} \leqslant T(t) \leqslant T_{max} \qquad (5)$$

We would like to find such optimal temperature profile which will maximize the criterion

$$I = E\Gamma_B(1) \qquad (6)$$

E( ) - expectation

Changing the system (1), (2) to Itó system and reducing it to the deterministic case we obtain the problem of the deterministic control

$$\frac{dx}{dt} = -(k_1+k_2)\left[1 - \frac{K^2}{2}(k_1+k_2)\right]x\,,\quad x(o) = 1 \qquad (7)$$

$$\frac{dy}{dt} = k_1\left[1 - \frac{K^2}{2}(k_1+k_2)\right]x\,,\qquad y(o) = 0 \qquad (8)$$

where

$$x = E\Gamma_A \quad , \quad y = E\Gamma_B \qquad (9)$$

Then the criterion will be the following

$$I = \max\ y(1) \qquad (10)$$

$$T_{min} \leqslant T(t) \leqslant T_{max}$$

This problem can be solved by using the Pontrygin Maximum Principle.

Main Results

It is easy to show that the sufficient condition of asymptotic stochastic stability with probability 1 of the system (1), (2) is the following inequality

$$1 - 2\frac{K^2}{2}(k_1+k_2) = 1 - 2\alpha(k_1+k_2) > 0 \qquad (11)$$

We introduce the function H (Hamiltonian)

$$H = -\lambda_A(k_1+k_2)\left[1-\alpha(k_1+k_2)\right] x + \lambda_B k_1 \left[1-\alpha(k_1+k_2)\right] x \qquad (12)$$

where λ_A and λ_B satisfy the following system of adjoint equations:

$$\frac{d\lambda_A}{dt} = \lambda_A(k_1+k_2)\left[1-\alpha(k_1+k_2)\right] - k_1\left[1-\alpha(k_1+k_2)\right], \lambda_A(1)=0 \qquad (13)$$

$$\frac{d\lambda_B}{dt} = 0 \quad , \quad \lambda_B(1) = 1 \qquad (14)$$

hence

$$H = -\lambda_A(k_1+k_2)\left[1-\alpha(k_1+k_2)\right] x + k_1\left[1-\alpha(k_1+k_2)\right] x \qquad (15)$$

In order to determine the optimal temperature profile, we shall show the applicability of several remarks, which considerably simplify the solution of the given problem.

Remark 1.

The Hamiltonian along the optimal path for $0 \leqslant t \leqslant 1$ is positive. It follows from the fact that $H(1) = k_1\left[1-\alpha(k_1+k_2)\right] \cdot x(1) > 0$ and from the fact that the value of Hamiltonian must be constant along the whole profile.

Remark 2

In the moment t=1 the optimal temperature is $T = T_{max}$ if

$$0 \leqslant \alpha < \frac{1}{2\,k_1 + \frac{E_1+E_2}{E_1}\cdot k_2} \qquad (16)$$

which follows from the fact that $H(1) = k_1\left[1- \ (k_1+k_2)\right]\,x(1)$ will be minimal for $T = T_{max}$, when the condition (16) is satisfied.

Remark 3

H may attain a stationary maximum when $E_2 > E_1$ (a stationary point at which $\frac{\partial H}{\partial T} = 0$)

$R\,T^2 \frac{\partial H}{\partial T}$ will be calculated from equation (1) and equated to zero

$$RT^2 \frac{\partial H}{\partial T}\,\frac{1}{x} = -\ \lambda_A(E_1k_1+E_2k_2)\left[1-\alpha(k_1+k_2)\right] + \lambda_A(k_1+k_2)(E_1k_1+$$
$$+\ E_2k_2) + E_1k_1\left[1-\alpha(k_1+k_2)\right] - \alpha k_1(E_1k_1+E_2k_2) = 0 \qquad (17)$$

hence

$$\lambda_A = k_1 \quad \frac{E_1\left[1-\alpha(k_1+k_2)\right] - \alpha(E_1k_1+E_2k_2)}{(E_1k_1+E_2k_2)\left[1-2\alpha(k_1+k_2)\right]} \qquad (18)$$

It can be showed, that the sign of the second derivative $\frac{\partial^2 H}{\partial T^2}$ will depend only on the expressions (E_1-E_2) and $1-2\alpha(k_1+k_2)$.

If $E_2 > E_1$ and $1-2\alpha(k_1+k_2) > 0$ then $\frac{\partial^2 H}{\partial T^2} < 0$ (19)

Remark 4

If $E_2 > E_1$ and α satisfies the condition (19), the optimal temperature profile is a non growing curve.

The equation (17) i.e. $\frac{\partial H}{\partial T} = 0$ is satisfied for every $T_{opt} \in \left[T_{min}, T_{max}\right]$.

Let us divide both sides of that equation by x and then differentiate wiht respect to t

$$-\ \frac{d\lambda_A}{dt}(E_1k_1+E_2k_2)\left[1-2\alpha(k_1+k_2)\right] - \lambda_A\,\frac{dT}{dt}\,\frac{1}{RT^2}\left[(E_1^2k_1 + E_2^2k_2)\right.$$

$$\left[1 - 2\alpha(k_1+k_2)\right] - 2\alpha(E_1k_1 + E_2k_2)^2\Big] + \frac{dT}{dt}\left[\frac{1}{RT^2} E_1^2k_1\left[1 - \alpha(k_1+k_2)\right] - \right.$$

$$\left. - 2E_1k_1\,\alpha(E_1k_1 + E_2k_2) - \alpha k_1\,(E_1^2k_1 + E_2^2k_2)\right] = 0 \qquad (20)$$

$\frac{d\lambda_A}{dt}$ is defined by equation (13) λ_A in turn is defined by equation (18) .

Completing these substitutions the following is obtained:

$$\frac{dT}{dt} = RT^2\,\frac{\left[1 - \alpha(k_1+k_2)\right]^2\left[E_1k_1 + E_2k_2\right]\left[1 - 2\alpha(k_1+k_2)\right]}{E_1E_2\left[1-2\alpha(k_1+k_2)\right]\left[1-\alpha(k_1+k_2)\right] + 2\alpha^2(E_1k_1+E_2k_2)^2} \qquad (21)$$

$\frac{dT}{dt}$ determined by this equation is always positive if the optimal profile is a rising curve. Moreover, it is a function of temperature only and of some constant coefficients.

Now we apply our remarks to constructions of optimal temperature profile.

<u>Construction of the optimal temperature profile when $E_1 < E_2$</u>

The required profile will be more conveniently constructed starting from the end, that is, from t=1 to t=0. In the sense of Remark 2, $T=T_{max}$ for t=1. Moving in the direction of decreasing T, the optimal temperature will continue to be $T=T_{max}$, up to the instant t=t at which the condition of a stationary maximum $\partial H/\partial T=0$ (Equation 17 is satisfied). Equation (17) is equivalent to Equation (18), which will be written as follows for $T=T_{max}$:

$$\lambda_A = k_{1max}\,\frac{E_1\left[1 - \alpha(k_{1max} + k_{2max})\right] - \alpha(E_1k_{1max} + E_2k_{2max})}{(E_1k_{1max} + E_2k_{2max})\left[1 - 2\alpha(k_{1max} + k_{2\,max})\right]} \qquad (22)$$

where

$$k_{1max} = k_{10}\exp(-\frac{E_1}{RT_{max}}),\quad k_{2max} = k_{20}\exp(-\frac{E_2}{RT_{max}}) \qquad (23)$$

In order to determine the coordinate $t = t_1$ "entry into the curve", further defined by Equation (21), we shall first integrate Equation (14) :

$$\frac{d\lambda_A}{dt} = \lambda_A(k_{1max} + k_{2max})\left[1 - \alpha(k_{1max} + k_{2max})\right] - $$
$$- k_{1max}\left[1 - \alpha(k_{1max} + k_{2max})\right] \quad \lambda_A(1) = 0 \tag{24}$$

Equating its solution for $t=t_1$ and (22) we find :

$$t_1 = 1 + \frac{1}{(k_{1max} + k_{2max})\left[1 - \alpha(k_{1max} + k_{2max})\right]}$$
$$\ln \frac{(E_2 - E_1)\, k_{2max}\left[1 - \alpha(k_{1\,max} + k_{2\,max})\right]}{(E_1 k_{1max} + E_2 k_{2max})\left[1 - 2\alpha(k_{1max} + k_{2max})\right]} \tag{25}$$

Starting from point t_1, we enter the curve defined by Equation (21), and the quantity $T(t_1)=T_{max}$ will be the missing initial condition for this differential equation.

However, the equation (21) is a differential equation with separated variables the analitic solution is very complicated. In this case, it is better to integrate by the approximation method, using numerical methods.

It can be shown that in the deterministic case the temperature profile is growing "more abruptly" than in the stochastic case i.e.

$$\left.\frac{dT}{dt}\right|_{det} > \left.\frac{dT}{dt}\right|_{stoch}$$

Integration of equation (21) is carried out until we reach t=0, unless we arrive earlier at the point $t=t_2$ at which $T=T_{min}$. In the second eventuality, we discontinue the calculation, and the constant temperature $T=T_{min}$ will be the optimal temperature until the point t=0 is reached.

Determination of changes in reagent concentrations in the course of the reaction with applied optimal temperature profile

In the most generalized case, the optimal temperature profile consists of three sections: the lower horizontal segment $T=T_{min}$ curve defined by Equation (21), and the value $T=T_{max}$ will be the missing initial condition of the equation.

Component (A)

The calculations are started in the segment encompassing the range $T=T_{min}$. The change in the concentration of component A is described here by the linear differential equation:

$$\frac{dx}{dt} = -(k_{1min}+k_{2min})\left[1-\alpha(k_{1min}+k_{2\,min})\right]x\ ,\ x(0) = 1 \tag{26}$$

where

$$k_{1min} = k_{10}\ \exp\left(-\frac{E_1}{RT_{min}}\right),\quad k_{2min} = k_{20}\ \exp\left(-\frac{E_2}{RT_{min}}\right) \tag{27}$$

The solution of equation (26) is

$$x(t) = \exp\left(-(k_{1min}+k_{2min})\left[1-\alpha(k_{1min}+k_{2min})\right]\ t\right) \tag{28}$$

From equation (28), we calculate the value of the concentration Γ_A at the point $t=t_2$ corresponding to the transition from the isotherm $T=T_{min}$ to the rising section of the temperature profile. Starting from this point, the temperature increases from $T=T_{min}$ to $T=T_{max}$, which is attained at $t=t_1$. In order to determine the change in the concentration Γ_A in the increasing temperature range $T_{min}\leqslant T \leqslant T_{max}$ we can use the formulae (15) and (18) with condition

$$x(T_{min}) = x(t_2) \tag{29}$$

then we obtain

$$x(T) = x(t_2)\ \frac{(E_1k_1 + E_2k_2)\left[1-2\alpha(k_1+k_2)\right]}{k_1k_2\left[1-\alpha(k_1+k_2)\right]^2 M_{min}} \tag{30}$$

where

$$M_{min} = \frac{(E_1k_{1min}+E_2k_{2min})\left[1-2\alpha(k_{1min}+k_{2min})\right]}{k_{1min}k_{2min}\left[1-\alpha(k_{1min}+k_{2min})\right]^2} \tag{31}$$

Equation (30) obtains in the region $t_2 \leqslant t \leqslant t_1$, where t_1 is a point corresponding to the transition from the rising section of the temperature profile to the isotherm $T=T_{max}$. In the end region

$t_1 \leqslant t \leqslant 1$ the temperature is always constant ($T=T_{max}$), and the change in the concentration of component A will be described by the linear differential equation

$$\frac{dx}{dt} = -(k_{1max} + k_{2max})\left[1-\alpha(k_{1max}+k_{2max})\right] x \qquad (32)$$

We calculate the initial condition from (30)

$$x(t_1) = x(T_{max}) \qquad (33)$$

From this equation we calculate the final mean concentration of the first component x(1)

$$x(1) = x(t_1)\exp\left(-(k_{1max}+k_{2max})\left[1-\alpha(k_{1max}+k_{2max})\right](1-t_1)\right) \qquad (34)$$

Component B

The calculations are started in the segment encompassing the range $T=T_{min}$. The change in the concentration of component B is described here by the differential equation

$$\frac{dy}{dt} = k_{1min}\left[1-\alpha(k_{1min} + k_{2min})\right] x, \quad y(0) = 0 \qquad (35)$$

The change in the concentration of component A in this segment is determined by equation (28). Hence we determine the concentration of component B at point $t=t_2$.

$$y(t) = \frac{k_{1min}}{k_{1min}+k_{2min}}\left[1 - \exp -(k_{1min}+k_{2min})\left[1-\alpha(k_{1min}+k_{2min})\right] t\right] \qquad (36)$$

Starting from this point, the temperature increases from $T=T_{min}$ up to $T=T_{max}$, which is attained at $t=t_1$. In order to determine the change in the concentration of component in this region, let us divide the kinetic equation (9) by equation (21) describing the temperature change as a function of t.

$$\frac{1}{E_1E_2}\frac{dy}{dT} = \frac{k_1\left[1-2\alpha(k_1+k_2)\right]\left[1-\alpha(k_1+k_2)\right] + 2\alpha^2(E_1k_1+E_2k_2)^2}{RT^2\left[1-\alpha(k_1+k_2)\right](E_1k_1+E_2k_2)\left[1-2\alpha(k_1+k_2)\right]}\, x \qquad (37)$$

x as a function of temperature T is known and defined by equation (30) hence

$$y(T_{max})=E_1E_2\int_{T_{min}}^{T_{max}} \frac{[1-\alpha(k_1+k_2)][1-2\alpha(k_1+k_2)]+2\alpha^2(E_1k_1+E_2k_2)^2}{k_2\ RT^2\ [1-\alpha(k_1+k_2)]^3\ M_{min}}\,x(t_2)dT \qquad (38)$$

At point $t=t_1$ the temperature $T=T_{max}$. The value of the concentration y at this point and further on in the region $t_1 \leqslant t \leqslant 1$ is calculated using equation (38). The change in the concentration of component B will be described by the differential equation corresponding to the isotherm $T=T_{max}$:

$$\frac{dy}{dt} = k_{1max}\left[1 - (k_{1max}+k_{2max})\right]x, \quad y(t_1) = y(T_{max}) \qquad (39)$$

Inserting the solution of equation (32) instead of x we can calculate the value of concentration of the second component at the moment $t=1$, which is the criterion in our problem.

$$y(1)=y(t_1)+x(t_1)\ \frac{k_{1max}}{k_{1max}+k_{2max}}\left[1 - \exp\left((k_{1max}+k_{2max})\left[1-\alpha(k_{1max} + k_{2max})\right](1-t_1)\right)\right] \qquad (40)$$

<u>The determination of change in the concentration of reagents in stochastic model with superimposed optimal temperature profile for deterministic model</u>

We shall now deal with the problem of application of optimal temperature profile which is determined for the deterministic model, in the stochastic model. It may happen that the designer of optimal temperature profile will not take into consideration disturbances, where as in real system they will appear.

Therefore, it is necessary to analyse such case and to determine the maximal concentration of the second component at the moment $t=1$. This concentration constitutes a criterion for the problem of optimization which has been brought forward.

First component (A)

In the first range $0 \leqslant t \leqslant d_2$ the temperature is constant at the same time d_2 denotes the moment t_2 of input in the range of rise in temperature for the deterministic model. The concentration of the first component is rendered then by formula (28) hence

$$x(d_2) = \exp\left(- (k_{1min}+k_{2min}) \left[1-\alpha(k_{1min}+ k_{2min})\right] d_2 \right) \tag{41}$$

In the second range $d_2 \leqslant t \leqslant d_1$ we determine temperature $T(t)$ from formula (22) assuming that $\alpha=0$, at the same time d_1 as previously d_2 denotes the moment t_1 of input in the range of constant temperature $T=T_{max}$ for deterministic model. We have then

$$\left.\frac{dT}{dt}\right|_{det} = RT^2 \frac{E_1k_1 + E_2k_2}{E_1E_2} \tag{42}$$

Dividing equation (7) by equation (42) we obtain

$$\frac{1}{E_1 E_2} \frac{dx}{dT} = \frac{-(k_1+k_2) \left[1 - \alpha (k_1+ k_2)\right]}{RT^2 (E_1k_1+ E_2k_2)} x \tag{43}$$

with initial condition (41) $x(T_{min}) = x(d_2)$.

Hence

$$x(T_{max})=x(d_2)\exp E_1E_2 \int_{T_{min}}^{T_{max}} \frac{-(k_1+k_2) \left[1 - \alpha(k_1+k_2)\right]}{RT^2 (E_1k_1+ E_2k_2)} dT \tag{44}$$

In the third range $d_1 \leqslant t \leqslant 1$ the concentration of the first component is rendered by formula (32) with initial condition determined in formula (44) $x(d_1) = x(T_{max})$.

From this formula we can determine the final concentration of the first component $x(1)$

$$x(1)=x(d_1)\ \exp\left(-(k_{1max}+k_{2max})\ \left[1-\alpha(k_{1max}+k_{2max})\right]\ (1-d_1)\right) \qquad (45)$$

<u>Second component</u> (B)

As in the case of the first component in the first range $0 \leqslant t \leqslant d_2$ the temperature is constant $T=T_{min}$ and the concentration of the second component is rendered by formula (36). Hence

$$y(d_2) = \frac{k_{1min}}{k_{1min}+k_{2min}}\left[1-\exp\left(-(k_{1min}+k_{2min})\left[1-\alpha(k_{1min}+k_{2min})\right]\ d_2\right)\right] \qquad (46)$$

In the second range $d_2 \leqslant t \leqslant d_1$ as in the case of the first component the range of temperature is rendered by formula (42).

Dividing equation(8)by equation (42) we obtain

$$\frac{1}{E_1\ E_2}\ \frac{dy}{dt} = \frac{k_1\ \left[1-\alpha(k_1+k_2)\right]}{RT^2\ (E_1k_1+E_2k_2)}\ x \qquad (47)$$

with initial condition (46) $y(T_{min}) = y(d_2)$. Hence

$$y(T_{max}) = \int_{T_{min}}^{T_{max}} \frac{E_1E_2}{RT^2}\ \frac{k_1\ \left[1-\alpha(k_1+k_2)\right]}{E_1k_1+E_2k_2}\ x\ (d_2)\cdot$$

$$\exp\left(\int_{T_{min}}^{T} -\frac{E_1E_2}{RT^2}\ \frac{(k_1+k_2)\ \left[1-\alpha(k_1+k_2)\right]}{(E_1k_1+E_2k_2)}\ dT_1\right) dT+y(d_2) \qquad (48)$$

In the third range $d_1 \leqslant T \leqslant 1$ the concentration of the second component is rendered by the solution of equation (32) with initial condition determined in formula (48). $y(d_1) = y(T_{max})$

Taking into consideration the initial condition $x(d_1) = x(T_{max})$ we can determine the maximum concentration of the second component.

$$y(1)=y(d_1)+x(d_1)\ \frac{k_{1max}}{k_{1max}+k_{2max}}\left[1-\exp\left(-(k_{1max}+k_{2max})\left[1-\right.\right.\right.$$

$$\left.\left.\left.-\ (k_{1max}+k_{2max})\right]\ (1-d_1)\right)\right] \qquad (49)$$

We shall try to illustrate the results presented in this paper by the numerical example inserted below.

Example:

We shall find optimal temperature profiles in the system of two parallel reactions of the first order for the following parametres

$k_{10} = 45.79\ e^{10}\ h^{-1}$	$x(0) = 1$
$k_{20} = 0.4579\ e^{20}\ h^{-1}$	$y(0) = 0$
$E_1 = 10\text{kcal/mol}$	$T_{max} = 500^\circ\ K$
$E_2 = 20\ \text{kcal/mol}$	$T_{min} = 400^\circ\ K$
$R = 1.98\ \text{cal/mol deg}$	

The coefficient, being the measure of intensity of noise according to formula (16). must fulfil the relation $0 \leqslant \alpha < \frac{1}{2k_{1max} + 3k_{2max}}$

The optimal temperature profiles determine for various accord- ing to the procedure considered in this paper, have been presented on Figure 1.

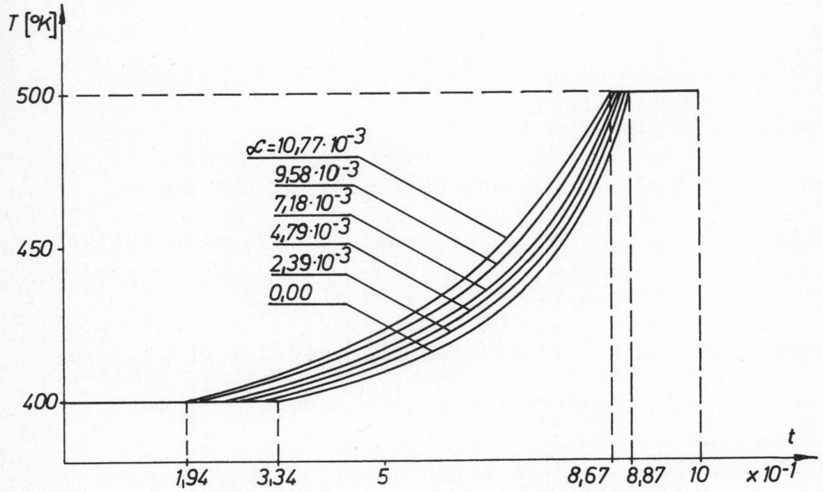

Fig. 1: Optimal temperature profiles T(t) determined for various α.

Dependence of maximum concentration of the first and the second component at the final moment t=1 in the function of intensity of disturbances defined by formulae (34), (40), (45), (49), has been illustrated on Figures 2 and 3.

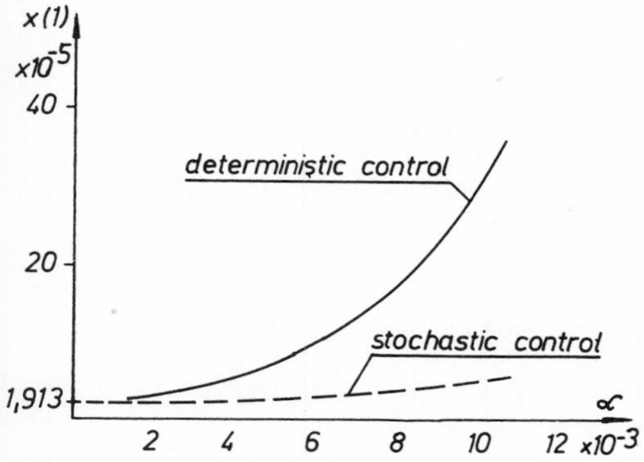

Fig. 2 : Comparison of maximum concentration corresponding to the open loop control calculated for both stochastic and deterministic models and applied to stochastic system - refering to first component.

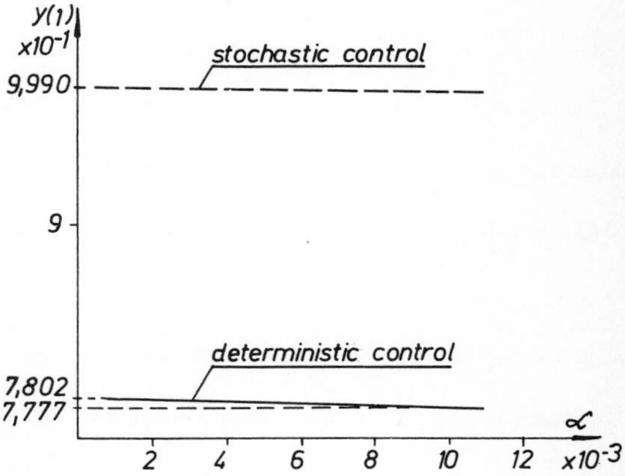

Fig. 3 : Comparison of maximum concentration corresponding to the open loop control calculated for both stochastic and deterministic models and applied to stochastic system - refering to second component.

Maximum concentration with the application of control corresponding to the deterministic model, has been denoted by the broken line. On the other hand maximum concentration with the application of control corresponding to the stochastic model, has been denoted by the continuous line.

General remarks and conclusions

From the theoretical consideration and from the example given, it follows that the increase of noise intensity causes the change of optimal temperature profiles.

They undergo the displacement to the left i.e. the moment t_2 of change-over is shorter and shorter for greater and greater value of the coefficient. At the same time, the rate of the increase of profile $\frac{dT}{dt}$ diminishes with insignificant change of the moment of change-over t_1.

From that one can draw a conclusion that the increase of the action of noise causes the same changes of optimal temperature profiles as the increase of coefficients $\frac{k_{20}}{k_{10}}$ in the case of deterministic control.

According to expectations the increase of the action of noise causes the decrease of the maximum mean value of concentration of the second component. This value constitutes the criterion of optimization. These changes are, however insignificant which permits to justify the application of open control.

Let us point out that in the case of the application of control corresponding with deterministic model, to stochastic system, much greater decrease of maximum concentration of the second component takes place. This points to the necessity of taking into consideration stochastic processes while determining optimal temperature profiles.

It is worth emphasizing here that if one assumed in the description of the system, the stochastic model connected with the differential one in the sense of Ito, the set of stochastic equations (11), (12) after carrying out the operation of calculating average value on it, would assume identical form as in the deterministic model (α =0).

The determination of optimal temperature profiles for systems considered in this paper in the deterministic case, has been presented in [2].

References

[1] A.Burghardt, J. Skrzypek Chem. Eng.Sci., 29, 1311 (1974)

[2] J. Skrzypek, Inżyniera Chemiczna, III, 2, 371 (1973)
International Chemical Engineering 14, 2, 214 (1974)

[3] Horn F., Troltenier U.: Chem. Ing.Techn., 32, 382 (1960)

[4] Horn F., Troltenier U.: Chem. Ing. Techn., 33, 413 (1961)

OUTPUT REGULATION IN PARTIALLY OBSERVABLE LINEAR DISTURBED SYSTEMS

Gy. Sonnevend

Dept. of Numerical Math. and Computer Science

Eötvös L. University, 1445. Budapest 8. Pf. 323.

Introduction.

Let us consider the time invariant /constant/ linear system

/1/ $\frac{dx(t)}{dt} = Fx(t) + Gu(t) + Dv(t), \qquad t > 0 ,$

/2/ $y(t) = Hx(t), \qquad z(t) = Cx(t) ,$

where $x(t) \in R^n$, is the state vector; $u(t) \in R^m$, the vector of the control input, by which we can influence the system; $v(t) \in R^d$, the vector of the unknown disturbance input; $y(t) \in R^s$, the vector corresponding to the value of the measured output; $z(t) \in R^p$, the output vector to be regulated. The numbers n, m, s, d, p and the matrices, $(F, G, D, H, C) = \Sigma$, of corresponding size, are assumed to be known exactly.

The main problem we study here is how to construct the control function u in order to guarantee the asymptotic stabilisation of the output z

/3/ $z(t) \to 0$, for $t \to \infty$, for all unknown x_o and v, where the unknown function $v : [0, \infty) \to R^d$ is assumed to be locally Lebesque integrable. The control function u should be non anticipating, i.e.

/4/ $u(t) = U(t, y(s), \quad 0 \le s \le t , \quad y^k(t) , \quad 0 \le k \le n-1) .$

Variants of this problem are obtained when we require that the values of $u(t)$ cannot depend on the derivatives $y^k(t)$ in this case we say that u is strongly non anticipating; or when we require that the convergence in /3/ could be made "arbitrarily fast" /see below/. The problem /1/-/3/ can also be interpreted as a differential game,

/we refer to the work of Krasovskii and Subbotin [7] for a deep analysis of the related idealisation problems and results, for the case of "bounded" controls, see also [9], (a unique specification of the optimal controls u, v could be given in the case of quadratic constraints leading to a cost functional: $\int_0^\infty (\|Cx(t)\|^2 + \|Pu(t)\|^2 - \|Qv(t)\|^2)dt = \inf_u \sup_v$, see [1], [3]).

We give a constructive necessary and sufficient condition for the solvability of this problem, showing that it has a solution if and only if it has one of the following special form, /"general stabilisator structure"/:

/5/ $u(t) = Ky(t) + K_o \eta(t) + \sum_{k=1}^{n-1} K_k y^k(t)$,

where $\eta(t) \in R^a$, is the state vector of an auxiliary, constant linear dynamic system "driven" by the measured output and its derivatives

/6/ $\frac{d\eta(t)}{dt} = A\eta(t) + \sum_{i=0}^{n-1} B_k y^k(t)$, $\eta(0) = 0$.

Here K, A, B_i, K_i, $i=0,1,\ldots,n-1$, are appropriately chosen constant matrices, which then assure exponential convergence in /3/ i.e. for some $c, \alpha > 0$

/7/ $\|z(t)\| \leq c\|x(0)\| e^{-\alpha t}$, for all $t > 0$, x_o and v.

Similar necessary and sufficient conditions will be given for the solvability of the variants of the problem, e.g. when $B_i, K_i = 0$, $i > 1$, and in (7) we require that the number α can be made arbitrarily large by suitable choice of A, B and K_k, $k \leq n-1$, see Theorems 2,3.

It turns out that the above restriction: $k \leq n-1$, is justified in the sense that the knowledge of $y^k(t)$, for $k \geq n$ does not add any more information about $x(t)$ and $v(t)$.

The system /6/ is a dynamic observer, whose task is to give - together with $y(t)$ - asymptotically, /exponentially fastly/ - those "coordinates" of the state vector, $x(t)$, which can be determined from $y(s)$, $0 \leq s \leq t$. Thus in the controller /5/ the first two terms always correspond to a "feedback", which is strongly non anticipating if $B_k=0$ for $k \geq 1$, while the last one is equal to the value of $S D_i v(t)$, $R(D_i) \subseteq R(D)$, /where S is a constant matrix and $D_i v(t)$ is the result of a disturbance inversion procedure/, it is a "feedforward" term, see [4], [8].

It is useful to investigate the algebraically analoguous discrete time problem, see [1]

/8/ $x(k+1) = Fx(k) + Gu(k) + Dv(k), \quad k=0,1,2,\ldots,$

/9/ $y(k) = Hx(k), \quad z(k) = Cx(k),$

the deep algebraic analogy /arising from the correspondence of the operators $x(k) \to x(k+1)$, $x(t) \to x'(t)$, leads to exactly parallel interpretations and results/, e.g. asymptotic and strongly non anticipating, arbitrary fast output stabilisation is possible in the continuous time system /1/-/2/, if and only if it is possible to guarantee, in the system /8/,/9/, for all $j > 0$, $z(N+j) = 0$, for some fixed N by some non anticipating control $u(k) = U(y(j),\ 0 \leq j \leq k)$, $0 \leq k \leq N$, and then N can be taken to be at most $2n$, see [9].

The output stabilisation problem - in special cases $D=0$ or $H=I$: identity - has been studied extensively, see [1], [3], [10]. The closely related problems of state observation and disturbance inversion will be here also considered. The main tools in our investigations are the notions of /conditioned and controlled/ invariant subspaces. They were introduced - following Kalman's pioneering works - by Basile and Marro and used extensively by Wonham, see [2], [3], and the closely related works by Silverman on the inversion problem.

In order to get the solutions of the above output stabilisation problems we had to develop the results of the works [2], [3] into an algebraically more complete treatment, the main difficulty was to satisfy the constraints arising from the partial observability, /rank $H < n$/, for this reason a precise analysis of the problems of best state reconstruction and observation was needed. In the proposed method there are analogies with the stategy building methods from the theory of differential games proposed by Pontrjagin, /see [8] for the notion of superiority invariant subspace introduced here/, Nikolskii and others.

The author acknowledges the help he received from professor R.E. Kalman.

§.1. Invariant subspaces associated to the system

Here we give refined definitions for the important notions of "controllability" and "observability", see [1], together with algorithms for state determination and disturbance inversion in the partially observed and disturbed system /1/,/2/.

The linear subspaces of R^n form a lattice under addition $+$ and intersection, $\cap$. Let us denote the range, /kernel/ of a matrix $G \in R^{nxm}$, $(H \in R^{sxn})$, by $R(G)$, $(\mathrm{Ker}\ H)$.

The set, V_i of those initial states x_o, which cannot be reconstructed from the output valus $y(0)$, $y(1),\dots,y(i-1)$ is given by the following recursion

/11/ $V_{i+1} = F^{-1}(V_i + R(D)) \cap \mathrm{Ker}\ H, \qquad V(0) = X$.

The set V_i of those initial states x_o which can be controlled in i-steps to the zero state, $x(i) = 0$, conditionally, i.e. inside $\mathrm{Ker}\ C$, (if $D=0$, $H=I$), is given by the similar recursion:

/12/ $\tilde{V}_{i+1} = F^{-1}(\tilde{V}_i + R(G)) \cap \mathrm{Ker}\ C, \qquad \tilde{V}(0) = 0$.

The set W_i of those states, $x(i)$, which can be conditionally reached from the zero state $x_o=0$, $x(j) \in \mathrm{Ker}\ C$, $0 \le j \le (i-1)$, i.e. inside $\mathrm{Ker}\ C$, (if $D=0$ $H=I$), can be computed by the recursion

/13/ $W(i+1) = R(G) + F(W(i) \cap \mathrm{Ker}\ C), \qquad W(0) = 0$.

The set W_i of those states, $x(i)$, which are unobservable from the values of the output $y(0),\dots,y(i-1)$, is given by

/14/ $\tilde{W}(i+1) = R(D) + F(\tilde{W}(i) \cap \mathrm{Ker}\ H), \qquad \tilde{W}(0) = X$.

It is easy to check that we have for all $i \geq 0$:

/15/ $V(i) \supseteq V(i+1)$, $\tilde{W}(i) \supseteq \tilde{W}(i+1)$,
$\tilde{V}(i) \subseteq \tilde{V}(i+1)$, $W(i) \subseteq W(i+1)$.

Because these sets are subspeces, we obtain that, for $k \geq 0$

/16/ $V(n+k) = V(n) = M(F,D,H)$, $\tilde{V}(n+k) = \tilde{V}(n) = Mp(F,G,C)$,

/17/ $W(n+k) = W(n) = m(F,G,C)$, $\tilde{W}(n+k) = \tilde{W}(n) = mp(F,D,H)$.

Thus we have defined four subspaces for every triple of matrices $A \in R^{pxp}$, $B \in R^{pxq}$, $C \in R^{rxp}$. They are invariants of the corresponding linear system, or its transfer function, $C(Is - A)^{-1} B$. Here either (F, G, C):"control part", or (F, D, H): "observation part".

As a consequence of the linearity of the system, and /11/, the value of $H_o x(t)$, where $\text{Ker } H_o = M(F, D, H)$-, i.e. the vector $x(t)$ modulo the last subspace - can be, for all $t \geq 0$ computed as a linear, constant combination of the first $(n-1)$ derivates of the output, and in fact the component of $x(t)$ in $\text{Ker } H_o$ is the largest unreconstructible part of the state at moment t, /from the values $y(s)$, $s \geq t$/. Similarly, for all $t > 0$ the states $x(0)$ could be "reconstructed", at most, modulo $Mp(F, D, H)$ from the knowledge of $x(t)$, and $y(s)$, $0 \leq s \leq t$, in fact from $x(t)$ and $y^k(t)$, $0 \leq k \leq n-1$. Both of the last two subspaces have also an other interpretation: they can be made invariant by appropriately chosen "feedback" control.

We say that a subspace V is $(F, R(G))$, /controlled/ invariant if

/18/ $FV \subseteq V + R(G)$, or equivalently, if $(F + GK)V \subseteq V$

for some matrix $K \in R^{mxn}$.

Proposition 1. The largest $(F, R(G))$ invariant subspace in Ker C is the subspace $M(F, G, C)$, see [2], [3].

We say that a subspace W is $(F, \text{Ker } H)$ /conditioned/ invariant if

/19/ $F(W \cap \mathrm{Ker}\, H) \subseteq W$, or equivalently, if $(F + LH)W \subseteq W$ for some matrix $L \in R^{nxs}$.

<u>Proposition 2.</u> The smallest (F, Ker H) invariant subspace containing R(D) is the subspace m (F, D, H), in /17/.

We introduce the subspaces $M(F, G, C) \cap m(F, C, G) = R(F, G, C)$ and $Mp(F, D, H) \cap mp(F, H, D) = Q(F, H, D)$.

Then as a consequence of the well known theorem on <u>spectral assignability</u>, - /stating that, if

/20/ $R(F,G) = R(G) + FR(G) + \dots + F^{n-1}R(G) = R^n, \quad Q(F,H) = \bigcap_{i=0}^{n-1} \mathrm{Ker}(HF^i)$

then for any real coefficient polinomial, $p(s) = s^n + \dots$, of degree n, there exists a matrix K, (L), such that the characteristic polynomial of (F+GK), ((F+LH)), is just p, see [3]/, - one obtains the following result:

<u>Proposition 3.</u> The spectrum of the matrix K in /18/ for $V=R=R(F,G,C)$ is arbitrarily assignable over R, moreover

/21/ $R(F, G, C) = R(F + GK, M \cap R(G)), \quad M = M(F, G, C)$,

for all K such that $(F + GK)M \subseteq M$. The invariant factors of $\bar{F}$, the restriction of (F + GK) on M/R, are then independent of K, /and identical with the numeratiors of the diagonal elements in the Smith - McMillan canonical form of the transfer function, $T(s) = C(Is-F)^{-1}G$, s=complex number, see [3], [4], [1]/. We do not make use of this last fact, but remark that all the problems westudy here with the "state space representation" of the system, could be also studied by "frequency domain" methods, then the various invariant subspaces appear as polinomials associated with the transfer functions T(s) and $W(s) = H(Is - F)^{-1}D$.

At this point it is useful to point out the <u>duality</u> relations among the introduced subspaces. /Which clearly show the duality between "observation" and "control"./

Let us define the dual system of /1/-/2/ as

/22/ $$\frac{-da(t)}{dt} = F^{*}a(t) + C^{*}b(t) + H^{*}e(t) , \quad t > 0 ,$$

/23/ $$c(t) = D^{*}a(t) , \quad d(t) = G^{*}a(t) ,$$

where the vectors here have the same interpretations as before; /or dropping the minus term in /22/, we could define backward time systems, i.e. the dual of discrete time system /8/-/9/ is defined by

$$a(k-1) = F^{*}a(k) + C^{*}b(k) + H^{*}e(k) , \quad k \geq 0 .$$

Here we assumed that the matrix M^{*} denotes the dual matrix, below we denote the dual subspace of $V \subseteq X = R^{n}$, by V^{*},

$$V^{*} = \{f \mid f \in X^{*}, \quad f(x) = 0 , \quad \text{for all} \quad x \in V\} .$$

The following duality relations hold: if V is $(F, R(G))$ invariant, then V^{*} is $(F^{*}, \text{Ker } G^{*})$ invariant.

$$m(F, H, D)^{*} = M(F^{*},D^{*},H^{*}), \quad Mp(F^{*},D^{*},H^{*}) = mp(F, H, D)^{*} .$$

<u>Proposition 4.</u> The spectrum of the matrix L /in 19/, for $W = Q = Q(F, H, D)$ is arbitrarily assignable outside Q, moreover

/24/ $$Q(F, H, D) = Q(F + LH, m + \text{Ker } H), \quad m = m(F, H, D) ,$$

for all L, such that $(F + LH)m \subseteq m$. The invariant factors of the restriction of $(F + LH)$ on $Q/_m$ are then indepenedent of L.

We say that an $F, R(G)$ invariant subspace V is <u>internally stabilizable</u>, if in /18/ K can be chosen so that the matrix $(F+GK)$ is stable on V.

As a consequence of proposition 3, the subspace $M(F, G, C)$ is internally stabilisable iff $\overline{F}$ is stable. The <u>largest internally stabilisable</u> $(F, R(G))$ <u>invariant subspace</u> in $\text{Ker } C$ is

/25/ $$Ms(F, G, C) = R(F, G, C) + \chi^{-}(F + GK) \cap M(F, G, C)) .$$

We say that an $(F, \text{Ker } H)$ invariant subspace is <u>externally stabilisable</u> if in /19/ the matrix L can be chosen so that $(F+LH)$ is stable on the factor space X/W. From proposition 4. follows, that $Q(F, H, D)$ is externally stabilisable, and that

/26/ $Q(F, H, D) \cap [\mathcal{X}^{+}(F + LH) + m(F, H, D)] = ms(F, H, D)$,

is the smallest externally stabilisable (F, Ker H) invariant subspace containing R(D). Here we used the usual notations for the subspaces $\mathcal{X}^{-}(M)$, $\mathcal{X}^{+}(M)$, of /asymptotically/ stable and unstable nodes of a matrix M.

The value of $x(t)$ can be determined in the system /1/-/2/ from the values of $y(s)$, $s \le t$, i.e. "observed", modulo $Q(F, H, D)$, with the help of a dynamic observer

$$\frac{d\eta(t)}{dt} = F\eta(t) + L(H\eta(t) - y(t)) , \qquad \eta(0) = 0 ,$$

in the sense that if $\text{Ker } H_1 = Q(F, H, D)$, then

/27/ $\| H_1 \eta(t) - H_1 x(t)\| \le c\|x(0)\| e^{-\beta t}$, for $t > 0$,

and here the convergence can be made arbitrarily fast i.e., β to be arbitrarily large. In fact the component of $x(t)$ in $\text{Ker } H \cap Q(F,H,D)$ is the largest unobservable part of $x(t)$.

If we are interested only in some exponential, but not arbitrarily fast convergence, then /27/ can be guaranteed also for all H_2, such that $\text{Ker } H_2 = ms(F, H, D)$, therefore $\text{Ker } H_2 \cap \text{Ker } H$, is the subspace of asymptotically unobservable states in the system /1/-/2/.

It should be noted that output differentiation, which is required for state reconstruction in the continuous time system, /1/-/2/ is, in practice often a very difficult, if not impossible, task,/a not "well set" problem/. Yet if it is allowed, /e.g. when the disturbance function is "impulse free", see /28/, and we can measure $y(t)$ exactly/, then the most information about $x(t)$, included in $y(s)$, $-\infty < s < \infty$, can be obtained by applying first the best /i.e. sharpest/ state reconstruction procedure, /to get $H_0 x(s)$, i.e. $x(s)$ modulo $M(F, D, H)$ from $y^k(s)$, $0 \le k \le n-1$, and secondly by applying the sharpest dynamic observer driven by $H_0 x(s)$, as the newly "observed" output, /we think that the proper word is: "reconstructed"/.

Summarising we have thus the following theorem:

Theorem 1. As a result of the consecutively performed best state reconstruction and observation procedures, the state $x(t)$ will be determined modulo an $(F, R(D))$ invariant subspace, $\mathrm{Ker}\ H^{o}$, $\mathrm{Ker}\ H^{o} = \mathrm{Ker}\ H_{o} \cap Q(F, H_{o}, R(D))$, $\mathrm{Ker}\ H_{o} = M(F, D, H)$.
If only /arbitrarily fast/ dynamic observers are allowed, then $x(t)$ can be determined modulo $\mathrm{Ker}\ H^{1} = \mathrm{Ker}\ H \cap \mathrm{Ker}\ H_{1}$, $\mathrm{Ker}\ H_{1} = Q(F,H,D)$.
The corresponding result holds also for the discrete time system, there we obtain specially, that the values $y(k+s)$, $|s| \geq n$ are unrevelent, /in addition to those for $|s|<n$/, for the computation of $x(k)$.

The disturbance inversion problems are solved through the state determination methods. We remind again that disturbance inversion is an allowed tool in our problem only if the disturbance is impulse free:

$$/28/ \qquad \int_{t}^{t+h} \|v(s)\| ds \leq c_{2}h , \quad \text{for all} \quad t > 0 \quad \text{and} \quad v.$$

for some constant c_2, /because then the map $T_h : v \to \overline{v}$, $\overline{v}(t) = v(t-h)$, continued to $x \to \overline{x} := x(\overline{v})$, will be continuous in the space of trajektories endowed with the uniform norm topology/.

Proposition 5. The maximal "univertible" part of disturbance, $Dv(t)$, in the system /1/, /2/ is given by $\overline{D}_1 v(t)$, where

$$R(\overline{D}_1) = \max\{R(D_1) \mid FV = V + R(D_1),\ R(D_1) \subseteq R(D)\} ,$$

where $V = M(F, D, H)$. The value of $D_i v(t)$, where $D_i = D - \overline{D}_1$, i.e. $Dv(t)$ modulo $\overline{D}_1$, can be computed as a linear combination of $y^k(t)$, $0 \leq k \leq n-1$. maximal uninvertible part of $Dv(t)$, in the

If we know the initial state $x(0) = x_o$, say $x_o = 0$, then the system /1/, /2/ is given by $\overline{D}_1^o v(t)$, where

$$R(\overline{D}_1^o) = \max\{R(D_1^o) \mid FV = V + R(D_1^o) ,\ R(D_1^o) \subseteq R(D)\} ,$$

where $V = R(F, D, H)$. The value of $D_i^o v(t)$, where $D_i^o = D - \overline{D}_1^o$, i.e. $Dv(t)$ modulo $\overline{D_1^o}$, can be obtained from the inverse system

$$\frac{dx^1(t)}{dt} = F_{11}x^1(t) + F_{12}w(t) + D_i^o v(t) \ , \qquad x^1(0) = 0 \ ,$$

/29/ $$\frac{dw(t)}{dt} = F_{21}x^1(t) + F_{22}w(t) \ , \qquad w(0) = 0 \ ,$$

where $w \in M(F, D, H)/R(F, D, H)$, $x^1 \in ({}^X/FR(F, D, H))$.

We see that the inverse system, i.e. F_{22}, is stable iff the subspace $M(F, D, H)$ is internally stable, $F_{22} = \overline{F}$, see Proposition 3

Using only dynamic observers and one differentiation "delay-free" inverses could be computed similarly.

§.2. The solution of the output stabilization problem

The first step in the following solution of the output stabilizing control is the observation, that it is necessary and sufficient to be able to localise the disturbance, $Dv(t)$, into Ker C, by an appropriate control, $u_1(t)$. This implies, under the assumption of complete knowledge of the state, that $R(D)$ should be contained in an $(F, R(G))$ invariant subspace, V, in Ker C, i.e.

/30/ $R(D) \subseteq V \subseteq \text{Ker } C$, $FV \subseteq V + R(G)$, i.e. $(F + GK)V \subseteq V$.

Further it is also both necessary and sufficient that we could stabilise the "disturbance decoupled" system on the factor space ${}^X/_V$ by a control $u_2(t)$, this is possible /in a completely observable system/, if and only if, see [3]

/31/ $V + R(F, G) = X$,

or if only asymptotic, but not arbitrarily fast stabilization is needed, if and only if

/32/ $\mathcal{X}^+(F) \subseteq V + R(F, G)$.

Finally we have to find out to what extent, /Ker $\overline{H}$ = ?/, the state vector, $x(t)$, should be known in order that the above disturbance localisation and stabilization be possible.

If we allow to use the derivatives of the output, i.e. a state reconstruction procedure, see §.2., then $\text{Ker } \bar{H} = \text{Ker } H^{o}$ is an $(F, R(D))$ invariant subspace and therefore we can simple factor out $\text{Ker } H^{o}$, i.e. consider the system /1/, /2/ only on the known state factor space $X/\text{Ker } H^{o}$.

Disturbance localisation is possible if and only if

/33/ $F(V \cap \text{Ker } \bar{H}) \subseteq V$, i.e. $(F + L\bar{H})V \subseteq V$,

i.e. V in /30/ is $(F, \text{Ker } \bar{H})$ invariant, where $x(t)$ is known modulo $\text{Ker } \bar{H}$, /e.g. as the result of a state observation, by some dynamic observer see §.2./. For the solvability of the stabilisation, by $u_2(t)$, $u = u_1 + u_2$, it is necessary and sufficient that the $(F, \text{Ker } \bar{H})$ invariant subspace V be externally stabilisable:

/34/ $\mathcal{X}^{+}(F + L\bar{H}) \subseteq V$ for some L , /specially $\text{Ker } \bar{H} \cap \text{Ker } H \subseteq V$, which means to know $x(t)$, at least, modulo V/, and here, in /34/, again we must require the same kind of stabilization /arbitrarily fast, or not/. The conditions /33/, /34/ and $R(D) \subseteq V$ imply that V must contain the smallest externally stabilisable $(F, \text{Ker } \bar{H})$ invariant subspace containing $R(D)$, and the conditions /30/ imply that V must be contained in the largest $(F, R(G))$ invariant subspace in $\text{Ker } C$, $M(F, G, C)$.

It is easy to show that /30/ and /33/ imply that the matrices K, L can be "factorized" by the same matrix, $\bar{K}$, $K = \overline{KH}$, $L = G\bar{K}$.

We can thus summarize the following solutions of the output stabilisation problems formulated in the introduction.

Theorem 2. The necessary and sufficient conditions for the possibility of arbitrarily fast output regulation, if we allow only strongly non anticipating control, are given by the following two conditions:

/35/ $\text{Ker } H_1 = Q(F, H, D) \subseteq M(F, G, C)$,

/36/ $M(F, G, C) + R(F, G) = X = R^{n}$.

the control $u(t)$ can be chosen as $\bar{K}_{o}y(t) + K_1H_1\eta(t)$, where

$H_1 \eta(t)$ is the "output" of a dynamic observer, (27).

Theorem 3. If we have to guarantee asymptotic stabilization only, i.e. to satisfy /7/ only for some $\alpha > 0$, by a strongly non anticipating control, then the necessary and sufficient conditions are given by

/37/ $\quad \text{Ker } H_2 = ms(F, H, D) \subseteq M(F, G, C)$,

/38/ $\quad \mathcal{X}^+ F \subseteq M(F, G, C) + R(F, G)$,

the control $u(t)$ can be chosen as $\bar{K}_o y(t) + \bar{K}_2 H_2 \xi(t)$, where $H_2 \xi(t)$ is the "output" of a dynamic observer.

A number of methods, /spectrum assignement procedures, see e.g. [3]/, are known for the choices of the stabilising matrices $\bar{K}$, L, see §1. Thus with the finite step algorithms for the computation of the above subspaces, (see §.1), we obtained by theorems 2 and 3 a constructive solution of our problems.

If we can allow output differentiation also, /i.e. state reconstruction, see Theorem 1, then in the above two theorems H could be substituted by H_o, $\text{Ker } H_o = M(F, D, H)$, see /5/,/6/.

It is easy to see that for the solvability of the output stabilisation problem it would be enough to use only those combinations of Gu, $u = (u^1, \ldots, u^m)$, which do not belong to the subspace M(F, G, C). Under this assuption, /i.e., if $R(F, G, C) = 0$, see Proposition 3/, the arising "closed loop system", /in which $v(t) \equiv 0$/, will be stable if and only if, - for both of the above theorems - the subspace M(F, G, C) is internally stable, i.e. the map $\bar{F}$, in Proposition 3, is stable. Thus if we would impose the stability of the closed loop system as a necessary requirement of control design, then in Theorems 2,3 the subspace M(F, G, C) should be substituted by the largest internally stabilisable (F, R(G)) invariant subspace in Ker C, Ms(F, G, C), constructed in /25/.

If disturbance inversion is allowed, i.e. we are able to compute $D_i v(t)$, see §.2., then in the above theorems we could take $(D-D_i)$ instead of D, whenever $R(D_i) \subseteq M(F, G, C) + R(G)$ is valid. To see this we introduce a further type of invariant subspaces. A subspace V is said to be $(F, R(G), R(D_i))$, /superiority/, invariant, if

/39/ $FV + R(D_i) \subseteq V + R(G)$, or equivalently iff

$(F + GK_1)V + R((I + GK_2)D_i) \subseteq V$, for some K_1, K_2 .

There exist an $(F, R(G), R(D_i))$ invariant subspace in $\operatorname{Ker} C$ iff $R(D_i) \subseteq M(F, G, C) + R(G)$, and then the $M(F, G, C)$ is the largest such subspace. Now the partially invertible disturbance, $Dv(t)$ can be localised into $\operatorname{Ker} C$ by a control $u(t)$ - if we know $x(t)$ only mod $\operatorname{Ker} \bar{H}$ - if and only if /39/, together with

$R(D-D_i) \subseteq V$, $F(V \cap \operatorname{Ker} \bar{H}) \subseteq V$,

is valid for $V = M(F, G, C)$. Then we can chose

$u(t) = \bar{K}_1 \bar{H} x(t) + \bar{K}_2 D_i v(t)$, see /6/ .

Again for the stabilisation problem it is enough to know $\bar{H}x(t)$ with an exponentially small error.

References

[1] R.E. Kalman, M.A. Arbib, P.L. Falb: "Topics in Mathematical System Theory", McGraw Hill, New York, 1969.

[2] G. Basile, G. Marro: "A new characterisation of some structural properties of linear systems", Int. J. on Control, /1973/ 17, 931-943.

[3] M.W. Wonham: "Linear Multivariable Control", Lect. Notes in Economics and Math. Systems, vol. 101. Springer, Berlin-Heidelberg - New York, 1974.

[4] E.J. Davison: "The robust Control of a Servomechanism Problem for Linear Time Invariant Multivariable Systems", I.E.E.E. Trans. Automat. Contr. vol. AC-21, No.1. February, 1976. 25-34.

[5] B.P. Molinari: "Extended Controllability and Observability for Linear Systems", I.E.E.E. Trans. Automat. Contr., vol. AC-21, No.1., February, 1976. 136-137.

[6] M.S. Nikolski: "Linear Theory of Observation", /in Russian/, Publications of the Computing Centre of the Academy of Sciences USSR, Nauka, Moscow, 1973.

[7] N.N. Krasovskii, A.I. Subbotin: "Positional Differential Games", /in Russian/, Nauka, Moscow, 1974.

[8] L.S. Pontrjagin: "On linear differential games II.", /in Russian/ Dokl. Akad. Nauk, 175. No.4. /1975/.

[9] G. Sonnevend: "Structural problems in the theory of bounded control", Proceedings of the 9th Symposium on Math. Programming Budapest, 1976, Akadémiai Kiadó - North Holland Publ. Compl., 1978, Budapest.

[10] M.F. Chang, J.B. Rhodes: "Disturbance localisation in linear systems with simultaneous decoupling pole assigument and stabilisation, I.E.E.E. Trans. Aut. Control, AC-20, /1975/, 518-523.

[11] L.M. Silverman: "Discrete Riccati Equations, Alternative Algorithms Asymptotic Properties and System Theory Interpretations" in "Advances in Control and Dynamic Systems: Theory and Applications", vol. 12, Academic Press, 1976.

"Reaction of continuos dynamic systems with complex form under time-space random fields".

Aleksander Waberski
Department of Mathematics and Physics
Silesian Polytechnical Institute
ul. Katowicka 16, Gliwice, 44-100, POLAND

Introduction.

Correlational and spectral analysis methods are widely applicable within the scope of examining reactions of discrete and continuous dynamic systems subjected to random excitations. [2], [5], [6], [7], [8], [9], [10], [11], [12], [13], [15], [18], [19], [20]. They examine interrelations between the probabilistic characteristics /average value/, the correlative function of random excitation processes conveniently referred to as the "input" and random processes describing the state of the system referred to as "output". In the case of continuous dynamic systems the random fields constitute the random processes. It is the aim of this work to present a certain new mathematical method for the determination of probabilistic characteristics of continuous dynamic systems with a complex form subject to non-stationary excitations by random fields. These problems arise in various technical problems e.g. by examining starting vibrations of air and rocket constructions. The demonstration of the methods has been carried out for the case of vibrating plates having a complex form and fixed stiff along the whole edge. The method discussed due to the applications of certain special functions introduced by V.L. Rvatschev [17] called the R-functions enabled solutions of the problem in the form of closed analytical formulas. Applying the R-functions one can obtain the equation of the outline of the edge of the area. This area describes practically free geometry. The characteristic feature of the R-function is that each of them corresponds to a definite logical function, the arguments of the latter are two discrete values 1 and 0. This quality enables to apply the contemporary methods of the algebra of logic to the solution of the boundary problems of mathematical physics in the fields of complex form. The logic function which corresponds to the R-function takes the value 1 when the point under investigation lies within the area or on its edge and the value 0 when it is outside the area. A free area with complex form can be set together by means of the multiplicity operations of addition, multiplication etc. upon the areas with simple form the equations of which are known. In the case of the R-functions the multiplicity operations will correspond then to the logical operations such as the OP /alternative/ operation and the AND /conjunctive/ operation. Thus the logical functions corresponding to the R-functions will posses in itself a certain coded information about the change - over /switch over, commutation/ of the sign in the equations describing the edge of the areas with complex form.

1. The formulation of the problem.

The stochastic boundary problems of mathematical physics, connected with linear continuous dynamic systems, can be written in the following operational form:

$$A u = f(\bar{x}, t, \gamma) \qquad (1)$$

$$B_j u \big|_{\partial\Omega} = 0 \qquad (2)$$

$$D_t^k u \big|_{t=+0} = 0 \qquad (3)$$

in the area $Q = \Omega \times [0, T] \quad (\bar{x} \in \Omega, \; t \in R_+^1)$
where $\Omega \subset R^n$

$\partial\Omega$ - the edge of the area Ω

f/x, t, γ/ - the exciting measurable random fields induced by probabilistic space /Γ, F, P/.

$\gamma \in \Gamma$, γ is the set of elementary sentences, F is the σ - algebra,
P - is the probabilistic measure in the probabilistic space /Γ ,F ,P/.
A, B_j - are the linear differential operators in the equation / 1 / and in the boundary conditions / 2 /.
D_t^k - the multi indicatory symbol of the differentiation of the relative time t.
In the case of the linear, continuous dynamic system which random vibrate the operator A appears most often in the shape of:

$$A u = a_0 \cdot \frac{\partial^2 u}{\partial t^2} + a_1 \frac{\partial u}{\partial t} + L u \qquad (4)$$

where

a_0, a_1 - are certain positive constans.

$$L u = \sum_{|p||q| \leq m}' (-1)^{|p|} D^p (a_{pq}(\bar{x}) D^q u) \qquad (5)$$

$$a_{pq}(\bar{x}) \in C^\infty(\Omega)$$

$$D^q = \frac{\partial^q}{\partial x_1^{q_1} \partial x_2^{q_2} \dots \partial x_n^{q_n}} \quad , \quad q = q_1 + q_2 + \dots q_n$$

The equation of random vibrations of thin linear plates is the special case of the equation /1/. In the case of variable thickness it has the shape of

$$\Delta(D\Delta u) - (1-\nu_c)\cdot\left(\frac{\partial^2 D}{\partial x^2}\cdot\frac{\partial^2 u}{\partial y^2} - 2\cdot\frac{\partial^2 D}{\partial x \partial y}\cdot\frac{\partial^2 u}{\partial x \partial y} + \frac{\partial^2 D}{\partial y^2}\cdot\frac{\partial^2 u}{\partial x^2}\right) + \varrho\cdot h\,\frac{\partial^2 u}{\partial t^2} + 2\cdot\varrho\cdot h\cdot n\cdot\frac{\partial u}{\partial t} = f(x,y,t,\gamma) \qquad (6)$$

and in the case of constant thickness

$$D\Delta\Delta u + \varrho\cdot h\cdot\frac{\partial^2 u}{\partial t^2} + 2\cdot\varrho\cdot h\, n\,\frac{\partial u}{\partial t} = f(x,y,t,\gamma) \qquad (7)$$

at the same time

$$\Delta\Delta = \frac{\partial^4}{\partial x^4} + 2\cdot\frac{\partial^4}{\partial x^2 \partial y^2} + \frac{\partial^4}{\partial y^4}$$ - biharmonic operator

$$\Delta = \frac{\partial^2}{\partial x^2} + \frac{\partial^2}{\partial y^2}$$ - Laplace`s harmonic operator

h - the thickness of the plate

ϱ - the density of the material of the plate

n - the coefficient of damping

E - Young`s module

ν_c - Poisson`s number

$$D = \frac{E\cdot h^3}{12(1-\nu_c^2)}$$ - the cylindrical stiffness of the plate.

In this paper we shall consider the stochastic vibrations of the plates with complex form, which are stiffly fixed on the whole edge. Their boundary conditions are expressed in the form of:

$$u\big|_{\partial\Omega} = 0 \qquad \frac{\partial u}{\partial \nu}\Big|_{\partial\Omega} = 0 \qquad (8)$$

Where ν - normal to the edge $\partial\Omega$

We shall assume the initial conditions as zero, thus

$$u(x,y,0) = \frac{\partial}{\partial t}u(x,y,0) = 0 \qquad (9)$$

2. R-function and boundary value problem of mathematical physics.

The analysis of continuous dynamic systems with complex geometry in the presented method requires the description of the complex geometrical form of these systems by means of an equation. It is possible to solve this problem positively applying the so-called R-function [17] . Let us consider the function $S_2/x/$ defined in the following way:

$$S_2/x/ \quad \underset{=}{df} \quad \begin{cases} 0 & \text{if} \quad x < 0 \\ 1 & \text{if} \quad x \geqslant 0 \end{cases} \qquad /10/$$

Definition 1. /of the R-function/

The function $y = f/x^1 \ldots\ldots\ldots x^n/$ to which corresponds such Boole's function $F/X_1 \ldots\ldots\ldots X_n/$ that the following relation:

$$S_2 \left[f/x^1 \ldots\ldots\ldots x^n/ \right] = F \left[S_2/x^1/, \ldots\ldots\ldots S_2/x^n/ \right] \qquad /11/$$

is fulfilled, is called the R-function.

Theorem 1. [17]

Let $F/X_1 \ldots\ldots\ldots X_n/$ be a certain Boole's function and $f(x^1, \ldots x^n)$ the R-function corresponding to it. If the considered closed area consisting of the elementary areas Ω_i $i = 1,2,\ldots.n$ defined by the continuous function $\varphi_i(x^1 \ldots\ldots\ldots x^n)$ is described by the equation

$$F[S_2(\varphi_1), S_2(\varphi_2), \ldots, S_2(\varphi_n)] = 1 \qquad (12)$$

this area may be defined by the inequality

$$f(\varphi_1, \varphi_2, \ldots \varphi_n) \geqslant 0 \qquad /13/$$

Making use of Theorem 1 the area Ω with practically free /optional/ geometry can by written in the form of an equation. Thus for instance if area Ω is an intersection of two areas Ω_1 and Ω_2 defined by the inequations:

$$\varphi_1 \geqslant 0, \quad \varphi_2 \geqslant 0, \quad \Omega = \Omega_1 \cap \Omega_2$$

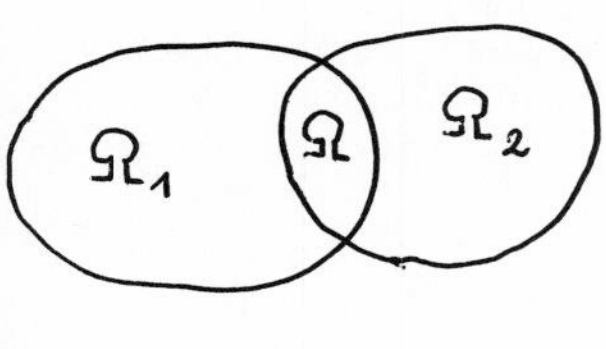

Fig. 1.

this area may be defined by the inequality

$$S_2(\varphi_1) \wedge S_2(\varphi_2) = 1$$

According to theorem 1 the area will be defined by the inequality: $\varphi_1 \wedge^* \varphi_2 \geqslant 0$ where $\wedge^*$ is a symbol of a certain R-function. In case the area Ω is two-dimensional it is convernient to introduce the notion the plan on the plane.

Definition 2.

The point pattern of the intersection of surface Ω and the plane xoy, is called a plan on the plane xoy. It has the equation f/x,y/=0 and intersets with the plane along a certain line $\partial\Omega$ the equation of which has the form f/x,y/ = 0.

The plan corresponds to a certain curve on the plane. The following relations are most often taken as the R-function corresponding to various logical operations:

[17] :

$$X \wedge_\alpha Y = \frac{1}{1+\alpha}\left(X+Y-\sqrt{X^2+Y^2-2\alpha \cdot X \cdot Y}\right) \quad (14)$$

$$X \vee_\alpha Y = \frac{1}{1+\alpha}\left(X+Y+\sqrt{X^2+Y^2-2\alpha XY}\right) \quad (15)$$

$$\sim X = -X \quad , \quad -1 < \alpha \leq 1 \quad (16)$$

Example 1.

A rectangle can be written as the section of

$\Omega_1 = (a^2-x^2) \geqslant 0$ - vertical band

$\Omega_2 = (b^2-y^2) \geqslant 0$ - horizontal band

thus $$\Omega = \Omega_1 \cap \Omega_2 = \Omega_1 \wedge \Omega_2 = 0$$

$$\Omega = (a^2-x^2) \wedge_0 (b^2-y^2)$$

Applying the R-functions for conjuction /AND operaton/ we obtain the equation of the contour of the area Ω in shape of:

$$\omega(x,y) = a^2-x^2+b^2-y^2-\sqrt{(a^2-x^2)^2+(b^2-y^2)^2} = 0$$

The equation of the contour should be normalized to I-st order i.e. fullfil the following conditions

$$\omega\big|_{\partial\Omega} = 0 \quad (17)$$

$$\omega \geqslant 0 \quad \text{where } \bar{x} \in \Omega \cup \partial\Omega \quad (18)$$

$$\frac{\partial\omega}{\partial\nu}\Big|_{\partial\Omega} = 1 \quad (19)$$

The equation of the contour can be normalized making use of the formula

$$\omega_1 \equiv \frac{\omega}{\sqrt{\omega^2 + |\operatorname{grad}\omega|^2}} = 0 \quad (20)$$

Example 2.

The equation of the circle with the radius R and the centre $O_1/a,b/$ which is normalized to the I-st order has the form:

$$\omega(x,y) = \frac{1}{2R}\cdot\left[R^2-(x-a)^2-(y-b)^2\right]=0$$

We shall now introduce the notion of the structure of solution.

Definition 3.

Let φ_1 be the element of the linear space R, φ_0, the known function belonging to the domain D_A of the operator A in equation /1/. Let K be the known operator projecting the space $R \to \Omega\times[0,T]$ The expression $u=K(\varphi_1)+\varphi_0$ is called the structure fullfiling the boundary conditions /2/ if with an optional φ_1 the function u fullfils the boundary conditions /2/. We shall now introduce operators D_K, D_K^{τ} as the extension of differential operators in relation to the normal ν and the tangent τ on the boundary of the area $\partial\Omega$ to the inside of the area Ω. In the case of two variables these operators have the form:

$$D_k f = \left(\frac{\partial\omega}{\partial x}\cdot\frac{\partial}{\partial x}+\frac{\partial\omega}{\partial y}\cdot\frac{\partial}{\partial y}\right)^k f \qquad (21)$$

$$D_k^{\tau} f = \left(\frac{\partial\omega}{\partial x}\cdot\frac{\partial}{\partial x}-\frac{\partial\omega}{\partial y}\cdot\frac{\partial}{\partial y}\right)^k f \qquad (22)$$

If the equation of the contour ω must be normalized to higher orders i.e.

$$\left.\frac{\partial^2\omega}{\partial\nu^2}\right|_{\partial\Omega} = \ldots\ldots = \left.\frac{\partial^k\omega}{\partial\nu^k}\right|_{\partial\Omega}=0 \qquad (23)$$

it is possible to employ the following recurrence formula /when normalized/:

$$\omega_k = \omega_{k-1} - \frac{1}{k!}\cdot\omega_1^{\,k} D_k\cdot\omega_{k-1} \qquad (24)$$

in order to select the structure of the solution in the considered boundary problems, we shall develop the solution into so-called Taylor's series around the contour described by the equation $\omega(\bar{x})$.

Theorem 2. [17]

If the function $\omega(\bar{x})$ is normalized to the order of /n + 1/, than

$$u(\bar{x},t,\gamma) = u_0^*(\bar{x},t,\gamma)+\sum_{k=1}^{n}\frac{1}{k!}\cdot u_k^*(\bar{x},t,\gamma)\cdot\omega^k(\bar{x})+O\left[\rho^{n+1}(\bar{x})\right] \qquad (25)$$

where $u^*(\bar{x},t,\gamma) = u\left[\bar{x}-\omega(\bar{x})\cdot\nabla\omega(\bar{x}),\, t,\gamma\right]$ (26)

is called normalizant of the function $u(x,t,\gamma)$ in relation to the function $\omega(\bar{x})$. . Making use of the above expansion and of the boundary conditions it is easy to educe the structure of the solutions for various boundary problems of mathematical physics. We shall show it in the case a plate fixed stiffly on the whole edge. The boundary conditions for such a plate have the form:

$$u\big|_{\partial\Omega} = 0 \qquad \frac{\partial u}{\partial \nu}\Big|_{\partial\Omega} = 0$$

If u_0^* and u_1^* are normalizants in relations to the contour normalized to I-st order, then

$$u\big|_{\partial\Omega} = u_0\big|_{\partial\Omega}\,, \qquad \frac{\partial u}{\partial \nu}\Big|_{\partial\Omega} = u_1\big|_{\partial\Omega}$$

Since u has to fullfil the boundary conditions i.e. $u_0=0, u_1=0$ and hence $u_0^*=0$ and $u_1^*=0$ so the structure of the solution is expressed by the formula

$$u(x,y,t,\gamma) = \omega^2(x,y)\cdot\frac{1}{2!}\cdot u_2^*(x,y,t,\gamma) = \omega^2(x,y)\cdot\psi(x,y,t,\gamma) \qquad (27)$$

The unknown functions $\psi(x,y,t,\gamma)$ can be presented in the form

$$\psi(x,y,t,\gamma) = \sum_{l+m=0}^{n} c_{lm}(t,\gamma)\cdot\Psi_{lm}(x,y) \qquad (28)$$

where

$$\Psi_{lm}(x,y) = T_l(\alpha x)\cdot T_m(\beta y)$$

T_l, T_m Tschebyshev`s multinomials of I-st type, $\alpha, \beta$ - normalizing coefficients. They depend on the localization of the origin of coordinates x·y and on the dimension of area $\Omega$. The unknown random functions $c_{lm}(t,\gamma)$ are sought by means of Galerkin-Ritz`s method.

3. Method of solution.

Applying Galerkin-Ritz`s method we shall present the solution of the boundary problem /7 - 9/ in the form:

$$u_n(x,y,t,\gamma) = \sum_{l+m=0}^{n} c_{lm}(t,\gamma)\cdot V_{lm}(x,y) \qquad (29)$$

The base functions $V_{lm}(x,y)$ in the case of a stiffly fixed plate have the form:

$$V_{lm}(x,y) = \omega^2(x,y)\cdot\Psi_{lm}(x,y) \qquad (30)$$

where

1. $\displaystyle\sum_{l+m=0}^{n} = \sum_{l=0}^{n}\sum_{m=0}^{n-l}$
2. $\omega/x,y/$ is the equation of the contour of the plate, normalized to the first order, written by means of the R-function.
3. $\Psi_{lm}(x,y) = T_l(\alpha\cdot x)\cdot T_m(\beta\cdot y)$

T_l, T_m are Tschebyshev's multinomials of the first type, while α and β are normalizing coefficients depending on the localization of the origin of coordinates xoy. The random functions $c_{lm}(t,\gamma)$ for a plate of constant thickness are calculated on the basis of the following system of ordinary differential equations:

$$\sum_{l+m=0}^{n}\left\{c_{lm}\iint_{\Omega} D\cdot\left[\Delta V_{lm}\cdot\Delta V_{kj}-(1-\nu_0)\cdot\left(\frac{\partial^2 V_{lm}}{\partial x^2}\cdot\frac{\partial^2 V_{kj}}{\partial y^2}+\frac{\partial^2 V_{kj}}{\partial x^2}\cdot\frac{\partial^2 V_{lm}}{\partial y^2}-\right.\right.\right.$$
$$\left.\left.-2\cdot\frac{\partial^2 V_{lm}}{\partial x\,\partial y}\cdot\frac{\partial^2 V_{kj}}{\partial x\,\partial y}\right)\right]dx\,dy+\left[2\cdot\rho\cdot h\cdot n\cdot\frac{dc_{lm}}{dt}+\rho\cdot h\cdot\frac{d^2c_{lm}}{dt^2}\right]\cdot\iint_{\Omega} V_{lm}\cdot V_{kj}\,dx\,dy=$$
$$=\iint_{\Omega} f(x,y,t,\gamma)\cdot V_{kj}\,dx\,dy \tag{31}$$

This system in the case of a plate stiffly fixed on the edge, resolves itself into the form:

$$\sum_{l+m=0}^{n}\left\{c_{lm}\iint_{\Omega} D\cdot\Delta V_{lm}\cdot\Delta V_{kj}\,dx\,dy+\left[2\cdot\rho\cdot h\cdot n\cdot\frac{dc_{lm}}{dt}+\rho\cdot h\cdot\frac{d^2c_{lm}}{dt^2}\right]\iint_{\Omega} V_{lm}\cdot V_{kj}\,dx\,dy=\right.$$
$$=\iint_{\Omega} f(x,y,t,\gamma)\cdot V_{kj}\,dx\,dy \tag{32}$$

The eigenvalues of the considered boundary problem can be calculated on the basis of the following symmetrical matrix determinant

$$\det\left|\left[V_{lm},V_{kj}\right]-\lambda\cdot\langle V_{lm},V_{kj}\rangle\right|=0 \tag{33}$$

where

$$\left[V_{lm},V_{kj}\right]=\iint_{\Omega}\frac{D}{\rho\cdot h}\left[\Delta V_{lm}\cdot\Delta V_{kj}-(1-\nu_0)\cdot\left(\frac{\partial^2 V_{lm}}{\partial x^2}\cdot\frac{\partial^2 V_{kj}}{\partial y^2}+\frac{\partial^2 V_{kj}}{\partial x^2}\cdot\frac{\partial^2 V_{lm}}{\partial y^2}-\right.\right.$$
$$\left.\left.-2\cdot\frac{\partial^2 V_{lm}}{\partial x\,\partial y}\cdot\frac{\partial^2 V_{kj}}{\partial x\,\partial y}\right)\right]dx\,dy \tag{34}$$

$$\langle V_{lm},V_{kj}\rangle=\iint_{\Omega} V_{lm}\cdot V_{kj}\,dx\,dy \tag{35}$$

and for plates stiffly fixed on the whole edge:

$$\left[V_{lm},V_{kj}\right]=\iint_{\Omega}\frac{D}{\rho\cdot h}\cdot\Delta V_{lm}\cdot\Delta V_{kj}\,dx\,dy \tag{36}$$

Knowing the eigenvalues λ_{lm} it is possible to orthogonalize the base functions $V_{lm}(x,y)$ in accordance with scalar product:

$$\left\{V_{lm},V_{kj}\right\}\overset{df}{=}\left[V_{lm},V_{kj}\right]-\lambda_{lm}\cdot\langle V_{lm},V_{kj}\rangle \tag{37}$$

and to normalize in accordance with the norm $|V_{lm}|=\sqrt{\langle V_{lm},V_{lm}\rangle}$

In the case of multiple eigenvalues λ_{lm} the secondary orthogonalization of the function V_{lm} according to scalar product $\langle\tilde{\cdot}\rangle$ should be carried out. Obtained in this way new base functions $\tilde{V}_{kj}(x,y)$ have the property that the system of ordinary differential equation /31-32/ has the canonic form i.e.

$$\ddot{c}_{kj}+2n\cdot\dot{c}_{kj}+\tilde{\omega}_{kj}^2\cdot c_{kj}=f_{kj}(t,\gamma) \tag{38}$$

where

$$\ddot{c}_{kj}=\frac{d^2c_{kj}}{dt^2}\ ,\quad \dot{c}_{kj}=\frac{dc_{kj}}{dt}\ ,\quad \tilde{\omega}_{kj}^2=\lambda_{kj}=\left[\tilde{V}_{kj},\tilde{V}_{kj}\right]$$

$$f_{kj}(t,\gamma) = \iint\limits_{\Omega} \frac{1}{\varsigma \cdot h} \cdot f(x,y,t,\gamma) \cdot \tilde{V}_{kj}(x,y)\, dx\, dy$$

The functions $\tilde{V}_{kj}(x,y)$ have the form $V_{kj}(x,y)$ after orthogonalization. In the case of weak damping when $\tilde{\omega}_{kj} \gg n$ only this sort of damping will be analyzed the final solution is defined by the formula:

$$u_n(x,y,t,\gamma) = \sum_{l+m=0}^{n} \int_0^t \iint\limits_{\Omega} \frac{e^{-n(t-\tau)} \cdot \sin p_{lm}(t-\tau)}{\varsigma \cdot h \cdot p_{lm}} \cdot f(\xi,\zeta,\tau,\gamma) \cdot \tilde{V}_{lm}(\xi,\zeta) \cdot d\xi\, d\zeta\, d\tau \cdot \tilde{V}_{lm}(x,y) \qquad (39)$$

where $p_{lm} = \sqrt{\tilde{\omega}_{lm}^2 - n^2}$

Green's function or the impulse function of transition for the above solution has the form:

$$G(x,\xi,y,\zeta,t,\tau) = \sum_{l+m=0}^{n} \frac{e^{-n(t-\tau)} \cdot \sin p_{lm}(t-\tau)}{\varsigma \cdot h \cdot p_{lm}} \cdot \tilde{V}_{lm}(\xi,\zeta) \cdot \tilde{V}_{lm}(x,y) \qquad (40)$$

The variance of translocation of a plate is expressed by the formula:

$$\sigma_n^2(x,y,t) = v_n(x,y,t) = \sum_{l+m=0}^{n} \sum_{k+j=0}^{n} \int_0^t \int_0^t \iint\limits_{\Omega} \iint\limits_{\Omega} K_f(\xi_1,\zeta_1,\tau_1,\xi_2,\zeta_2,\tau_2) \cdot \tilde{V}_{lm}(\xi_1,\zeta_1) \cdot \tilde{V}_{kj}(\xi_2,\zeta_2) \cdot G_{lm}(t-\tau_1) \cdot G_{kj}(t-\tau_2)\, d\xi_1\, d\zeta_1\, d\tau_1\, d\xi_2\, d\zeta_2\, d\tau_2 \cdot \tilde{V}_{lm}(x,y) \cdot \tilde{V}_{kj}(x,y) \qquad (41)$$

where

$$G_{lm,kj}(t-\tau) = \frac{e^{-n(t-\tau)} \cdot \sin p_{lm,kj}(t-\tau)}{\varsigma \cdot h \cdot p_{lm,kj}} \qquad (42)$$

The correlation function of the non-stationary random load is expressed by means of the formula:

$$K_f(t_1,t_2) = \sum_{q=1}^{N} \sigma_q^2 \cdot \cos\beta_q(t_1^2 - t_2^2) \qquad (43)$$

Substituting the correlation function expressed by the formula /43/ to the relation /41/ we obtain after enumerating the integrals appearing in it, complicated formulas on the basis of which it is possible to calculate the variance of translocation of the plates.

4. Numerical example.

Numerical calculations were carried out for the stiffly fixed trapezial plate with a circular hole cut out in the middle of it /as is shown of the ilustration 2/.

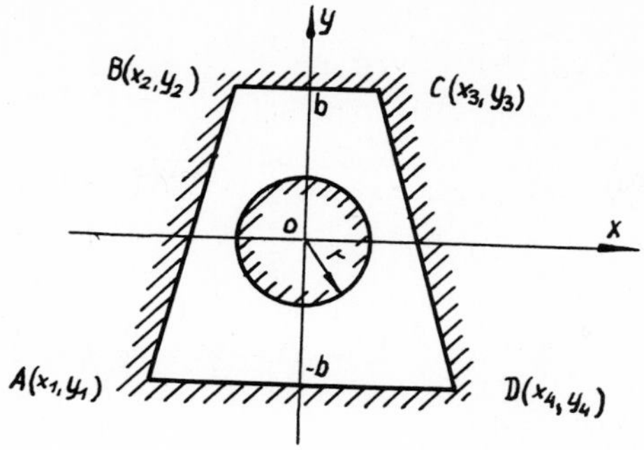

Fig.2

The equation of the contour of the plate written by means of the R-functions has the form:

$$\omega(x,y) = f_1 \wedge_0 f_2 \wedge_0 f_3 \wedge_0 f_4 \tag{44}$$

where

$$f_1 = \frac{-x\cdot(y_2-y_1)+y(x_2-x_1)-y_1\cdot x_2+y_2\cdot x_1}{\sqrt{(y_1-y_2)^2+(x_1-x_2)^2}} \geq 0$$

$$f_2 = \frac{-x\cdot(y_4-y_3)+y(x_4-x_3)-y_3\cdot x_4+x_3\cdot y_4}{\sqrt{(y_3-y_4)^2+(x_3-x_4)^2}} \geq 0$$

$$f_3 = \frac{1}{2b}(b^2-y^2) \geq 0 \qquad f_4 = -\frac{1}{2r}(r^2-x^2-y^2) \geq 0$$

and $\wedge_0$ is the operation of the R-function expressed by the formula

$$X \wedge_0 Y = X + Y - \sqrt{X^2+Y^2}$$

The coordinates of the corners of the plate have been defined

$/x_1 , y_1/ = A/-1; -1/$
$/x_2 , y_2/ = B/-0,5; 1/$
$/x_3 , y_3/ = C/0,5; 1/$
$/x_4 , y_4/ = D/1; -1/$

The following physical and geometrical parameters have been assumed:

$E = 2\cdot 10^{11} \frac{N}{m^2}$

$h = 0,02$ m

$\rho = 7800 \frac{kg}{m^3}$

$n = 5$

$\nu_0 = 0,3$
b = 1 m
r = 0,25 m
The parameters in the correlation functions /43/ have been defined
N = 1
$\delta_1 = 2$
$\beta_1 = 1$
The algorithm of the calculations on a digital computer can be devided into three stages:

1/ The numerical calculation of double integrals appearing in Galerkin-Ritz's system of ordinary differential equations.
2/ The determination of the frequency of free vibrations and the orthogonalization of the base function of the solution.
3/ The calculation of the variances according to the educed analytic formulas.

The numerical calculations have been carried out mainly in order to investigate the space distribution concerning the variances of the translocation of the plates, in order to localize its maximal value. It has also been investigated whether or not this localization is variable in time. Further or not this localization is variable in time. Further investigations of the variation of variances in time have been omitted as less interesting. The reason of this fact is that such an analysis by work /in the qualitative sense/ would not give much new; it would reduce the problem to the investigation of the reaction of the reactions of the harmonic oscillator upon the excitation with non-stationary random processes. The investigations of this type have already been carried out in the works [9], [18], [19]. The variance of the translocation of the plate has been shown in the shape of contour lines for the times $t_1 = 0,01$ s, $t_2 = 0,02$ s, $t_3 = 0,03$ s and $t_4 = 5$ s on the figure 3-6 for the random load with the correlation function /43/. The maximum values of the variance are localized in the middle upper part of the plate and are distributed symmetrically in relation to the vertical axis of coordinate system, in the lower part of the plate. This variance is the starting point for further studies concerning the problems of reliability of working plates and other constructional elements exposed to random loads.

S U M M A R Y

In this paper there has been presented the new mathematical

method of calculation of the probabilistic characteristic of mechanical systems with complex geometry. This method has been demonstrated on the example of random vibrating plates. This method is based on the application of certain special functions called the R-functions. In order to demonstrate this method there have been presented numerical calculations of probabilistic characteristics for concrete plates with complex geometry which have been fixed stiff on the whole edge.

R E F E R E N C E S

1/ A. Bensoussan - "Filtrage optimal des systemes lineaires", Dunod, Paris, 1971.

2/ W.W. Bołotin - "The statistic methods in structural mechanics" /in russ./, J.L.S. Moskwa, 1965.

3/ J. Bothroyd - "Eigenvalues and eigenvectors of the symmetric systems /A - λ B/ • X = 0", Communication of the ACM, vol. 10, N° 3, March 1967, p. 181-182.

4/ P.A. Businger - "Eingenvalues of a real symmetric martix by the QR method" , Communication of the ACM, vol. 8, No 4, April 1965, p. 217-219.

5/ T.K. Caughey - "Response of monlinear string to random loading", J.Appl. Mech. vol. 26, No 3, 1959.

6/ T.K.Caughey, H.J. Strumpf - "Transient response of a dynamic system under random excitation", J.Appl. Mech. No 28, 1961.

7/ S.H. Crandall cd. - "Random vibration", vol. 1, MIT Press Cambridge, Mass 1959.

8/ S.H.Crandall cd. - "Random vibration", vol. 2, MIT Press Cambridge, Mass 1963.

9/ E.Czogała - "Reaction of the continuous dynamical systems, to space - time random fields" /in polish/ 1974, Gliwice Zeszyty Naukowe Pol. Śl. No 29.

10/ M.F.Dimentberg - "Forced vibrations of the plates under load showing time-space random process" /in russ./.

11/ J.Dyer - "Response of plates to a decaying and convecting random pressure field", J.Acoust. Soc. Amer. 35, 7, 1963.

12/ A.C.Eringen, Lafayette ind "Response of beams and plates to random loads", J.Appl.Mech, March 1957, 1.

13/ J.K.Knowles - "On the dynamic response of a beam to a randomly moving load", J.Appl. Mech. ASME, vol. 35, No 1, March 1968.

14/ J.L.Lions, E.Magenes - "Problemes aux limites non homogenes at applications", Dunod, Paris, 1968.

15/ W.A.Palmov - "A thin plate under the influence of random load" /in russ./, P.M.M., No 26, w. 3, 1962.

16/ L.W.Rakova, N.G.Sklepus, E.E.Bezludnyj - "Approximated calculation of double integrals by Rvatschev method" /in russ./, Izviestia A.N.BSSR, ser. FMN, No 1, 1969, p.129-132.

17/ W.L.Rvatschev and others - "The method of R-function in the problem of translocation and vibration of plates with complex form" , Naukowa Dumka, 1973, Kiew.

18/ B.Skalmierski, A.Tylikowski - "Stochastic processes in dynamics" PWN, Warszawa, 1972.

19/ A.Tylikowski - "Non-stationary processes in mechanical systems caused by certain random perturbation", disertation, Gliwice, 1969, /in polish/.

20/ E.Wong - "Stochastic processes in information and dynamical systems", Mc Graw-Hill, 1971.

The contour lines of the variance of the plate`s translocation under the excitation of the non-stationary random field with the correlation function

$$K_f(t_1,t_2)=\sum_{q=1}^{N} \sigma_q^2 \cos\beta_q (t_1^2-t_2^2)$$

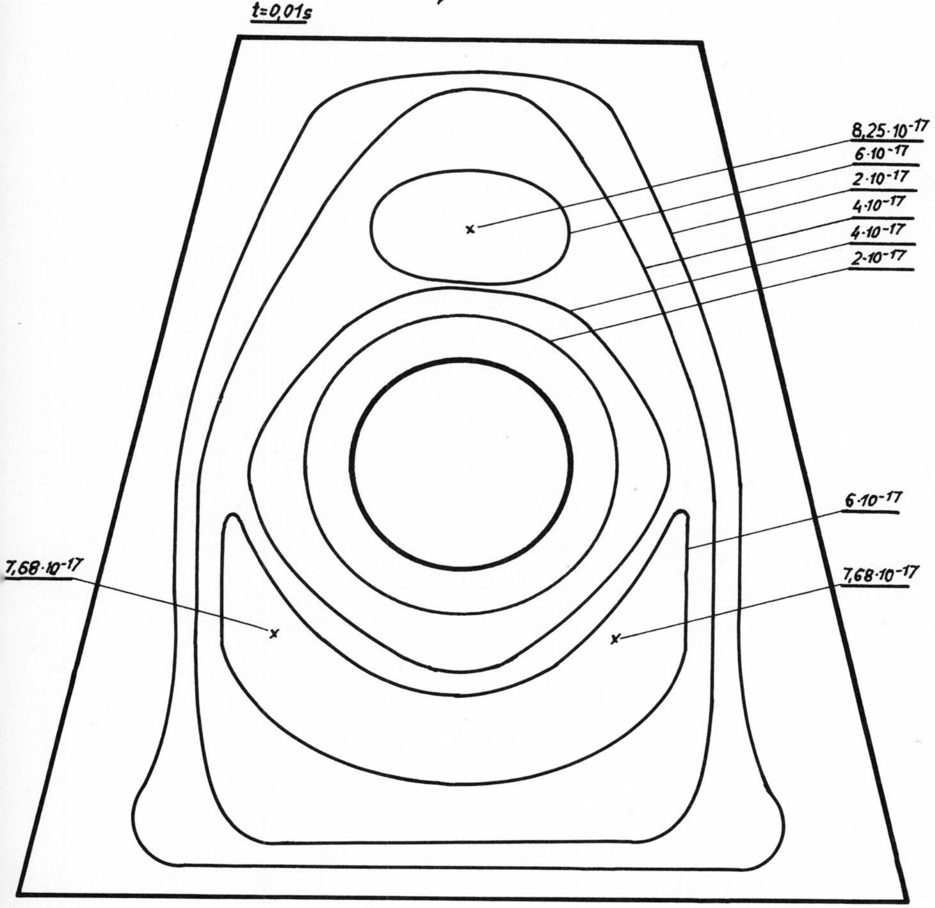

Fig. 3

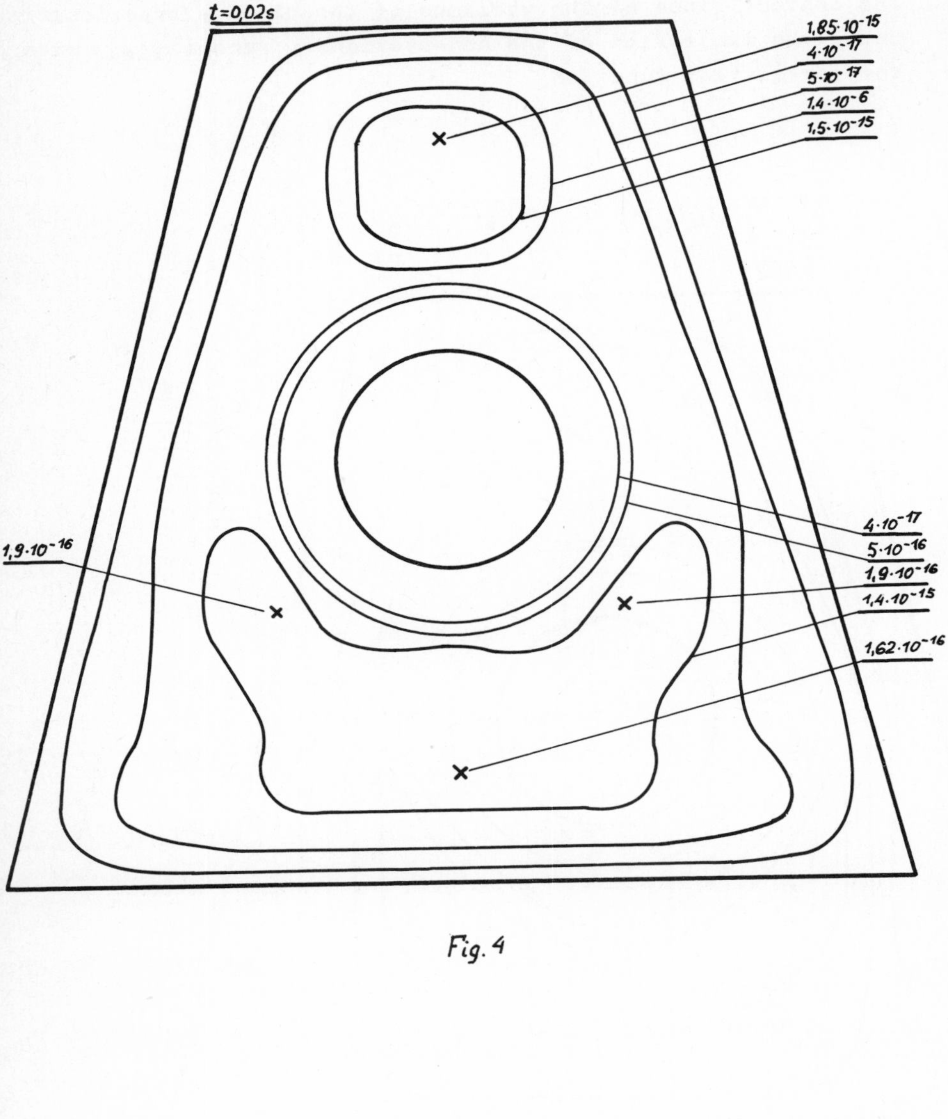
t=0,02s
1,85·10⁻¹⁵
4·10⁻¹⁷
5·10⁻¹⁷
1,4·10⁻⁶
1,5·10⁻¹⁵
4·10⁻¹⁷
5·10⁻¹⁶
1,9·10⁻¹⁶
1,4·10⁻¹⁵
1,62·10⁻¹⁶
1,9·10⁻¹⁶

Fig. 4

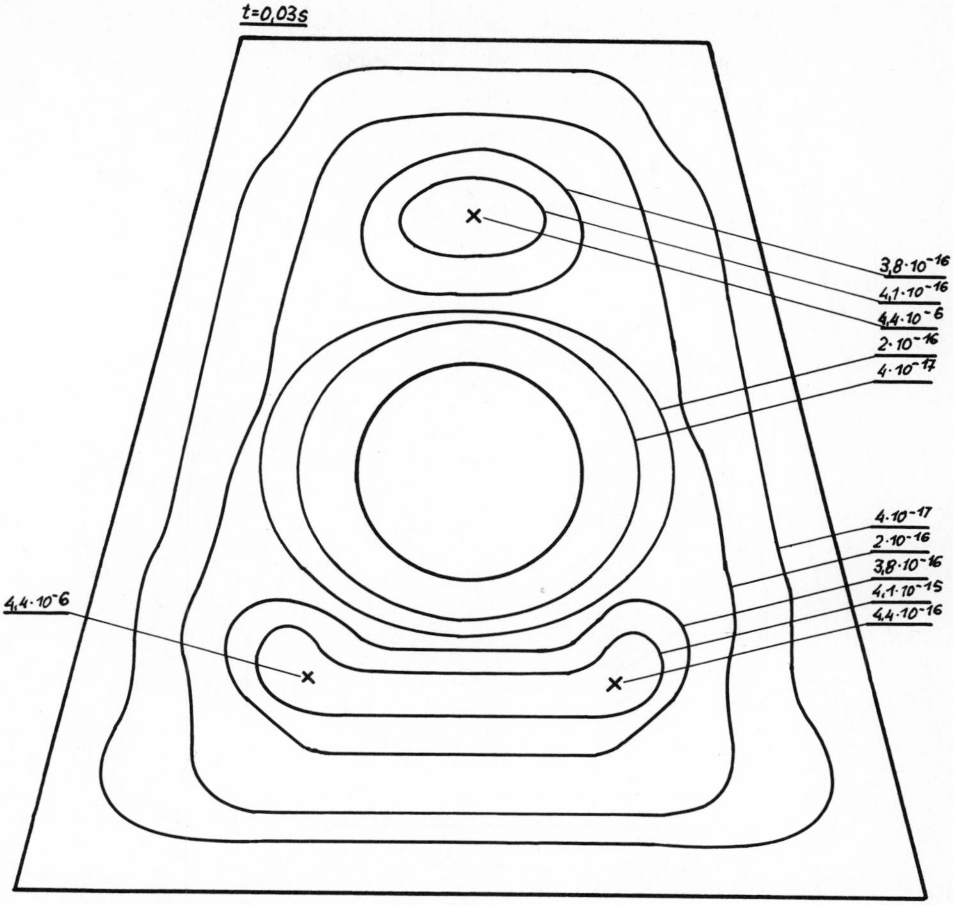
t=0,03s
3,8·10⁻¹⁶
4,1·10⁻¹⁶
4,4·10⁻6
2·10⁻¹⁶
4·10⁻¹⁷
4·10⁻¹⁷
2·10⁻¹⁶
3,8·10⁻¹⁶
4,1·10⁻¹⁵
4,4·10⁻¹⁶
4,4·10⁻6

Fig. 5

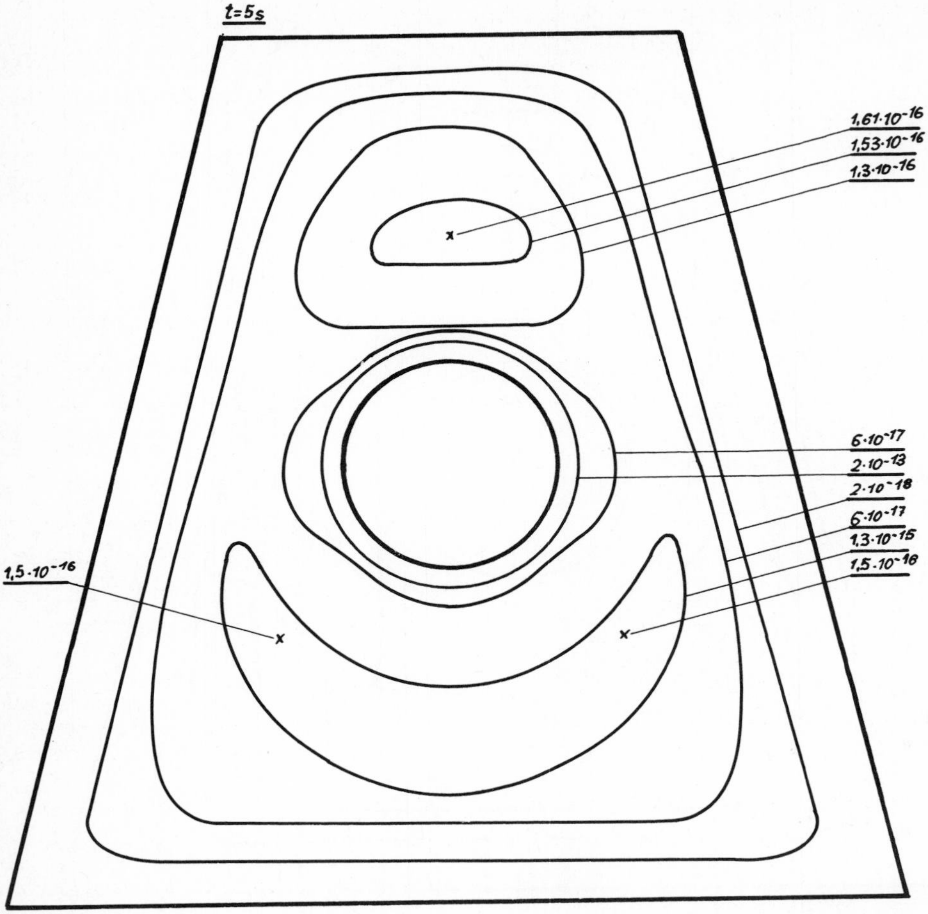
t=5s
1,61·10^-16
1,53·10^-16
1,3·10^-16
6·10^-17
2·10^-13
2·10^-18
6·10^-17
1,3·10^-15
1,5·10^-16
1,5·10^-16

Fig. 6

ON THE CHARACTERIZATION AND THE COMPUTATION OF THE CHARACTERISTIC FUNCTION OF A GAME WITHOUT SIDE PAYMENTS

J.L. Goffin
Faculty of Management
McGill University
1001 Sherbrooke ST. West
Montreal, H3A 1G5 Quebec

A. Haurie
INSEA
B.P. 406
Rabat, Morocco

1. INTRODUCTION

A game in normal form is defined by a set of m players $M = \{1,2,\ldots,m\}$ and one real-valued cost criterion for each of the players: $\psi_j(x_1,\ldots,x_m)$, $j \in M$ where x_k denotes the decision of the kth player.

The solution concepts for such games always involve assumptions about the behaviour of coalitions (or subsets) of players.

The core is a cooperative concept, and is composed of the cost vectors which are such that no coalition can achieve a lower cost for each of its members. Aumann [1] has ascribed a precise meaning to the vague notion of what it is that a coalition can guarantee to its members; this is not trivial which are not part of the coalition might do. A more complete and much clearer description of this problem can be found in Scarf [14]. More formally, a vector payoff is in the core if:

(i) it is Pareto minimum for the set of all players (M).

(ii) no coalition $S \subset M$ can guarantee itself a payoff which is lower for all the members of S and this whatever the decisions of the anticoalition (M/S).

The characteristic functions associates to each coalition a set V(S), called value, which describes the costs that the coalition S can achieve by itself. The somewhat imprecise definition that has been given here is that of α-value, and the α-core, in which it is assumed that a coalition must announce its choice of strategies before its anticoalition.

It can be shown [7,8] that the α-value can be subsumed by the set of Pareto-minima. The α-value is the set of attainable costs for modified costs functions, which can be computed from the costs $\psi_j, j \in M$, but involve a maximum over the strategies of the anticoalition. Thus the characterization and the computation of the α-value for a given coalition reduces to the problem of finding Pareto-minima of a vector-valued function, which is, as a max function, non differentiable.

The exciting works of Scarf [14,15] and Shapley [16] have defined computational procedures which, <u>given</u> the characteristic function will produce a point in the core, under some conditions. On the other hand the study of the optimization of the non-differentiable functions has given rise to a number of usable computational procedures, due to Wolfe [17], Lemaréchal [12], Shor [19], Mifflin [13] and others. The study Non-differentiable Optimization [18] is a good reference on that subject.

It does appear that the technology required to compute a point in the core of a game given in normal form is available; at least conceptually, as the complexity and the amount of computations would be enormous.

We will review some conditions which characterize the α-value for a given coalition. The original part of this paper is that an example of these conditions will be given.

In this introduction, we refrained from giving any notation; the reason for this is that it is rather messy, and as, in what follows, we will deal with the α-value for one given coalition, rather unnecessary.

2. DEFINITION AND NOTATIONS.

Throughout this paper, one will study one coalition only (denoted by S)

Let:

X the set of feasible (joint) strategies for S,

x a (joint) strategy for S,

Y the set of feasible (joint) strategies for M/S

y a (joint) strategy for M/S

$f(x,y) = (f_j(x,y))_{j \in S}$

a vector valued function listing the costs to the players in S, if the coalition plays x and the anticoalition plays y.

The costs for the players not in S will be irrelevant for the definition of the α-value V(S). It is assumed that X and Y are subsets of finite dimensional Euclidian spaces.

To show that the notation given above originates from a game in normal form, we will give three examples.

(i) Pure strategies:

x_j, $j \in M$ is a strategy for player $j \in M$

X_j, $j \in M$ is the feasible set for player $j \in M$

$\psi_j\ (x_1,\ldots,x_m)$ is the cost to player $j \in M$

then

$$x = (x_j)_{j \in M}$$

$$X = \prod_{j \in S} X_j$$

$$y = (x_j)_{j \in M/S}$$

$$Y = \prod_{j \in M/S} X_j$$

$f_j(x,y) = \psi_j(x_1,\ldots,x_m) \quad \forall\, j \in S.$

(ii) Mixed independent strategies (polymatrix games: it is assumed that all X_j are finite sets).

$$x = (p_j(x_j))_{x_j \in X_j, j \in S} \qquad X = \prod_{j \in S} T_j$$

where T_j is the $|X_j| - 1$ simplex, which is the convex hull of the unit vectors in in R^{X_j} ; this simply means that p_j is a probability distribution on X_j.

$$y = (q_j(x_j)_{x_j \in X_j, j \in M/S} \qquad Y = \prod_{j \in M/S} T_j \,.$$

$$f_j(x,y) = \sum_{k \in S} \sum_{\ell \in M/S} \sum_{x_k \in X_k} \sum_{x_\ell \in X_\ell} [\prod_{k \in S} p_k(x_k)\psi_j(x_1,\ldots,x_m) \prod_{\ell \in M/S} q_\ell(x_\ell)]$$

(iii) Mixed correlated stragegies:

$$x = (P(x_j)_{j \in S})_{(x_j)_{j \in S}} \in \prod_{j \in S} X_j \qquad X = T$$

where T is the $|\prod_{j \in S} X_j| - 1$ simplex, which is the convex hull of the unit vectors in $R^{\prod_{j \in S} X_j}$; P is a joint probability distribution on the joint strategies

$$y = (Q(x_j)_{j \in M/S})_{(x_j)_{j \in M/S}} \in \prod_{j \in M/S} X_j \qquad Y = T'$$

where T' is similarly defined.

$$f_j(x,y) = \sum_{(x_k)_{k \in S}} \sum_{(x_\ell)_{\ell \in M/S}} P((x_k)_{k \in S})\psi_j(x_1,\ldots,x_m)\; Q((x_\ell)_{\ell \in M/S})$$

where the sums are taken, respectively, on

$$\prod_{k \in S} X_k \quad \text{and} \quad \prod_{\ell \in M/S} X_\ell \,.$$

One could think of defining 9 cases, by allowing the coalition and anticoalition to play separately according to any of the three concepts. Fortunately, the α-value is the same whether the anticoalition plays (i),(ii),(iii); thus only 3 cases need to be studied.

It can be seen that (i) and (ii) could give rise to the same notation, as the strategy space for the coalition is a Cartesian product of the strategy space for each player in the coalition. Case (iii) would give a different notation as the strategy space is not a Cartesian product (the probability constraint is a joint constraint); this creates quite a few problems where one wants to extend Scarf's results on the nonemptiness of the core. In case (iii) the function $f_j(x,y)$ is linear in x, and linear in y. In case (ii) the function $f_j(x,y)$ is not, in general, a quasi-convex function of x and y.

And thus in neither case (ii) nor (iii) is the core guaranteed to be non-empty (except, I think, in case (iii) when $m \leq 3$).

3. THE MODIFIED COSTS

Definition: The modified cost is the vector function

$$\varphi(x) = (\varphi_j(x))_{j \in S}$$

where

$$\varphi_j(x) = \max_{y \in Y} f_j(x,y).$$

The definition of the α-value given by Aumann [1], reduces to (see [7,8]):

$$V(S) = \{\omega \in R^M : \exists\, x \in X \text{ such that } \varphi_j(x) \leq \omega_j, \forall\, j \in S\}.$$

This definition implicitly assumes that the coalition expects the anticoalition to play a different stratey against each player of the coalition; it also assumes the free accrual (disposal) of costs. The set of Pareto minima of V(S) represents the best that the coalition can do on its own. It contains the complete, and minimal, information needed in order to compute the core. In case (iii), $\varphi_j(x)$ is a convex (piecewise linear function) of x, and thus V(S) is a convex set. When the functions $\varphi_j(x)$ are convex, then a point $x^* \in X$ as Pareto minimum if it minimizes an all positive scalarization of the φ_j, and furthermore, if $x^* \in X$ is a Pareto minimum, then it minimizes a non-negative (not zero) scalarization of the φ_j. In case (ii) or case (i), in general, other conditions for Pareto minima are required.

4. NECESSARY CONDITIONS FOR OPTIMALITY:

Assumptions:

(1) $f_j(x,y)$ is continuous in x and y, and $\nabla_x f_j(x,y)$ is continuous in x and y.

(2) Y is compact

(3) $x \in X \subset R^n$

(4) $X = \{x \in R^n : g_i(x) \geq 0,\ i = 1,\dots,r\}$,

where all g_i are C^1. Also the constraints satisfy a constraint qualification (Kuhn-Tucker).

Various results by Demyanov [5], Bram [2], Danskin [4] give information about $\varphi_j(x)$.

Let

$$Y_j(x) = \{y \in Y : \varphi_j(x) = f_j(x,y)\}$$

$$V_j(x) = \{u \in R^n : u = \nabla_x f_j(x,y), \text{ for some } y \in Y_j(x)\}$$

$$W_j(x) = \mathrm{Conv}(V_j(x)), \text{the convex hull of } V_j(x)$$

then $Y_j(x)$ is compact

φ_j is C^0

$V_j(x)$ is compact (and so is $W_j(x)$)

the directional derivative $\varphi'_j(x;h)$ exists,

and

$$\varphi'_j(x;h) = \max_{v \in V_j(x)} \langle v,h \rangle = \max_{w \in W_j(x)} \langle w,h \rangle$$

also

$$\lim_{h \to 0} \frac{\varphi_j(x+h) - \varphi_j(x) - \varphi'_j(x;h)}{\|h\|} = 0$$

The set $W_j(x)$ is the subdifferential at $h = 0$ of $\varphi'_j(x;h)$ seen as a function of h; it is the generalized gradient of φ_j at x (see Clarke [3]). Under the previously stated assumptions, a Kuhn-Tucker type necessary condition can be shown [7 , 8 , 9 ,10] the proof of which is simply a pasting of proofs for vector valued C^1 functions, and scalar maximum functions.

Theorem:

If $x^* \in X$ is a Pareto minimum for $\varphi(x)$, $x \in X$, then there exists

$w_j \in W_j(x^*)$ $\quad j \in S$

$\alpha_j \geq 0$ $\quad j \in S$ $\quad$ with $\sum_{j \in S} \alpha_j = 1$

$\lambda_i \geq 0$ $\quad i = 1,\ldots,r$

such that

$$\sum_{j \in S} \alpha_j w_j = \sum_{i=1}^{r} \lambda_i \nabla_x g_i(x^*) \tag{4.1}$$

and

$$\sum_{i=1}^{r} \lambda_i g_i(x^*) = 0 \tag{4.2}$$

Remarks:

(i) Sufficiency follows if $\alpha_j > 0$ $\forall j \in S$, all g_i are quasiconcave, and all φ_j are pseudoconvex

(i.e.: $\varphi'_j(x;h) \geq 0 \rightarrow \varphi_j(x+h) \geq \varphi_j(x)$).

(ii) Using Caratheodory;s theorem, (4.1) could be rewritten as:

$$\sum_{j \in S} \sum_{k_j=1}^{n_j} \beta_{jk_j} v_{jk_j} = \sum_{i=1}^{r} \lambda_i \nabla_x g_i(x^*)$$

with

$$\beta_{jk_j} \geq 0 \qquad \sum_{j \in S} \sum_{k_j=1}^{n_j} \beta_{jk_j} = 1$$

$$v_{jk_j} \in V_j(x^*) \quad \forall k_j \quad \forall j \in S \quad ; \qquad n_j \leq n+1 .$$

(iii) A more sophisticated use of Caratheodory's theorem would imply that the total number of non zero β_{jk_j} and λ_i can be taken to be no more than n+1.

(iv) (4.1) can be written as:

$$\sum_{j \in S} \alpha_j \varphi'_j(x^*;h) \geq \sum_{i=1}^{r} \lambda_i \langle \nabla_x g_i(x^*),h \rangle$$

(v) (4.1) and (4.2) can be written as saying that x^* is a fixed point of a point to set map [20].

(vi) The necessary condition simply says that x^* is a stationary point of a scalarization of the φ_j's.

One result mentioned in paragraph 2 will be used in the study of an example.

Theorem: Whether the anticoalition plays pure, mixed independent, or mixed correlated strategies the same function $\varphi(x)$ (and thus the same α-value $V(S)$) ensues.

Proof: For mixed correlated strategies $f_j(x,y)$ is linear in y, and thus attains its maximum at a vertex of Y. Y is a simplex whose vertices represent pure strategies. Thus the two definitions of $\varphi_j(x)$ are identical. Mixed independent strategies are caught in between. Q.E.D.

5. AN EXAMPLE:

This was not a trivial task as to give an example of the usefulness of the necessary conditions, one needed a point such that (i) the function φ is not differentiable, (ii) it is a Pareto minimum and (iii) it is not a global minimum of a scalarization of the φ_j.

We needed a coalition of at least two players to get a non scalar problem; an anticoalition of at least one player to get non-differentiability; mixed independent strategies to avoid convexity. Two pure strategies per player were enough though.

Let $X_1 = \{1,2\}$, $X_2 = \{1,2\}$, $X_3 = \{1,2\}$, and the costs be given by:

$$\begin{bmatrix} (-1,-3) & (-2,-4) \\ (0,2) & (-3,0) \end{bmatrix} \qquad \begin{bmatrix} (-2,-1) & (2,1) \\ (1,0) & (-3,-2) \end{bmatrix}$$

where player 1 plays row, player 2, column, and player 3, matrix; the first (second) number is the cost to the first (second) player. As we will compute $V(\{1,2\})$, the costs of the third player are irrelevant; mixed independent strategies will be played.

Now $x = (X1(1), X1(2), X2(1), X2(2))$

and $x \in X \subset R^4$ if

$$\begin{array}{llll} X1(1) + X1(2) & & = 1 & \lambda_1 \\ & X2(1) + X2(2) & = 1 & \lambda_2 \\ X1(1) & & \geq 0 & \mu_1 \\ X1(2) & & \geq 0 & \mu_2 \\ & X2(1) & \geq 0 & \mu_3 \\ & X2(2) & \geq 0 & \mu_4 \end{array}$$

where X1(2) is the probability that player 1 plays strategy 2. The λ's and μ's represent the Lagrange multipliers.

Let $f'(x,y')$ be the cost function if {3} plays pure strategies: $y' \varepsilon Y' = \{1,2\}$.

Let $f(x,y)$ be the cost function if {3} playes mixed (independent or correlated makes no difference) stragegies:

$$y \varepsilon Y = \{(Y(1),Y(2)): Y(1) + Y(2) = 1,\ Y(1) \geq 0,\ Y(2) \geq 0\}\ .$$

Then

$$\varphi_j(x) = \underset{y' \varepsilon Y'}{\text{Max}}\ f'_j(x,y') = \underset{y \varepsilon Y}{\text{Max}}\ f_j(x,y) \quad \text{for } j = 1,2.$$

A complete plot of $\varphi(X)$ was drawn , using increments of X1 and X2 of .01, (i.e. 10201 points); the data given earlier is the result of quite a bit if trial and error. See Figure (1).

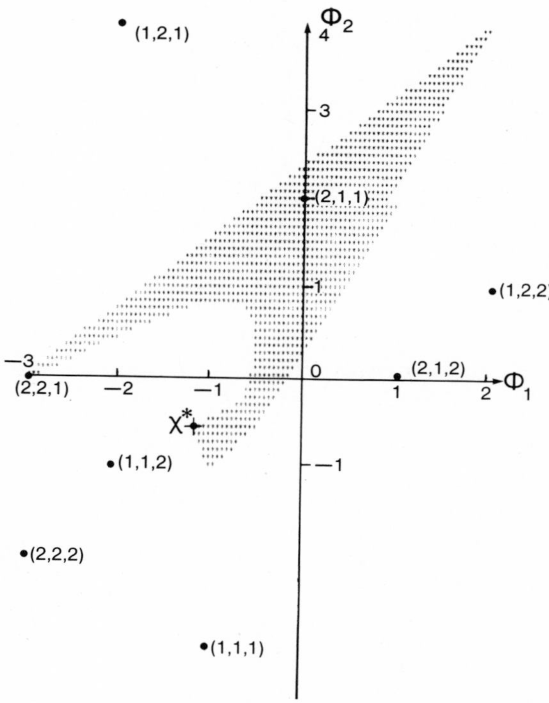

The point $\varphi_1 = -1.2$, $\varphi_2 = -.6$ is a Pareto minimum. It can be checked that it is achieved for

$$x^* = (X1(1) = 1,\ X1(2) = 0,\ X2(1) = .8,\ X2(2) = .2)$$

and that

$$Y'_1(x^*) = \{1,2\} \qquad Y'_2(x^*) = \{1\}$$

if {3} plays pure strategies and

$$Y_1(x^*) = \text{Conv}\{(0,1),(1,0)\} \quad Y_2(x^*) = \{(0,1)\}$$

if {3} plays mixed strategies.

First it should be clear that $\mu_1 = \mu_3 = \mu_4 = 0$ by (4.2) and also that λ_1 and λ_2 are not restricted in sign.

We will write the necessary conditions for the case where {3} plays pure. Details will be given only for the derivative with respect to X1(1); the first component of the set $V'_1(x^*)$ is

$$\{-1\ X2(1) - 2\ X2(2),\ -2\ X2(1) + 2\ X2(2)\}$$

and for the set $V'_2(x^*)$ it is:

$$\{-3\ X2(1) - 4\ X2(2)\}.$$

Note that if {3} played mixed then the first component of the set $V_1(x^*)$ would be:

$$\{(-1\ X2(1) - 2\ X2(2))\ Y(1) + (-2\ X2(1) + 2\ X2(2))\ Y(2): \quad Y(1) + Y(2) = 1,\ Y(1) \geq 0,\ Y(2) \geq 0\}\ .$$

So when one replaces x^* by its value, one gets for the necessary conditions (we use the version of the theorem given in remark (11)):

$$\begin{aligned}
-1.2\ \beta_{11} - 1.2\ \beta_{12} - .6\ \beta_2 &= \lambda_1 \\
-\ .6\ \beta_{11} + \ .2\ \beta_{12} - .4\ \beta_2 &= \lambda_1 + \mu_2 \\
-\ \beta_{11} - 2\ \beta_{12} - \beta_2 &= \lambda_2 \\
\beta_{11} + \beta_{12} + \beta_2 &= 1 \\
\beta_{11}, \beta_{12}, \beta_2, \mu_2 &\geq 0\ .
\end{aligned}$$

The general solution of that linear system is:

$$\begin{aligned}
&\beta_{11} &\qquad &\beta_2 = 2 - \tfrac{5}{2}\beta_{11} \\
&\beta_{12} = -1 + \tfrac{3}{2}\beta_{11} &\qquad &\mu_2 = -1 + .6\ \beta_{11} \\
&\lambda_2 = -\tfrac{3}{2}\beta_{11} &\qquad &\lambda_1 = -\tfrac{3}{2}\beta_{11}
\end{aligned}$$

where

$$\frac{2}{2} \leq \beta_{11} \leq \frac{4}{5}\ .$$

Note $\alpha_1 = \beta_{11} + \beta_{12}\ \varepsilon\ [\frac{2}{3}, 1]$; $\alpha_2 = 1 - \alpha_1$, are weights which give a local minimum of $\alpha_1\varphi_1(x) + \alpha_2\varphi_2(x)$.

The system is a linear system (given x^*) with a finite number of variables (it is essentially an infinite system in the case where {3} plays mixed).

Caratheodery's theorem would say that a solution exists for which no more than 5(n=4) β, λ and μ are non-zero: if $\beta_{11} = \frac{4}{5}$ then $\beta_2 = 0$, and that condition is satisfied.

This research has been supported by the NRC of Canada, the DGES of Quebec, and by a Faculty Research leave of the Faculty of Management of McGill.

ACKNOWLEDGEMENT

The computer program used to draw Figure 1 has been written by Prasad Padmanabhan.

REFERENCES

[1] Aumann, R.J., The Core of a Cooperative Game without side Payments, Trans. Amer. Math. Soc. 98 (1961), 539-552.

[2] Bram, J., The Lagrange Multiplier Theorem for Max-Min with Several Constraints, J. SIAM App. Math. Vol. 14, (1966), 665-667.

[3] Clarke, F.H., Generalized Gradients and Applications, Trans. Amer. Math. Soc., 205 (1975), 247-262.

[4] Danskin, J.M., The Theory of Max-Min, Springer Verlag (New York) 1967.

[5] Demyanov, V. Malozemov, V., On the Theory of Non-linear Minimax Problems Russian Mathematical Surveys, Vol. 26, No.3, May-June, 1971.

[6] Goffin, J.L. Conditions for Pareto Optimality in a Multicriterion Systems with Non-differentiable Functions; Proceedings of the eleventh annual Allerton Conference on Circuit and System Theory, 1973.

[7] Goffin, J.L., Haurie, A. Necessary Conditions and Sufficient Conditions for Pareto Optimality in a Multicriterion Perturbed System; in 5th Conference on Optimization Techniques, edited by R. Conti and A. Ruberti, Springer Verlag, 1973.

[8] Goffin, J.L. Haurie, A., Pareto Optimality with Non-differentiable Cost Functions, with Alain Haurie, in Multiple Criteria Decision-Making, Proceedings of a Conference on Multiple Criteria Decision-making, Jouy-enJosas, France 1975, edited by H. Thiriez and S. Zionts, Lecture notes in Economics and Mathematical Systems (130), Springer Verlag (1976).

[9] Haurie, A. On Pareto Optimal Decisions for a Coalition of a Subset of Players, IEEE Trans. on Automatic Control, April 1973.

[10] Hurwicz, L. Programming in Linear Spaces, in Studies in Linear and Non-Linear Programming, by Arrow, Hurwicz and Uzawa, Stanford University Press, 1958.

[11] Kuhn, H.W., and Tucker, A.W., Non-linear Programming, 2nd Berkeley Symposium of Mathematical Statistics and Probability University of California Press, Berkeley, 1951.

[12] Lemarechal, C., Note on an Extension of "Davidon" Methods to Non-differentiable Functions, Mathematical Programming, Vol. 7, (1974), No. 3, North-Holland, Amsterdam.

[13] Mifflin, R., An Algorithm for Constrained Optimization with Semismooth Functions, Math. of O.R. (1977)

[14] Scarf, H. E., The Core of an N Person Game, Econometrica, 35, No. 1, 1967, 50-69.

[15] Scarf, H.E., On the Existence of a Cooperative Solution for a General Class of N-Person Games, Journal of Economic Theory, 3, 169-181 (1971)

[16] Shapley, L.S., On Balanced Games without side Payments, Mathematical Programming, ed. by T.C. Hu and S.M. Robinson, Academic Press, New York, 1973.

[17] Wolfe, P. Note on a Method of Conjugate Subgradients form Minimizing Non-differentiable Functions, Mathematical Programming, Vol.7, (1974), No. 3, North-Holland, Amsterdam.

[18] Non-differentiable Optimization, Mathematical Programming , Study 3, edited by M.L. Balinski and P. Wolfe (1975).

[19] Shor, N.Z., A Class of Almost-differentiable Functions and a Minimization Method for Functions of this Class, Cybernetics, July (1974), 599-606; Kibernetika, 4 (1972), 65-70.

[20] Goffin, J.L. Haurie, A. On the Computation of the Value of a Game without side Payments, IEEE. Proceedings in Decision and Control, (Dec. 1976).

Evasion in the Plane[†]

by

G. Leitmann and H. S. Liu

University of California
Berkeley, California

Abstract

We consider dynamical systems subject to control by two agents, one of whom desires that no trajectory of the system, emanating from outside a given set, intersects that set no matter what the admissible actions of the other agent. Constructive conditions sufficient to yield a feedback control for the agent seeking avoidance were given earlier. These are employed here to deduce an evader control for the planar pursuit-evasion problem with bounded normal accelerations.

[†]Based on research supported by NASA, Ames Research Center.

1. Introduction

A problem of collision avoidance arises whenever two, or more, objects move in space. Here we consider the case of two objects moving in the same plane, e.g., two ships. One object (evader) is capable of determining the relative position and velocity of the other object (pursuer). The pursuer may be active or passive; that is, he may desire collision, or he may be unable to measure the evader's relative position and velocity and thereby cause collision through inadvertence. Each controls his motion by means of his normal acceleration whose values are constrained. The evader desires to maneuver so as to avoid collision no matter what the actions of the pursuer.

The problem outlined above belongs to the following class of problems. There is given a dynamical system subject to control by two agents, one of whom desires that no trajectory of the system, emanating from outside a prescribed set, intersects that set no matter what the admissible actions of the other agent. Such problems have been discussed in Refs. 1-9, among others. There the treatment is within the framework of differential games, Refs. 6 and 7, either as games of kind (qualitative games) or games of degree (quantitative games). In the former approach, the players seek a saddlepoint for time of collision or for miss-distance, Refs. 1-3 and 8, 9, and in the latter, barriers are sought which separate regions in which collision can be brought about from regions in which avoidance can be assured, Refs. 4 and 5. These techniques usually require numerical integration. Furthermore, only *necessary* conditions are employed so that avoidance cannot be assured. In Ref. 10 we propose an alternative approach, namely, the *constructive* utilization of conditions *sufficient* to guarantee avoidance. Before discussing the planar avoidance problem in Sec. 3, we state the general avoidance problem and the results of Ref. 10. The more general case in which each player has his own target on which he desires termination is treated in Ref. 11.

2. General Problem Statement and Results

Let

$$p^i(\cdot) : R^n \times R \to \text{the nonempty subsets of } R^{d_i}, \quad i = 1, 2$$

be feedback controls (strategies) belonging to given classes of possibly set-valued functions, $\mathcal{U}_i$, with control values u^i ranging in prescribed sets, U_i (which may depend on state and time); that is, given $(x, t) \in R^n \times R$

$$u^i \in p^i(x, t) \subseteq U_i \subseteq R^{d_i} \quad , \quad i = 1, 2 \quad .$$

Let

$$f(\cdot) : R^n \times R \times R^{d_1} \times R^{d_2} \to R^n$$

be a prescribed function, and for given $p^i(\cdot) \in \mathcal{U}_i$, $i = 1, 2,$ define a set-valued function $F(\cdot)$ by

$$F(x, t) \overset{\Delta}{=} \{z \in R^n \mid z = f(x, t, u^1, u^2),\ u^i \in p^i(x, t)\}$$

$$= f(x, t, p^1(x, t), p^2(x, t)) \quad .$$

Then a dynamical system, e.g. Refs. 12 and 13, is defined by the relation

$$\dot{x} \in F(x, t) \quad . \tag{1}$$

Given $(x_0, t_0) \in \Delta \times R$, where Δ is an open set (or the closure of an open set) in R^n , solutions of (1) are absolutely continuous functions on intervals of R

$$x(\cdot) : [t_0, t_1] \to R^n \quad , \quad x(t_0) = x_0$$

such that

$$\dot{x}(t) \in f(x(t), t, p^1(x(t), t), p^2(x(t), t)) \tag{2}$$

a.e. $[t_0, t_1]$.

Now let there be given an *anti-target*, T , in Δ , that is a given set into which no solution of (1) must enter for some $p^1(\cdot) \in \mathcal{U}_1$ and all $p^2(\cdot) \in \mathcal{U}_2$. Consider a closed subset, A , of Δ such that $A \supset T$ and consider also the closure, Δ_ε , of an open subset of Δ such that

$$\Delta_\varepsilon \supset A \quad \text{and} \quad \partial\Delta_\varepsilon \cap \partial A \cap \text{int}\,\Delta = \emptyset \quad .$$

We call A the *avoidance set* and

$$\Delta_A \overset{\Delta}{=} \Delta_\varepsilon \setminus A$$

the *safety zone*. If a solution avoids A then it cannot enter T , and if a strategy $p^1(\cdot)$ is used in Δ_A that guarantees avoidance of A for all $p^2(\cdot)$, then a solution originating outside of A cannot reach A .

For given $p^1(\cdot) \in \mathcal{U}_1$, let K denote the set of all trajectories of (1) for all $(x_0, t_0) \in \Delta_A \times R$ and all $p^2(\cdot) \in \mathcal{U}_2$. Then, given system (1) and sets $\mathcal{U}_1$ and $\mathcal{U}_2$, a prescribed set A is *avoidable* if there is a $p^1(\cdot) \in \mathcal{U}_1$ and $\Delta_A \times R \neq \emptyset$ such that

$$K \cap A = \emptyset \quad . \tag{3}$$

Note that (3) implies global avoidance, that is, avoidance for all $(x_0, t_0) \in (\Delta \setminus A) \times R$. Avoidance set A may be any set containing anti-target T ; often it is different from T .

The following theorem and corollary are proved in Ref. 10.

Theorem A given set A is avoidable if there exist a nonempty set Δ_A and two functions, a strategy $p^1(\cdot) \in \mathcal{U}_1$ and a C^1 function $V(\cdot) : S \to R$, $S(\text{open}) \supset \overline{\Delta_A \times R}$, such that for all $(x, t) \in \Delta_A \times R$

(i) $V(x, t) > V(x', t') \quad \forall\, x' \in \partial A \,, \quad \forall\, t' \geq t \,,$

and $\forall\, u^1 \in \tilde{p}^1(x, t)$

(ii) $\dfrac{\partial V(x, t)}{\partial t} + \nabla_x V(x, t) \;\; f(x, t, u^1, u^2) \geq 0 \quad \forall\, u^2 \in U_2 \,,$

where $\tilde{p}^1(\cdot)$ is the restriction of $p^1(\cdot)$ to $\Delta_A \times R$ ∎

Let

$$H(x, t, u^1, u^2) \triangleq \frac{\partial V(x, t)}{\partial t} + \nabla_x V(x, t) f(x, t, u^1, u^2) \;.$$

Then the Theorem has a

<u>Corollary</u> Given $(x, t) \in \Delta_A \times R$, if there is a $(\tilde{u}^1, \tilde{u}^2) \in U_1 \times U_2$ such that

$$\text{(i)}\; H(x, t, \tilde{u}^1, \tilde{u}^2) = \max_{u^1 \in U_1} \; \min_{u^2 \in U_2} \; H(x, t, u^1, u^2)$$

and

$$\text{(ii)}\; H(x, t, \tilde{u}^1, \tilde{u}^2) \geq 0$$

then condition (ii) of the Theorem is met. Furthermore, $\tilde{u}^1 \in \tilde{p}^1(x, t)$, provided the resulting $p^1(\cdot) \in \mathcal{U}_1$ ∎

Note that the Corollary is constructive in that it may permit construction of $\tilde{p}^1(\cdot)$. Usually, $\tilde{u}^1 = \tilde{p}^1(x, t)$ a.e., that is, except on discontinuity manifolds.

3. <u>Avoidance in the Plane</u>

Consider two agents, called pursuer P and evader E , moving in a plane. Let $\vec{v}_P$ and $\vec{v}_E$ be the velocities (relative to an inertial reference frame) of P and E , respectively. We suppose that their speeds $v_P = |\vec{v}_P|$ and $v_E = |\vec{v}_E|$, are constants and that $v_E > v_P$.

Referring to Figure 1, the kinematic equations of motion are (note that here, unlike in Ref. 5, position is relative to the pursuer)

$$\begin{aligned} \dot{r} &= v_E \cos \beta - v_P \sin \theta \\ \dot{\theta} &= \frac{1}{r} (v_E \sin \beta - v_P \cos \theta) - u_P \\ \dot{\beta} &= \frac{-1}{r} (v_E \sin \beta - v_P \cos \theta) + u_E \end{aligned} \tag{4}$$

where

$$u_E \triangleq \dot{\theta}_E \quad , \quad u_P \triangleq \dot{\theta}_P$$

are the controls of E and P , respectively; that is, E and P control their motions by means of their normal acceleration components. These are constrained; namely,

$$|\, u_E \,| \leq \bar{u}_E \text{ (given)} \quad , \quad |\, u_P \,| \leq \bar{u}_P \text{ (given)} \quad . \tag{5}$$

Evader, E , wishes to avoid having pursuer, P , approach more closely than a given distance $\bar{r}$; that is, the anti-target

$$T = \{(r, \theta, \beta) \in \Delta \mid r \leqslant \bar{r}\} \tag{6}$$

with

$$\Delta = \{(r, \theta, \beta) \mid r \in R_+ , \theta \in R , \mid \beta \mid \leqslant \pi\} .$$

As will be seen subsequently, it suffices to consider $\mid \beta \mid \leqslant \pi$.

There arises now the question of selecting an avoidance set, A . To allow E maneuverability, one wants r "sufficiently" large when $\dot{r} = \dot{r}_{min}$, but when $\dot{r} = \dot{r}_{max}$ one can allow $r = \bar{r}$, where

$$\dot{r}_{min} = - v_E - v_P \quad (\theta = \frac{\pi}{2} \pm 2n\pi , \beta = \pm \pi)$$

$$\dot{r}_{max} = v_E + v_P \quad (\theta = - \frac{\pi}{2} \pm 2n\pi , \beta = 0) .$$

This is accomplished, for instance, by

$$A = \{(r, \theta, \beta) \in \Delta \mid r - \bar{r} \leqslant b(1 + \sin \theta) + c\beta^2\} \tag{7}$$

for given constants $b \geqslant 0$, $c > 0$.

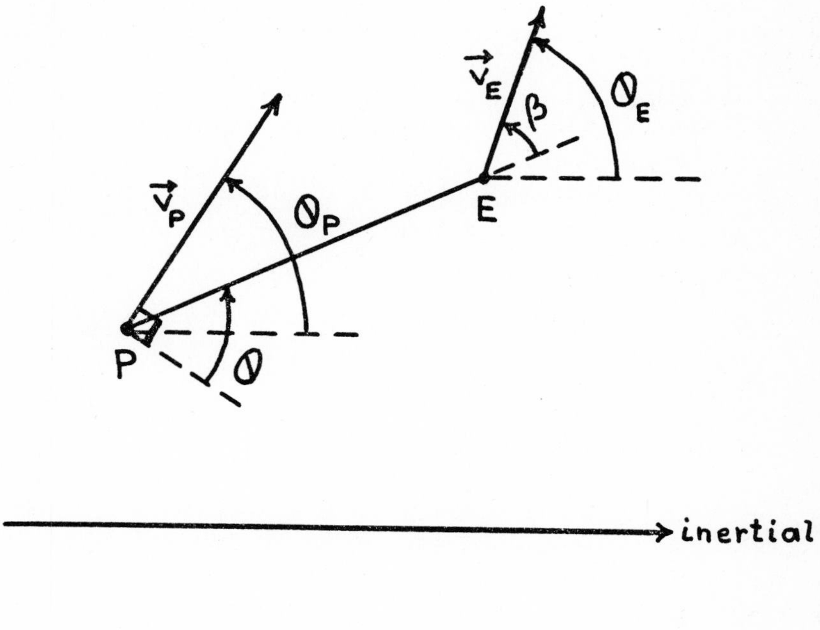

Figure 1, Coordinate System

To satisfy condition (i) of the Theorem we choose $V(\cdot)$ such that

$$V(r, \theta, \beta, t) = r - \bar{r} - b(1 + \sin\theta) - c\beta^2 \quad . \tag{8}$$

To apply condition (ii) of the Theorem we form

$$\begin{aligned} H(r, \theta, \beta, t, u_E, u_P) &= (v_E \cos\beta - v_P \sin\theta) \\ &- b \cos\theta \left[\frac{1}{r} (v_E \sin\beta - v_P \cos\theta) - u_P \right] \\ &- 2c\beta \left[\frac{1}{r} (v_P \cos\theta - v_E \sin\beta) + u_E \right] \quad . \end{aligned} \tag{9}$$

First we check to see what is required to assure $H \geqslant 0$ for all u_P, satisfying constraints (5), and for all $(r, \theta, \beta) \in \Delta_A$. Δ_A has not been defined yet; it depends on Δ_ε. For instance, one might let
$\Delta_\varepsilon = \{(r, \theta, \beta) \in \Delta \mid r \leqslant \bar{r} + b(1 + \sin\theta) + c\beta^2 + \varepsilon \ , \ \varepsilon = \text{constant} > 0\}$.

At $\beta = 0$

$$\begin{aligned} H &= (v_E - v_P \sin\theta) + \frac{b\, v_P}{r} \cos^2\theta + b\, u_P \cos\theta \\ &\geqslant v_E - v_P \sin\theta + b\, u_P \cos\theta \geqslant v_E - \sqrt{v_P^2 + b^2 u_P^2} \quad . \end{aligned}$$

Thus, to satisfy condition (ii) of the Theorem we make the conservative choice

$$b < \frac{\sqrt{v_E^2 - v_P^2}}{\bar{u}_P} \quad . \tag{10}$$

Next we impose the conditions of the Corollary. Since u_E and u_P are separated in H, it follows readily that

$$\begin{aligned} \tilde{u}_E &= - \bar{u}_E && \text{for } \beta > 0 \\ \tilde{u}_E &= \bar{u}_E && \text{for } \beta < 0 \\ \tilde{u}_E &\in [-\bar{u}_E, \bar{u}_E] && \text{for } \beta = 0 \end{aligned} \tag{11}$$

and

$$\begin{aligned} \tilde{u}_P &= \bar{u}_P && \text{for } \cos\theta < 0 \\ \tilde{u}_P &= - \bar{u}_P && \text{for } \cos\theta > 0 \\ \tilde{u}_P &\in [-\bar{u}_P, \bar{u}_P] && \text{for } \cos\theta = 0 \quad . \end{aligned} \tag{12}$$

Now we investigate conditions on $\bar{u}_E$ which assure satisfaction of (ii) of the Corollary; namely, in view of (11) and (12), for all $(r, \theta, \beta) \in \Delta_A$ and all (u_E, u_P) satisfying (5)

$$\min_{u_P} \max_{u_E} H = v_E \cos\beta - v_P \sin\theta + \frac{b}{r}(v_P \cos\theta - v_E \sin\beta)\cos\theta$$
$$+ \frac{2c\beta}{r}(v_E \sin\beta - v_P \cos\theta) \tag{13}$$
$$+ 2c|\beta|\bar{u}_E - b|\cos\theta|\bar{u}_P \geq 0 .$$

To obtain a conservative estimate for the required value of $\bar{u}_E$ we rewrite (13) as

$$\bar{u}_E \geq \max_{(r,\theta,\beta)\in\Delta_A} \frac{1}{2c|\beta|}\Big[v_P \sin\theta - v_E \cos\beta$$
$$+ \frac{b}{r}(v_E \sin\beta - v_P \cos\theta)\cos\theta$$
$$+ \frac{2c\beta}{r}(v_P \cos\theta - v_E \sin\beta) + b|\cos\theta|\bar{u}_P\Big]$$

whence

$$\bar{u}_E \geq \max_{\beta\in[0,\pi]} \Big[\frac{v_P}{\bar{r}} + \frac{1}{2c\beta}\Big(\sqrt{v_P^2 + b^2 \bar{u}_P^2}$$
$$- v_E \cos\beta + \frac{b\, v_E}{\bar{r}} \sin\beta\Big)\Big] . \tag{14}$$

Letting

$$g(\beta) \overset{\Delta}{=} \sqrt{v_P^2 + b^2 \bar{u}_P^2} - v_E \cos\beta + \frac{b\, v_E}{\bar{r}} \sin\beta$$
$$h(\beta) \overset{\Delta}{=} \frac{1}{2c\beta}$$

we rewrite (14) as

$$\bar{u}_E \geq \frac{v_P}{\bar{r}} + \max_{\beta\in[0,\pi]} g(\beta)\, h(\beta) . \tag{15}$$

An even more conservative bound is then found by replacing

$$\max_{\beta\in[0,\pi]} g(\beta)\, h(\beta)$$

by

$$\max g(\beta) \ \max h(\beta) \quad \text{for} \quad \beta \in \{\beta\in[0,\pi] \mid g(\beta) \geq 0\} .$$

In this connection we utilize condition (10) whence

$$g(0) = -v_E + \sqrt{v_P^2 + b^2 \bar{u}_P^2} < 0$$

so that $\hat{\beta} \in (0, \pi]$ for $g(\hat{\beta}) = 0$. We arrive at the very conservative bound

$$\bar{u}_E \geq \frac{v_P}{\bar{r}} + \frac{1}{2c}\left[v_E \sqrt{1 + (b/\bar{r})^2} + \sqrt{v_P^2 + b^2 \bar{u}_P^2} \right] \left[\cos^{-1} \frac{1}{v_E} \sqrt{\frac{v_P^2 + b^2 \bar{u}_P^2}{1 + (b/\bar{r})^2}} - \cos^{-1} \frac{1}{\sqrt{1 + (b/\bar{r})^2}} \right]^{-1} . \tag{16}$$

To reiterate, given pursuer and evader speeds v_P and v_E, respectively, pursuer control bound $\bar{u}_P$, missdistance $\bar{r}$, and constants $b \geq 0$, $c > 0$, with b subject to (10), the use of evader control (11) with $\bar{u}_E$ satisfying (16) guarantees collision avoidance, provided, of course, $\mathcal{U}_1$ admits piecewise continuous functions. Of course, the evader needs to implement such a control only on Δ_A.

As assumed in the definition of Δ, $\beta \in [-\pi, \pi]$ since

$$\dot{\beta}|_{\beta = \pm \pi} = \frac{v_P \cos \theta}{r} \pm \bar{u}_E$$

and by (16)

$$\bar{u}_E \geq \frac{v_P}{\bar{r}} \quad , \quad \bar{r} < r$$

whence it follows that

$$\dot{\beta}|_{\beta = \pi} < 0 \quad , \quad \dot{\beta}|_{\beta = -\pi} > 0 \ .$$

To illustrate the aforegoing results consider

$$v_E = 300 \text{ m/s} \quad , \quad v_P = 225 \text{ m/s}$$

$$\bar{u}_P = 1 \text{ rad/s} \quad , \quad \bar{r} = 3000 \text{ m} \ .$$

Then (10) becomes

$$b \in [0, 198.4) \ .$$

For example, with $c = 6 \times 10^4$ and

(i) $b = 100$, $\bar{u}_E \geq 0.083$

(ii) $b = 10$, $\bar{u}_E \geq 0.081$

(iii) $b = 1$, $\bar{u}_E \geq 0.081$.

Finally, we can draw these conclusions:

(i) The bound on $\bar{u}_E$ given by (16) is quite insensitive to changes in the value of b, and it can be decreased by increasing the value of c (that is, by increasing the size of the avoidance set A).

(ii) For given β , the contour of A is "nearly" circular (more so at $\beta = \pi$ than at $\beta = 0$).

References

1. Hagedorn, P. and Breakwell, J. V., A Differential Game with Two Pursuers and One Evader, Journal of Optimization Theory and Applications, Vol. 18, No. 1, 1976.
2. Breakwell, J. V. and Merz, A., Toward a Complete Solution of the Homicidal Chauffeur Game, Proceedings of First International Conference on the Theory and Applications of Differential Games, University of Massachusetts, Amherst, 1969.
3. Foley, M. A. and Schmitendorf, W. E., A Class of Differential Games with Two Pursuers versus One Evader, IEEE Transactions on Automatic Control, Vol. AC-19, No. 3, 1974.
4. Merz, A., Optimal Evasive Maneuvers in Maritime Collision Avoidance, Navigation, Vol. 20, No. 2, 1973.
5. Vincent, T. L., Avoidance of Guided Projectiles, The Theory and Application of Differential Games (edited by J. D. Grote), Reidel Publishing Company, Dordrecht, 1975, p. 267.
6. Isaacs, R., Differential Games, John Wiley and Sons, New York, New York, 1965.
7. Blaquiere, A., Gerard, F. and Leitmann, G., Quantitative and Qualitative Games, Academic Press, New York, New York, 1969.
8. Krasovskiy, N. N., A Differential Game of Approach and Evasion. 1., Engineering Cybernetics, No. 2, 1973.
9. Krasovskiy, N. N., A Differential Game of Approach and Evasion, 2., Engineering Cybernetics, No. 3, 1973.
10. Leitmann, G. and Skowronski, J., Avoidance Control, Journal of Optimization Theory and Applications, to appear.
11. Getz, W. and Leitmann, G., Qualitative Differential Games with Two Targets, Journal of Mathematical Analysis and Applications, to appear.
12. Filippov, A. F., Classical Solutions of Differential Equations with Multi-Valued Right-Hand Side, SIAM Journal of Control, Vol. 5, No. 4, 1967.
13. Roxin, E., On Generalized Dynamical Systems Defined by a Contingent Equation, Journal of Differential Equations, Vol. 1, p. 188 f., 1965.

A DIFFERENTIAL GAME APPROACH TO COLLISION AVOIDANCE OF SHIPS

Geert Jan Olsder, Jan L. Walter, Twente University of Technology,
P.O. Box 217, Enschede, The Netherlands.

1. Introduction.

Given two ships in each other's neighborhood in the open sea, the critical question is whether a collision can be avoided. It is assumed that the helmsman of ship 1 has complete information on the state of ship 2, but the helmsman of ship 2 is not aware of the presence of the other. This lack of information makes the situation hazardous. He may actually perform a manoeuvre to cause a collision which might otherwise not occur. In this paper we will focus on worst-case situations, that is, one ship tries to avoid and the other ship tries to cause a collision.
If we assume that a stalemate cannot arise, the state space (which is three dimensional: relative position and relative heading) can be divided into a region where ship 1 can avoid a collision and a region where it cannot. The division of the state space will be discussed quantitatively for the idealized case in which both ships move in the open, flat, sea at constant (not necessarily the same) speeds and with bounded turnrates. Within these bounds the turnrates are the controls of the ships. The ships themselves are given by a line segment of a given length. A collision is defined to be an overlap of the two line segments.
A method for the exact construction of the dividing surface, to be called Σ, which is semi-permeable [1], will be given. This surface will consist of only those states from which "optimal" play by both helmsmen leads to a touch or graze-situation. There are 18 essentially different kinds of those situations. The construction of Σ will be accomplished by the computation of the section of all possible pieces in hyperplanes in the state space defined by relative heading = constant. A numerical example will be given. For other approaches to problems of collision avoidance one can consult for instance [2], [3], [4]. In both [2] and [3] a 'collision' takes place if the ships come closer to each other than a given distance. W.r.t. our approach this would mean that the ships are round disks, as viewed from above, instead of the more shipshaped line segments. In [4] the problem treated as a "game of degree" instead of "game of kind" [1]. Both ships try to maximize the minimum distance between the two ships. The approach taken in this paper is similar to the one in [5]. In [7] some more practical questions of a real implementation of a collision avoidance equipment have been considered.
Because of space limitations many arguments and details have been omitted; they can be found in [8].

2. The model and the general form of the analysis.

Two ships move in a horizontal plane which represents the sea surface. There are no obstacles, in the form of, for instance, other ships or land. The ships will be denoted by P_1 and P_2 respectively. Ship P_i (i = 1,2) is represented by a line segment of length l_i. Ship P_i has a constant speed v_i and a maximum angular velocity w_i. The velocity vector of P_i points in the direction of its line segment. The turning point of the ship is its bow. This means that the ship (i.e. the line segment) is always tangential to the trajectory described by the bow. Hence the ship can sweep with its stern. Apart from their being positive, no other assumptions are made on v_i, w_i, l_i.

Ship P_1 tries to avoid a collision whereas P_2 tries to cause it, whereby a collision is defined as a crossing or overlap of the two line segments. If $v_1 > v_2$, P_1 will face a critical situation for a finite time. If, however, $v_1 < v_2$, P_2 can come back all the time. It then depends on the other parameters whether ship P_1 can avoid a collision (if $w_1 >> w_2$, P_1 may escape at every attack by a simple side-stepping; compare the toreador and the bull). The analysis, however, applies to all cases.

If (x,y) are the position coordinates of P_2 relative to P_1, with the y-axis aligned with ship P_1's velocity vector and the x-axis to the right of P_1, and if θ is the angle measured clockwise from P_1's velocity vector to P_2's velocity vector, then the equations of motion in the three dimensional state space (x,y,θ) are [1]

$$\begin{aligned} \dot{x} &= - w_1 u_1 y + v_2 \sin\theta , \\ \dot{y} &= w_1 u_1 x - v_1 + v_2 \cos\theta , \qquad (2.1) \\ \dot{\theta} &= - w_1 u_1 + w_2 u_2 \end{aligned}$$

where the controls u_i are normalized turnrates, i.e. $|u_i(t)| \le 1$ and $u_i = 0$ corresponds to a straight motion, $u_i = +1$ to a full right turn and $u_i = -1$ to a to a full left turn, all in "real" space.

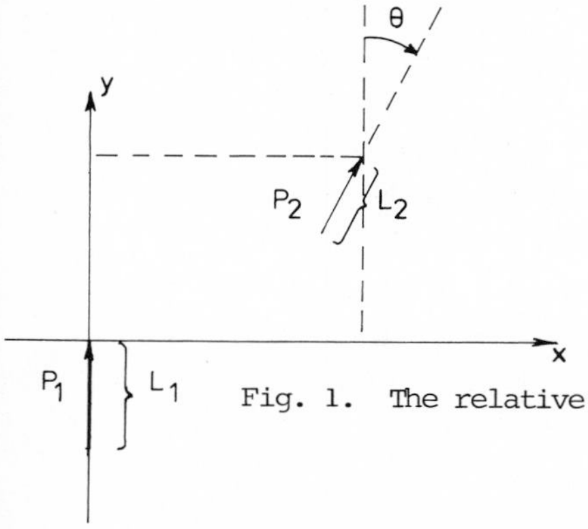

Fig. 1. The relative coordinates x, y and θ.

The dividing semi-permeable surface Σ will consist only of those states in the three-dimensional state space from which "optimal" play by both ships leads to a graze-situation, i.e. the ships just touch each other. Though optimal play has not yet exactly been defined by means of a cost function, a deviation from optimal play by P_1 will cause a collision and a deviation from optimal play by P_2 results in the ships not colliding.
The surface Σ will consist of many portions, each belonging to a particular graze-situation. Each portion of Σ is obtainable by consideration of a particular local differential game with a terminal payoff [1], only defined in the immediate neighbourhood of the graze-situation. In each local game the terminal state x_f, y_f, θ_f, describing a particular graze-situation, together with the terminal values of the adjoint equations, which can also be obtained, are given in terms of a single parameter with a bounded range. The adjoint variables will be denoted by V_x, V_y and V_θ, and the terminal values of these variables are provided with a subscript f. Backwards integration of the optimal paths in state space is now possible, where, at each time instant, u_1 and u_2 are chosen such as to satisfy Isaac's equation [1]. These paths, for each local game, thus yield a surface in the state space of which the parameters are time and the above mentioned single parameter. To each local game corresponds such a surface. The determination of Σ from all these surfaces will be illustrated in section 4 by a numerical example. It will turn out that not every part of a particular surface forms part of Σ. Notice that the semi-permeable surface is intrinsic to the system (2.1) and has nothing to do with the particular terminal payoff chosen.

3. The local games.

There are essentially eighteen types of graze-manoeuvres, which will be indicated by the numbers 1 - 18, as shown in fig. 2. Another eighteen cases exist by left-right reflections. These new graze-situations will be denoted by primes. For each type the terminal values of the adjoints can be determined, from which, together with the terminal states of the system, the optimal paths can be traced backwards in time and simultaneously the optimal strategies can be obtained. As an example consider case 1 (corresponding to case 1 in fig. 2). Near the end both ships turn right. The final state is $x_f = 0$, $\dot{x}_f = 0$, so that θ_f is given by $\sin \theta_f = w_1 y_f / v_2$. The parameter is y_f, with the obvious limits $\max(-v_2/w_1, -l_1) \le y_f \le 0$. When one integrates backwards in time with these endconditions it turns out that P_1 has a switch. In fig. 3 it is shown how, for varying parameter y_f, all the paths form a semi-permeable surface in the state-space.

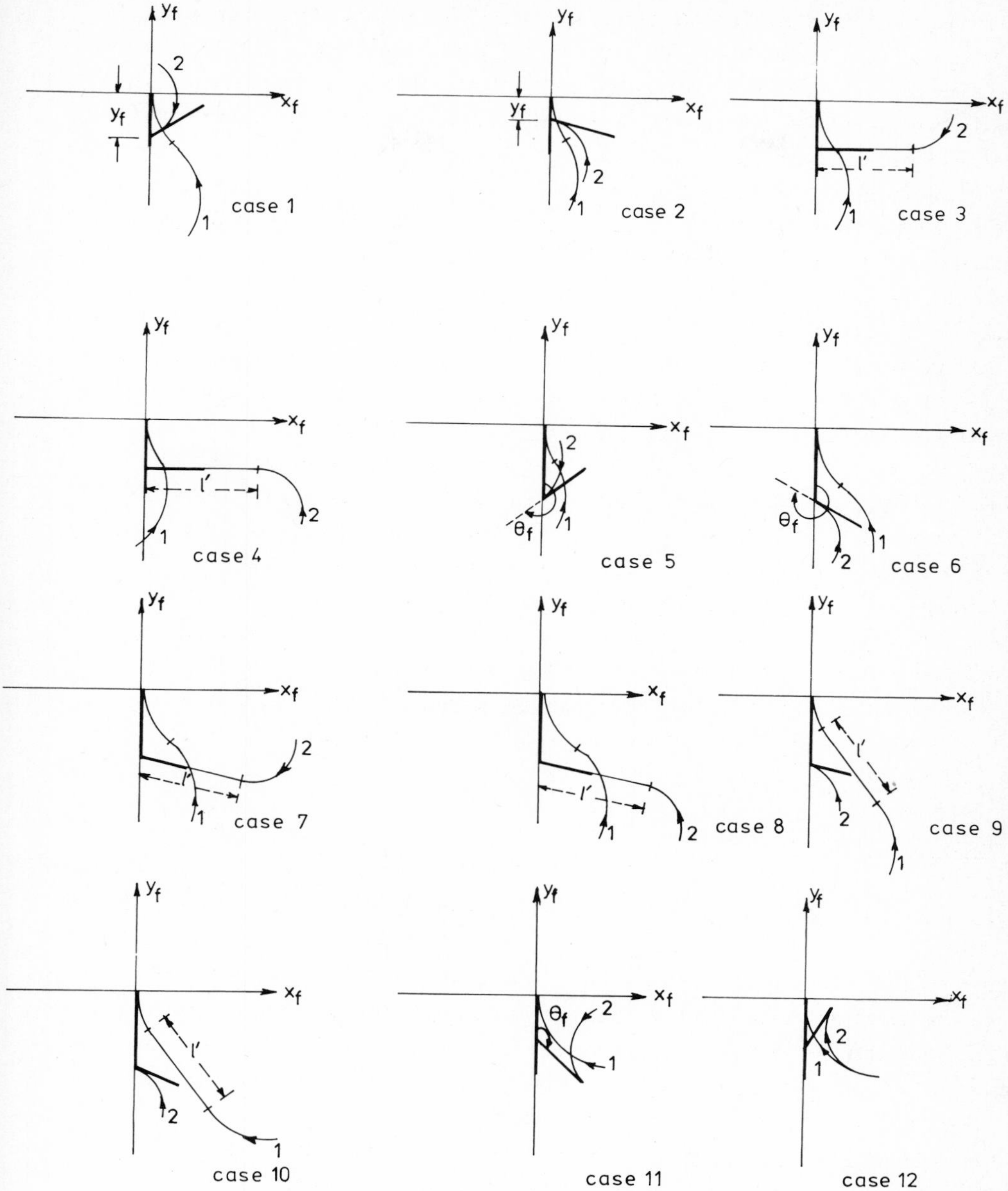

y_f
x_f
1
2
l'
θ_f
case 1
case 2
case 3
case 4
case 5
case 6
case 7
case 8
case 9
case 10
case 11
case 12

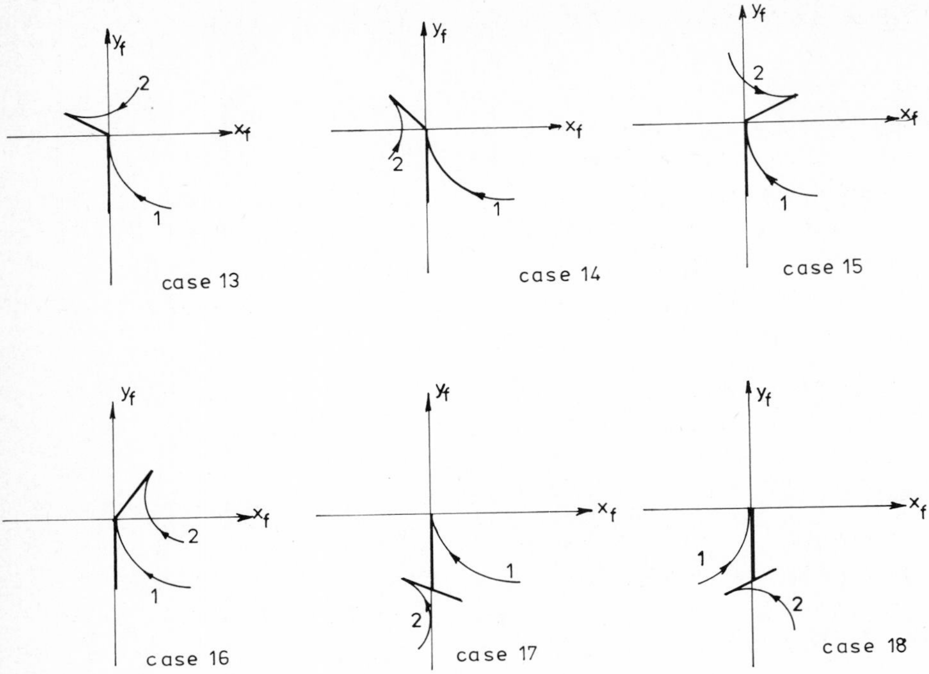

Fig. 2. The local games.

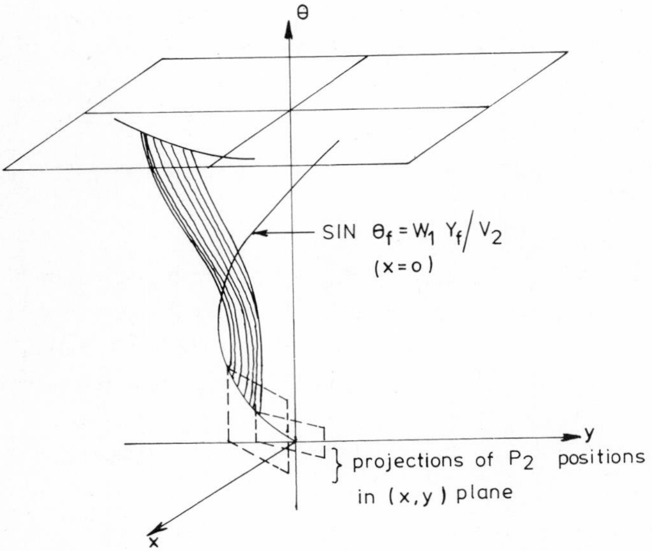

Fig. 3. The semi-permeable surface corresponding to case 1.

4. Synthesis of the dividing surface Σ for some numerical cases.

A computer program has been written which constructs the semi-permeable surfaces of all the 2 × 18 local games. The composite surface Σ is made up out of pieces of these semi-permeable surfaces. This surface Σ qualifies for the dividing surface between collision/escape only if 1° it separates the state space into two regions, and 2° the junction of two semi-permeable surfaces does not leak. The second condition actually means that two semi-permeable surfaces which intersect form a composite semi-permeable surface.Conditions for this to hold are discussed in [6].

In the basis example we assume that $v_1 = 1$, $v_2 = 0,75$, $w_1 = 0,5$, $w_2 = 0,75$, $l_1 = 0,8$, $l_2 = 0,6$. These quantities correspond to the following real data: velocity P_1:50 km/hours; velocity P_2: 37,5 km/hours; minimum turn radius P_1: 1 km; minimum turn radius P_2: 0,5 km; length P_1: 400 m; length P_2: 300 m. The demonstration of a region, to be called R, entirely surrounded by a composite semi-permeable surface of course requires the construction for all values of θ. This has been carried out for $\theta = k\,\frac{\pi}{6}$, k = 0, 1, ... ,6 and the sections of R and its boundary in the planes $\theta = k\,\frac{\pi}{6}$ are plainly discernable. In figs. 4.1 - 4.7 these sections are shown, with all the unnecessary arcs deleted. The solution gives that if P_2 belongs to R, ship P_1 cannot avoid a collision if P_2 really wants a collision. Notice that no "optimal" (e.g. time-optimal) strategies are given for P_1 and P_2 for resp. postponing the collision as long as possible resp. achieve a collision as soon as possible. Only if P_2 is initially situated in R near Σ, then the optimal strategies belonging to a point near and on Σ, will yield a strategy which leads to a collision.

Conclusions.

A differential game method has been developed which divides the state-space into two regions. In one region ship P_2 can always cause a collision, in the other region ship P_1 can prevent a collision, provided it uses an appropriate strategy. In a future publication slightly different sets of ships will be considered such as to get a feeling for the sensitivity of the collision avoidance capacities w.r.t. the three parameters which characterize a ship (i.e. length, velocity, maximum turnrate).

The ship models considered were very simple; the speeds are constant. More realistic is that the speed decreases during manoeuvring. This, however, will make the analysis much more complicated. Another feature which has not been considered is what to do if three ships are in each other's neighborhood instead of two.

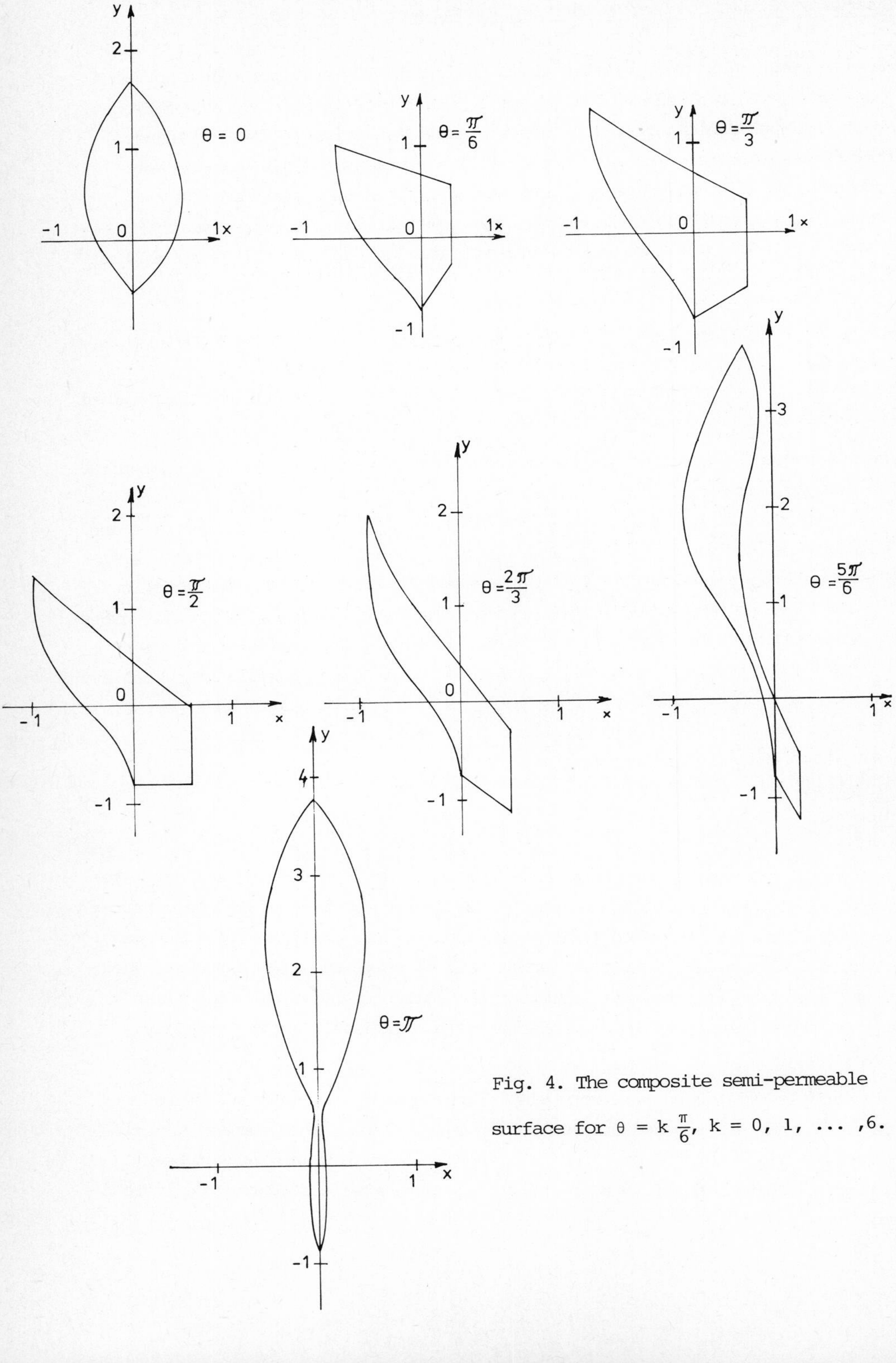

Fig. 4. The composite semi-permeable surface for $\theta = k\,\frac{\pi}{6}$, $k = 0, 1, \ldots ,6$.

References:

[1] R. Isaacs, Differential Games, 2nd. ed., Wiley, 1976.

[2] T. Miloh, "Determination of Criterical Manoeuvres for Collision Avoidance Using the Theory of Differential Games", Institut für Schiffbau, Hamburg, Bericht Nr. 319 (1974).

[3] T.L. Vincent, W.Y. Peng, "Ship Collision Avoidance", Workshop on Differential Games, Naval Academy, Annapolis, USA, 1973.

[4] A.W. Merz, "Optimal Evasive Manoeuvers in Maritime Collision Avoidance", Navigation, vol. 20, Nr. 2, pp. 144-152, 1973.

[5] G.J. Olsder, J.V. Breakwell, "Role Determination in an Aerial Dogfight", Int. Journal of Game Theory, vol. 3, Nr. 1, pp. 47-66, 1974.

[6] P. Bernhard, "Conditions de Coin pour les jeur differentiels" Seminaire sur les Jeux Differentiels, Centre d'Automatique, Paris, 1971.

[7] J.S.Karnarkar, J.A. Sorenson, J.F. Tyler, "Analysis of a Novel Marine Collision Avoidance System", in: "Ship Operation Automation", Pitkin, Roche, Williams (eds.), North Holland, 1976.

[8] G.J. Olsder, J.L. Walter "Collision Avoidance of Ships", Internal report Dept. of Applied Mathematics, Twente University of Technology, 1977.

A METHOD FOR COMPUTING NASH EQUILIBRIA FOR NON-ZERO-SUM DIFFERENTIAL GAMES

B. Tołwiński
Systems Research Institut
Polish Academy of Sciences
Newelska 6, 01-447 Warsaw/Poland

1. Introduction

The aim of this paper is to discuss the problem of computing open-loop, unconstrained Nash equilibrium point of a non-zero-sum differential game, defined as follows:
Let $\{I = 1, \ldots, N\}$ be a set of players and $T = [t_o, t_f]$ a given time interval. We assume, that for all $i \in I$ the i-th player chooses a measurable function $u_i: T \rightarrow R^{m_i}$, trying to minimize a cost functional

$$J_i(u_1, \ldots, u_N) = \int_{t_o}^{t_f} L_i(x, u_1, \ldots, u_N, t)dt + K_i(x(t_f)) \quad (1)$$

subject to the n-dimensional state equation

$$\dot{x} = f(x, u_1, \ldots, u_N, t), \quad x(t_o) = x_o \quad (2)$$

A point $u^* = (u_1^*, \ldots, u_N^*)$ is called Nash equilibrium solution of the game (1)-(2), if for all $i \in I$ and all admissible strategies u_i it satisfies the system of inequalities

$$J_i(u^*) \leqslant J_i(u_1^*, \ldots, u_{i-1}^*, u_i, u_{i+1}^*, \ldots, u_N^*) \quad (3)$$

Define a Hamiltonian function for each player from the set I as

$$H_i(x, u_1, \ldots, u_N, p_i, t) = L_i + p_i' f \quad (4)$$

where p_i is a vector of dimension n. As it was shown in [11], if the standard regularity assmmptions about functions L_i, K_i and f are satisfied, then the following necessary conditions for an open-loop Nash solution must hold:

$$\dot{x} = f(x, u, t), x(t_o) = x_o \quad (5)$$

$$\dot{p}_i = -(\partial/\partial x) H_i(x, u, p_i, t), p_i(t_f) = (\partial/\partial x(t_f)) K_i(x(t_f)) \quad (6)$$

$$0 = (\partial/\partial u_i) H_i(x, u, p_i, t) \quad (7)$$

where $u = (u_1, \ldots, u_N)$ and $i \in I$.
From conditions (5)-(7) various numerical procedures for computing Nash solutions can be derived [1,3,7,8,10,12,13]. Some of these

techniques will be discussed in the next Sections.

2. Some comments on existing techniques for computing Nash controls

Several authors have been concerned with iterative techniques for computing Nash equilibria for general, non linear-quadratic non--zero-sum differential games. Starr [10] proposed a simple "cycling" procedure, where, if initial solution $u^o = (u_1^o, \ldots, u_N^o)$ is given and $k = 1,2,\ldots,$ then controls u_i^k for $i \in I$ are defined as

$$u_i^k = \arg\min_{v_i} J_i\left(u_1^{k-1}, \ldots, u_{i-1}^{k-1}, v_i, u_{i+1}^k, \ldots, u_N^k\right)$$

For some problems this method may converge, but even then the convergence will be slow, since near the Nash solution each functional is insensitive to u_i but sensitive to u_j for $j \neq i$. Sage [8] and Elsner and Mukundan [1] proposed a gradient technique, where u_i^k for $i \in I$ and $k = 1,2,\ldots$ are defined as

$$u_i^k = u_i^{k-1} - s_i^k (\partial/\partial u_i) H_i(x^{k-1}, u_1^{k-1}, \ldots, u_N^{k-1}, p_i^{k-1}, t).$$

s_i^k denotes here a step length, x^{k-1} - the trajectory obtained from equation (5), where $u^{k-1} = (u_1^{k-1}, \ldots, u_N^{k-1})$ was substituted for u, and p_i^{k-1} - co-state variables obtained from equations (6), where u^{k-1} and x^{k-1} were substituted for u and x respectively. A serious drawback of this method follows from the fact, that in general one has no indication how to choose the quantities s_i^k to make the procedure move towards an equilibrium solution. In [1] the suggestion was made to define s_i^k as numbers satisfying the system of equations

$$J_i\left(u_1^{k-1} - s_1^k (\partial H_1/\partial u_1), \ldots, u_N^{k-1} - s_N^k (\partial H_N/\partial u_N)\right) = 0, \quad i \in I \tag{8}$$

However such an approach does not seem very promising in the situation, where first, it is in general impossible to compute second derivatives $J''_{iu_iu_j}$ of the cost functionals (1) and, in consequence it is rather difficult to compute s_i^k satisfying (8) even approximatly (as Newton technique for solving non-linear equations cannot be applied), and second, even the exact computation of s_i^k satisfying (8) does not guarantee, that the method will converge to a Nash solution. An easily implementable method, called "ping-pong" algorithm, was proposed in [3]. For some problems it turned out to be quite efficient, but unfortunatly, like the gradient technique, the ping-pong algorithm may fail to converge even for simple linear-quadratic games, no matter how close an initial reference solution u^o would be to the actual one. This fact will be illustrated by the following example:

Let $I = \{1,2\}, T = [0,1]$

$$J_1 = (1/2) \int_0^1 u^2(t)dt + x(1)y(1)$$

$$J_2 = (1/2) \int_0^1 v^2(t)dt - x(1)y(1)$$

$$\dot{x} = u, \quad \dot{y} = v, \quad x(0) = y(0) = 0,$$

where u is the control of the first player, v is the control of the second player, x and y are state variables, and u, v, x, y are scalar functions defined on T. In the ping-pong algorithm controls u^k, $k = 1,2,\ldots,$ are obtained from the system of equations

$$(\partial/\partial u_i) H_i(x^{k-1}, u^k, p_i^{k-1}, t) = 0, \quad i \in I \tag{9}$$

(x^{k-1} and p_i^{k-1} **are** defined as in the gradient method) or equivalently are defined as

$$u^k(t) = \arg\min_u (1/2)\left(\sum_{i=1}^{N} \left| (\partial/\partial u_i) H_i(x^{k-1}, u, p_i^{k-1}, t)\right|^2\right) \tag{10}$$

for $t \in T$.

Using this algorithm to solve the game defined above one computes the controls u^k and v^k as

$$u^k = -y^{k-1}(1), \quad v^k = x^{k-1}(1),$$

what follows from the fact, that the Hamiltonians of the problem are of the form

$$H_1 = (1/2)u^2 + y(1)u + x(1)v$$

$$H_2 = (1/2)v^2 - y(1)u - x(1)v$$

Now, let c be any constant number not equal to 0 and assume, that $u^o(t) = v^o(t) = c$ for all $t \in T$. Obviously one has

$$(u^1,v^1) = (-c,c), \; (u^2,v^2) = (-c,-c), \; (u^3,v^3) = (c,-c),$$

$$(u^4,v^4) = (c,c) \quad \text{and} \quad (u^{4l},v^{4l}) = (c,c) \quad \text{for } l = 1,2,\ldots$$

and the procedure fails to converge to the unique Nash solution $(u^*,v^*) = (0,0)$ of the game for every $(u^o,v^o) = (c,c)$. Note, that for the given game and the given initial solution the cycling procedure and the gradient method with step length s_i^k satisfying exactly the system (8) also fail to converge.

3. Differential game as an optimal control problem

To overcome the difficulties exposed in Section 2 one can propose the following approach. Observe first, that if the system of equations (5)-(7) has a solution, then it is equivalent to the following optimal control problem:

Minimize the functional

$$J(u) = (1/2) \int_{t_o}^{t_f} \left(\sum_{i=1}^{N} |(\partial/\partial u_i) H_i(x,u,p_i,t)|^2 \, dt \right. \tag{11}$$

subject to the state equations

$$\mathring{x} = f(x,u,t), \quad x(t_o) = x_o \tag{12}$$

$$\mathring{p}_i' = -(\partial/\partial x) H_i(x,u,p_i,t), \quad p_i(t_f) = (\partial/\partial x(t_f)) K_i(x(t_f)), \quad i \in I \tag{13}$$

where $u = (u_1, \ldots, u_N)$.

The Hamiltonian of this problem has the form:

$$H(x,p,u,\eta,\delta,t) = (1/2) \sum_{i=1}^{N} |\partial H_i/\partial u_i|^2 + \eta' f - \sum_{i=1}^{N} \delta_i (\partial H_i/\partial x)' \tag{14}$$

where

$$\mathring{\eta} = -\partial H/\partial x, \quad \eta(t_f) = -\sum_{i=1}^{N} \delta_i(t_f)(\partial^2/\partial x^2(t_f)) K_i(x(t_f)) \tag{15}$$

$$\mathring{\delta}_i = -\partial H/\partial p_i = -(\partial H_i/\partial u_i)(\partial f/\partial u_i) + \delta_i'(\partial f/\partial x), \quad \delta_i(t_o) = 0, \quad i \in I \tag{16}$$

$$p = (p_1, \ldots, p_n), \quad \delta = (\delta_1, \ldots, \delta_N).$$

Observe, that the TPBVP (15)-(16) decomposes into two seperate:

initial value problem (16), which does not depend on the co-state variable η, and terminal value problem (15), which can be solved after co-state variables δ_i are found from (16). In consequence, the problem (11)-(13) can be solved by an arbitrary optimization scheme developed for the standard optimal control problems, i.e. gradient techniques, strong variation methods etc.

The Hamiltonian (14) gives a new interpretation of the ping-pong algorithm. If one compares the definition (10) with expression (14) it will become clear, that the computation of the control u^k in the ping-pong method is equivalent to the minimization of the first term of the Hamiltonian (14) and, in consequence, the ping-pong method can be considered as an approximation to the so-called min-H, or strong variation techniques applied to the solution of the optimal control problem (11)-(13). Recall, that the basic operation of such techniques is the minimization of the Hamiltonian of the problem in $u(t)$ for all $t \in T$ and for given state and co-state variables. In our case one has to compute

$$\bar{u}^k(t) = \arg \min_{u(t)} H(x^{k-1}(t), p^{k-1}(t), u(t), \eta^{k-1}(t), \delta^{k-1}(t), t) \tag{17}$$

for $t \in T$ and trajectories $x^{k-1}, p^{k-1}, \eta^{k-1}, \delta^{k-1}$ corresponding to the control u^{k-1}, and then to put simply $u^k = \bar{u}^k$ [2], or, in more sophisticated methods, to define u^k as some combination of $\bar{u}^k$ and u^{k-1} in the way to assure that the inequality

$$J(u^k) < J(u^{k-1}) \tag{18}$$

is satisfied, i.e. the algorithm has the descent property [4, 5, 6].

4. Algorithm

We are able now to define a general algorithm for computing a Nash solution for the differential game (1)-(2), or more strictly speaking - a point satisfying necessary conditions for such a solution. As it has been demonstrated, such a point can be found by solving the optimal control problem (11)-(13) by means of any standard computational technique. However, taking into consideration the good performance of the ping-pong procedure demonstrated on a number of examples and the fact, that if it converges it can save a considerable computation effort, as avoiding the integration of the differential equations (15) and (16), one can propose an approach, where the co-state variables η and δ are computed only, when the ping-pong procedure fails.

Consider the following algorithm

Step 1. Select an initial reference solution $u^o = (u_1^o, \ldots, u_N^o)$, where

$u^o: T \rightarrow R^{m_1} \times \ldots \times R^{m_N}$.

Set $k = 1$.

Step 2. Compute $J(u^{k-1})$ according to (11). If $J(u^{k-1}) = 0$ then stop, otherwise compute $\bar{u}^k$ using the ping-pong method (see definition (10), where $\bar{u}^k$ is substituted for u^k).

Step 3. If $J(\bar{u}^k) \leqslant s \cdot J(u^{k-1})$ for some $0 < s < 1$, then set $u^k = \bar{u}^k$, $k = k + 1$ and go to Step 2. Otherwise go to Step 4.

Step 4. Solve equations (16) and (15) to find δ_i^{k-1}, $i \in I$, and η^{k-1} corresponding to u^{k-1}. Define u^k using some strong variation technique with respect to the optimal control problem (11)-(13), set $k = k + 1$ and return to Step 2.

It is easy to see, that if for example the first order methods of Jacobson and Mayne or Mayne and Polak are used in Step 4 of the algorithm, then for the game considered in Section 2 it converges in two steps, with variables η and δ_i, $i \in I$, computed only once. More general convergence results can be obtained with u^k in Step 4 defined by means of Mayne's and Polak's method [6]. These results are given in Appendix, where also the detailed version of the algorithm is presented.

5. Conclusion

The problem of computing open-loop Nash solution of a general non-zero-sum differential game has been reduced to an optimal control problem tractable by any standard computational technique. Next, an algorithm has been proposed, which takes advantage of the special form of the problem and, in consequence, may save a considerable computational effort. The algorithm (see Appendix) converges to a point satisfying necessary conditions for Nash equilibrium at least for a class of differential games, which includes linear-quadratic problems.

Computation of open-loop Nash solutions may be in many practical cases intersting in itself, but it seems also, that the development of efficient techniques for obtaining such solutions is basic, if one is interested in computing sampled data Nash controls for general non linear-quadratic games [9], which is important in view of possible applications of the Game Theory in Economics.

Appendix

We are going to present now a more detailed version of the algorithm from the Section 4, where the strong-variation procedure of Mayne and Polak is used as the optimization technique in Step 4.

Assume first, that the problem is to find an open-loop control $u\colon T\to R^m \left(m=\sum_{i=1}^{m_N} m_i\right)$ satisfying conditions (5)-(7), which is bounded in L_∞ metric, and suppose, that if such a control exists, then Ω is a closed, convex and bounded subset of R^m such, that $u(t)\in \in \operatorname{int}\Omega$ a.e. on T. The following definitions are analogous to those given in [6]:

$$\bar{H}(u,t) = \min_{w\in\Omega} H\left(x^u(t),p^u(t),w,\ \eta^u(t),\ \delta^u(t),t\right) \tag{19}$$

$$\Theta(u) = (1/(t_f-t_o)) \int_{t_o}^{t_f} \bar{H}(u,t) - H\left(x^u(t),\ p^u(t),u(t),\ \eta^u(t),\ \delta^u(t),t\right)dt \tag{20}$$

$$I_u^H = \left\{t\in T \mid \bar{H}(u,t) - H\left(x^u(t),\ p^u(t),u(t),\ \eta^u(t),\ \delta^u(t),t\right)\leqslant \Theta(u)\right\} \tag{21}$$

for $\alpha\in[0,\ t_f-t_o]$ $I_{\alpha u}\subset T$ and

$$\mu(I_{\alpha u})=\alpha,$$

$$\alpha\in[0,\ \mu(I_u^H)]\Rightarrow I_{\alpha u}\subset I_u^H$$

$$\alpha\in[\mu(I_u^H),t_f-t_o]\Rightarrow I_{\alpha u}\supset I_u^H$$

$$\forall\alpha\in[0,\ \mu(I_u^H)],\ \{t\in I_u^H,\ t'\in I_{\alpha u},\ t<t'\}\Rightarrow\{t\in I_{\alpha u}\} \tag{22}$$

$$\forall\alpha\in[(I_u^H),\ t_f-t_o]\ ,\ \{t\in T,t'\in I_{\alpha u}\setminus I_u^H,\ t<t'\}\Rightarrow\{t\in I_{\alpha u}\}.$$

$$G = \{u\colon T\to \Omega \mid u \text{ measurable}\} \tag{23}$$

<u>Algoritm</u>

Step 0. Select $u^o\in G$, $\beta\in(0,1)$ and $s\in(0,1)$.
Step 1. Set $k = 0$ and $u = u^o$.

Step 2. Compute x by solving (5).
Step 3. Compute p_i, $i \in I$, by solving (6).
Step 4. Compute $J = J(u)$ according to (11).
Step 5. If $J = 0$ then stop, otherwise go to Step 6.
Step 6. Compute

$$\bar{u}(t) = \arg\min_{v \in \Omega} V(v,t) \text{ for } t \in T, \text{ where}$$

$$V(v,t) = \sum_{i=1}^{t} \left| (\partial/\partial u_i) H_i(x(t), v, p_i(t), t) \right|^2$$

If $V(\bar{u}(t), t) = 0$ a.e. on T then go to Step 7, otherwise go to Step 9.

Step 7. Compute state and co-state variables $\bar{x}$ and $\bar{p}_i$, $i \in I$, corresponding to the control $\bar{u}$. Compute $\bar{J} = J(\bar{u})$.
Step 8. If $\bar{J} \leqslant s\, J$ then set $k = k + 1$, $u^k = \bar{u}$, $u = \bar{u}$, $x = \bar{x}$, $J = \bar{J}$ and go to Step 5, otherwise go to Step 9.
Step 9. Compute δ_i, $i \in I$, by solving (16).
Step 10. Compute η by solving (15).
Step 11, Compute

$$\hat{u}(t) = \arg\min_{v \in \Omega} H\left(x^u(t),\ p^u(t),\ v,\ \eta^u(t),\ \delta^u(t),\ t\right)$$

for $t \in T$

Step 12. Compute $\Theta(u)$. If $\Theta(u) = 0$ then stop, otherwise go to Step 13.
Step 13. Set $\alpha = t_f - t_o$.
Step 14. Set $u_\alpha = \begin{cases} \hat{u}(t), & t \in I_{\alpha u} \\ u(t), & t \in T \setminus I_{\alpha u} \end{cases}$
Step 15. Compute x_α, $p_{i\alpha}$ $i \in I$ and J_α corresponding to the control u_α.
Step 16. If $J_\alpha - J \leqslant \alpha\,\Theta(u)/2$, then set $k = k + 1$, $u^k = u_\alpha$, $x = x_\alpha$, $p_i = p_{i\alpha}$ $i \in I$, $J = J_\alpha$ and go to Step 5. Otherwise go to Step 17.
Step 17. Set $\alpha = \beta \cdot \alpha$ and go to Step 14.

Note, that the algorithm stops in any point u satisfying the minimum principle for the optimal control problem (11) - (13) with additional condition, that controls belong to the set G. If $J(u) > 0$ then one can try to restart the algorithm with a new initial solution u^o, or, in some cases, one may conclude, that the game has no Nash solution u, for which $u(t) \in \operatorname{int}\Omega$ for $t \in T$.

Assume now, that functions f and L_i, $i \in I$, are of the following form:

$$f(x,u_1, \ldots, u_N, t) = Ax + \sum_{j=1}^{N} B_j u_j \tag{24}$$

$$L_i(x,u_1, \ldots, u_N, t) = x'Qx + \sum_{j=1}^{N} x'R_{ij}u_j + g_i(u,t) \tag{25}$$

where A, B_j, Q_i, R_{ij} for i, $j \in I$ are matrix functions defined and continuous on T, $g_i : R^m \times T \to R$, $i \in I$, and their partial derivatives g'_{iu}, g''_{iuu}, g'''_{iuuu} are continuous on $\Omega \times T$. Suppose also, that the expression

$$\sum_{i=1}^{N} |(\partial/\partial u_i) g_i(u,t)|^2 \tag{26}$$

is a strictly convex function of variable u.

<u>Theorem</u>

If the assumptions stated above are satisfied (they are satisfied in particular for a linear-quadratic game), then

1. There exists a unique solution $\bar{u}$ of the optimal control problem (11) - (13) with the additional constraint $u(t) \in \Omega$ a. e. on T.
2. Let U be a set of Nash solutios u of the game (1)-(2), with the property, that $u(t) \in \text{int}\,\Omega$ a. e. on T. If the set U is not empty, then it has only one element u^* and $u^* = \bar{u}$.
3. The sequence $\{u^k\}$ generated by the algorithm converges to $\bar{u}$ in L_2 metric.

The proof of this theorem follows from the slightly generalized results of [6] and from the fact, that under our assumptions the functional (11) is convex on G [13].

References

1. Elsner W.B, Mukundan R. - Linear feedback strategies in non-zero-sum differential games, Int.J.Systems Sci., <u>6</u>, 6 (1975).
2. Gotlieb R.G. - Rapid convergence to optimum using a Min-H strategy, AIAA J., <u>5</u>, 2 (1967).
3. Holt D., Mukundan R. - A Nash algorithm for a class of non-zero-sum differential games, Int.J.Systems Sci., <u>2</u>, 4 (1972).
4. Jacobson D.H., Mayne D.Q. - Differential dynamic programming, Elsevier, N.Y. 1970.

5. Krylow I.A., Tchernoushko F.L. - On sequential approximations method for optimal control problem, J.Vytch.Mat. i Mat.Fiz.,2,6 (1962) (in Russian).

6. Mayne D.Q., Polak E. - First-order strong variation algorithms for optimal control, J.Opt.,TH.Appl.,16,3/4 (1975).

7. Pau L.F. - Differential games and a Nash equilibrium searching algorithm, SIAM J. Control, 13,4 (1975).

8. Sage A.P. - On gradient methods for optimization of differential game strategies, SWIEEE CO Rec. Techn. Pap. 22nd Annual Southwest IEEE Conf. and Exhibit., Dallas 1970, N.Y. 1970, pp. 188-191.

9. Simaan M., Cruz J.B. - Sampled data Nash controls in non-zero--sum differential games, Int.J.of Control, 17,6 (1973).

10. Starr A.W. - Computation of Nash equilibria for non-linear differential games, Proc. First Int. Conf. on the Theory and Appl. of Diff. Games, Amherst, Mass., pp.IV-13 - IV-18.

11. Starr A.W.,Ho Y.C. - Non-zero-sum differential games, J. Opt. Th. Appl., 3,3 (1969).

12. Tolwiński B. - Numerical solution of N-person, non-zero-sum differential games, Control and Cybernetics, 6,4 (1977).

13. Tolwiński B. - Computational methods in the theory of non-zero-sum differential games, Techn. Report of Internetional Scientific Group, Institut for Systems Studies, Moscow USSR, 1976-1977 in Russian .

NUMERICAL APPROXIMATION AND IDENTIFICATION IN A 1-D PARABOLIC DEGENERATED NON-LINEAR DIFFUSION AND TRANSPORT EQUATION

by

G. Chavent(*) and G. Cohen(**)

ABSTRACT

The physical motivation of this study is the recovering of oil from a porous medium by injection of water[1]. For high water injection rate, the transport term is preponderant, thus we give an approximation by discontinuous finite elements, which also remains valid for the pure transport equation. Identification of non-linearities appearing in the state equation is then performed by minimizing an output error criterion. Numerical examples are given.

1. THE PHYSICAL PROBLEM AND ITS MODELLIZATION.

Let us consider a core sample of homogeneous porous material, which we suppose to be *initially filled with oil*. We inject (cf. fig. 1) some water through its left hand side end with a *known rate*, and we *observe the rates of oil and water recovered* through the right hand side end [2]:

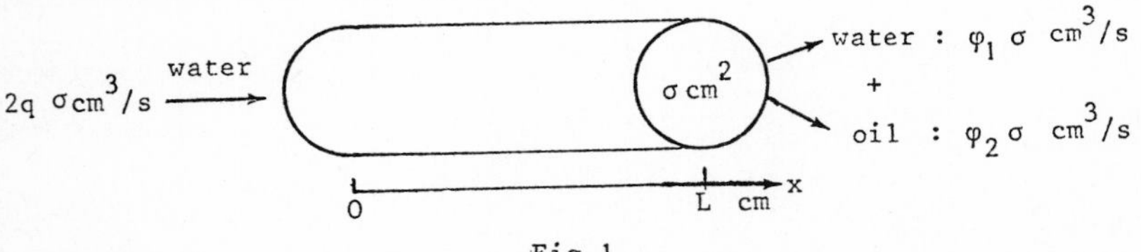

Fig.1

The water does not appear instanteanously at the right hand side end, but only after the *break-through* time (B.T.), and then in growing quantities.

The state of the system is the *(reduced) water saturation* $u(x,t)$ at point x and time t, ranging between 0 and 1. The saturation is not easily accessible to measurement.

---

(1) This study has been made in the frame of a research contract n° 05/77 between IRIA and the IFP (Institut Français du Pétrole) which we thank for allowing as to publish those numerical results.

(2) This experience corresponds to the laboratory study of core samples obtained during the drilling of wells in the oil industry ; but the two-phase flow process involved here has other applications, as the study of unsaturated soils for instance.

(*) IRIA, Domaine de Voluceau, POB. 105, F-78150 Le Chesnay and University of PARIS-IX, 75016 Paris, France.

(**) IRIA, same address.

The state equation is (cf [1] [2] [3]) :

$$(1) \qquad \emptyset \frac{\partial u}{\partial t} - \frac{\partial}{\partial x}\left(K \frac{\partial}{\partial x} \alpha(u)\right) + q(t) \frac{\partial}{\partial x}(b(u)) = 0 \qquad \text{in }]0,L[\times]0,T[$$

with the boundary conditions :

$$(2) \quad \begin{cases} \text{Dirichlet at } x = 0 : \\ u(o,.) \overset{\Delta}{=} u_e \\ u_e \equiv 1 \end{cases} \qquad (3) \quad \begin{cases} \text{Unilateral condition at } x = L : \\ u(L,.) \overset{\Delta}{=} u_s \\ u_s \leq 1 \, , \; \varphi_1 \geq 0 \, , \; (1-u_s)\varphi_1 = 0 \end{cases}$$

and an initial condition :

$$(4) \qquad u(x,o) = u_o(x)$$

where the <u>water output rate</u> φ_1 is related to the saturation u by :

$$(5) \qquad \varphi_1(t) = (1+b(u))q - K \frac{\partial}{\partial x} \alpha(u) \; .$$

The parameters appearing in (1) to (5) are the porosity $\emptyset$ and the permeability K, that are, in our case, easily determined by preliminary experiences, and the <u>two functions of saturation</u> $\alpha(u)$ <u>and</u> $b(u)$, related to the relative permeabilities and capillarity pressure curves <u>that are not precisely known</u>. Their typical forms are given in Fig. 2. <u>One may notice that the derivative</u> α' <u>of</u> α <u>vanishes for</u> $u = 0$ <u>and</u> $u = 1$, so that the <u>diffusion term of equation</u> (1) <u>is degenerated</u>. The mathematical study of system (1) to (5) under the above conditions, has been made in [4], where the existence of a solution has been proved.

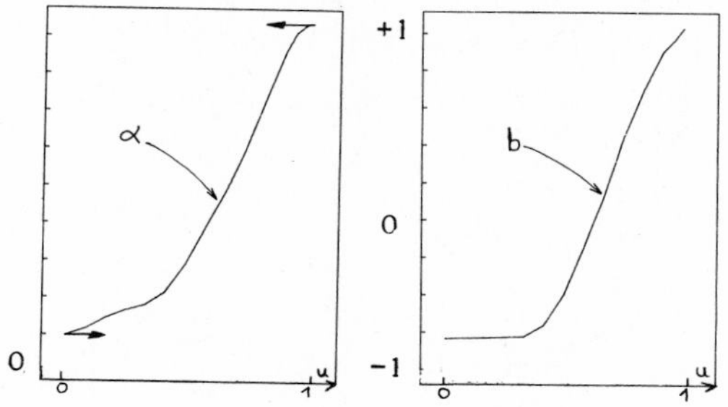

<u>Fig. 2</u>

<u>The identification problem then consists in trying to determine</u> α <u>and/or</u> b <u>from the measurement</u> $\varphi_{1\,ob}$ <u>of the water output rate</u> $\varphi_1(t)$.
This can be achieved in minimizing the observation error functional :

$$(6) \qquad J(\alpha,b) = \int_o^T (\varphi_1(t) - \varphi_{1ob}(t))^2 dt = \text{Min}$$

over a suitable set of functions α and/or b. A proof of the existence of a solution to that optimization problem and a calculus of the Gâteaux-derivative of J may be

found in [3], for a non-degenerated diffusion term and condition (3) being replaced by a Dirichlet condition.
The choice of a good approximation scheme for the state equations (1) to (5) is a necessary preliminary to the numerical resolution of the optimization problem (6). We present in § 2 such an approximation, and in § 3 its use for the optimization problem.

2. NUMERICAL APPROXIMATION OF THE STATE EQUATIONS

Divide $[0,L]$ into N intervals of length $h = \frac{L}{N}$, and set $x_i = ih$ $i = 0,1\ldots N$.
The two ideas which guided us in the choice of the approximation are the following :

i) for high water injection rate $q(t)$, equation (1) behaves quite as a first order hyperbolic equation, and the solution u can present very stiff fronts. So u was approximated by discontinuous finite elements, following the ideas introduced by Lesaint [5] for the linear first order hyperbolic equation :

$$(7)\quad \begin{cases} u \simeq u_h \in V_h \\ V_h = \{v \in L^2(0,L) \mid v|_{]x_i,x_{i+1}[} \in \mathcal{P}^k \ i=0,1..N-1\} \end{cases}$$

where $\mathcal{P}^k$ is the set of polynomials in x of degree $\leq k$.

ii) we need a good numerical approximation for the water output rate $\varphi_1(t)$ (as it appears in the functional (6) to be minimized !), which includes a derivative of u. As we have already chosen a discontinuous u_h, the only way out is to use a mixed approximation, following the ideas of [6] : besides u we must approximate the quantity $r = -K\frac{\partial}{\partial x}(\alpha(u)) = -K\alpha'(u)\frac{\partial u}{\partial x}$ (and not $\frac{\partial u}{\partial x}$ itself, that is not square integrable as α' vanishes for $u = 0$ and $u = 1$) :

$$(8)\quad \begin{cases} r = -K\frac{\partial}{\partial x}(\alpha(u)) \simeq r_h \in Q_h \\ Q_h = \{v \in \mathcal{C}([0,L]) \mid v|_{]x_i,x_{i+1}[} \in \mathcal{P}^{k+1} \ i=0,1..N-1\} \end{cases}$$

We shall use the straightforward notations :

$$(8\text{bis})\qquad b_h = b(u_h)\ ,\quad b^+_{hi} = b(u^+_h(x_i))\ ,\quad b^-_{hi} = b(u_h x_i))\ , \text{etc}\ldots$$

Following the above guideline, we are led to the following semidiscretization in space of the state equations (1) to (5) :

$$(9)\quad \left\{ \int_{x_i}^{x_{i+1}} \emptyset \frac{du_h}{dt}v + \int_{x_i}^{x_{i+1}} \frac{\partial r_h}{\partial x}v + q(t)\left\{(b^+_{h_i} - b^-_{h_i})v(x_i) + \int_{x_i}^{x_{i+1}} \frac{\partial b_h}{\partial x}v\right\} = 0 \right.$$
$$\forall v \in \mathcal{P}^k\ ,\ \forall\, i = 0,1\ldots N-1\ ,\ \forall t \in^1 [0,T]$$

with

(10) $$u_{ho}^{-} = u_e \qquad \forall\, t \in [0,T]$$

and

(11) $$\int_0^L \frac{r_h s_h}{K} = \int_0^L \alpha_h \frac{\partial s_h}{\partial x} + \alpha(u_e)s_h(o) - \alpha(u_s)s_h(L) \qquad \forall\, s_h \in Q_h \,,\quad \forall\, t \in [0,T]$$

(12) $$u_e \equiv 1 \,,\quad u_s \leq 1 \,,\quad \varphi_{1h} \geq 0 \,,\quad (1-u_s)\varphi_{1h} = 0$$

(13) $$u_h(x,o) = u_{ho}(x)$$

(14) $$\varphi_{1h}(t) = (1+b_{hN}^{-}(t))q(t) + r_{hN}(t)\ .$$

The time discretization is simply made by replacing $\frac{\partial u_k}{\partial t}$ by $\frac{u_h^{n+1} - u_k^n}{\tau}$ (τ being the time step) and using an explicit scheme in (9).

This approximation main features are the following :

- it is valid from the pure parabolic case ($q \equiv 0$: imbibition process) to the pure hyperbolic case ($\alpha \equiv 0$: Buckley-Leverett approximation)
- it satisfies exactly the mass balance for each of the two fluids separately ; for water for instance, one checks easily that

(15) $$\begin{cases} \frac{d}{dt}\int_0^L \emptyset\, u_h(t)dx = \phi_{1h}(t) - \varphi_{1h}(t) \quad \text{where} \\ \phi_{1h}(t) = \left[1+b_{ho}^{-}(t)\right] q(t) + r_{ho}(t) = \text{water input rate at } x=0. \end{cases}$$

- it gives a precise determination of the water break-through time : it is the first time when u_s becomes equal to 1.
- for $k = 0$, u_h is a piecewise constant function, and the equations (9) to (11) turn out to be identical, for an adequate choice of quadrature formulas, to the usual finite difference scheme. The equations (12)(14) give however a new formulation of the B.C. at $x = L$. We shall refer to that case when using the term "finite differences".
- for $k = 1$, u_h is a discontinuous piecewise linear function, and we get new schemes depending on the choice of the quadrature formulas we use. The best choice turned out to be to use the Simpson formula in all integrals appearing in (9) and in (11). We shall refer to that case when using the term "discontinuous finite elements".

Numerical results of the resolution of the state equations :

The datas : $L = 24\text{cm}$, $\emptyset = .156$, $K = 1.3\ 10^{-8}$, $\sigma = 18\text{cm}^2$, $q = 4.629\ 10^{-4}$ cm/s, the functions α and b as in Fig. 2. These datas correspond to the rather high water injection rate of 1 cm^3/mn, and a stiff water saturation front is expected. The actual saturations (ranging from .15 to .63) are used in the numerical computations instead of the reduced saturations (ranging from 0 to 1).

The water saturation front : the saturation distribution after 1000 s. are plotted in Fig. 3 for k = 0 and 1 and for a space step size growing from 1/4 to 1 cm.

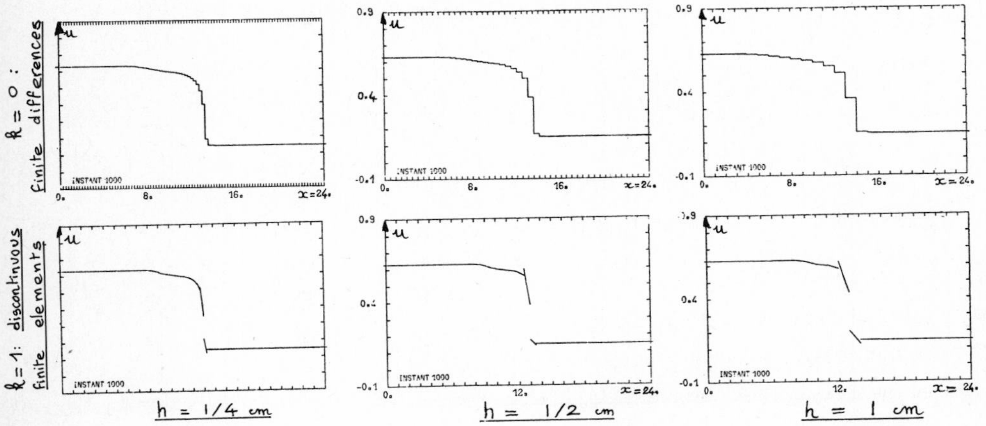

Fig. 3

More instructive is the superposition of the curves for the steps h = 1/4 and h = 1.

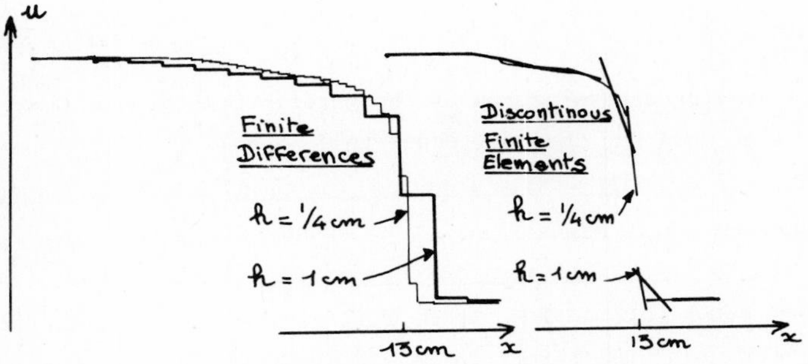

Fig. 4: Comparison of the water front obtained for different schemes and step size

One sees that the position of the front is, for discontinuous finite elements, practically independent from the space step size, which is the case, as it is well known, for the finite difference.

The overall process : we have plotted on Fig. 5 the successive positions of the front every 200 s., for finite elements with a step of 1/4 cm.

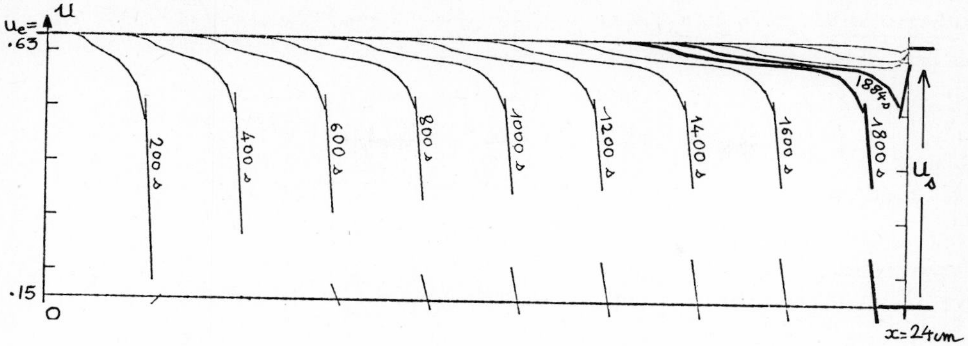

Fig. 5

We have reinforced the two curves at time $t = 1884$ s. and $t = 1800$ s., corresponding respectively to the water break-through time and to the last curve plotted before that time. One sees that the "surface" saturation u_s has grown up from .15 to .63 during the time interval between the two curves.

The water output rate $\varphi_1(t)$ and the break through time (B.T.) : we have plotted on Fig. 6, instead of the water output rate $\varphi_1(t)$, the output water/oil ratio (WOR) more commonly used by oil engineers :

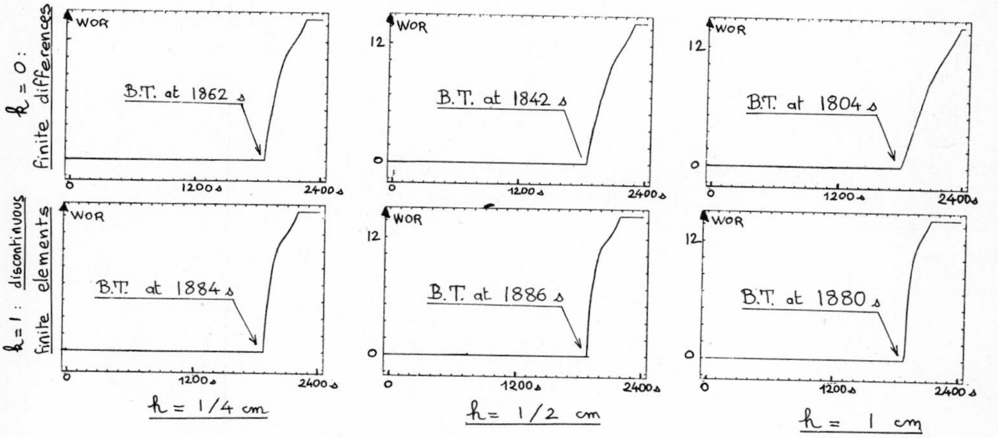

Fig. 6

Here too, the superiority of finite elements over finite difference is appearent : the break-through time is practically independent from the discretization for the finite elements, and the WOR-curves form remains very similar, which is not the case with the finite differences.

The computation time : for the simulation of the flow up to 2400 s , the computation time required on an IRIS 80 CII is :

	(h,τ) in (cm,s)	(1 , 4)	(1/2,2)	(1/4,1)
computer time in minute and hundreth of minute	Finite differences	.38	.79	2.29
	Finite elements	0.6	1.71	5.53

For the same number of unknowns in space, for instance finite differences with h = 1/4 and finite elements with h = 1/2, the maximum admissible time steps τ are about 5s in both cases, so that the corresponding computing time is 0.46 s for finite differences and 0.68 s for finite elements, i.e. an increase of about 50 % which is not prohibitive on account of the gain in accuracy observed.

The limiting cases : we have drawn on Fig. 7 the solution u obtained with the pure parabolic case (left) corresponding to the imbibition process, and the pure hyperbolic case (right). In the latter case, the physical minimum entropy solution is obtained.

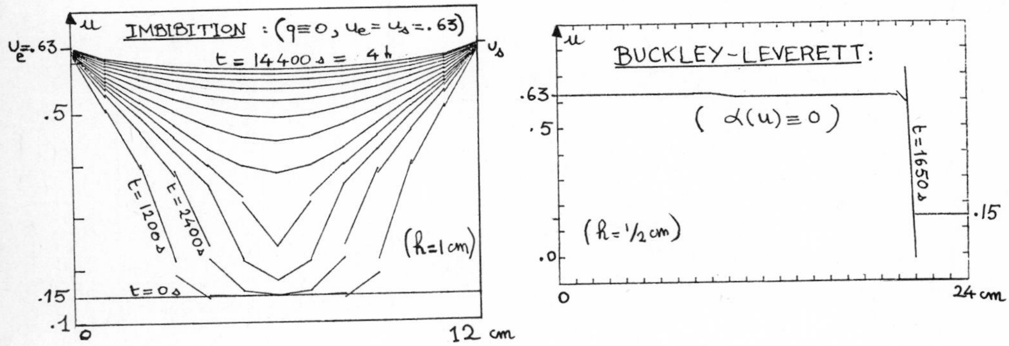

Fig. 7

CONCLUSION :

The use of discontinuous finite elements, together with the unilateral condition at the right-hand side of the slab, gives a good approximation of the saturation front and of the water output rate, even for large step sizes : it is this scheme that we shall use in the identification procedure.

3. NUMERICAL RESOLUTION OF THE IDENTIFICATION PROBLEM.

Because of the unilateral condition in (12), the state equations (9) to (14) are not differentiable. Thus, we approximate them first, by penalizing (cf. [4]), the corresponding variational inequality, this leading to replace the boundary condition (12) by :

$$\text{(12bis)} \qquad u_e \equiv 1 \ , \qquad \varphi_{1h} - \frac{1}{2\varepsilon}\left[(u_s-1)^+\right]^2 = 0$$

where $\varepsilon > 0$ is given "small enough".

We define the -only semi-discretized in space for sake of (relative) simplicity - approximated criterion J_h by

$$\text{(16)} \qquad J_h(\alpha,b) = \int_0^T (\varphi_{1h}(t) - \varphi_{1ob}(t))^2 dt$$

We define then a Lagrangian with the criterion (16) and the state equations (9)(10)(11)(12bis)(13) which leads us, after some meticulous and tedious calculus, to define a discrete adjoint state $p_h(t) \in V_h$, $\theta_h(t) \in Q_h$, $t \in [0,T]$; (think of θ_h as being an approximation of $-K \frac{\partial p_h}{\partial x}$). With the system of notation (8bis), the adjoint equations are (compare with (9)(10)(11)(12bis) and (13)) :

$$\text{(17)} \qquad \left\{ \begin{array}{l} -\int_{x_{i-1}}^{x_i} \emptyset \frac{dp_h}{dt} v + \int_{x_{i-1}}^{x_i} \alpha'_h \frac{\partial\theta_h}{\partial x} v - q(t) \left\{ b'^{-}_{hi}(p^+_{hi}-p^-_{hi})v(x_i) + \int_{x_{i-1}}^{x_i} b'_h \frac{\partial p_h}{\partial x} v \right\} = 0 \\ \forall\ v \in \mathcal{P}^k, \quad \forall\ i=1\ldots N\ , \quad \forall\ t \in [0,T] \end{array} \right.$$

$$\text{(18)} \qquad p^+_{hN} = p_s \qquad \forall\ t \in [0,T]$$

$$\text{(19)} \qquad \int_0^L \frac{\theta_h s_h}{K} = \int_0^L p_h \frac{\partial s_h}{\partial x} + p_e s_h(o) - p_s s_h(L) \qquad \forall\ s_h \in Q_h \quad \forall\ t \in [0,T]$$

$$\text{(20)} \qquad p_e \equiv 0\ ,\ \alpha'(u_s)\theta_{hN} - \frac{1}{\varepsilon}(u_s-1)^+[p_s + 2(\varphi_{1h}-\varphi_{1ob})] = 0 \qquad \forall\ t \in [0,T]$$

$$\text{(21)} \qquad p_h(x,T) = 0 \ . \qquad \forall\ x \in [0,L].$$

The first order variation δJ_k of criterion J_h associated to variations $\delta\alpha$ and δb of the parameters is then given by :

$$\text{(22)} \qquad \left\{ \begin{array}{l} \delta J_h = -\int_0^T q(t) \left\{ \sum_{i=0}^{N} \delta b^-_{hi}(p^+_{hi}-p^-_{hi}) + \sum_{i=0}^{N-1} \int_{x_i}^{x_{i+1}} \delta b_h \frac{\partial p_h}{\partial x} \right\} \\ - \int_0^T \left\{ \delta\alpha(u_e)\theta_{ho} - \delta\alpha(u_s)\theta_{hN} + \int_0^L \delta\alpha_h \frac{\partial\theta_h}{\partial x} \right\} \ . \end{array} \right.$$

The functions α and b are represented as continuous piecewise linear functions of the saturation u :

Fig.8 The discretization of the parameter functions α and b

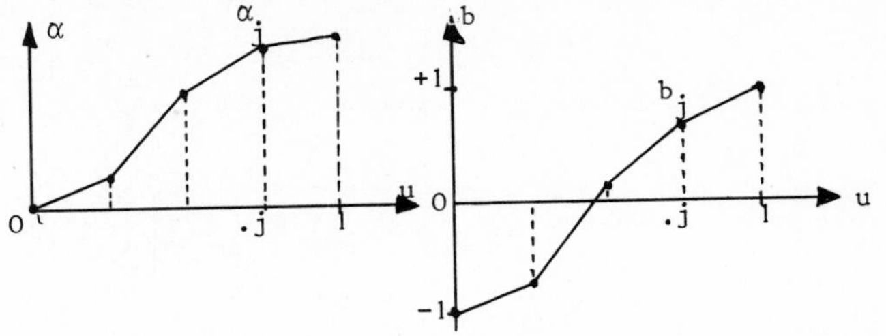

It is then easy to derive from (22) the exact gradient of J_h with respect to the numerical unknowns α_j and b_j of the optimization problem.
The monotonicity (α and b are known to be the increasing functions of u) and regularity constraints ($|\alpha''|$ and $|b''|$ bounded, in order to avoid oscillating non linearities) are taken into account simply by penalization of the objective function (16).
A conjugate gradient method is used.

Numerical results for the identification problem.

The datas : The same datas as in § 2 are used for the state equation, excepted that a shorter slab (6 cm) is used, in order to save computation time, as no information is gained before the water break-through time. Discontinuous finite elements (k=1) are used, with a space step of h=1/2 cm.
The actual saturations (ranging from .15 to .63) are used, and the interval [.15, .63] is divided, as shown in Fig. 8, into 13 intervals for the discretization of α and b , so that there are 14 numerical unknown values α_j and/or 14 values b_j. The functions α and b of Fig. 2 have been used for the simulation of the observation φ_{1ob} (t). We shall refer to these as to the "true" parameters α^* , b^* .

The "no model-error" case : The observation φ_{1ob}.(t) is computed, using the "true" parameters α^* and b^* , with exactly the same numerical model as used for the identification ((9)(10)(11)(12bis)(13) with k=1 and h = 1/2) hence $J(\alpha^*,b^*)= 0$. The results obtained when we try to identify b from that observation, are shown in Fig. 9 , where the water saturation repartition, u after 300s is also shown. Though the minimization of J_h has been very effective (computed and observed water output rates are identical up to the drawing precision), the computed b turns out to be very different from the "true" one, namely for the saturations less than .4, which seems to have absolutely no influence on the observation. In correspondence with that, one observes a discrepancy between the computed and the "true" water front.

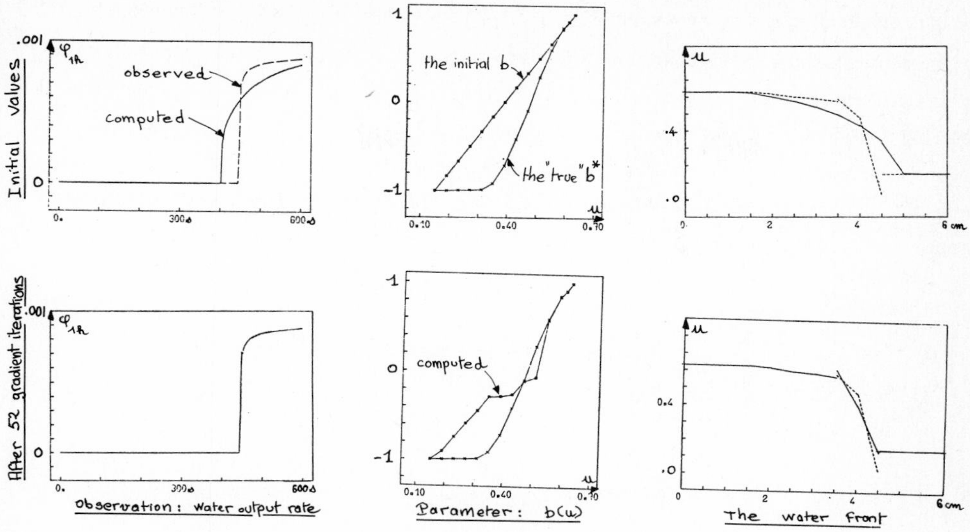

Fig. 9

The "model-error" case : Assuming the physical laws underlying equations (1) to (5) to be exact, and our approximation scheme to be converging when $h \to 0$, a (noiseless) "actual" observation $\varphi_{1ob.}(t)$ can be simulated in using the discretized equations with a small h, chosen here equal to 1/4 cm, so that $J(\alpha^*, b^*) \neq 0$. The results for the identification of b are shown on Fig.10. A small discrepancy between computed and observed water output rate is still visible, but the model error has not affected the computed b, which is very similar to the one shown in Fig. 8.

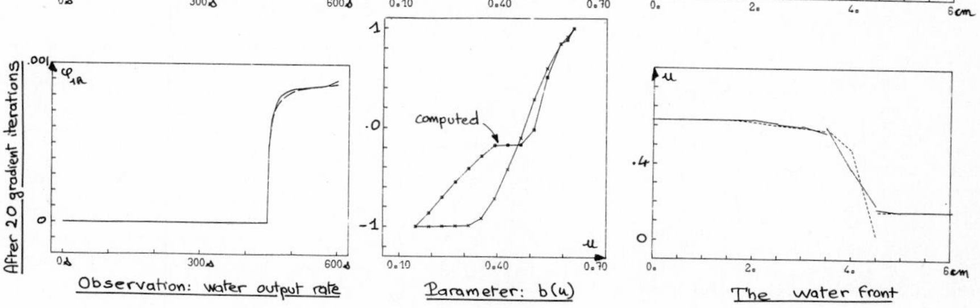

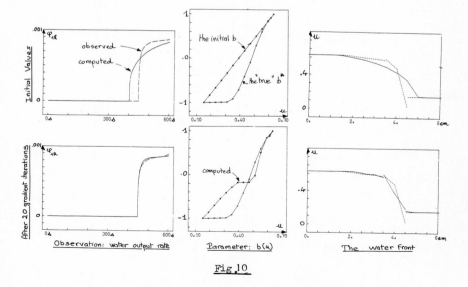

Fig.10

The identification of α is shown on Fig.11 . The initial value of J is very small which means that for the chosen water injection rate q , the diffusion parameter α has very little influence on the water output rate φ_{1h} , i.e. the state equations are nearly of hyperbolic type. The initial discrepancy between φ_{1h} and $\varphi_{1ob.}$ also results from the model-error, which, in our case, could be as much important as the error $\alpha - \alpha^*$ on the parameter. Nevertheless, the optimization algorithm works, the output error is still reduced, but of course, the computed α does not look at all like α^*.

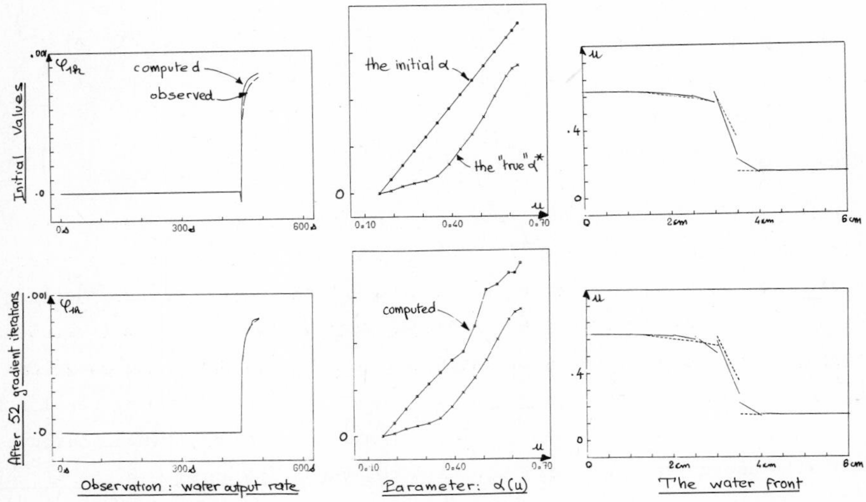

Fig. 11

About the simultaneous identifications of α and b : The above numerical results, showing the poor influence of α and b on φ_{1h}, do not encourage the simultaneous identification of those two functions. Engineers, however, did not know condition (12) and used the Dirichlet B.C. :

(12 ter) $\qquad u_e \equiv 1 , \quad u_s \equiv Y(t-t_{BT})$ where Y is the Heavyside function,

and where t_{BT} is the experimental break-through time, which gives a less realistic water output rate $\varphi_{1h}(t)$, having some parasit oscillation just after the B.T. However numerical experimentations (cf. [7]) have shown that this condition artificially increased the influence on φ_{1h} of the values of α and b for saturations less than .4. The results of simultaneous identification of α and b in that case, all the remaining conditions being unchanged, are shown in Fig. 12.

Conclusion of the numerical results of the identification problem :

The above results show that the proposed optimization procedure works satisfactorily, but that the observation of the water output rate $\varphi_{1h}(t)$ is not sufficient, when the realistic B.C.(12) are used, for the determination of α and b over the whole range of saturations. It is suggested to ask the experimentators more observations, for instance the value, along time, of the saturation u at a given point of the slab.

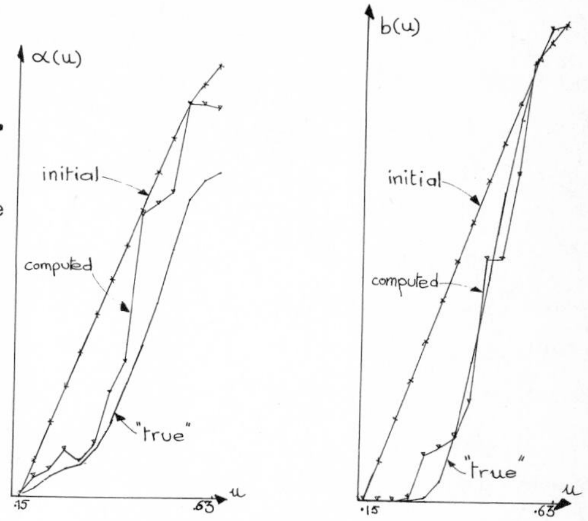

Fig. 12

3. CONCLUSION

We have given an adequate discretization of a 1-D rather general non-linear degenerated diffusion and transport equation, valid under conditions ranging from the pure parabolic case to the pure hyperbolic case. We have calculated the corresponding adjoint state, and the exact gradient of the approximated output error criterion, and applied it to the resolution of an identification problem. Numerical results have been shown.

REFERENCES

[1] M. MUSKAT The Flow of Homogeneous Fluids through porous Media, McGrawHill, New-York, (1937).

[2] G. CHAVENT A new formulation of diphasic incompressible flows in porous media, in Lecture Notes in Mathematics, Vol. 503, Springer, (1976).

[3] G. CHAVENT About the identification and modelling of miscible or immiscible displacement in porous media, in Lecture Notes in Control and Information Sciences, Vol. 1, Springer, (1978).

[4] G. CHAVENT Un théorème d'existence dans une inéquation variationnelle parabolique dégénérée modélisant le déplacement d'un fluide par un autre non miscible, LABORIA Report No. 260, IRIA, POBox 105, 78150 Le Chesnay, France.

[5] P. LESAINT and P.A. RAVIART On a finite element method for solving the neutron transport equation. Symposium on Mathematical aspects of finite elements in P.D.E. Math. Research Center, University of Wisconsin, Madison,(1974).

[6] P.A. RAVIART and J.M. THOMAS A mixed finite element method for second order elliptic problems.

[7] G. COHEN Thesis, Paris, (to appear).

OPTIMIZATION OF THE DESIGN OF AN IN-FLIGHT REFUELING SYSTEM

J. Eichler
Ben Gurion University of the Negev
Beer Sheva, Israel

ABSTRACT

The design goals of an in-flight refueling system are to maintain drogue motion within specified limits for given flight conditions of speed and altitude, and in the presence of disturbances caused by wind gusts and wing vibration. Thus, a set of desired flight conditions are given and a model of maximum wind gust disturbances and wing vibration (frequency and magnitude) are specified and the designer must choose design variables such as drogue weight and drag, hose length, width and weight per unit length, that will obtain a dynamic condition of drogue motion within tolerance (usually the diameter of the drogue traversed in a specified time). This problem requires the development and solution of non-linear partial differential equations of the hose drogue motion.

The problem is treated by first solving the static position partial differential equations and subsequently developing linearized perturbation equations about the nominal static solution. Simplifying assumptions are made that permit a linear set of hyperbolic partial differential equations with variable coefficients to be obtained. These are solved numerically many times as part of a parameter search over the range of design variables for the optimum set. The paper reports on the model used and the solution of the equations within the optimization of the design.

SYMBOL LIST

$\bar{c}$ mean aerodynamic chord (inches)

p lift per unit length of hose (lbs./ft.)

P drogue lift (lbs.)

q drag per unit length of hose (lbs./ft.)

Q drogue drag (lbs.)

s distance along the hose (ft.)

s_o length of the hose (ft.)

T tension (lbs.)

t time (sec.)

V velocity of the aircraft (knot)

w input wind disturbance (ft.)

W weight of the drogue (lbs.)

x axis along horizontal plane at the level of the pod (ft.)

y vertical axis from the pod down, positive down (ft.)

ρ_H hose weight per unit length (slugs/ft.)

INTRODUCTION

The dynamic problem of in-flight refueling [1] contains a number of phases: the tanker attaining flight conditions for fueling; letting out the hose-drogue system; the transient motion of the hose-drogue system until it is fully out and stabilized; subsequent disturbances due to vertical wind gusts, wing vibration etc.; hook up to the drogue; transfer of fuel and so on. This study deals with that phase beginning after the transient due to set-up of the system has passed and continuing until the phase of hook-up. To be considered are the effects on a set-up hose-drogue system, of the disturbances due to: vertical wind gusts and wing vibration. (see Figure 1).

The tanker plane is assumed to be in level flight at a constant velocity. The main design question is whether the vertical motion of the drogue will allow hook-up, to be done easily. Ideally, if the drogue's vertical motion does not exceed ± its own diameter, the fueler plane will be able to insert its probe into a section of the cone of the drogue and attain hook-up.

The possibility of attaining this desired dynamic response to disturbances will vary for different flight conditions of altitude and speed, and will of course depend on the design parameters of the hose-drogue system. It is assumed that the wing vibration spectrum can be represented by 2 frequencies (obtained from a structural analysis of the aircraft) and that a worst case vertical wind gust is defined for the design specification. The flight conditions are represented by 2 conditions: sea level, 265 knots and 25,000 ft., 346 knots. The changeable design parameters and their nominal values are:

1.	weight/length of the hose full (lbs./ft.)	1.858
2.	length of the hose (ft.)	50.0
3.	drogue weight (lbs.)	52.8
4.	drogue drag ($C_D A$) (ft.^{2})	2.2

In reference 1 a tradeoff between two designs is described. The desire is expressed at the conclusion of reference 1 to further optimize the selected design. " A more complete performance optimization could now be conducted. This would use the programs developed in this report as subroutines to a merit function reflecting the design goal and using a search routine such as gradient, simplex, or a grid search to solve for the optimum design parameters." This paper presents the implementation of this performance optimization task.

Optimization of a design can never really be considered complete without an analysis of the sensitivity of the design. Consequently, an analysis of the sensitivity of the optimum design was conducted and is also presented.

The paper follows the topical headings:

1. Problem definition.
2. Solution of the partial differential equations.
3. Optimization method.
4. Optimization and sensitivity results.
5. Acknowledgement.
6. Future work.

1. PROBLEM DEFINITION

Considering the tanker plane to be in level flight at a constant velocity, we seek the optimum set of design variables within a specified region of allowed variation about the nominal design set. "Optimum" is defined by specifying a cost function, the minmum of which will be obtained by the optimum design.

Two flight conditions are considered: sea level, 265 knots and 25,000 ft., 346 knots. For each condition a worst case vertical wind gust was considered, with a mangnitude w_{de} of 66 ft./sec. at sea level and 61.3 ft./sec. at 25000 ft. altitude. The disturbance was modelled by:

$$w = \frac{w_{de}}{2}\left[\sin \frac{2\pi}{25\,\bar{c}} Vt\right]$$

where V is the speed of the tanker plane, t is time and $\bar{c}$ is the mean aerodynamic chord of the wing (taken as 283 inches). The basic input disturbance thus has a 50 halfchord wavelength.

For each flight condition, wing vibration disturbances were also considered. Based on structural analysis results, it was determined that two modes are important, one at 1.3 Hz. and the other at 2.1 Hz.. Their relative magnitudes were in the ratio of about 2:1, the lower frequency having the larger magnitude. The model for wing vibration was taken as a unit sinusoid input at the pod end of the hose-drogue system. (see Figure 1). Consequently, at each flight condition the dynamic response to three separate inputs were considered: one worst case vertical wind gust and two different frequency, wing vibrations. This was considered to be an adequate representation of the dynamic performance of the in-flight refueling system. Naturally what was of most interest in the dynamic response, was the amplitude and velocity of the motion of the drogue.

A cost function was constructed as a weighted function of the steady state dynamic performance. The function was:

$$YP = \sum_{i=1}^{2}\left[y_{v1.3} + 0.5\, y_{v2.1} + \dot{y}_{wg}\right]_i$$

where i varies over the

two flight conditions and $y_{v1.3}$, $y_{v2.1}$ are the steady state wing vibration amplitudes at the drogue, for a 1.3 Hz. and 2.1 Hz disturbance, respectively. $\dot{y}_{wg}$ is the steady state motion of the drogue per second due to the vertical wind gust disturbance. Thus, the cost function will attain a minimum at the optimum setting of the design parameters. There are certain natural limits to the variation of each design variable, that are possible. The weight/length, for example cannot be decreased without changing the hose, but may easily be increased by adding weights along the length of the hose. Similarly, the length of the hose may not be decreased below about 45 ft. without running the risk of a collision between the plane receiving fuel and the tanker plane. Based on these types of considerations the following range of variation of the design variables was considered:

	nominal	range
1. weight/length (lbs./ft.) full	1.858	1.858 to 2.158
2. length (ft.)	50.0	45 to 60
3. drogue weight (lbs.)	52.8	48 to 58
4. drogue drag ($C_D A$) (ft.^{2})	2.2	2.0 to 2.6

In summary, the problem was to find a set of design variables, within the specified range that minize the cost function; where the cost function represents the steady state motion of the drogue in the flight operation envelope and under disturbances due to vertical wind gust and wing vibration.

2. SOLUTION OF THE PARTIAL DIFFERENTIAL EQUATIONS

The mathematical model for the in-flight refueling hose-drogue system [1] is derived as follows. Referring to figure 1, we summarize the forces acting on a differential of hose ds and obtain:

eq. 1) $$\rho_H \frac{\partial^2 y}{\partial t^2} ds + p ds + T\frac{\partial y}{\partial s} - \rho_H g ds - \left(T + \frac{\partial T}{\partial s} ds\right)\left(\frac{\partial y}{\partial s} + \frac{\partial^2 y}{\partial s^2} ds\right) = 0$$

eq. 2) $$\rho_H \frac{\partial^2 x}{\partial t^2} ds - q ds + T\frac{\partial x}{\partial s} - \left(T + \frac{\partial T}{\partial s} ds\right)\left(\frac{\partial x}{\partial s} + \frac{\partial^2 x}{\partial s^2} ds\right) = 0$$

The boundary conditions of these equations will be described at the pod end to be zero if we consider wind gusts alone, and sinrt if we consider wing vibration alone (where r is the frequency of the vibration in radians/sec.). At the drogue end we will require that the resultant forces at the end of the hose be in the same direction as the resultant forces acting on the drogue.

The assumptions in deriving the model and further details are given in reference 1. The method of solution is also given in reference 1. Basically, the non-linear partial differential equations (PDE) shown in equations 1 and 2 are solved in a two step procedure. The first step is to consider that static solution, and then to develope linear perturbation equations about the static solution. The static PDE can be solved by approximating several terms by first order Taylor Series thus obtaing equations that allow a closed form solution. Perturbation equations are developed about the static solution. The perturbation equations yield a PDE for y with coefficients that are a function of the independent variable s distance along the hose. Two solution methods are described in reference 1. For the optimization study described in this paper, only the numerical method solution (using finite differences) was used. The solutions were obtained on a CDC Cyber 173 computer. The numerical method solution was permitted to run until the transient passed and the steady state solution obtained. The cost function was calculated based on the steady state solution values.

3. OPTIMIZATION METHOD

Prior to and as a first step in optimizing, it is most important to determine the texture of the cost function surface. The cost function can be thought of as a surface in 4 design variables plus 1 cost variable (= 5 dimensional) space. At each point specified by a set of design variable values there is a cost. A smooth texture would mean that small variations of design produce only small changes in cost, and a rough texture would be evidenced by large cost changes occuring for small design value variations. Of course we mean both positive and negativs cost changes. Rough texture would signify that there may be local minima. We, naturally seek the global minimum. The method and resolution of our search for the optimum must be dictated by this phenomenon of rough ness of texture. The above consideration led to a first step of conducting a coarse grid search within the range of the design variables. The cost was calculated for 192 different designs.

The designs considered were all combinations of:

	values	number of values
1. weight/length	1.858,2.008,2.158	3
2. length	45, 50, 55, 60	4
3. drogue weight	48, 51, 54, 58	4
4. drogue drag	2.0, 2.2, 2.4, 2.6	4

This gives 3X4X4X4 = 192 designs. The results of this coarse search were costs that varied from 9.24 at design 3444 (where 3 indicates the third value of design parameter one, i.e. 2.158; 4 indicates the fourth value of design parameter 2, i.e. 60 etc.) and a high value of 22.19 at design 2212. The texture was reasonably smooth but with three local minima indicated by this coarse search. They were at 1124, cost 10.44; at 1244, cost 10.35 and at 3444, cost 9.24.

A finer grid search was conducted at these points. At 1124 the four designs, combinations of 1.858; 45, 47; 50, 52 and 2.6 were considered. They all resulted in a higher cost than the 10.44 obtained at 1124. At 1244, four designs, combinations of 1.858,1.929; 49, 51; 58 and 2.6 were considered. The design 1.929, 49, 58, 2.6 yielded a cost of 10.34 which is just 0.01 under the 10.35 obtained at 1244. All other costs were higher.

At 3444 which gave the lowest cost, a search of 16 designs was made. These were combinations of 2.140, 2.158; 58, 60; 57, 58; and 2.5, 2.6. A low cost of 9.23 was obtained at 2.140; 60; 58; 2.6.

Further grid searches could have been made but it seemed reasonable to switch to an optimizing method that would grind down to the minimum at the best grid point so far obtained. It was decided to use the Simplex method of Nelder and Mead (see references 2 and 3). The starting simplex was at design 3444 with four other points at: 2.10, 59.0, 57.0, 2.5; 2.15, 59.5, 57.5, 2.55; 2.14, 58.9, 57.4, 2.570; and 2.130, 59.80, 57.20 2.510. Because the design was near the edge of the range of design variables, penalty functions were added to the cost function. The penalty function terms were:

$$VC1*(ROP/2)^4 + VC2*(s_o/50)^4 + VC3*(W/50)^4 + VC4*(C_DA/2.3)^4$$

where VC1 through VC4 were zero if the variable range was not exceeded and 1 if it was exceeded. ROP is weight/length; s_o is length; W is drogue weight and C_DA is drogue drag. Each variable is normalized by its approximate average value to give equal importance to all variables irrespective of their natural numerical range.

The Simplex method reached a minimum within the design range and after five steps. The minimum found was at: 2.1535, 59.725, 57.850, 2.5925 and with a cost of 9.19 .

4. OPTIMIZATION AND SENSITIVITY RESULTS

The optimization method is a coarse then a fine grid search followed by the Simplex method [2,3] at the best local minimum. This yielded an optimum design near the edge of the region of permitted design variation. Three local minima were investigated and the "global" minimum was found at design: 2.1535, 59.725, 57.850, 2.5925 with a cost of YP = 9.19. The cost of the nominal design was 17.3 and the worst design evaluated had a cost of 22.19.

Sensitivity of the design to variations of design parameters was estimated at the conclusion of the coarse grid search. Sensitivity S, is defined:

$$S = \frac{YP(i+deli,j) - YP(i,j)}{\overline{YP}} \Big/ \frac{deli}{\overline{i}}$$

where YP(i + deli,j) is the cost of the perturbed design, YP(i,j) is the cost of the unperturbed design, $\overline{YP}$ is the average cost in the interval; deli is the perturbation of the design (grid size step) and $\overline{i}$ is the average value of the design parameter. Thus S is normalized to both the cost and the size of the design parameter. Calculations of sensitivity based on the coarse grid values in the vicinity of 3444 gave (in absolute value):

S(wt./length) = 0.646
S(hose length) = 2.45
S(drogue weight) = 0.766
S(drogue drag) = 0.988

We notice that the system is most sensitive to hose length.

The sensitivity of the design was also calculated for changes in flight conditions. The nominal conditions of sea level, 265 knots; 25,000 ft., 346 knots, were perturbed and the costs calculated. The sensitivity results were:

altered condition	sensitivity
1. Increase high altitude velocity from 346 to 380.6	-1.40
2. " low " " " 265 to 291.5	0.04
3. Decrease high " " " 346 to 311.4	-1.83
4. " low " " " 265 to 238.5	1.25
5. Increase high altitude from 25,000 to 27,500	0.84
6. Decrease " " " " to 22,500	-1.03

These results indicate that an increase of velocity at high altitude lowers the cost, whereas a decrease in velocity increases the cost. The opposite is true at low velocity. Any change in the high altitude valuc causes the cost to increase.

5. ACKNOWLEDGEMENT

I wish to acknowledge the assistance of Mr. S. Yosephon who ran the Simplex program, and I thank him for his contribution to this paper. I also wish to thank Mr. K. Winnikoff for help in programing and my wife Hedy who did all the programming.

6. FUTURE WORK

The system considered in this study is taken to be fixed, i. e. none of the design parameters may be varied in flight. However since we notice a great sensitivity to hose length it leads us to consider the possibility of a dynamic design where hose length is the variable parameter. That is we seek $s_o(h,V)$ (hose length as a function of the tanker's altitude and velocity) that gives us an optimum design. This would be usefull when refueling at altitudes and velocities other than the nominal ones, and would give us a design with optimum performance at each flight condition.

It would also be usefull to study the problem from the frequency and resonance point of view. This would perhaps give some insight into what is happening physically. One can guess that there is resonance interaction between the "natural" frequencies of the hose-drogue system and the input disturbance frequencies of wing vibration and wind gust. The optimum solution arrived at in this study appears to straddle the nodes of the combined reaction so as to minimize the resulting amplitude at the drogue. Such a complete study, however, might require many solutions of the partial differential equations and therefore be very costly.

REFERENCES

1. Eichler, J. "Dynamic Analysis of an In-Flight Refueling System" accepted for publication in the A.I.A.A. Journal of Aircraft.

2. Kuester, J. L.; Mize, J. H. "Optimization Techniques with Fortran" McGraw Hill, Inc. 1973 pages 298-308.

3. Nelder, J.A.; Mead, R. "A Simplex Method for Function Minimization" The Computer Journal, Vol. 7 (1964) pages 308-313.

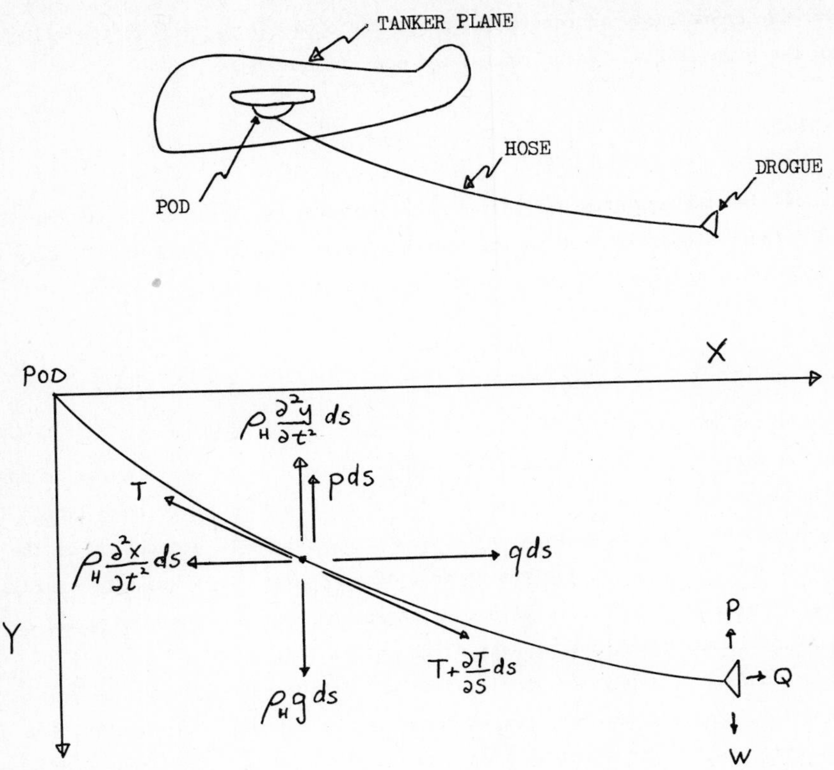

FIGURE 1   Geometry

THE FE (FINITE ELEMENT) & LP (LINEAR PROGRAMMING) METHOD AND THE RELATED METHODS FOR OPTIMIZATION OF PARTIAL DIFFERENTIAL EQUATION SYSTEMS

Tanehiro Futagami
Department of Civil Engineering
Hiroshima Institute of Technology
Miyake, Itsukaichi, Saiki, Hiroshima, 738, Japan

Abstruct

By combining finite element method with linear programming, a new optimization technique (Finite Element & Linear Programming Method, or, the FE&LP Method) has been developed in order to control differential equations with both equality or inequality constraints and an objective function. The generalization of the FE&LP method and the related methods are studied. The tractability in the initial and boundary conditions and the equality or inequality constraints makes sure that the proposed methods becomes useful techniques for several new types of boundary value problems.

Introduction

Finite element & linear programming method (the FE&LP method) (1, 2, 3, 4) has been developed in order to control systems of partial differential equations with both equality or inequality constraints and an objective function. Such systems are frequently encountered in various engineering and scientific problems of control and optimal design and, especially, are of interest in control of field problems (heat conduction, diffusion-convection, electric or magnetic potential, etc.). In the development of the FE&LP method, the combined use of finite element method and linear programming has been adopted. The finite element method, originated in structural mechanics, is powerful numerical method for the solution of differential equations because of its generality with respect to geometry and material properties (5, 6, 7). Linear programming is one of the most frequently used mathematical methods of operations research (8, 9). In the development of the FE& LP method the concepts of the decision variable and the state variable are adopted as in Bellman's dynamic programming (10, 11) and Pontryagin's maximum principle (12, 13). The FE&LP method utilizes the advantages of the numerical techniques of both finite element method and linear programming. Aguado and Remson (14) have suggested the

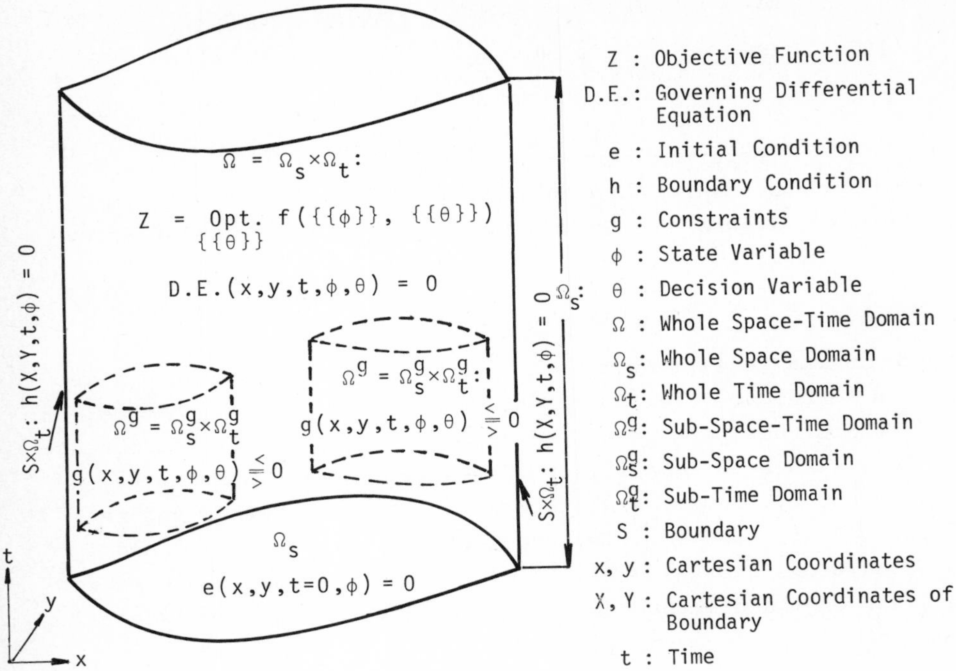

Fig. 1 General Concepts of FE&LP Method and Related Methods

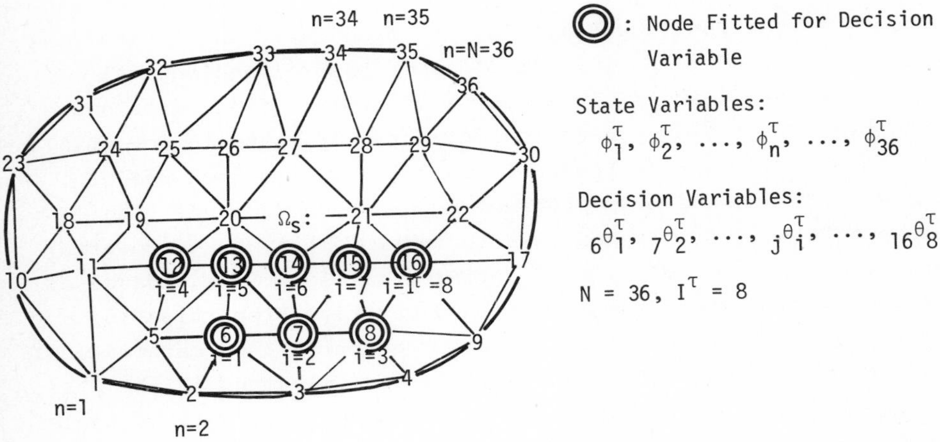

Fig. 2 Whole Space Domain Ω_s Devided into FE&LP Method and Related Methods

combined use of finite element method with linear programming in the study of ground-water management, in which finite difference method is used instead of finite element method.

1. Systems of Basic Differential Equations

The FE&LP method and the related methods are developed and systematized in order to control the following systems of partial differential equations.

Objective Function (throughout the whole domain $(\Omega = \Omega_s \times \Omega_t)$)

$$Z = \underset{\{\{\theta\}\}}{Opt.}\ f(\{\{\phi\}\}, \{\{\theta\}\}) = \begin{cases} \underset{\{\{\theta\}\}}{Max.}\ f(\{\{\phi\}\}, \{\{\theta\}\}) \\ \underset{\{\{\theta\}\}}{Min.}\ f(\{\{\phi\}\}, \{\{\theta\}\}) \end{cases} \quad (1)$$

subject to:

Equilibrium Equations

Governing Differential Equation (in the whole domain $(\Omega = \Omega_s \times \Omega_t)$)

$$D.E.\ (x,\ t,\ \phi,\ \theta) = 0 \quad (2)$$

Initial Condition (in the whole space domain Ω_s)

$$e\ (X,\ t=0,\ \phi) = 0 \quad (3)$$

Boundary Conditions (on the boundaries S)

$$h\ (X,\ t,\ \phi) = 0 \quad (4)$$

Constraints (in the subdomains $(\Omega^g = \Omega^g_s \times \Omega^g_t)$)

$$g\ (x,\ t,\ \phi,\ \theta) \lesseqgtr 0 \quad (5)$$

in which ϕ = the state variable (temperature, concentration, potential, etc.); θ = the decision variable (controllable load, controllable charge, etc.); $\{\{\phi\}\}$ = vector of the state variables in Ω; $\{\{\theta\}\}$ = vector of the decision variables in Ω; $\Omega = \Omega_s \times \Omega_t$ = the whole space-time domain; Ω_s = the whole space domain; Ω_t = the whole time domain; $\Omega^g = \Omega^g_s \times \Omega^g_t$ = sub-space time domain; Ω^g_s = sub-space domain; Ω^g_t = sub-time domain; S = the boundary; x = Cartesian coordinates (x, y, z); X = Cartesian coordinates of the boundary (X, Y, Z); and t = time.

2. Formulation of the FE&LP Method and the Related Methods

The combining use of finite element method and linear programming (or non-linear programming, etc.) in Eqs. 1-5 yields the following matrix-vector forms of the FE&LP method (or finite element & non-linear programming method, etc.).

Objective Function

$$Z = \underset{\{\{{}_j\theta_i^\tau\}\}}{Opt.}\; f\,(\{\{\phi_n^\tau\}\},\ \{\{{}_j\theta_i^\tau\}\}) \tag{6}$$

subject to:

Equilibrium Equations ((T×N)-Eqs.)

$$[A + \frac{1}{\Delta t^1} C]\{\phi_n^1\} + [D^1]\{{}_j\theta_i^1\} = \{Q_n^1\} + [\frac{1}{\Delta t^1} C]\{\Phi_n^0\} \tag{7-1}$$

$$-[\frac{1}{\Delta t^\tau} C]\{\phi_n^{\tau-1}\} + [A + \frac{1}{\Delta t^\tau} C]\{\phi_n^\tau\} + [D^\tau]\{{}_j\theta_i^\tau\} = \{Q_n^\tau\} \quad (\tau = 2 \sim T) \tag{7-τ}$$

Constraints (L-Eqs.)

$$[G_\phi]\{\{\phi_n^\tau\}\} + [G_\theta]\{\{{}_j\theta_i^\tau\}\} \lesseqgtr \{\{B_l\}\} \tag{8}$$

Nonnegative Conditions

$$\phi_n^\tau \geqq 0 \;(\tau = 1 \sim T,\ n = 1 \sim N),\quad {}_j\theta_i^\tau \geqq 0 \;(\tau = 1 \sim T,\ i = 1 \sim I^\tau) \tag{9}$$

in which [A] = the state matrix derived from finite element method, $(N\times N)$ matrix; $[D^\tau]$ = the decision matrix at τth time step, $(N\times I^\tau)$ sparse matrix; $[G_\phi]$ = the state-constraint matrix, $(L\times(T\times N))$ matrix; $[G_\theta]$ = the decision-constraint matrix, $(L\times(\sum_{\tau=1}^{T} I^\tau))$ matrix; $\{\phi_n^\tau\}$ = vector of the state variables at τth time step; $\{{}_j\theta_i^\tau\}$ = vector of the decision variables (controllable load vector, controllable charge vector, etc.) at τth time step; $\{Q_n^\tau\}$ = constant vector (uncontrollable load vector, uncontrollable charge vector, etc.) at τth time step; $\{\Phi_n^0\}$ = initial state vector; $\{\{B_l\}\}$ = constant vector in the constraints; $\tau = 1 \sim T$ = time step number; Δt^τ = increment of time in τth time step; $n = 1 \sim N$ = state variable number at each time step (node number in finite elements); $i = 1 \sim I^\tau$ = decision variable number at τth time step; j = node number fitted for ith decision variable; $l = 1 \sim L$ = constraint number; and [C] = the capacity matrix derived from finite element method, $(N\times N)$ matrix.

The FE&LP method is one that optimizes the objective function under the conditions of the equilibrium equations and the constraints. Since all of the variables in linear programming have to be nonnegative because of the limitation in the computational algorithm based on the simplex method, the nonnegative conditions (Eq. 9) are required. In the FE&LP method the number of the variables is $(T\times N + \sum_{\tau=1}^{T} I^\tau)$, the number of the equilibrium equations is $(T\times N)$, and the number of the constraints is L, respectively. In the sense of general linear programming, the equilibrium equations of the FE&LP method are also the constraints. Thus, the FE&LP method is a kind of linear programming in which the number of the variables is $(T \times N + \sum_{\tau=1}^{T} I^{\tau})$ and the number of the constraints is $(T \times N + L)$, respectively. In the EE&LP method the solution for the state variables and the solution for the decision variables are obtained simultaneously by the simplex method.

The equilibrium equations (Eqs. 7) are obtained the following procedures. At first, the Galerkin finite element method based on the weighted residual process (5) is applied to the governing equation (Eq. 2). The application yields the following algebraic equations.

$$[A]\{\phi_n^{\tau}\} + [C]\{\frac{\partial\phi}{\partial t}\Big|_n^{\tau}\} - \{\theta_n^{\tau}\} = \{Q_n^{\tau}\} \quad (\tau = 1 \sim T) \tag{10}$$

Although several time stepping scheme in finite element method have been presented, the following backward differencing (15) is used in this research.

$$\{\frac{\partial\phi}{\partial t}\Big|_n^{\tau}\} = \frac{1}{\Delta t^{\tau}}\,(\{\phi_n^{\tau}\} - \{\phi_n^{\tau-1}\}) \quad (\tau = 1 \sim T) \tag{11}$$

In order to reduce the number of the decision variables, θ_n^{τ} should be dropped at the nodal points where the controllable load does not exist. Therefore, the following expression for the decision variables is used instead of $\{\theta_n^{\tau}\}$ and the number of the decision variables at each time step is reduced from N to I^{τ}.

$$-\,[D^{\tau}]\{{}_j\theta_i^{\tau}\} = -\,[d_{ni}^{\tau}]\{{}_j\theta_i^{\tau}\} \tag{12}$$

in which $[D^{\tau}] = [d_{ni}^{\tau}]$ = the decision matrix, $(N \times I^{\tau})$ sparse matrix composed of zero factors with the exceptions of '-1' in I^{τ} factors whose row number is j and whose column number is i.

Substitution of Eqs. 11 and 12 into Eqs. 10 yields the equilibrium equations (Eqs. 7).

2. FE&LP Method in Control of Field Problems

In order to clarify the features of the FE&LP method, the application of the method to field problems (heat conduction, diffusion-convection, electric or magnetic potential, etc.) is studied. In the field problems the basic differential equation systems are as follows:

Objective Function (throughout the whole domain $(\Omega = \Omega_s \times \Omega_t)$)

$$Z = \underset{\{\{\theta\}\}}{Opt.}\, f(\{\{\phi\}\}, \{\{\theta\}\}) = \begin{cases} \underset{\{\{\theta\}\}}{Max.}\, f(\{\{\phi\}\}, \{\{\theta\}\}) \\ \underset{\{\{\theta\}\}}{Min.}\, f(\{\{\phi\}\}, \{\{\theta\}\}) \end{cases} \tag{13}$$

subject to:

Equilibrium Equations

Governing Differential Equation (in the whole domain ($\Omega = \Omega_s \times \Omega_t$))

$$\underbrace{\omega \frac{\partial \phi}{\partial t} = div\ (K\ grad\ \phi)}_{\phi\text{-terms}} + \underbrace{\theta}_{\theta\text{-term}} + \underbrace{Q}_{const} \tag{14}$$

Initial Condition (in the whole space domain Ω_s)

$$\phi(x,y,t=0) = \Phi^0(x,y) \tag{15}$$

Boundary Conditons (on the boundaries S)

$$\phi(X,Y,t) = \Phi_b(X,Y,t), \quad \frac{\partial \phi}{\partial n}\Big|(X,Y,t) = 0 \tag{16}$$

Constraints (in the subdomains ($\Omega^g = \Omega_s^g \times \Omega_t^g$))

$$\underline{\Phi} \leq \phi \leq \overline{\Phi}, \text{ or } \begin{cases} \phi \geq \underline{\Phi} \\ \phi \leq \overline{\Phi} \end{cases}$$
$$\underline{\Theta} \leq \theta \leq \overline{\Theta}, \text{ or } \begin{cases} \theta \geq \underline{\Theta} \\ \theta \leq \overline{\Theta} \end{cases} \tag{17}$$

in which ϕ = the state variable (temperature, concentration, potential, etc.); θ = the decision variable (controllable load, controllable rate of production, controllable charge, etc.); Q = constant (uncontrollable load, uncontrollable rate of production, uncontrollable charge, etc.), K = thermal conductivity, diffusion coefficient, etc.; ω = constant ($\omega = \rho c$ in heat conduction equation, $\omega = 1$ in diffusion-convection equation, etc.); Φ^0 = initial state; Φ_b = prescribed boundary value; $\underline{\Phi}$ = the lower limit of the state variable, $\overline{\Phi}$ = the upper limit of the state variable; $\underline{\theta}$ = the lower limit of the decision variable; $\overline{\Theta}$ = the upper limit of the decision variable.

As for the constraints, although only the lower and upper limits of the state variable and decision variable are imposed in the above equation systems, we can impose other constraints, if necessary.

The application of the FE&LP method to the above differential equation systems yilds the matrix-vector forms. The formulation of the FE&LP method for a simple model in heat conduction problem are shown in Eqs. 18-26 (see Fig. 3). In the model all of the boundaries are nonconvective ones, or $\partial\phi/\partial n = 0$ on the all boundaries.

Objective Function

$$Z = \underset{\{\{_j\theta_i^\tau\}\}}{Opt.}\ f\ (\{\{\phi_n^\tau\}\}, \{\{_j\theta_i^\tau\}\}) \simeq \underset{\{\{_j\theta_i^\tau\}\}}{Opt.}\ \sum_{\tau=1}^{3}\left(\sum_{n=1}^{5} \alpha_n^\tau \phi_n^\tau + \sum_{i=1}^{2} \beta_i^\tau\ {}_j\theta_i^\tau\right) \tag{18}$$

subject to:

Equilibrium Equations ((3×5)-Eqs.)

$$[A + \frac{1}{\Delta t^1} C]\{\phi_n^1\} + [D^1]\{{}_j\theta_i^1\} = \{Q_n^1\} + [\frac{1}{\Delta t^1} C]\{\Phi_n^0\} \quad (\tau = 1) \qquad (19\text{-}1)$$

$$-[\frac{1}{\Delta t^2} C]\{\phi_n^1\} + [A + \frac{1}{\Delta t^2} C]\{\phi_n^2\} + [D^2]\{{}_j\theta_i^2\} = \{Q_n^2\} \quad (\tau = 2) \qquad (19\text{-}2)$$

$$-[\frac{1}{\Delta t^3} C]\{\phi_n^2\} + [A + \frac{1}{\Delta t^3} C]\{\phi_n^3\} + [D^3]\{{}_j\theta_i^3\} = \{Q_n^3\} \quad (\tau = 3) \qquad (19\text{-}3)$$

Constraints ((3×(4+2))-Eqs.)

$$\phi_4^1 \geqq \underline{\Phi}_4^1 \ (= 6.0),\ \phi_4^1 \leqq \overline{\Phi}_4^1 \ (= 20.0),\ \phi_5^1 \geqq \underline{\Phi}_5^1 \ (= 6.0),\ \phi_5^1 \leqq \overline{\Phi}_5^1 \ (= 20.0),$$
$${}_1\theta_1^1 \leqq {}_1\overline{\Theta}_1^1 \ (= 30.0),\ {}_3\theta_2^1 \leqq {}_3\overline{\Theta}_2^1 \ (= 30.0) \qquad (20\text{-}1)$$

$$\phi_4^2 \geqq \underline{\Phi}_4^2 \ (= 6.0),\ \phi_4^2 \leqq \overline{\Phi}_4^2 \ (= 20.0),\ \phi_5^2 \geqq \underline{\Phi}_5^2 \ (= 6.0),\ \phi_5^2 \leqq \overline{\Phi}_5^2 \ (= 20.0),$$
$${}_1\theta_1^2 \leqq {}_1\overline{\Theta}_1^2 \ (= 30.0),\ {}_3\theta_2^2 \leqq {}_3\overline{\Theta}_2^2 \ (= 30.0) \qquad (20\text{-}2)$$

$$\phi_4^3 \geqq \underline{\Phi}_4^3 \ (= 6.0),\ \phi_4^3 \leqq \overline{\Phi}_4^3 \ (= 20.0),\ \phi_5^3 \geqq \underline{\Phi}_5^3 \ (= 6.0),\ \phi_5^3 \leqq \overline{\Phi}_5^3 \ (= 20.0),$$
$${}_1\theta_1^3 \leqq {}_1\overline{\Theta}_1^3 \ (= 30.0),\ {}_3\theta_2^3 \leqq {}_3\overline{\Theta}_2^3 \ (= 30.0) \qquad (20\text{-}3)$$

Nonnegative Conditions

$$\phi_n^\tau \geqq 0 \ (\tau = 1 \sim 3,\ n = 1 \sim 5),\ {}_j\theta_i^\tau \geqq 0 \ (\tau = 1 \sim 3,\ i = 1 \sim 2) \qquad (21)$$

with

$$[A] = \begin{bmatrix} 0.84 & -0.66 & 0.00 & -0.16 & 0.00 \\ -0.66 & 2.01 & -0.66 & -0.33 & -0.33 \\ 0.00 & -0.66 & 0.84 & 0.00 & -0.16 \\ -0.16 & -0.33 & 0.00 & 1.18 & -0.66 \\ 0.00 & -0.33 & -0.16 & -0.66 & 1.18 \end{bmatrix} \qquad (22)$$

$$[\frac{1}{\Delta t^\tau} C] = \begin{bmatrix} 0.01 & 0.01 & 0.00 & 0.01 & 0.00 \\ 0.01 & 0.04 & 0.01 & 0.01 & 0.01 \\ 0.00 & 0.01 & 0.01 & 0.00 & 0.01 \\ 0.01 & 0.01 & 0.00 & 0.03 & 0.01 \\ 0.00 & 0.01 & 0.01 & 0.01 & 0.03 \end{bmatrix} \quad (\tau = 1 \sim 3) \qquad (23)$$

$$[D^\tau] = \begin{bmatrix} -1 & 0 \\ 0 & 0 \\ 0 & -1 \\ 0 & 0 \\ 0 & 0 \end{bmatrix} \quad (\tau = 1 \sim 3) \qquad (24)$$

$$\begin{Bmatrix} \Phi_1^0 \\ \Phi_2^0 \\ \Phi_3^0 \\ \Phi_4^0 \\ \Phi_5^0 \end{Bmatrix} = \begin{Bmatrix} 5.0 \\ 5.0 \\ 5.0 \\ 5.0 \\ 5.0 \end{Bmatrix} \qquad (25) \qquad \begin{Bmatrix} Q_1^\tau \\ Q_2^\tau \\ Q_3^\tau \\ Q_4^\tau \\ Q_5^\tau \end{Bmatrix} = \begin{Bmatrix} 0 \\ 2.0 \\ 0 \\ 0 \\ 0 \end{Bmatrix} \quad (\tau = 1 \sim 3) \qquad (26)$$

The units of the state variable and the decision variable are not described in the simple model, because the method is applicable to general physical problems irrespective of the variables considered.

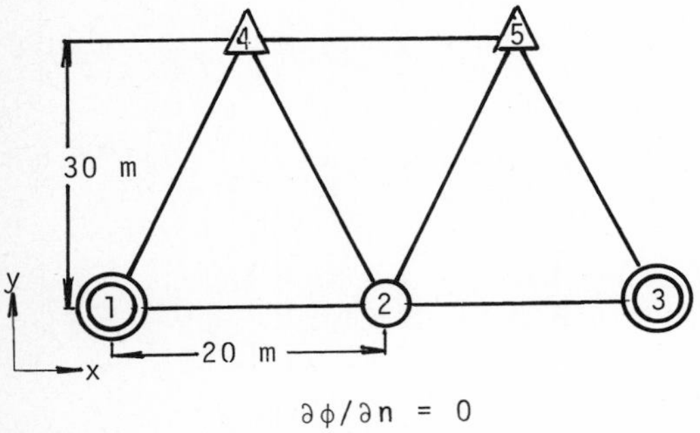

$T = 3,\ N = 5$

$I^1 = I^2 = I^3 = 2$

$K = 1.0\ m^2/sec$

$\Delta t^\tau = 3600\ sec\ (\tau = 1 \sim 3)$

$\Phi_n^0 = 5.0\ (n = 1 \sim 5)$

$\underline{\Phi}_4^\tau = \underline{\Phi}_5^\tau = 6.0\ (\tau = 1 \sim 3)$

$\overline{\Phi}_4^\tau = \overline{\Phi}_5^\tau = 20.0\ (\tau = 1 \sim 3)$

${}_1\underline{\Theta}_1^\tau = {}_3\underline{\Theta}_2^\tau = 0.0\ (\tau = 1 \sim 3)$

${}_1\overline{\Theta}_1^\tau = {}_3\overline{\Theta}_2^\tau = 30.0\ (\tau = 1 \sim 3)$

◎ : Node Fitted for Location of Controllable Rate of Product

○ : Node Fitted for Location of Uncontrollable Rate of Product

△ : Regulated Node in State Constraints

Fig. 3 Simple Model of FE&LP Method in Field Problem

3. Conclusions

The generalization of the FE (Finite Element) & LP (Linear Programming) method and the related methods is studied. A formulation of the FE&LP method in field problem is presented. The tractability in both the initial and boundary conditions and the equality or inequality constraints makes sure that the FE&LP method and the related methods become useful techniques for several new types of boundary value problems in systems optimization. An efficient computational algorithm for the FE&LP method in water pollution control has been presented (3, 4). The development of the efficient computational algorithm makes it possible to attack large scale problems. As for the related method, a stochastic finite element & linear (or non-linear) programming could be developed.

References

1. Futagami, T., "Development of Finite Element & Linear Programming Method and Its Application to Problems of Water Pollution Control," Proceedings, 19th Hydraulics Conference, Japan Society of Civil Engineers, Feburuary, 1975, pp. 133-138, (in Japanese).
2. Futagami, T., Tamai, N., and Yatsuzuka, M., "FEM Coupled with LP for Water Pollution Control," Journal of Hydraulics Divisison, ASCE, 102, HY7, 1976, pp. 881-897.
3. Futagami, T., "Several Mathematical Methods in Water Pollution Control — The Finite Element & Linear Programming Method and the Related Methods,"Thesis Presented to Kyoto University, in Partial Fullfillment of the Requirements for the Degree of Doctor of Phylosophy, February, 1976.
4. Futagami, T., Fukuhara, T., and Tomita, M., "Transient Finite Element & Linear Programming Method in Environmental Systems Control — The Efficient Computational Algorithm, Preprints, IFAC Environmental Systems Symposium, Kyoto, August, 1977, Pergamon Press, pp. 143-150.
5. Zienkiewicz, O. C., The Finite Element Method in Engineering Science, 2nd ed.,McGraw-Hill, 1971.
6. Bruch, J. C., Jr., and Zyvoloski, G., "Transient Two-Dimensional Heat Conduction Problems Solved by the Finite Element Method," International Journal for Numerical Methods in Engineering, Vol. 8, 1974, pp. 481-494.
7. Comini, D. S., Guidice, Lewis, W. R., and Zienkiewicz, "Finite Element Solution of Non-Linear Heat Conduction Problems with Special Reference to Phase Change," International Journal for Numerical Methods in Engineering, Vol. 8, 1974, pp. 613-624.
8. Dantig, B. G., "Linear Programming and Extensions," Princeton University Press, 1963.
9. Gass, I. S., "Linear Programming, Methods and Applications," 3rd ed., McGraw-Hill, Kogakusha, Ltd., 1969.
10. Bellman, R., "Dynamic Programming," Princeton University Press, 1957.
11. Futagami, T., "Dynamic Programming for a Sewage Treatment System," Proceedings, 5th International Water Pollution Research Conference, Pergamon Press, Spring 1971, pp. II-21/1-II-21/12.
12. Pontryagin, L S., Boltyanskii, R. V., Gamkrelidze, R. V. and Mischenko, E. F., "The Mathematical Theory of Optimal Processes," Wiley-Interscience, 1962, (English Translation by Trirogoff).
13. Fan, L. T., "The Continuous Maximum Principle," John Wiley & Sons, 1966.
14. Aguado E., and Remson, I., "Ground-Water Hydraulics in Aquifer

Management," Journal of the Hydraulic Division, ASCE, Vol. 100, No. HY1, January, 1974, pp. 103-108.

15. Wilson, L. E., "The Determination of Temperatures within Mass Concrete Structures," Report No. 68-17, Structural Engineering Laboratory, University of California, December, 1968.

Table of Contents

A GENERALIZATION OF THE DUALITY IN OPTIMAL CONTROL AND SOME NUMERICAL CONCLUSIONS

R. Klötzler, Leipzig / GDR
Karl-Marx-University, Sektion Mathematik

In some recent papers [4], [5] and lately in more detailed form on the conference EQUADIFF 4 in Prague [6] the author stated a new duality principle in optimal control, which is effectiv also without any assumptions of convexity. For its description we consider general problems of optimal control of the type

$$(1)\qquad J(x,u) := \int_{\partial\Omega} l(t,x)\,do + \int_{\Omega} f(t,x,u)\,dt \longrightarrow \text{Min}$$

subject to all vector-valued state functions $x \in X$ and control functions $u \in U(x)$ by side conditions

$$x^i_{t_\alpha} = g^i_\alpha(t,x,u) \qquad (i = 1,\dots,n;\ \alpha = 1,\dots,m) .$$

Here Ω is a strongly Lipschitz domain of R^m,

$X = \{ x \in W^{1,n}_p(\Omega) \mid (t,x(t)) \in \bar{G}$ on Ω, $b(t,x(t)) = \sigma$ on $\partial\Omega\}$ for $p > m$, and

$U(x) = \{ u \in L^r_p(\Omega) \mid u(t) \in V(t,x(t)) \subset R^r$ a.e. on $\Omega\}$ for every $x \in X$.

Moreover we state the following basic assumptions :

a) G is an open set of R^{n+m} ;
b) $V(.,.)$ is a normal map from $\bar{G}$ into $2^{(R^r)}$ in the sense of Joffe/Tichomirow [3] p. 338 ;
c) l and b are real continuous functions on $\partial\Omega \times R^n$;
d) f as well as g^i_α are real functions on $\bar{G} \times R^r$ satisfying the Carathéodory condition in the following meaning: they are (Lebesgue-) measurable functions in respect of the first argument t and continuous functions for almost every fixed $t \in \Omega$;
e) The set $\mathcal{P}$ of all processes $\langle x,u \rangle$ (i.e. admissible pairs

x,u of problem (1)) is not empty ;

f) The function $f(.,x(.),u(.))$ is minorized by a function $\psi \in L_1^1(\Omega)\ \forall\ \langle x,u\rangle \in \mathcal{P}$.

Further we use the following denotations:

H let be the Pontryagin function, defined by

$$H(t,\xi,v,y) := -f(t,\xi,v) + y_i\, g^i\,(t,\xi,v) \quad \text{on } \bar{G}\times R^r\times R^{mn} \ ;$$

$\mathcal{H}$ let be the Hamiltonian function, defined by

$$\mathcal{H}(t,\xi,y) := \sup_{v\in \bar{V}(t,\xi)} H(t,\xi,v,y) \quad \text{on } \bar{G}\times R^{nm} \ ;$$

Q(t) denotes the following cuts of $\bar{G}$

$$Q(t) := \begin{cases} \{\xi\in R^n \mid (t,\xi)\in \bar{G}\} & \forall\ t\in\Omega \\ \{\xi\in R^n \mid (t,\xi)\in \bar{G},\, b(t,\xi) = 0\} & \forall\ t\in\partial\Omega ; \end{cases}$$

$\mathcal{Y}$ let be the set of all functions $S = (S^1,\dots,S^m) \in W^{1,m}_\infty(G)$ having the following properties:

*) each class of distribution derivatives of S^α ($\alpha = 1,\dots,m$) contains a bounded representative S^α_j ($j = 1,\dots, n+m$) - we denote $S^\alpha_i = S^\alpha_{t^i}$ for $i = 1,\dots,m$ and $S^\alpha_{i+m} = S^\alpha_{\xi^i}$ for $i = 1,\dots,n$,

**) there are uniformly bounded sequences of functions $z^\alpha_k \in C^1(R^{n+m})$ and their derivatives satisfying pointwise the conditions

$$\lim_{k\to\infty} z^\alpha_k = S^\alpha \quad\text{and}\quad \lim_{k\to\infty} (\partial z^\alpha_k/\partial \xi^j) = S^\alpha_j \qquad \text{on } \bar{G} \ ;$$

$$\delta_S(t,\xi) := S^\alpha_{t^\alpha}(t,\xi) + \mathcal{H}(t,\xi,S_\xi(t,\xi))$$

denotes the "defect of the Hamilton-Jacobi equation" $\forall\ S\in\mathcal{Y}$;

$$\Lambda_S(t) := \sup_{\xi\in Q(t)} \delta_S(t,\xi) \quad \forall\ t\in\Omega \text{ and } S\in\mathcal{Y} \ .$$

Then according to [6] the following duality theorem holds.

Theorem 1 . Let $\langle x,u\rangle$ be a process and $S\in\mathcal{Y}$, then $J(x,u) \geqq L(S)$,if we define the expression L(S) by

$$(2) \qquad L(S) := -\int_{\Omega} \Lambda_S(t)\, dt + \int_{\partial\Omega} \inf_{\xi \in Q(t)} \left[S^{\alpha}(t,\xi) n_{\alpha}(t) + l(t,\xi) \right] do,$$

where $n_{\alpha}(t)$ ($\alpha = 1,\dots,m$) denote the components of the unit vector of the exterior normal on $\partial\Omega$ at the point t .The equality $J(x,u) = L(S)$ occurs if and only if the following conditions are fulfilled:

$$H(t,x,u,S_{\xi}(t,x)) = \mathcal{H}(t,x,S_{\xi}(t,x)) \text{ a.e. on } \Omega \quad ,$$

$$\delta_S(t,x(t)) = \Lambda_S(t) \text{ a.e. on } \Omega \ ,$$

$$S^{\alpha}(t,x)n_{\alpha}(t) + l(t,x) = \inf_{\xi \in Q(t)} \left[S^{\alpha}(t,\xi)n_{\alpha}(t) + l(t,\xi) \right] \text{ a.e. on } \partial\Omega \ .$$

We remark that in consequence of our basic assumptions the integrands in formula (2) are summable (in the broad sense). Furthermore it is easily seen,that in virtue of this Theorem 1 a duality is defined between the original problem (1) and its dual problem

$$(3) \qquad L(S) \longrightarrow \text{Max on } \mathcal{Y} \ .$$

If we write formula (2) under the special statement

$$S^{\alpha}(t,\xi) = y_0^{\alpha}(t) + y_i^{\alpha}(t)\,\xi^i \quad (\alpha = 1,\dots,m) \quad ,$$

then this duality changes over into the well-known form,which was presented through additional assumptions of convexity on problem (1) by K.O.Friedrichs [2] , M.M.Zwetanow [9] ,R.T.Rockafellar [8] Ekeland/Temam [1] and other authors. About further general conclusions from Theorem 1 compare also the paper [6] .

Now we shall discuss and applicate this Theorem 1 only for the class of parametric variational problems

$$(4) \qquad J(x) = \int_{\sigma}^{T} f(x,\dot{x})\, dt \longrightarrow \text{Min} \quad \text{on } W_p^{1,n}(\sigma,T)$$

among boundary conditions $x(\sigma) = x_o$, $x(T) = x_T$ and state restrictions $x(t) \in \overline{G}_o \subset R^n \ \forall \ t \in [\sigma,T]$ by a given domain G_o with $\partial G_o \in C_1^o$.
We do it in the conciousness that each optimal control problem con-

cerning simple integrals is reducible into a problem of type (4) under more general boundary conditions and analytical properties of f (comp. [8]).

Here we suppose again the above-mentioned basic assumptions by $\Omega = (\sigma,T)$, $m = 1$, $g_1^i(t,\xi,v) \equiv v^i$ ($i = 1,\dots,n$), $G = G_o \times (\sigma,T)$, $U = L_p^n(\sigma,T)$. Furthermore we suppose the property $f \geqq \sigma$ as well as the positiv homogeneity of degree one in respect of the function $f(x,.)$. Thus we obtain by the additional statement $S_t \equiv \sigma$,as is easily verified,

$$\delta_S(t,\xi) = \mathcal{H}(t,\xi,y) = \begin{cases} \sigma & \text{if } y \in \mathfrak{F}(\xi) \\ \infty & \text{if } y \notin \mathfrak{F}(\xi) \end{cases}$$

and

$$\Lambda_S = \begin{cases} \sigma & \text{if for every } \xi \in \bar{G}_o \quad S_\xi(\xi) \in \mathfrak{F}(\xi) \\ \infty & \text{otherwise} \end{cases} ,$$

where $\mathfrak{F}(\xi)$ is the bounded convex figuratrix set in the sense of Carathéodory,Minkowski and Hadamard,defined by

$$\mathfrak{F}(\xi) := \left\{ z \in R^n \;\middle|\; z_i v^i \leqq f(\xi,v) \;\; \forall \, v \in R^n \right\} .$$

Therefore Theorem 1 leads in the special case of problem (4) to the estimate

$$J(x) \geqq S(x_T) - S(x_o) \tag{5}$$

for every admissible state function x of problem (4) and each $S \in \mathcal{T}$ satisfying the constraint

$$S_\xi(\xi) \in \mathfrak{F}(\xi) \quad \forall \, \xi \in G_o \; . \tag{6}$$

We denote the class of these functions S by $\mathfrak{A}(G_o)$.

Theorem 2 . Formula (5) holds too for every $S \in \bar{\mathfrak{A}}(G_o)$,where

$$\bar{\mathfrak{A}}(G_o) = \left\{ S \in W^{1,1}_\infty(G_o) \;\middle|\; S_\xi(\xi) \in \mathfrak{F}(\xi) \text{ a.e. on } G_o \right\} .$$

Proof. In consequence of our goodness of G_o we may consider only this case,without limitation of generality,where $x_o, x_T \in G_o$

and $x(t) \in G_o \; \forall \; t \in [0,T]$. Namely, if we have proved Theorem 2 for every $x \in X_o := \{ x \in W_p^{1,n}(G_o) \mid x(t) \in G_o \; \forall \; t \in [0,T] \}$, then each admissible function x^* of problem (4) allows to construct a sequence $x^{(k)} \in X_o$ having the properties

$$\lim_{k\to\infty} J(x^{(k)}) = J(x^*) \, , \quad \lim_{k\to\infty} x^{(k)}(t) = x^*(t) \quad \forall \; t \in [0,T] \; .$$

Thus we obtain from $J(x^{(k)}) \geqq S(x^{(k)}(T)) - S(x^{(k)}(0))$ for $k \to \infty$ in the limit $J(x^*) \geqq S(x^*(T)) - S(x^*(0)) = S(x_T) - S(x_o)$.

Now let be x an admissible state function of problem (4) belonging to X_o , and S should be an element of $\bar{\mathfrak{A}}(G_o)$. Then also $\bar{S} := \gamma S \in \bar{\mathfrak{A}}(G_o)$ for each constant $\gamma \in (0,1)$. Through mollified functions of S we can construct a sequence $\sigma_k \in C^1(\bar{G}_o)$ with the following properties:

on each domain $D \subset\subset G_o$ satisfying $x(t) \in D$ on $[0,T]$

σ_k converges uniformly to $\bar{S}$,

$\partial \sigma_k / \partial \xi^i$ converges uniformly a.e. to $\bar{S}_{\xi^i}$,

$\exists \; K(\gamma, D)$ such that $\partial \sigma_k / \partial \xi \in \mathfrak{F}(\xi) \quad \forall \; k > K(\gamma, D)$.

Because of formula (5) - applicated on D -

$$J(x) \geqq \sigma_k(x_T) - \sigma_k(x_o) \quad \forall \; k > K(\gamma, D) \, ,$$

and in the limit for $k \to \infty$ we obtain

$$J(x) \geqq \bar{S}(x_T) - \bar{S}(x_o) = \gamma \, (S(x_T) - S(x_o)) \qquad \forall \; \gamma \in (0,1) \; .$$

Since γ is arbitrary, it follows for $\gamma \to 1$

$$J(x) \geqq S(x_T) - S(x_o) \, ,$$

and the proof is finished.

In correspondence with Theorem 2 a modified dual problem to problem (4) is stated by

(7) $$S(x_T) - S(x_o) \longrightarrow \text{Max} \quad \text{on} \quad \bar{\mathfrak{A}}(G_o) \quad .$$

Theorem 3 . Suppose that the integrand of problem (4) satisfies the supplementary condition

$$M = \sup \left\{ f(\xi,v) \;\middle|\; \xi \in G_o \,,\; v \in R^n \,,\; |v| = 1 \right\} \quad .$$

Then $\inf_{x \in X} J(x) = \operatorname{Max}_{S \in \bar{\mathcal{A}}(G_o)} \left[S(x_T) - S(x_o) \right]$.

Proof. We denote

$$X(x_o,\xi) := \left\{ x \in W_p^{1,n}(\sigma,T) \;\middle|\; x(t) \in \bar{G}_o \;\forall\, t \in [\sigma,T], x(\sigma)=x_o,\; x(T)=\xi \right\}$$

such that $X = X(x_o,x_T)$ as the set of admissible state functions of problem (4). In consequence of our basic assumptions

$$\exists \inf_{x \in X(x_o,\xi)} J(x) > -\infty \quad \forall\, \xi \in \bar{G}_o \quad .$$

We choose for S the Bellman function

$$S^*(\xi) = \inf_{x \in X(x_o,\xi)} J(x) \tag{8}$$

as a continuous function on $\bar{G}_o$.

Further $\forall\, \xi \in G_o \,,\; v \in R^n$

$$S^*(\xi+\varepsilon v) - S^*(\xi) = \inf_{x \in X(x_o,\xi+\varepsilon v)} J(x) - \inf_{x \in X(x_o,\xi)} J(x)$$

$$\leqq \left[\inf_{x \in X(x_o,\xi)} J(x) + \inf_{x \in X(\xi,\xi+\varepsilon v)} J(x) \right] - \inf_{x \in X(x_o,\xi)} J(x)$$

$$\leqq \int_\sigma^\varepsilon f(\xi+\tau v,\, v)\, d\tau$$

for sufficiently small $\varepsilon > \sigma$,if we observe the invariance of the integral $J(x)$ by parameter transformations $\tilde{t} = \phi(t)$ with $\phi \in C^1$ and $\phi'(t) > \sigma$ everywhere.

Because of the positiv homogeneity of $f(\xi,v)$ in respect of **v** we obtain finally the estimate

$$S^*(\xi+\varepsilon v) - S^*(\xi) \leqq |v| \cdot \int_\sigma^\varepsilon f(\xi+\tau v, v_o)\, d\tau \quad \text{with} \quad v = v_o\,|v| \tag{9}$$

and by use of the supplementary condition of Theorem 3

(10) $S^*(\xi+\varepsilon v) - S^*(\xi) \leqq \varepsilon M|v| \quad \forall \xi \in G_o, \varepsilon < \varepsilon_o(\xi), v \in R^n$.

From formula (10) we conclude the local Lipschitz continuity of S^* on G_o, such that by the famous theorem of Rademacher [7] S^* is total differentiable for almost every $\xi \in G_o$. This result leads through formula (9), divided by ε, for $\varepsilon \to 0$ to the property

$$S^*_{\xi^i}(\xi)\, v^i \leqq f(\xi,v) \quad \forall\, v \in R^n \text{ and a.e. } \xi \in G_o ,$$

respectivley to $S^*_\xi \in \mathfrak{F}(\xi)$ a.e. on G_o. Hence in regard of the supplementary condition of Theorem 3 $S^* \in W^{1,1}_\infty(G_o)$. By summarizing of those conclusions the Bellman function S^* according to (8) fulfils the conditions $S^* \in \bar{\mathfrak{A}}(G_o)$, $S^*(x_o) = 0$ and

$$\inf_{x \in X} J(x) = \inf_{x \in X(x_o,x_T)} J(x) = S^*(x_T) = S^*(x_T) - S^*(x_o) .$$

This means by Theorem 2 $\inf\limits_{x \in X} J(x) = \operatorname{Max}\limits_{S \in \bar{\mathfrak{A}}(G_o)} \left[S(x_T) - S(x_o)\right]$,

such that Theorem 3 is valid.

Finally we sketch a numerical application of Theorem 2 . Now G_o is assumed as a bounded polyhedral domain of R^n . Then in $\bar{G}_o$ we embed a network $\Gamma = (N,E)$ consisting of a finite set N of nodes ξ_i and of a set E of orientated edges (as pairs of elements of N). We suppose $x_o \in N$, $x_T \in N$ and require that every elements of N and E are subsets of G_o . If we now introduce the sets $X_\Gamma(x_o,\xi) \subset X(x_o,\xi)$ of all orientated ways x on Γ going from x_o to $\xi \in N$, and moreover $I_\Gamma(\xi) := \operatorname{Min}\limits_{x \in X_\Gamma(x_o,\xi)} J(x)$, then obviously $I_\Gamma(x_T) \geqq \inf\limits_{x \in X} J(x)$.

The values $I_\Gamma(\xi_i)$ and especially $I_\Gamma(x_T)$ we can compute very easy by well-known algorithms of the graph theory. Through this valuation of N we are able to find a positive constant $\beta \leqq 1$ depending of the speciell structure of Γ , such that for the function $S(\xi) := \beta\, I_\Gamma(\xi)$ on N exists an extension $S \in \bar{\mathfrak{A}}(G_o)$.

Hence by Theorem 2 we obtain the both-sided estimate

(11) $$I_\Gamma(x_T) \geqq \inf_{x \in X} J(x) \geqq \beta\, I_\Gamma(x_T) .$$

We illustrate this method for the problem of the shortest way in euclidean metric within a polyhedral plane labyrinth $\bar{G}_o$,which allows a triangulation by equilateral triangles. In this example $f(\xi, v) = |v|$ and $\mathfrak{F}(\xi)$ is represented by the unit circle. Using the linear extension of $S = \beta I_{\Gamma}$ from N to each triangle area of $\bar{G}_o$,we may choose $\beta = \frac{1}{2}\sqrt{3}$.

References

[1] I.Ekeland,R.Temam, Analyse convexe et problèmes variationnels, Gauthier-Villars,Paris 1974.

[2] K.Friedrichs, Ein Verfahren der Variationsrechnung das Minimum eines Integrals als das Maximum eines anderen Ausdrucks darzustellen, Göttinger Nachr. 1929,13-20.

[3] A.D.Joffe,W.M.Tichomirov, Theory of optimal problems,(Russian) Nauka,Moscow 1974 .

[4] R.Klötzler, Einige neue Aspekte zur Bellmanschen Differentialgleichung, Proceedings of the All-Union Symposium on Optimal Control and Differential Games,(Russian),Tbilisi 1976, 146-154.

[5] R.Klötzler, Weiterentwicklungen der Hamilton-Jacobischen Theoree, Sitzungsberichte d.Akad.d.Wiss.d.DDR (to appear).

[6] R.Klötzler, On a general conception of duality in optimal control, Proceedings of the conference EQUADIFF 4 , Prague 1977 (to appear).

[7] H.Rademacher, Über partielle und totale Differenzierbarkeit von Funktionen mehrerer Variablen und über die Transformation der Doppelintegrale, Math.Ann. 79 (1918),340-359.

[8] R.T.Rockafellar, Conjugate convex functions in optimal control and the calculus of variations, Journ.Math.Anal.Appl. 32 (1970),174-222.

[9] M.M.Zwetanov, On duality in problems of the calculus of variations,(Russian) Comptes Rendus de l'Acad.Bulg.Sci. 21 (1968),733-736.

On Optimal Damping of One-Dimensional Vibrating Systems

Werner Krabs
Technische Hochschule Darmstadt
Fachbereich Mathematik
Schloßgartenstr. 7
D-61oo Darmstadt
West Germany

o. Introduction

This note is concerned with vibrations that are governed by a wave equation in one space variable where the spatial differential operator L is linear of even order 2n, symmetric, and positive definite. The motion is controlled via the 2n corresponding boundary conditions of L within a given time interval $[o,T]$ where the 2n admissable control functions are taken from suitable weakly compact convex subsets of $H_2[o,T]$. Further, initial conditions are prescribed. The details of this dynamical model are given in Section 1.

The aim is to minimize the vibration energy at T under admissable controls. This problem is rephrased in Section 2 as a convex approximation problem in a suitable L^2-vector function space. By well known results from the theory of convex approximation problems in Hilbert spaces the existence of a unique solution is ensured which is characterized by a maximum principle. This turns out to be equivalent to a weak bang-bang principle for the corresponding optimal control vector functions.

In Section 3 it is sketched how by Fourier's method the given control problem can be approximated by a sequence of similar problems in finite-dimensional vector function spaces which have unique optimal control vectors that can be characterized by a strong bang-bang principle. Each weak cluster point of the sequence of these approximate solutions then solves the given problem.

For the special case of the vibrating string this analysis has been introduced in [5] and more detailed numerical calculations were carried out in [1]. Based on investigations in [6] which lead to a direct solution of the problem of minimizing the vibration energy also immediate numerical solutions are given in [1] and [2].

In Section 4 numerical results are presented for the vibrating string and beam where the approximate problems were solved by the conditional gradient method whose special form in the case of convex approximation problems was also developed by Gilbert [4]. The results of Section 4 are mostly taken from [1] and [3] where more details can be found. The conditional gradient method is described in [1] and [2] where also further references as to its development and possible improvements are given.

The problem of controllability does not seem to have been discussed in this general context. However, there are various investigations in this direction concerned with the wave equation (also in higher dimensions) and the vibrating beam. As representative of the one-dimensional case the paper [7] of Russell can be considered who made numerous contributions to this topic. Further references can be found in [2],[5], and [6].

1. The Dynamical Model

We consider a motion that is governed by a differential equation of the following kind

$$y_{tt}(x,t)+Ly(x,t)=o \qquad (1.1)$$

for all $x\in(o,1)$, $t>o$ where L is a linear differential operator of order 2n with respect to x whose coefficients do not depend on t. The domain D_L of L is assumed to consist of all functions $z\in C^{2n}[o,1]$ that satisfy 2n boundary conditions of the form

$$R_i^o z = \sum_{j=o}^{2n-1} \alpha_{ij} z^{(j)}(o) = o,$$
$$i=1,\ldots,n. \qquad (1.2)$$
$$R_i^1 z = \sum_{j=o}^{2n-1} \beta_{ij} z^{(j)}(1) = o,$$

Let L be symmetric and positive definite such that Friedrich's extension of L is a self-adjoint positive definite linear operator $\hat{L}$ on a dense subspace $D_{\hat{L}}$ of $L_2[o,1]$ that has a complete orthonormal sequence of eigenfunctions $z_1, z_2, \ldots \in D_{\hat{L}}$ with corresponding eigenvalues $\lambda_1, \lambda_2, \ldots \in R$ such that $o<\lambda_1 \leq \lambda_2 \leq \ldots$ and $\lim_{k\to\infty} \lambda_k = +\infty$. Further, each function $z \in L_2[o,1]$ can be expanded in a Fourier series

$$z(x) = \sum_{k=1}^{\infty} a_k z_k(x) \text{ with } a_k = \int_o^1 z(x) z_k(x)\,dx.$$

The motion, governed by (1.1) is assumed to be controlled on the boundary, within a given time interval $[o,T]$, $T>o$, in the form

$$R_i^o y(\cdot,t) = u_i(t),$$
$$t \in [o,T],\ i=1,\ldots,n, \qquad (1.3)$$
$$R_i^1 y(\cdot,t) = v_i(t),$$

where the u_i's and v_i's are taken from closed subspaces X_i^o and X_i^1, respectively, of the vector space

$$H_2[o,T] = \{v \in C^1[o,T] \mid v'' \text{ exists almost everywhere on } [o,T] \text{ and is in } L_2[o,T]\}.$$

Further, initial conditions

$$y(x,o) = y_o(x),\ y_t(x,o) = y_1(x),\ o<x<1, \qquad (1.4)$$

are prescribed where $y_o \in D_{\hat{L}}$ and $y_1 \in L_2([o,1])$ are given functions. Then the initial-boundary value problem (1.1), (1.2 with $u_i = v_i = o$ for $i=1,\ldots,n$), (1.4) has exactly one generalized solution of the form

$$\hat{y}(x,t) = \sum_{k=1}^{\infty} (a_k \cos\sqrt{\lambda}_k t + b_k \sin\sqrt{\lambda}_k t) z_k(x)$$

where

$$a_k = \int_o^1 y_o(x) z_k(x)\,dx,$$

$$b_k = \frac{1}{\sqrt{\lambda}_k} \int_o^1 y_1(x) z_k(x)\,dx,\ k=1,2,\ldots$$

For each $i=1,\dots,n$, let r_i^o and r_i^1 be the solution of

$$Lr_i^o = o \text{ on } (o,1), \quad \text{and} \quad Lr_i^1 = o \text{ on } (o,1),$$

$$\left.\begin{matrix} R_j^o r_i^o = \delta_{ji}, \\ R_j^1 r_i^o = o, \end{matrix}\right\} j=1,\dots,n, \text{ and } \left.\begin{matrix} R_j^o r_i^1 = o, \\ R_j^1 r_i^1 = \delta_{ji}, \end{matrix}\right\} j=1,\dots,n,$$

respectively. Then there is exactly one generalized solution $y^*=y^*(x,t,u_1'',\dots,u_n'',v_1'',\dots,v_n'')$ of the problem

$$y_{tt}^*+Ly^* = \sum_{i=1}^n r_i^o u_i'' + \sum_{i=1}^n r_i^1 v_i'' \text{ on } (o,1)\times(o,T),$$

$$R_i^o y^*(\cdot,t) = R_i^1 y^*(\cdot,t) = o \text{ for all } t\in[o,T],$$

$$y^*(x,o) = y_t^*(x,o) = o \text{ for } o<x<1.$$

This is given by

$$y^*(x,t,u'',v'') = \sum_{i=1}^n y_i^*(x,t,u_i'',v_i'')$$

with

$$y_i^*(x,t,u_i'',v_i'') = \sum_{k=1}^\infty \Big[\frac{h_k^{o,i}}{\sqrt{\lambda}_k}\int_o^t \sin\sqrt{\lambda}_k(t-\tau)u_i''(\tau)d\tau + \frac{h_k^{1,i}}{\sqrt{\lambda}_k}\int_o^t \sin\sqrt{\lambda}_k(t-\tau)v_i''(\tau)d\tau\Big]z_k(x)$$

where

$$h_k^{o,i} = \int_o^1 r_i^o(x)z_k(x)dx \text{ and } h_k^{1,i} = \int_o^1 r_i^1(x)z_k(x)dx.$$

On defining

$$y(x,t,u,v) = \hat{y}(x,t) - y^*(x,t,u'',v'') + \sum_{i=1}^n r_i^o(x)u_i(t) + r_i^1(x)v_i(t) \tag{1.5}$$

we obtain the unique generalized solution of (1.1),(1.3),(1.4). The aim now consists of minimizing the vibration energy at T under the control $(u,v)=(u_1,\dots,u_n,v_1,\dots,v_n)$ which is given by

$$E(T,u,v) = \frac{1}{2}\int_o^1 y_t(x,T,u,v)^2 + My(x,T,u,v)^2 dx \tag{1.6}$$

where M denotes the square-root of $\hat{L}$. M can be computed by using the identity

$$\int_o^1 z(x)\hat{L}z(x)dx = \int_o^1 Mz(x)^2 dx \text{ for all } z\in D_{\hat{L}}.$$

On minimizing $E(T,u,v)$ the control functions u_i and v_i are only allowed to vary in

$$U_i=\{u_i\in X_i^o \mid u_i(o)=o,\ u_i'(o)=o \text{ and } |u_i''|\le 1 \text{ a.e.on}[o,T]\}$$

and (1.7)

$$V_i=\{v_i\in X_i^1 \mid v_i(o)=o, v_i'(o)=o \text{ and } |v_i''|\le 1 \text{ a.e.on}[o,T]\}.$$

2. The Minimization of Energy as a Convex Approximation Problem

From (1.5) we obtain

$$y_t(x,T,u,v)=\hat{y}_t(x,T)-y_t^*(x,T,u'',v'') + \sum_{i=1}^n r_i^o(x)u_i'(T)+r_i^1(x)v_i'(T)$$

$$My(x,T,u,v)=M\hat{y}(x,T)-My^*(x,T,u'',v'') + \sum_{i=1}^n Mr_i^o u_i(T)+Mr_i^1 v_i(T)$$

where

$$u_i'(T) = \int_o^T u_i''(t)dt, \quad v_i'(T) = \int_o^T v_i''(t)dt,$$

$$u_i(T) = \int_o^T (T-t)u_i''(t)dt, \quad v_i(T) = \int_o^T (T-t)v_i''(t)dt.$$

If we define, for each $i=1,\dots,n$,

$$u_i(t) = \int_o^t (t-\tau)w_i^o(\tau)d\tau \text{ and } v_i(t) = \int_o^t (t-\tau)w_i^1(\tau)d\tau, \tag{2.1}$$

we get a one-to-one correspondence between U_i and V_i, respectively, and

$$W_i^{o,1}=\{w\in L_2[o,T] \mid |w|\le 1 \text{ a.e. on } [o,T], \int_o^t (t-\tau)w(t)dt\in X_i^{o,1}\} \tag{2.2}$$

such that $w_i^o=u_i''$ and $w_i^1=v_i''$ a.e.. Therefore we define

$$S(w^o,w^1)(x)=\begin{pmatrix} y_t^*(x,T,w^o,w^1)-\sum_{i=1}^{n} r_i^o(x)\int_o^T w_i^o(t)dt \\ -r_i^1(x)\int_o^T w_i^1(t)dt \\ My^*(x,T,w^o,w^1-\sum_{i=1}^{n} Mr_i^o\int_o^T (T-t)w_i^o(t)dt \\ -Mr_i^1\int_o^T (T-t)w_i^1(t)dt \end{pmatrix}$$

and obtain a linear mapping from $L_2[o,T]^{2n}$ into $Y=L_2[o,1]^2$ such that the vibration energy (1.6) is given by

$$E(T,u,v)=\frac{1}{2}\|\hat{g}-S(w^o,w^1)\|_Y^2$$

with

$$\hat{g}(x)=\begin{pmatrix}\hat{y}_t(x,T)\\ M\hat{y}(x,T)\end{pmatrix}$$

and $u_i,v_i,i=1,\ldots,n$ defined by (2.1).
Thus minimizing $E(T,u,v)$ on $U_1\times\ldots\times U_n\times V_1\times\ldots\times V_n$ (1.7) is equivalent with finding $(\hat{w}^o,\hat{w}^1)\in W=W_1^o\times\ldots\times W_n^o\times W_1^1\times\ldots\times W_n^1$ (2.2) such that

$$\|\hat{g}-S(\hat{w}^o,\hat{w}^1)\|_Y\leq\|\hat{g}-S(w^o,w^1)\|_Y \qquad (2.3)$$

$$\text{for all } (w^o,w^1)\in W.$$

This is a convex approximation problem.
As W is weakly compact in $L_2[o,T]^{2n}$ and $S:L_2[o,T]^{2n}\to L_2[o,1]^2$ is continuous, hence weakly continuous, the image S(W) is weakly compact in $L_2[o,1]^2$ and therefore closed. As it is also convex there exists exactly one $S(\hat{w}^o,\hat{w}^1)\in S(W)$ such that (2.3) holds and $S(\hat{w}^o,\hat{w}^1)$ is characterized by the maximum principle

$$\langle\hat{g}-S(\hat{w}^o,\hat{w}^1),\ S(\hat{w}^o,\hat{w}^1)\rangle_Y$$

$$=\max_{(w^o,w^1)\in W}\langle\hat{g}-S(\hat{w}^o,\hat{w}^1),\ S(w^o,w^1)\rangle_Y \qquad (2.4)$$

where $\langle\cdot,\cdot\rangle_Y$ denotes the scalar product in $Y=L_2[o,1]^2$.
Let S^* denote the adjoint operator of S mapping $L_2[o,1]^2$ into $L_2[o,T]^{2n}$. Then, for each $(w^o,w^1)\in W$, we have

$$<\hat{g}-S(\hat{w}^o,\hat{w}^1),S(w^o,w^1)>_Y = S^*(\hat{g}-S(\hat{w}^o,\hat{w}^1)),(w^o,w^1)>_Y$$

$$= \int_o^T <\hat{S}_o^*(t),w^o(t)>_{R^n} + <\hat{S}_1^*(t),w^1(t)>_{R^n} dt$$

where

$$\hat{S}_{oi}^*(t) = \hat{a}_{oi}^o + \hat{b}_{oi}^o(T-t) + \sum_{k=1}^{\infty} \hat{a}_{ki}^o \cos\sqrt{\lambda}_k(T-t) + \hat{b}_{ki}^o \sin\sqrt{\lambda}_k(T-t),$$

$$\hat{S}_{1i}^*(t) = \hat{a}_{oi}^1 + \hat{b}_{oi}^1(T-t) + \sum_{k=1}^{\infty} \hat{a}_{ki}^1 \cos\sqrt{\lambda}_k(T-t) + \hat{b}_{ki}^1 \sin\sqrt{\lambda}_k(T-t),$$

$$\hat{a}_{oi}^{o,1} = -\int_o^1 r_i^{o,1}(x) y_t(x,T,\hat{u},\hat{v})dx,$$

$$\hat{a}_{ki}^{o,1} = h_k^{o,1,i} \int_o^1 z_k(x) y_t(x,T,\hat{u},\hat{v})dx,$$

$$\hat{b}_{oi}^{o,1} = -\int_o^1 Mr_i^{o,1}(x) My(x,T,\hat{u},\hat{v})dx, \tag{2.5}$$

$$\hat{b}_{ki}^{o,1} = \frac{h_k^{o,1,i}}{\sqrt{\lambda}_k} \int_o^1 Mz_k(x) My(x,T,\hat{u},\hat{v})dx$$

and $\hat{u}_i(t) = \int_o^t (t-\tau)\hat{w}_i^o(\tau)d\tau$, $\hat{v}_i(t) = \int_o^t (t-\tau)\hat{w}_i^1(\tau)d\tau$.

Therefore the maximum principle (2.4) is equivalent with the weak bang-bang-principle

$$w_i^o(t) = \operatorname{sgn} S_{oi}^*(t) \text{ for all } t\in[o,T] \text{ such that } S_{oi}^*(t)\neq o,$$

$$w_i^1(t) = \operatorname{sgn} S_{1i}^*(t) \text{ for all } t\in[o,T] \text{ such that } S_{1i}(t)\neq o, \tag{2.6}$$

$$i=1,\ldots,n,$$

if not $\hat{S}_o^*(t)=\Theta_n$ and $\hat{S}_1^*(t)=\Theta_n$ for almost all $t\in[o,T]$ in which case (2.3) holds for all $(w^o,w^1)\in L_2[o,T]^{2n}$.

3. Approximate Solution of the Problem

For each $N\in\mathbb{N}$ we define

$$y_i^N(x,t,w_i^o,w_i^1) = \sum_{k=1}^{N} \frac{h_k^{o,i}}{\sqrt{\lambda}_k} \int_o^t \sin\sqrt{\lambda}_k(t-\tau)w_i^o(\tau)d\tau + \frac{h_k^{1,i}}{\sqrt{\lambda}_k} \int_o^t \sin\sqrt{\lambda}_k(t-\tau)w_i^1(\tau)d\tau$$

for $w_i^{o,1}\in W_i^{o,1}$ (2.2), $i=1,\ldots,n$, and

$$S^N(w^o,w^1)(x)=\begin{pmatrix} y_t^N(x,T,w^o,w^1)-\sum_{i=1}^{n} r_i^o(x)\int_o^T w_i^o(t)\,dt \\ -r_i^1(x)\int_o^T w_i^1(t)\,dt \\ My^N(x,T,w^o,w^1)-\sum_{i=1}^{n} Mr_i^o\int_o^T (T-t)\,w_i^o(t)\,dt \\ -Mr_i^1\int_o^T (T-t)\,w_i^1(t)\,dt \end{pmatrix}$$

where

$$y^N(x,t,w^o,w^1) = \sum_{i=1}^{n} y_i^N(x,t,w_i^o,w_i^1).$$

Then there is exactly one $S^N(w^{N,o},w^{N,1})\in S(W)$ such that

$$\|\hat{g}-S^N(w^{N,o},w^{N,1})\|_Y \leq \|\hat{g}-S^N(w^o,w^1)\|_Y \qquad (3.1)$$

$$\text{for all } (w^o,w^1)\in W.$$

Furthermore, $(w^{N,o},w^{N,1})$ is characterized by the bang-bang-principle

$$w_i^{N,o}(t)=\operatorname{sgn} S_{oi}^{N*}(t) \text{ for almost all } t\in[o,T],$$
$$w_i^{N,1}(t)=\operatorname{sgn} S_{1i}^{N*}(t) \text{ for almost all } t\in[o,T] \qquad (3.2)$$

for $i=1,\ldots,n$ (and hence is unique) where

$$S_{oi}^{N*}(t)=a_{oi}^{N,o}+b_{oi}^{N,o}(T-t) + \sum_{k=1}^{N} a_{ki}^{N,o}\cos\sqrt{\lambda}_k(T-t)+b_{ki}^{N,o}\sin\sqrt{\lambda}_k(T-t),$$

$$S_{1i}^{N*}(t)=a_{oi}^{N,1}+b_{oi}^{N,1}(T-t) + \sum_{k=1}^{N} a_{ki}^{N,1}\cos\sqrt{\lambda}_k(T-t)+b_{ki}^{N,1}\sin\sqrt{\lambda}_k(T-t),$$

$i=1,\ldots,n$ have only finitely many zeroes or vanish identically. If this is the case for all $i=1,\ldots,n$, then (3.1) holds for all $(w^o,w^1)\in L_2[o,T]^{2n}$. The coefficients $a_{ki}^{N,o}, a_{ki}^{N,1}, b_{ki}^{N,o}, b_{ki}^{N,1}$ are obtained analogously to (2.5). Finally, each weak cluster point of $(w^{N,o},w^{N,1})$ is an optimal admissable control vector and

$$\lim_{N\to\infty} \|S^N(w^{N,o},w^{N,1})-S(\hat{w}^o,\hat{w}^1)\|_Y=o.$$

4. Special Cases

4.1. The Vibrating String

We consider the motion of a homogeneous string of length 1 such that L in (1.1) is given by

$$Ly(x,t) = -y_{xx}(x,t)$$

for all $x \in (o,1)$, $t>o$. In this case we have n=1 and the boundary conditions (1.3) are assumed to be of the form

$$y(o,t)=o \text{ and } y(1,t)=v(t), \quad t\in[o,T],$$

i.e., we choose X_1^o={zero function} and $X_1^1=H_2[o,T]$.
As initial conditions (1.4) we consider

$$y(x,o)=\sin\pi x, \; y_t(x,o)=o, \; o<x<1. \qquad (4.1)$$

The energy to be minimized at T>o is then of the usual form

$$E(T,v) = \frac{1}{2}\int_o^1 y_t(x,T,v)^2+y_x(x,T,v)^2dx$$

where v variies in the set

$$V=\{v\in H_2[o,T] \,|\, v(o)=v'(o)=o, |v''|\leq 1 \text{ a.e.on } [o,T]\}. \qquad (4.2)$$

The following table shows the location of the jumps of the approximate optimal bang-bang control

$$w_1^{N,1}(t)=\text{sgn}S_{11}^N(t) \text{ for almost all } t\in[o,T]$$

according to (3.2) and the corresponding energy values E(T,v) where $v''=w_1^{N,1}$ for N=2o and T=1,2,3,4.

T	jumps of v"		E(T,v)
1	o.1523o9 o.99718o		2.o16697513
2	o.169571 o.946o96	1.998365	1.47355o47o
3	o.152781 o.9o1836	2.o7o752 2.998o46	1.o35921927
4	o.156977 o.92o762 2.o71264	3.o33317 3.9967o1	o.65245o5o5

The initial value of the energy is given by $\frac{\pi^2}{4}$ = 2.4674o11oo.
The value of $v''=w_1^{N,1}$ between o and the first jump is always equal to +1.
For comparison we mention that for T=2 the jumps of v" of the control $v\in V$ that minimizes E(T,v) on V (which is unique

in this case) are located at o.169619 and o.946114 and that E(T,v)=1.473563490 (see [1] and [3]).

4.2. The Vibrating Beam

We consider the motion of the so called Euler beam of length 1 where L in (1.1) is given by

$$Ly(x,t)=a^2y_{xxxx}(x,t)$$

for all $x\in(o,1)$, $t>o$, with $a=(EI/A\rho)^{1/2}$ where Young's modulus E, the moment of inertia I of the cross section area A with respect to the middle axis of the beam, the cross section area A, and the density ρ are assumed to be constants such that a=o.6. In this case we have n=2 and as boundary conditions (1.3) we take

$$y(o,t)=y_{xx}(o,t)=o,$$
$$y(1,t)=v_1(t),y_{xx}(1,t)=v_2(t)$$

for $t\in[o,T]$, i.e. $X_1^o=X_2^o=\{\text{zero function}\}$ whereas $v_1\in V_1$ and $v_2\in V_2$ according to (1.7) where X_1^1 and X_2^1 will be suitably chosen. As initial conditions (1.4) we again consider (4.1). The energy to be minimized at T>o is now given by

$$E(T,v_1,v_2)=\frac{1}{2}\int_o^1 y_t(x,T,v_1,v_2)^2+a^2y_{xx}(x,T,v_1,v_2)^2dx$$

where for the choice of $v_i\in V_i$, i=1,2, according to (1.7) we consider two cases:

a) $V_1=\{\text{zero function}\}$, $V_2=V$ by (4.2),

b) $V_1=V$ by (4.2), $V_2=\{\text{zero function}\}$.

The following tables show the location of the jumps of the approximate optimal bang-bang control

a) $w_2^{N,1}(t)=\mathrm{sgn}S_{12}^{N*}(t)$ for almost all $t\in[o,T]$,

b) $w_1^{N,1}(t)=\mathrm{sgn}S_{11}^{N*}(t)$ for almost all $t\in[o,T]$

according to (3.2) and the corresponding energy values a) $E(T,o,v_2)$ where $v_2''=w_2^{N,1}$, b) $E(T,v_1,o)$ where $v_1''=w_1^{N,1}$ for N=2o and a) T=1,2,3,4, b) T=1,2,4,8. σ denotes the value of a) v_2'' and b) v_1'' between zero and its first jump. The value of the initial energy is 8.766818193.

a)

T	jumps of v_2''		σ	$E(T,o,v_2)$
1	o.995859		+1	8.184350885
2	1.981748		+1	7.053318717
3	1.894394	2.950521	+1	7.083335342
	1.930213			

T	jumps of v_2''			$E(T,0,v_2)$
4	0.061707	2.071369	−1	8.072573180
	0.531876	2.781223		
	1.067106	3.069484		
	1.651752			

b)

T	jumps of v_1''			$E(T,v_1,0)$
1	0.580615		−1	7.490836292
2	0.617098	1.678131	−1	6.063920740
	0.942593			
4	0.647791	2.769857	−1	3.193348414
	0.925077	3.049052		
	1.708824	3.830890		
	1.986109			
8	0.005536	4.249668	+1	1.732460921
	0.557721	4.801853		
	1.066569	5.310701		
	1.618754	5.862886		
	2.127602	6.371734		
	2.679787	6.923919		
	3.188635	7.432767		
	3.740820	7.984852		

References

[1] Eichenauer,W. and W.Krabs: On the Numerical Solution of Certain Control-Approximation Problems:I.Application to the Vibrating String. Preprint Nr.358 des Fachbereichs Mathematik der TH Darmstadt, Juni 1977.

[2] Eichenauer,W. and W.Krabs: On an Application of the Conditional Gradient Method to a Problem in Optimal Control.To appear in the Proceedings of the International Conference on Methods of Mathematical Programming in Zakopane,Poland,September 1977.

[3] Eichenauer,W.and W.Krabs: On the Numerical Solution of Certain Control-Approximation Problems:II.Application to the Vibrating Beam. Preprint Nr. 366 des Fachbereichs Mathematik der TH Darmstadt, Juli 1977.

[4] Gilbert,E.G.: An Iterative Procedure for Computing the Minimum of a Quadratic Form on a Convex Set. SIAM J.on Control 4,61-80(1966).

[5] Krabs,W.: Ein Kontroll-Approximationsproblem für die schwingende Saite. In: Numerische Methoden der Approximationstheorie, Band 3,ISNM 30,257-275,Birkhäuser 1976.

[6] Krabs,W.: Über die einseitige Randsteuerung einer schwingenden Saite in einen Zustand minimaler Energie. Computing 17, 351-359 (1977).

[7] Russell,D.L.: Nonharmonic Fourier Series in the Control of Distributed Parameter Systems. J.Mathem.Anal.Appl. 18,542-560(1967)

STABILITY AND STABILIZABILITY OF LINEAR CONTROL SYSTEMS ON HILBERT SPACE VIA OPERATOR DILATION THEORY*

by

N. Levan
Department of System Science
University of California
Los Angeles, CA 90024
USA

L. Rigby
Department of Computing & Control
Imperial College
London SW7 2BZ / UK

SUMMARY

We study in this work the stability and stabilizability problem of linear control systems on Hilbert space, using the so called Dilation Theory of Hilbert space operators.

Let (A,B) denote the discrete-time system described by the state-input equation

$$x_{n+1} = Ax_n + Bu_n, \; n>0, \; x_n \text{ in } H, \; u_n \text{ in } U \tag{1}$$

Similarly, let $(\mathcal{A},B)$ be the continuous-time system:

$$\dot{x} = \mathcal{A}x + Bu, \; t>0, \; x \text{ in } H, \; u \text{ in } U \tag{2}$$

In both cases the state space H and the control space U are Hilbert spaces. The operator B is bounded linear from U to H, while A is taken to be a contraction operator on H, $||A|| \leq 1$, and similarly $\mathcal{A}$ is assumed to be the generator of a C_o contraction semigroup $T(t); t \geq 0$, on H.

The system (A,B) (resp. $(\mathcal{A},B)$) is said to be s(strong)-stable if

$$A^n x \to 0, \; n \to \infty, \text{ for all } x \text{ in } H \tag{3}$$

$$(\; T(t)x \to 0, \; t \to \infty, \text{ for all } x \text{ in } H \;)$$

If the system is not s-stable and if a bounded linear operator F : H → U can be founded so that the feedback system (A + BF,B) (resp. $(\mathcal{A} + BF,B)$) is s-stable. Then the original system is said to be s-stabilizable.

Now let C and D be bounded linear operators on H and $\mathcal{H}$ respectively. If H is a proper subspace of $\mathcal{H}$, and if

$$C^n = P_H \, D^n \, P_H, \; n>0 \tag{4}$$

where P_H is the orthogonal projection from $\mathcal{H}$ onto H. Then D is called a (strong) dilation of C [1].

It is clear from (3) and (4) that, for the discrete-time system (A,B): $x_n = A^n x = P_H \, D^n \, P_H \, x$, for x in H. Hence one can study stabi-

* Work supported in part by the National Science Foundation under Grant No. ENG75-11876

lity of the system using the dilation D of A. In particular if A is a contraction, then D can be chosen to be either an isometric or an unitary operator. Necessary and sufficient conditions for stability as well as the subspace of stable states can be found using the dilations of A.

For the continuous-time case, since the semigroup T(t) is contractive, it cogenerator $C = (A + I)(A - I)^{-1}$ is also a contraction. Hence dilations of T(t) can be gotten from those of C. Furthermore, since

$$\lim_{t\to\infty} ||T(t)x|| = \lim_{n\to\infty} ||C^n x|| , \quad x \text{ in } H$$

stability of T(t) can again be investigated using the dilation theory.

The s-stabilization problem was also investigated. In this case we rely heavily on the notion of a completely nonunitary operator, and on a canonical decomposition of contractions due originally to Nagy and Foias [1]. Here it was shown that, for a large class of distributed parameter systems which occurred in many physical problems, s-stabilizability is equivalent to controllablity of the set of unstable states of the system. This, of course, is the analog of the wellknown result of Wonham [2] in finite dimensional case.

REFERENCES

[1] B. Sz-Nagy & C. Foias, Harmonic Analysis of Operators on Hilbert Space, American Elsevier, New York, 1970.

[2] W.M. Wonham, On Pole Assignement in Multi-Input Controllable Linear Systems, IEEE Trans. Automat. Contr., Vol. AC-12 (1967), pp 660-665.

ON DISCRETE-TIME RITZ-GALERKIN APPROXIMATION OF CONTROL CONSTRAINED OPTIMAL CONTROL PROBLEMS FOR PARABOLIC EQUATIONS

I. Lasiecka, K. Malanowski
Systems Research Institute
Polish Academy of Sciences
ul.Newelska 6, 01-447 Warszawa

1. Statement of optimal control problem

Let $\Omega \subset R^n$ be a bounded domain (an open set) with the boundary Γ regular enough, and let T be a fixed time.

Let us consider a system described in the cylinder $Q=\Omega \times (0,T)$ by the following parabolic equation

$$\frac{\partial y(x,t)}{\partial t} - A(x)y(x,t) = f(x,t) \tag{1}$$

where A is a self-adjoint elliptic operator given by

$$A(x)y(x) = \sum_{i,j=1}^{n} \frac{\partial}{\partial x_j}\left(a_{ij}(x) \frac{\partial y(x)}{\partial x_i}\right) - a_0(x)y(x)$$

and $a_0(\cdot)$ and $a_{ij}(\cdot)$ are properly regular functions.

It is assume the homogeneous boundary conditions of Neumann type are satisfied :

$$\frac{\partial y(\sigma,t)}{\partial \eta_A} = \sum_{i,j=1}^{n} a_{ij}(\sigma) \frac{\partial y(\sigma,t)}{\partial x_j} \cos(\eta, \sigma_i) = 0 \quad \text{in } \Gamma \times (0,T) \tag{1a}$$

where η is the unite outward normal to Γ.

Moreover an initial condition

$$y(x,0) = y^p(x) \tag{1b}$$

is satisfied.

It is well known [3] that for $f \in L^2(Q)$ and $y^p \in H^1(\Omega)$ equation (1) has a unique solution

$$y \in H^{2,1}(Q) \subset C(0,T;H^1(\Omega)) \tag{2}$$

In optimal control problem we assume that

$$f(t) = B\, u(t) \tag{3}$$

where

$$B \in \mathcal{L}(L^2(\Omega); L^2(\Omega)) \cap \mathcal{L}(H^1(\Omega); H^1(\Omega)) \qquad (4)^{1/}$$

$$u \in U_{ad} = \{u \in L^2(Q) : |u(x,t)| \leqslant 1 \quad \text{almost everywhere}\} \qquad (5)$$

Moreover there is given the cost functional

$$J(u,y) = \int_0^T \int_\Omega \varphi(u(x,t), y(x,t))\, dx\, dt \qquad (6)$$

where $\varphi(.,.)$ is a two times differentiable function, convex with respect to both arguments and strictly convex with respect to u.

We consider the following

Problem P-1

Find $u^0 \in U_{ad}$ such that

$$J(u^0, y(u^0)) \leqslant J(u, y(u)) \qquad \forall u \in U_{ad} \qquad (7)$$

where $y(u)$ is the solution of (1), where the right hand side is given by (3).

Due to the assumptions of strict convexity of functional (6) with respect to u and of convexity of U_{ad}, Problem P1 has a unique solution.

2. Lagrange formalism and regularity of optimal solution

It is convenient to write the constraints (5) in the form

$$\psi(u(x,t)) \leqslant 0 \qquad \text{a.e. in } Q \qquad (8)$$

where

$$\psi^T(u) = [u-1, \; -u-1]$$

Optimal control problem P-1 can be treated as the problem of minimizing of functional (6) on appropriate Hilbert space subject to constraints of equality type in the form of state equation and inequality type (8).

1/ $H^r(\Omega)$ denotes here the Sobolev space of order r of functions square integrable together with their (weak) derivatives up to the order r with the norm denoted by $\|.\|_r$.
Similarly $H^{r,s}(Q)$ is the Sobolev space of orders r and s with respect to space and time variables respectively [4] with the norm denoted by $\|.\|_{r,s}$.
$L^2(0,T;X)$ and $C(0,T;X)$ denote respectively the spaces of square integrable and continuous functions from $(0,T)$ into X.

To such a problem we can apply Lagrange formalism.
Namely we introduce a Lagrangian

$$L(u,y,\lambda,p) : L^2(Q)\times H^{2,1}(Q)\times L^2(Q)\times L^2(Q)$$

$$L(u,y,\lambda,p) \overset{def}{=} J(u,y) + \left(\left(p, \frac{dy}{dt} - Ay - Bu\right)\right) + ((\lambda, \psi(u))) \quad (9)$$

The solution of Problem P-1 can be characterized as the saddle point of Lagrangian (9) as follows [2] :

Lemma 1

There exist Lagrange multipliers

$$\lambda^o \in L^2(Q) \quad (10)$$

$$\lambda^o \geqslant 0 \quad (10a)$$

and $p^o \in H^{2,1}(Q)$ satisfying the adjoint equation

$$\frac{\partial p^o(x,t)}{\partial t} + A(x)p^o(x,t) = \partial_y J(u^o,y^o)(x,t) \quad \text{in } Q \quad (11)$$

$$\frac{\partial p^o(\sigma,t)}{\partial \eta_A} = 0 \quad \text{in } \Gamma\times(0,T) \quad (11a)$$

$$p^o(x,T) = 0 \quad \text{in } \Omega \quad (11b)$$

such that Lagrangian L assumes its saddle point at (u^o,y^o,λ^o,p^o), i.e.

$$L(u^o,y^o,\lambda,p) \leqslant L(u^o,y^o,\lambda^o,p^o) \leqslant L(u,y,\lambda^o,p^o) \quad (12)$$

$$\forall u \in L^2(Q)$$

$$\forall y \in H^{2,1}(Q),\ y(0) = y^p$$

$$\forall \lambda \in L^2(Q),\quad \lambda \geqslant 0$$

$$\forall p \in L^2(Q)$$

Remark

In Lemma 1 the only non-trivial result is that the multiplier λ^o satisfies regularity condition (10). To obtain this result the fact that constraints (5) are of local type is used.
In [2] there was also shown the following

Lemma 2

Optimal control u^o and Lagrange multiplier p^o satisfy the following regularity conditions

$$u^o \in H^{1,1}(Q) \quad (13)$$

$$p^o \in H^{3,2}(Q) \quad (14a)$$

If additionally

$$y^p \in H^3 \quad \text{and} \quad \frac{\partial y^p}{\partial \eta_A} = 0 \tag{15}$$

then also

$$y^o \in H^{3,2}(Q) \tag{14b}$$

Remark

In the proof of (13) the form (5) of constraints, conditions (12) and already known fact that $y^o, p^o \in H^{2,1}(Q)$ are used. Conditions (14) are obtained from (13), (15) and some regularity results for parabolic equations [1,3] .

3. Finite-dimensional approximation

Problem P-1 will be approximated by a finite-dimensional one using discrete-time Ritz-Galerkin method.

To use this method it is more convenient to rewrite state equation (1) in variational form [4] . Namely let us take the scalar product in $L^2(\Omega)$ of both sides of (1) by any arbitrary function $g \in H^1(\Omega)$. Using Green formula and taking advantage of (1a) we get

$$\left(\frac{dy(t)}{dt}, g\right) + a(y(t), g) = (f(t), g) \qquad \forall g \in H^1(\Omega) \tag{16}$$

$$y(0) = y^p \tag{16a}$$

where

$$a(y,z) = \int_\Omega \left(\sum_{i,j=1}^{n} a_{ij}(x) \frac{\partial y(x)}{\partial x_i} \frac{\partial z(x)}{\partial x_j} + a_0(x) y(x) z(x)\right) dx$$

is a bilinear form continuous and coercive on $H^1(\Omega)$.

In the space $G = H^1(\Omega)$ we introduce the family $\{G_h\}$ of finite dimensional subspaces G_h depending on parameter h destinated to tend to zero.

It is assumed that the following condition is satisfied

$$\exists c > 0 \quad \forall g \in G \quad \forall h > 0 \quad \exists g_h \in G_h \qquad \text{such that}$$

$$\| g - g_h \|_0 + h \| g - g_h \|_1 \leq c\, h^s \| g \|_s \qquad s = 1,2 \tag{17}$$

The element g_h is called an interpolate of g .

Assumption (17) is satisfied for example by the spaces of linear finite elements [6].

The functions of time will be approximated by step functions depending on parameter τ . To this end let us divide the interval $[0,T]$ into $M(\tau)$ subintervals of the length $\tau = \frac{T}{M(\tau)}$. Let χ_i be the characteristic function of $[i\tau, (i+1)\tau)$.

By $E_\tau(0,T;Z)$ we denote the space of step functions defined on $[0,T)$ with the range in Z , which are given by

$$z_\tau(t) = \sum_{i=0}^{M(\tau)-1} z_\tau(i\tau)\chi_i(t) \qquad z(i\tau)\in Z \tag{18}$$

For elements $z\in L^2(0,T;Z)$ we define the operator P_τ of projection on $E_\tau(0,T;Z)$ putting in (18)

$$z_\tau(i\tau) = \frac{1}{\tau}\int_{i\tau}^{(i+1)\tau} z(t)\,dt \tag{19}$$

In particular for functions $z\in L^2(0,T;H^1(\Omega))$ we introduce the following interpolate $z_{h,\tau}\in E_\tau(0,T;G_h)$

$$z_{h,\tau}(t) = \sum_{i=0}^{M(\tau)-1} z_{h,\tau}(i\tau)\chi_i(t) \tag{20}$$

where $z_{h,\tau}(i\tau)\in G_h$ is an interpolate in G_h of the element $z_\tau(i\tau)$.

The state equation (1) is approximated by the following difference equation

$$(\nabla y_{h,\tau}(t),g_h) + a(y_{h,\tau+\theta}(t),g_h) = (\varphi_\tau(t),g_h) \quad \forall g_h\in G_h \tag{21}$$

$$y_{h,\tau}(0) = y_h^p \tag{21a}$$

where

$$\nabla y_{h,\tau}(t) = \sum_{i=0}^{M(\tau)-1} \frac{1}{\tau}\big(y_{h,\tau}((i+1)\tau) - y_{h,\tau}(i\tau)\big)\chi_i(t) \tag{22a}$$

$$y_{h,\tau+\theta}(t) = \sum_{i=0}^{M(\tau)-1} \big((1-\theta)y_{h,\tau}((i+1)\tau) + \theta\, y_{h,\tau}(i\tau)\big)\chi_i(t) \qquad 0\leqslant\theta\leqslant\tfrac{1}{2} \tag{22b}$$

$y_h^p\in G_h$ is an interpolate of y^p.

For every $\varphi_\tau\in E_\tau(0,T;L^2(\Omega))$ equation (21) has a unique solution $y_{h,\tau}\in E_\tau(0,T;G_h)$. It is well known [7] that the convergence of the solutions of (21) to that of (1) can be estimated as follows

<u>Lemma 3</u>

<u>If</u> $\quad y,\ \frac{dy}{dt}\in L^2(0,T;H^2(\Omega))$

<u>then</u>

$$\|y_{h,\tau} - z_{h,\tau}\|_{1,0} \leqslant c\left[\|f-\varphi_\tau\|_{0,0} + (\tau+h)\right] \tag{23}$$

<u>where</u> $z_{h,\tau}$ <u>is an interpolate of the solution y of (1)</u>.

In order to introduce a finite dimensional approximation of

controls we define a family $\{V_k\}$ of spaces V_k of step functions defined on Ω, and depending on a parameter k destinated to tend to zero. For the elements v of the space $V = L^2(\Omega)$ we define the operators P_k of projection on V_k in the same way as in (19).

It is assumed that the subspaces V_k are constructed in such a way, that

$$\exists c > 0 \quad , \quad \forall v \in H^1(\Omega)$$
$$\| v - P_k v \|_0 \leqslant c\, k \| v \|_1 \tag{24}$$

For control functions $u \in L^2(Q)$ we introduce interpolants $u_{k,\tau} \in U_{k,\tau} = E_\tau(0,T;V_k)$ given by

$$u_{k,\tau} = P_k P_\tau u \tag{25}$$

Now we can formulate problem of optimization P-2 approximating P-1

<u>Problem P-2</u>

<u>Find</u> $u^0_{k,\tau} \in U_{ad} \cap U_{k,\tau}$ <u>such that</u>

$$J(u^0_{k,\tau}, y_{h,\tau}(u^0_{k,\tau})) \leqslant J(u_{k,\tau}, y_{h,\tau}(u_{k,\tau})) \quad \forall u_{k,\tau} \in U_{ad} \cap U_{k,\tau} \tag{26}$$

Problem P-2 has a unique solution which can be characterized as the saddle point of the following Lagrangian

$$L_d(u_{k,\tau}, y_{h,\tau}, \lambda_{k,\tau}, p_{h,\tau}) \overset{\text{def}}{=}$$
$$= J(u_{k,\tau}, y_{h,\tau}) + ((\nabla y_{h,\tau}, p_{h,\tau})) + a(y_{h,\tau+\theta}, p_{h,\tau}) - ((B\, u_{k,\tau}, p_{h,\tau})) +$$
$$+ ((\lambda_{k,\tau}, \psi(u_{k,\tau}))) \quad \text{i.e.} \tag{27}$$

$$L_d(u^0_{k,\tau}, y^0_{h,\tau}, \lambda_{k,\tau}, p_{h,\tau}) \leqslant L_d(u^0_{k,\tau}, y^0_{h,\tau}, \lambda^0_{k,\tau}, p^0_{h,\tau}) \leqslant$$
$$\leqslant L_d(u_{k,\tau}, y_{h,\tau}, \lambda^0_{k,\tau}, p^0_{h,\tau}) \tag{28}$$

$$\forall u_{k,\tau} \in E_\tau(0,T;V_k) \; ; \; \forall y_{h,\tau} \in E_\tau(0,T;G_h) \; , \quad y_{h,\tau}(0) = y^p_h \; ;$$
$$\forall \lambda_{k,\tau} \in E_\tau(0,T;V_k) \; ; \quad \lambda_{k,\tau} \geqslant 0 \; ; \quad \forall p_{h,\tau} \in E_\tau(0,T;G_h).$$

The element $p^0_{h,\tau}$ satisfies the following adjoint equation

$$(\nabla p^0_{h,\tau}(t), g_h) - a(p^0_{h,\tau+}(t), g_h) = (\partial_y J(u^0_{k,\tau}(t), y^0_{h,\tau}(t)), g_h)$$
$$\forall g_h \in G_h \tag{29}$$
$$p^0_{h,\tau}(T) = 0 \tag{29a}$$

4. Convergence of finite dimensional approximation

Saddle points conditions (12) and (28) will be used to find the estimation of the norm of difference $u^o - u_k^o$.

Note that the Lagrangians L and L_d are different and defined on different spaces. Hence the respective saddle points conditions can not be used directly and to obtainded the needed estimation instead of elements u^o, y^o, λ^o, p^o their interpolates $u_{k,\tau}, y_{h,\tau}$, $\lambda_{k,\tau}$ and $p_{h,\tau}$ will be used.

Expanding L_d into Taylor series with respect to u and y in the vicinity of the point $u_{k,\tau}, y_{h,\tau}, \lambda_{k,\tau}, p_{h,\tau}$, taking advantage of the strict convexity of φ and using (12) and (28) after some tedious transformations we get [2] :

$$\begin{aligned}
\|u^o_{k,\tau} - u_{k,\tau}\|^2_{0,0} \leqslant c\ & \|u^o_{k,\tau} - u_{k,\tau}\|_{0,0} \Big(\|u^o - u_{k,\tau}\|_{0,0} + \\
& + \|y^o - y_{h,\tau}\|_{0,0} + \|p^o - p_{h,\tau}\|_{0,0} \Big) + \|y^o_{h,\tau} - y_{h,\tau}\|_{0,0} \Big(\|u^o - u_{k,\tau}\|_{0,0} + \\
& + \|y^o - y_{h,\tau}\|_{0,0} + \Big\|\frac{dp^o}{dt} - \frac{dp_h}{dt}\Big\|_{0,0} \Big) + \|y^o_{h,\tau} - y_{h,\tau}\|_{1,0} \|p^o - p_{h,\tau+(1-\theta)}\|_{1,0} + \\
& + \|p^o_{h,\tau} - p_{h,\tau}\|_{0,0} \Big(\|u^o - u_{k,\tau}\|_{0,0} + \Big\|\frac{dy^o}{dt} - \frac{dy_h}{dt}\Big\|_{0,0} \Big) + \qquad (30) \\
& + \|p^o_{h,\tau} - p_{h,\tau}\|_{1,0} \|y^o - y_{h,\tau+\theta}\|_{1,0} + \|\lambda^o_{k,\tau} - \lambda^o\|_{0,0} \|u^o - u_{k,\tau}\|_{0,0}
\end{aligned}$$

Note that each of the components at the right-hand side of (30) is a product of two factors the second of which contains the norms of respective differences of optimal controls, states and Lagrange multipliers and their interpolates.

These differences can be estimated using regularity conditions (13) and (14) , definition (19) as well as assumptions (17) and (24). Namely we have [2] :

$$\|u^o - u_{k,\tau}\|_{0,0} \leqslant c\ (\tau + k) \qquad (31)$$

$$\|y^o - y_{h,\tau}\|_{0,0} \leqslant c\ (\tau + h^2) \qquad (32a)$$

$$\|y^o - y_{h,\tau+\theta}\|_{1,0} \leqslant c\ (\tau + h^2) \qquad (32b)$$

$$\Big\|\frac{dy^o}{dt} - \frac{dy_h}{dt}\Big\|_{0,0} \leqslant c\ h \qquad (32c)$$

$$\|p^o - p_{h,\tau}\|_{0,0} \leqslant c\ (\tau + h^2) \qquad (33a)$$

$$\left\| p^o - p_{h,\tau+(1-\theta)} \right\|_{1,0} \leqslant c\,(\tau + h^2) \tag{33b}$$

$$\left\| \frac{dp^o}{dt} - \frac{dp_h}{dt} \right\|_{0,0} \leqslant c\,h \tag{33c}$$

Substituting these estimations to (30) we get

$$\left\| u^o_{k,\tau} - u_k \right\|^2_{0,0} \leqslant c \Big[\left\| u^o_{k,\tau} - u_{k,\tau} \right\|_{0,0} (\tau + h^2 + k) + \left\| y^o_{h,\tau} - y_{h,\tau} \right\|_{0,0} (\tau + h^2 + k) +$$

$$+ \left\| y^o_{h,\tau} - y_{h,\tau} \right\|_{1,0} (\tau + h) + \left\| p^o_{h,\tau} - p_{h,\tau} \right\|_{0,0} (\tau + h^2 + k) +$$

$$+ \left\| p^o_{h,\tau} - p_{h,\tau} \right\|_{1,0} (\tau + h) + \left\| \lambda^o_{k,\tau} - \lambda^o \right\|_{0,0} (\tau + k) \Big] \tag{34}$$

On the other hand using (1), (3), (11) and (21), (29) as well as conditions of regularity (13) and (14) we get from Lemma 3 :

Lemma 4

<u>If y^p satisfies conditions (15) then</u>

$$\left\| y_{h,\tau} - y^o_{h,\tau} \right\|_{1,0} \leqslant c \Big[\left\| u^o_{k,\tau} - u^o \right\|_{0,0} + (\tau + h) \Big] \tag{35a}$$

$$\left\| p_{h,\tau} - p^o_{h,\tau} \right\|_{1,0} \leqslant c \Big[\left\| u^o_{k,\tau} - u^o \right\|_{0,0} + (\tau + h) \Big] \tag{35b}$$

Using the differential form of conditions of minimum for both Lagrangians L and L_d with respect to u and $u_{k,\tau}$ respectively as well as (10) one can obtain [2] :

Lemma 5

<u>The following estimation takes place</u>

$$\left\| \lambda^o_{k,\tau} - \lambda^o \right\|_{0,0} \leqslant c \Big[\left\| u^o_{k,\tau} - u^o \right\|_{0,0} + (\tau + h) \Big] \tag{36}$$

Substituting (35) and (36) to (34) and taking into account (31) we obtain the following theorem being the main result of the paper:

Theorem 1

<u>If initial condition y^p satisfies (12) then the following estimation take place</u>

$$\left\| u^o_{k,\tau} - u^o \right\|_{0,0} = 0\,(\tau + h + k) \tag{37a}$$

$$\| y^o_{h,\tau} - y^o \|_{1,0} = 0\,(\tau + h + k) \tag{37b}$$

$$J\left(u^o_{k,\tau},\ y(u^o_{k,\tau})\right) - J\left(u^o, y(u^o)\right) = 0\left(\tau^2 + h^2 + k^2\right) \tag{37c}$$

Remark

The above results can be extended [5] to the case of more complicated control constraints using the results of convex programming stability obtained by W.W. Hager [1] . The case of state space constraints is much harder mostly due to the difficulties in obtaining regularity results.

References

[1] Hager W.W. : Lipschitz continuity for constrained processes, to appear .

[2] Lasiecka I.,Malanowski K. : On regularity of solutions to convex optimal control problems with control constraints for parabolic systems , to appear in "Control and Cybernetics" .

[3] Lasiecka I.,Malanowski K. : On discrete-time Ritz-Galerkin approximation of control constrained optimal control problems for parabolic systems , to appear in "Control and Cybernetics".

[4] Lions J.P.,Magenes E. : Problèmes aux limites non homogènes vol. 2, Dunod, Paris 1968.

[5] Malanowski K. : On convergence of finite dimensional approximation to control constrained optimal control problems for distributed systems , presented at the conference "Methods of Mathematical Programming", September 12-16,1977, Zakopane, Poland.

[6] Strang G.,Fix G. : An analysis of the finite element method. Prentice-Hall, New York 1973.

[7] Zlámal M. : Finite element methods for parabolic equations. Math. of Computation, 28 1974 , pp. 393-404.

OPTIMAL DESIGN AND EIGENVALUE PROBLEMS

Bernard ROUSSELET
I.M.S.P.
Parc Valrose
06034 NICE CEDEX - France

INTRODUCTION

We investigate optimisation problems where the control Ω is a domain of $\mathbb{R}^n$ (optimal design) and where the functional J involves eigenvalues of a transmission problem.

Problems of this king were treated already in particular cases by Lord Rayleigh (1877) [9] and Hadamard [6].

A model problem is to find an optimal domain Ω^* for the functional:

$J(\Omega) = \sum_{i=1}^{i=N} \left| \frac{1}{\lambda_i(\Omega)} - \frac{1}{\alpha_i} \right|^2$ where $\{\alpha_i\}_{i=1,\ldots,N}$ are prescribed numbers and $\{\lambda_i(\Omega)\}_{i=1,\ldots,N}$ are the N smaller eigenvalues of the boundary problem:

$$\begin{cases} -\Delta u = \lambda u \text{ in } \Omega \\ u = 0 \text{ on the boundary } \partial\Omega . \end{cases}$$

The regularity of the mapping $\Omega \to J(\Omega)$ is classical (Courant Hilbert) for regular deformations of regular bounded domains.

In the first part of this paper we extend this result to the case of lipschitz regularity and to equations arising in neutronic diffusion modelisation. The main tool to be used is perturbation theory.

In the numerical study of the second part, for simplicity, we only consider the model problem.

It is shown there how to use a gradient type method to reach an approximate optimal domain; we emphasize the interaction of the different methods: finite elements, variable domains, eigenvalues, computation of gradient.

The implementation of the optimisation method in a Fortran program and some computing results are then presented.

For a detailed discussion of all these topics the reader is referred to [12]; on the other hand [10] treats a simpler problem.

I - THEORETICAL RESULTS

§ 1 - Statement of the problem

Neutronic diffusion modelisation leads to the following problem:
find the eigenvalues and eigenfunctions of the partial derivative system :

$$(1.1.)\qquad -\nabla^* (a_i \nabla u_i) + (Lu)_i = \lambda (Ku)_i \quad \text{in } \Omega \qquad \text{for } i = 1,\dots,p$$

with the boundary conditions :

$$(1.2.)\qquad u_i + \ell_i \frac{\partial u_i}{\partial u} = 0 \quad \text{on } \partial\Omega \quad \text{for } i = 1,\dots,p \text{ ;}$$

we have made use of the following notations and assumptions:
∇ is the gradient operator; ∇^* is the divergence operator, $\frac{\partial}{\partial \nu}$ is the outward normal derivative, $u = (u_1,\dots,u_p)$ takes its values in $\mathbb{R}^p$; Ω has a subdomain partition where $\{a_i\}_{i=1,\dots,p}$ (resp. L and K) are C^1 and L^∞ scalar (resp. matrix) functions; K is in general non one to one; in this paper L,K are supposed self-adjoint (see [12] for what be done without this hypothesis).

We shall make use of the variational form of this eigenvalue problem.
We shall need some hypothesis:

H1. Ω is a "lipschitz manifold with boundary" (Ω is locally mapping of a half-space by uniformely bilipschitz homeomorphisms).

Definition 1.1. The collection of all the bounded open sets of $\mathbb{R}^n$ verifying H1 is denoted by VL_n.

H2. $\Omega \in VL_n$ is supposed to be equiped with a subdomain partition $\Omega_j \in VL_n$ $(j=1,\dots,m)$ such that: $\bar{\Omega}_j \subset \Omega$ $(j = 1,\dots,m)$, $\bar{\Omega}_i \cap \bar{\Omega}_j = \emptyset$.
We set: $\Omega_0 = \Omega \setminus (\bigcup_{j=1}^{m} \bar{\Omega}_j)$. Such a domain equiped with its subdomain partition will be denoted by $(\Omega; \Omega_0,\dots,\Omega_m)$.

H3. The p (resp.2) scalar (resp. matrix) functions a_i (resp.L and K) are supposed to be in $L^\infty(\Omega)$ (resp. $L^\infty(\Omega, \mathcal{L}(\mathbb{C}^p))$) and C^1 in each Ω_j $(j = 0,\dots,m)$; moreover the p_i scalar functions ℓ_i are in $C^0(\partial\Omega)$; it is supposed that there exists strictly positiv constants such that $a_i(x) \geq c > 0$ almost everywhere in Ω and $\ell_i(x) \geq c' > 0$ everywhere on $\partial\Omega$; $K(x)$ is not generally one to one.

Definition 1.2. Let $H(\Omega) = (L^2_{\mathbb{C}}(\Omega))^p$, $V(\Omega) = (H^1_{\mathbb{C}}(\Omega))^p$; under the hypothesis H1, H2,H3 we consider :

(i) the sesquilinear form on $H(\Omega)$ with domain $V(\Omega)$:

$$(1.3.)\qquad a[u,v] = \underline{\underline{a}}[u,v] + \underset{\sim}{a}[u,v] \quad \text{with}$$

(1.4.) $$\underline{\underline{a}}[u,v] = \sum_{i=1}^{i=p} \int_\Omega a_i(x)\,(\nabla u_i, \nabla v_i)_{\mathbb{C}^n}\, dx$$

(1.5.) $$\underset{\sim}{a}[u,v] = \int_\Omega (L(x)\,u, v)_{\mathbb{C}^p}\, dx + \sum_{i=1}^{i=p} \int_{\partial\Omega} \frac{a_i(x)}{\ell_i(x)}\, u_i(x)\, \bar{v}_i(x)\, d\sigma(x)$$

for all u and v in $V(\Omega)$

(ii) the boundet sesquilinear form on $H(\Omega)$:

(1.6.) $$b[u,v] = \int_\Omega (K(x)\,u, v)_{\mathbb{C}^p}\, dx$$

Then we consider the variational eigenvalue problem :

under H1,H2,H3 hypothesis, find the scalar $\lambda \in \mathbb{C}$ such that there exists at least one $u \in V(\Omega) \setminus \{o\}$ satisfying

(1.7.) $\forall v \in V(\Omega)\quad a[u, v] = \lambda\, b[u,v]$ where a and b are the two forms defined by (1.3.) and (1.6.).

Remark 1.3. As usual a solution of (1.1.) and (1.2.) is a solution of (1.7.) and a smooth enough solution of (1.7.) is solution of (1.1.) and (1.2.)(see [13]).

However the study of the regularity of the eigenvalues with respect to the domain will not need the operator formulation (1.2.),(1.3.) and a fortiori will not use more regularity than $u \in V(\Omega)$; nevertheless we shall need an abstract operator formulation that will be given in what follows.

Till the end of this chapter the standard reference is [12].

It can be shown that under H1,H2,H3 assumptions the form a is a densely defined symetric, closed one which is bounded from below.

The use of a representation theorem (see [7]) lets us associate to the form a, a selfadjoint bounded from below operator A such that :
$D(A) \subset D(a)$ and $\forall u \in D(A)$, $\forall v \in D(a)$, $a[u, v] = (A\,u, v)_H$; $D(A)$ (resp. $D(a)$) denotes the (functional) domain of the operator A (resp. the form a).

We denote by B the bounded operator associated to b .

Proposition - Definition 1.4. (abstract operator formulation)

If we assume:

H4. For almost all $x \in \Omega$, $L(x)$ is a positiv definit self-adjoint matrix, then A is invertible and the scalar $\lambda \in C$ such that there exists $u \in D(A) \setminus \{o\}$ satisfying $A u - \lambda B u = o$ are the inverses of the non zero eigenvalues of $A^{-1}B$; they will be denoted by generalised eigenvalues of (A,B) (g.e.); as A^{-1} is a compact operator they form a discrete set; the operators are self-adjoint bounded from below thus they can be ordered in a sequence of real numbers going to infinity exactly as for the laplacian spectrum, (see [12] for the proof).

Statement of the optimisation problem

As in the model problem we intend identifying a domain Ω^* that should be optimal for a functional of the type:

$$J_N(\Omega) = d\big(\lambda_1(\Omega),\dots,\lambda_N(\Omega)\ ;\ \alpha_1,\dots,\alpha_N\big)$$

where $\{\alpha_i\}_{i=1,\dots,N}$ are prescribed scalars ("desired values" for the N smallest eigenvalues); d is a distance between the N-tuples $\{\lambda_i\}_{i=1,\dots,N}$ and $\{\alpha_i\}_{i=1,\dots,N}$.

The way of associationg the operators (and thus the eigenvalues) to a domain Ω is complicated by the fact that the discontinuity surfaces of the coefficients have to be identified as well as the boundary $\partial\Omega$ (modelisation of the physical problem).

This association is specified by :

(1.11.) H5. We suppose to be given on the one hand p functions $\ell_i \in C^o(\mathbb{R}^n)$ such that $\ell_i(x) \geq c > o$ (for every $x \in \mathbb{R}^n$ and $i = 1,\dots,p$) on the other hand $p.(m+1)$ real functions $a_{ij} \in C^1(\mathbb{R}^n)$ such that $a_{ij}(x) \geq c > o$ $(i = 1,\dots,p\ ;\ j = o,\dots,m)$ and $2.(m+1)$ real, matrix, self-adjoint functions $L_j \in C^1(\mathbb{R}^n, \mathcal{L}(\mathbb{C}^p)), K_j \in C^1(\mathbb{R}^n, \mathcal{L}(\mathbb{C}^p))$ (for $j = o,\dots,m$).

To every $(\Omega ; \Omega_o,\dots,\Omega_m)$ satisfying H1 and H2 we associate p functions $a_i \in L^\infty(\bar{\Omega})$ defined by $a_i|_{\Omega_j} = a_{ij}$ $(j = o,\dots,m)$ and the two functions $L \in L^\infty(\mathbb{R}^n, \mathcal{L}(\mathbb{C}^p))$, $K \in L^\infty(\mathbb{R}^n, \mathcal{L}(\mathbb{C}^p))$ defined by $L|_{\Omega_j} = L_j$, $K|_{\Omega_j} = K_j$.

These functions thus defined are the coefficients of the forms (1.3.), (1.6.) set in $(\Omega ; \Omega_o , \dots , \Omega_m)$.

Remark 1.5. Hypothesis H5 is the choice of a process which enables us to associate to every $(\Omega ; \Omega_o,\dots,\Omega_m)$ satisfying H1,H2 , some forms which verify H3–H4 . As regard the topology of the family of domains, it is known ([2] p.420) that the eigenvalues are not continue for a too weak topology. As our "lipschitz manifold with boundary" are stable by bilipschitz homeomorphisms (change of variable in the local maps) we may set :

Definition 1.6. The set VL_n of all the bounded open sets of $\mathbb{R}^n$ which are "lipschitz manifolds with boundary" is equiped with the topology generated by the fundamental system of neighborhood :

$$v(\Omega^*, \varepsilon) = \{\Omega \in VL_n \mid \exists F \in W^{1,\infty}(\mathbb{R}^n, \mathbb{R}^n),\ \|F\| \leq \varepsilon,\ \Omega = (I+F)(\Omega^*)\}$$

§ 2 - Regularity of the functional

Orientation . In what follows we shall consider $(\Omega; \Omega_0, \dots, \Omega_m)$ satisfying H1 and H2 ; for every $F \in W^{1,\infty}(\mathbb{R}^n, \mathbb{R}^n)$ such that $\|F\|_{1,\infty} < c < 1$ $(I+F)$ is a bilipschitz homeomorphism of a neighborhood of Ω on a neiborhood of $\Omega_F = (I+F)(\Omega)$; we denote $\Omega_{j,F} = (I+F)(\Omega_j)$, a_F , b_F (resp. A_F , B_F) the forms (resp. the operators) associated to $(\Omega, \Omega_0, \dots, \Omega_m)$ according to the process of H5 and remark 1.5. ; a_F , b_F satisfy then hypothesis H3 , H4 .

We first proove that a_F , b_F , A_F , B_F are Frechet-derivable, in some sense, respectively to F ; we may only conclude the continuity and Gateau-derivability of eigenvalues.

I . Frechet-derivability of the operators

The problem is local, we thus restric the study to a neighborhood of a fixed domain Ω; the idea consists in transforming the problem in variable domains into a problem in a fixed one (thus in fixed functional spaces) with coefficients depending upon the deformation F.

For brevity we only sketch the proof and when it becomes technical we only go into further details for the form b (we refer the reader to [12] for a complete treatment).

Remark 2.1. In what follows we denote $f \circ (I+F)$ by $\tilde{f}$ for every function defined in a neighborhood of Ω_F .

We go back to fixed spaces by

Lemma 2.2. Let $\tilde{a}_F$ and $\tilde{b}_F$ denote the two sesquilinar forms deduced from a_F and b_F with a change of variable in the coefficients (for example with $\phi = I+F$

(2.1.) $\tilde{b}_F[u,v] = \int_\Omega (\tilde{K} u, v)_{\mathbb{C}^n} |D\phi|\, dx)$; let $\tilde{A}_F$, $\tilde{B}_F$ be the associated operators; then the generalised eigenvalues of (A_F, B_F) are g.e. of $(\tilde{A}_F, \tilde{B}_F)$ and conversely; moreover the eigenspaces are in correspondance through the change of variable.

The derivative of b_F involves integrals over the subdomains Ω and is stated in-terms of bounded perturbations.

The derivability of a_F must be expressed in terms of <u>relatively</u> bounded perturbations; we deduce (with a result of perturbation theory that we proove in [12]) the Frechet derivability of F A_F^{-1} runs in a neighborhood of zero in $W^{1,\infty}(\mathbb{R}^n, \mathbb{R}^n)$.

2. Continuity and derivability of the eigenvalues

It is known that we cannot deduce the Frechet derivability of eigenvalues (see [7] and [12]).

The continuity is expressed by :

THEOREM 2.4. Let $\{\lambda_1, \dots, \lambda_m\}$ be m g.e. of (A_o, B_o) counted up with their multiplicity; for every neighborhood in $\mathbb{C}$ of $\{\lambda_1, \dots, \lambda_m\}$ containing no other g.e., there exists a neighborhood v of zero in $W^{1,\infty}(\mathbb{R}^n, \mathbb{R}^n)$ such that for every $F \in v$, w contains exactly m g.e. (counted up with their multiplicity) of (A_F, B_F).

What can be said about Gateau-derivability is expressed by :

THEOREM 2.5. Under the hypothesis H1,H2,H3,H4, let be a g.e. λ_o of (A_o, B_o) of multiplicity m (such that $\frac{1}{\lambda}$ be a <u>semi-simple</u> eigenvalue of $A_o^{-1} B_o$), the group of m g.e. of (A_F, B_F) neighbor of λ_o, defined for $\|F\|$ small enough in Theorem 2.4. is Gateau-derivable in zero :

$$\mu_j(tF) = \lambda_o + t\,\mu_j^{(1)}(F) + o(t) \qquad j = 1, \dots, m \,;$$

$\mu_j(tF)$ are the m g.e. (counted up with their multiplicity) of (A_F, B_F) neighbor of λ_o.

Corollary 2.6. Under the hypothesis of theorem 2.5. if λ_o is a <u>simple</u> g.e. it remains simple for t small enough (Theorem 2.4.), its derivative for $t = o$ is given by :

$$\lambda'(\Omega, F) = a_1[u_o, u_o] - \lambda_o\, b_1[u_o, u_o] \tag{2.5.}$$

where $A_o u_o = \lambda_o B_o u_o$ and $(B_o u_o, u_o) = 1$

3. Expressions of the derivative

usable in the computing practice for a <u>simple</u> eigenvalue of the <u>model</u> problem.

A direct use of corollary 2.6. gives for the model problem:

(2.6.) $$\lambda'(\Omega, F) = -2\int_\Omega (DF\nabla u, \nabla u)\,dx + \int_\Omega \|\nabla u\|^2\,\nabla^*F\,dx - \lambda(\Omega)\int_\Omega u^2\,\nabla^*F\,dx$$

a. Transformation into boundary integrals

It is a widely used method (see [3], [5], [6], [8]).
If the data are smooth enough for using a Green formula we easily deduce from (2.6.) the well-known Hadamard Formula

(2.7.) $$\lambda'(\Omega, F) = -\int_{\partial\Omega} \left(\frac{\partial u}{\partial \nu}\right)^2 (F, \nu)\,d\sigma$$

b. Choice of a family of deformation

In the case of two dimensions it is convenient to use conformal transformations (cf. [4] chapter X); writing $F(x+iy) = P(x+iy) + i\,Q(x+iy)$ an easy computation gives for (2.6.) the very simple expression:

(2.8.) $$\lambda'(\Omega, F) = -2\lambda(\Omega)\int_\Omega u^2\,\frac{\partial P}{\partial x}(x, y)\,dx\,dy\,.$$

In this expression no derivative of u is used, this will be very usefull in the computing applications.

II - NUMERICAL STUDY

§ 1 - Numerical methods

1. Introduction

Let us emphasize that (as usual in non convex optimisation) we do not propose a method to find a global minimum but only a local one. However this answers generally to the engineer's preoccupation: he has got a form of domain in the neighborhood of which he wishes to find an optimum. See [10] and [12] for further details.

The direct problem is discretised with a finite element method. It furnishes a discrete eigenvalue problem of the form $Ax = \lambda Bx$ which is solved by a simultaneous iteration method generalising the power method ([12]).

2. Optimisation method

a. Choice of the method

It is well known for optimal problem in an hilbert space (a fortiori in $\mathbb{R}^n$) that the best local choice of descent is the gradient G of the functional J,

$$(1.1.)\quad \begin{cases} x_o \in H \quad \text{given} \\ x_{n+1} = x_n - \varphi_n \, G(x_n) \ , \varphi_n > o \end{cases}$$

In our problem the collection of geometrical domains where occurs the optimisation do not have any vectorial structure by itself.

But the domains mapped from a fixed one by diffeomorphisms get the properties of the space of diffeomorphisms. On the other hand a discreted domain may be considered as a point of $\mathbb{R}^{2.n_h}$ (n_h equals the number of nodes of the triangulation); the vonverse is false as the triangles associated to a point of $\mathbb{R}^{2.n_h}$ may overlap.

In both cases, a high computing difficulty, is the construction of a good triangulation in each comain generated by the algorithm. As we cannot use a screen, automatic triangulation algorithms were of no use; this leads us to transport the triangulation of a fixed reference domain upon the itarated domains (by the way of diffeomorphisms).

This choice gives a new problem: how to get correct triangulations (from the point of view of finite elements).

The use of holomorphic deformations ensure the conservation of the angles; it is convenient to use complex polynomials. We may also use for simplicity functions which are polynomials on every triangle but we need check the angle conservation.

b. Expression of the gradient

We refer the reader to ([12] § II, 2.2.a , or [10] § II. 4.a.) for a boundary expression of the gradient.

For what concerns the distributed expression we have :

Proposition 1.1. Let J be the functional defined in I. § 1. ("statement of the optimisation problem") and Ω be a polygonal domain (with a correct triangulation); if all the eigenvalues used in J are simple the gradient sought in the form $G(z) = \sum_{p \leq N} b_p z^p$ with $b_p = \gamma_p + i\,\delta_p$ is given by :

$$(1.2.)\quad \begin{cases} \gamma_p = -2p \sum_{i=1}^{i=m} \lambda_i \dfrac{\partial J}{\partial \lambda_i} \int_\Omega u_i^2 \, \mathrm{Re}\big((x - i\,y)^{p-1}\big)\, dx\, dy \\ \delta_p = 2p \sum_{i=1}^{i=m} \lambda_i \dfrac{\partial J}{\partial \lambda_i} \int_\Omega u_i^2 \, \mathrm{Im}\big((x + i\,y)^{p-1}\big)\, dx\, dy \end{cases}$$

where u_i is a normalised eigenfunction associated to λ_i.

The proof is a straightforward application of formula I.(2.8.)

Remark 1.2. The use of transformations affine on every triangle gives also a pleasant expression (see [12] its implementation is in progress.

§ 2 - Implementation

For technical details about domain triangulations and equations formation we refer the reader to [12] .

For computing the gradient in the case of the boundary expression: the main problem is evaluating the normal derivatives $\frac{\partial u}{\partial \nu}$ with the values of u at the nodes of the triangulations; it is tedicus and does not seem accurate and stable. In the case of the distributed expression and globally polynomials deformations we just need evaluate expressions of the kind :

$$\int_\Omega u^2(x+iy)^q \, dx\, dy = \sum_{t\in\mathcal{D}(\Omega)} \int_T u^2(x+iy)^q \, dx\, dy$$

we use that u is affine on the triangles T and use a change of variable to go back to the standard reference rectangle isosceles triangle to make the computations.

The method is implemented in a Fortran program of more than two thousands instructions.

The general structure of the program is as follow:

(i) reading of the data

(ii) eigenvalues computation

(iii) functional computation

(iv) gradient computation

(v) "φ determination"

(vi) stopping tests and as the case may be go to (iv) or (vii)

(vii) end.

Its general feature is very simple but we must not forget that behind (ii) there is a finite element routine!

The number of the numerical results is too short to assess exactly the area of applicability of the method. Shortage of time and money (as well as convenient hardware) explain this deficiency.

As usual in gradient methods the decrease of the functional is fast at the beginning and slower then.

With the use of the boundary expression of the gradient, the functional stops decreasing although the gradient is not very small (to working accuracy); this does not appear with the use of the distributed gradient.

The non correctness of the triangulation prevents the program to go further; this would not be the case while using an interactive screen which enables to construct a new triangulation.

The non unicity of the optimal domain is very clear numerically.

In conclusion the method proposed is a very general approach to eigenvalue optimal design problems; to be of practicle use it needs a high scale computer but it should be of higher interest in engineering areas where generally there is no analytical solution.

REFERENCES

[1] CEA J., Optimisation: théorie et algorithmes
Dunot (1971)

[2] COURANT R. and HILBERT D., Methods of Mathematical Physics,
Interscience Publishers (1953)

[3] DERVIEUX A. and PALMERIO B., Thèse de l'Université de Nice

[4] DIEUDONNE J. Calcul infinitésimal
Hermann (1968)

[5] GARABEDIAN P.R. and SCHIFFER M., Variational problems in the theory of elliptic partial differential equations.
J. of rat.mech. and analys. p.137-171 (1953)

[6] HADAMARD J., Mémoire sur le problème d'analyse relatif à l'équilibre des plaques élastiques encastrées.
Mémoire des Savants étrangers,33 (1908)

[7] KATO T., Perturbation theory for linear operators
Springer-Verlag (1966)

[8] MURAT F. and SIMON J., Thèses de l'Université de Paris VI

[9] RAYLEIGH J.W., The theory of Sound
(New York Dover Publications) (1945)

[10] ROUSSELET B., Problèmes inverses de valeurs propres;
Optimization Techniques: Lecture Notes in Computer Sciences,41, Tome 2,
p.77-85, Springer-Verlag (1976)

[11] ROUSSELET B., Comptes-Rendus de l'Académie des Sciences, 283, série A,1976,
p.507

[12] ROUSSELET B., Thèse de l'Université de Nice (1977).

Optimal Control of a Parabolic Boundary Value Problem

Ekkehard Sachs

Technische Hochschule Darmstadt
Fachbereich Mathematik
Schloßgartenstr. 7
D-6100 Darmstadt
West Germany

1. Introduction

In this paper we are dealing with a control problem which arises in heating processes where the heat transfer takes place by radiation. For a physical description of this process see Carslaw/Jaeger [2] and Butkovskiy [1].
The aim of controlling is to achieve at time T, T>o fixed, a certain temperature distribution by using bounded controls at the boundary.
In Yvon [9] and Lions [5] existence theorems and numerical methods of this problem can be found.
v. Wolfersdorf [7],[8] derives a maximum principle for a class of nonlinear parabolic control problems with quadratic cost-functionals, however, without existence theorems on solutions of the differential equation and the whole optimization problem.
In Sachs [6] a bang-bang-principle is proved for the nonlinear problem with convex Fréchet-differentiable cost-functionals. In this paper we extend these results to convex and continuous cost-functionals. Hence we are able to treat the case where the maximum-norm acts as the objective and we also obtain a bang-bang-principle for this type of problems.

2. Mathematical Formulation of the Problem

We are describing a one-dimensional diffusion process with the Stefan-Boltzmann boundary condition.

Let the temperature be denoted by $y(t,x)$ where $x \in [o,1]$ is the location and $t \in [o,T]$ the time. Then we have

$$y_t(t,x) = y_{xx}(t,x) \qquad t \in (o,T),\ x \in (o,1) \qquad (1)$$

$$y(o,x) = o \qquad x \in [o,1] \qquad (2)$$

$$y_x(t,o) = o \qquad t \in (o,T] \qquad (3)$$

$$y_x(t,1) = -y^4(t,1) + u(t) \qquad t \in (o,T]. \qquad (4)$$

For reasons of proving existence of solutions of (1)-(4) we rewrite (4) with $M > o$

$$My(t,1) + y_x(t,1) = My(t,1) - y^4(t,1) + u(t) \quad t \in (o,T] \quad (4_M)$$

We assume the controls to be bounded, i.e.

$$U = \{ u \in L_p[o,T] : o \leq u(t) \leq 1 \text{ a.e. on } [o,T] \}$$

with $2 < p < \infty$.

<u>Definition:</u> A function $y(u,\cdot,\cdot) \in C([o,T] \times [o,1])$ is called a solution of (1) - (4_M) if it fulfills

$$y(u,t,x) = \int_o^t G_M(t-s,x,1)(My(u,s,1) - y^4(u,s,1) + u(s))ds \quad (5)$$

where G_M is the corresponding Green's function.

G_M can be written by Fourier expansion as

$$G_M(t-s,x,\xi) = \sum_{k=1}^{\infty} e_k(x)e_k(\xi) \exp(-\mu_k^2(t-s))$$

where $e_k(\cdot)$ and μ_k are the orthonormalized eigenfunctions and eigenvalues of the following boundary value problem

$$e''(x) = \mu e(x) \qquad x \in (o,1)$$

$$e'(o) = o, \qquad Me'(1)+e(1) = o.$$

The uniqueness and existence proof in Sachs [6] can be modified easily to the case of nonnegative controls in $L_\infty[o,T]$:

<u>Theorem 1:</u> For each $u \in L_\infty[o,T]$ such that $u(\ ) \geq o$ and $M \geq 4\|u\|_\infty^3$ there exists a unique solution of (5) and some $K>o$ with

$$o \leq y(u,t,x) \leq K \tag{6}$$

for $t \in [o,T]$, $x \in [o,1]$. If $\|u\|_\infty \leq 1$, then $K=1$.

Uniqueness is assured only for the class of solutions which are physically relevant.

Let $\varphi : C[o,1] \to \mathbb{R}$ be a continuous convex functional and $z \in \mathbf{C}[o,T]$ a fixed temperature distribution,
Defining a nonlinear operator by

$$Ru = y(u,T,\cdot),$$

for $u \in L_\infty[o,T]$, $u(t) \geq o$ a.e. $[o,T]$ we consider the following problem.

(P) Find $\hat{u} \in U$ such that

$$\varphi(R\hat{u}-z) \leq \varphi(Ru-z)$$

for all $u \in U$.

<u>3. Bang-Bang-Principle</u>

We have proved the following existence theorem in Sachs [6]:

<u>Theorem 2:</u> There exists an optimal control $\hat{u} \in U$ of problem (P).

We are turning our attention to the differentiability of the operator R.

Lemma 3 (Sachs[6]): The operator $R: U \to C[o,1]$ is Fréchet-differentiable at each point $\hat{u} \in U$ and we have $(M \geq 4)$

$$R'_{\hat{u}}(u) = \int_{o}^{T} G_M(T-s,\cdot,1)\,[(M-4y(\hat{u},s,1)^3)\, y'_{\hat{u}}(u,s,1)+u(s)]\,ds$$

where $h = y'_{\hat{u}}(u,\cdot,1) \in C[o,T]$ is a solution of the Volterra integral equation

$$h - S(Mh-4y(\hat{u},\cdot,1)^3 h) = -Su$$

with

$$Su = \int_{o}^{\cdot} G_M(\cdot-s,1,1)\,u(s)\,ds.$$

For the cost-functional $f(u) = \varphi(Ru-z)$ let us define $D \subseteq L_q[o,T]$, $1/p+1/q=1$, by

$$D = (R'_{\hat{u}})^* (\partial\varphi(R\hat{u}-z)) \tag{7}$$

$$= \{\tilde{l} \in L_q[o,T] : \exists\, l \in \partial\varphi(R\hat{u}-z) \subseteq C[o,1]^* \text{ with } \tilde{l} = l \cdot R'_u = (R'_u)^*(l)\},$$

where $\partial\varphi(R\hat{u}-z)$ is the subgradient of φ at $R\hat{u}-z$.

Theorem 4: The objective functional

$$f(u) = \varphi(Ru-z)$$

is directionally differentiable at each $\hat{u} \in U$ in all directions $u \in L_p[o,T]$ and we have

$$f'_{\hat{u}}(u) = \max_{\tilde{l} \in D} \tilde{l}(u).$$

Proof: We set $\hat{w} = R\hat{u}-z$. Since φ is continuous and convex, the directional derivative $\varphi'_{\hat{w}}(w)$ at $\hat{w} \in C[o,1]$ exists for all directions and can be computed by the subgradient of φ

$$\varphi'_{\hat{w}}(w) = \max_{l \in \partial\varphi(\hat{w})} l(w).$$

Another property of φ is to be locally Lipschitz-continuous because of the assumptions on φ (cf. Ekeland/Temam [4], 2.4).

Let $\{\lambda_k\} \subseteq \mathbb{R}$ be such that $\lim_{k\to\infty} \lambda_k = o$.

Then we obtain for $u, \hat{u} \in U$

$$\left|\frac{1}{\lambda_k}(\varphi(R(\hat{u}+\lambda_k u)-z))-\varphi(R\hat{u}-z))-\varphi'_{\hat{w}}(R'_{\hat{u}}(u)))\right|$$

$$=\left|\frac{1}{\lambda_k}(\varphi(R\hat{u}+\lambda_k R'_{\hat{u}}(u)+\lambda_k e_k - z)-\varphi(R\hat{u}-z))-\varphi'_{\hat{w}}(R'_{\hat{u}}(u))\right|$$

$$\le\left|\frac{1}{\lambda_k}(\varphi(R\hat{u}+\lambda_k R'_{\hat{u}}(u)-z)-\varphi(R\hat{u}-z))-\varphi'_{\hat{w}}(R'_{\hat{u}}(u))\right| +$$

$$\frac{1}{\lambda_k}\left|\varphi(R\hat{u}+\lambda_k(R'_{\hat{u}}(u)+e_k)-z)-\varphi(R\hat{u}+\lambda_k R'_{\hat{u}}(u)-z)\right|$$

where $\lim_{k\to\infty}\|e_k\| = o$ by the Fréchet-differentiability of R (Lemma 3). The first term vanishes for $k\to\infty$ because φ is directionally differentiable and the second one because φ is locally Lipschitz-continuous.

We have

$$f'_{\hat{u}}(u) = \varphi'_{\hat{w}}(R'_{\hat{u}}(u)) = \max_{l \in \partial\varphi(\hat{w})} l(R'_{\hat{u}}(u))$$

$$= \max_{l \in \partial\varphi(\hat{w})} (R'^{*}_{\hat{u}} l)(u) = \max_{\tilde{l} \in D} \tilde{l}(u).$$

Concerning optimality we have the following theorem:

<u>Theorem 5:</u> If $\hat{u} \in U$ is optimal, then there exist $\hat{\tilde{l}} \in D$ such that

$$\hat{\tilde{l}}(u-\hat{u}) \ge o \quad \text{for all } u \in U. \qquad (8)$$

<u>Proof:</u> It is well known that for an optimal control $\hat{u}\in U$ we have for each $u\in U$

$$f'_{\hat{u}}(u-\hat{u}) = \max_{\tilde{l}\in D} \tilde{l}(u-\hat{u}) \geq 0, \tag{9}$$

or equivalently

$$\min_{u\in U} \max_{\tilde{l}\in D} \tilde{l}(u-\hat{u}) \geq 0. \tag{10}$$

Since $U\subset L_p, D\subset L_q$ $(2<p<\infty,\ p^{-1}+q^{-1}=1)$ are weakly compact and L_p, L_q are reflexive Banachspaces we can apply a min-max-theorem, cf. Göpfert[3], p.132, which gives

$$\max_{\tilde{l}\in D} \min_{u\in U} \tilde{l}(u-\hat{u}) = \min_{u\in U} \max_{\tilde{l}\in D} \tilde{l}(u-\hat{u}). \tag{11}$$

(11) and (10) imply the existence of $\hat{\tilde{l}}\in D$ such that

$$\min_{u\in U} \hat{\tilde{l}}(u-\hat{u}) \geq 0,$$

which is identical with assertion (8).

Using the adjoint operator of $R'_{\hat{u}}$ in the sense of integral operators we derive from Lemma 3 and Theorem 5:

<u>Theorem 6:</u> If $\hat{u}\in U$ is an optimal control, then there exists $\hat{l}\in\partial\varphi(R\hat{u}-z)$ such that

$$\int_0^T c(t)u(t)\,dt \geq \int_0^T c(t)\hat{u}(t)\,dt \tag{12}$$

for all $u\in U$, where $c\in L_q[0,T]\cap C[0,T)$ solves

$$c(t) = \int_t^T G_M(s-t,1,1)\ (M-4y(\hat{u},s,1)^3)c(s)\,ds + \hat{l}(G_M(T-t,1,\cdot)). \tag{13}$$

It can be shown that $(R'_{\hat{u}})^*\hat{l}(u)$ can be represented by (12) with (13) (cf. Sachs[6]).

We shall prove the bang-bang-principle for our problem

Theorem 7: If $\hat{u}\in U$ is optimal and the problem (P) is well-posed, i.e. $\hat{u}$ is not a solution of the unconstrained problem (P_u)

$$\inf\{f(u) \mid u\in L_\infty[o,T],\ u(t)\geq o \text{ a.e. on } [o,T]\}$$

then $\hat{u}$ has the property $\hat{u}(t)\in\{o,1\}$, a.e.on $[o,T]$, except on a set which is nowhere dense in $[o,T]$.

Proof: If $\hat{u}$, optimal, is not of the described type, there exists a set $L\subset[o,T]$ and an interval $[t_o,t_1]$, $t_1<T$, such that the Lebesgue measure $\mu(L)$ is greater than zero, $[t_o,t_1]\subseteq\bar{L}$ and

$$o<u(t)<1 \qquad \text{for } t\in L.$$

From this statement it is not hard to deduce in connection with (13) that $c(t)=o$ for $t\in L$ and by continuity of c on $[o,T)$, $c(t)=o$ on $\bar{L}$ or

$$c(t)=o \quad \text{on } [t_o,t_1]. \tag{14}$$

$c(\cdot)$ is the unique solution of the linear Volterra integral equation (13). Furthermore we introduce $d(\cdot,\cdot)$ as the solution of

$$d(t,x) = \int_t^T G_M(s-t,x,1)(M-4y(\hat{u},s,1)^3)d(s,1)ds$$
$$-\hat{l}(G_M(T-t,x,\cdot)). \tag{15}$$

We obtain a solution of (15) by setting $d(s,1)=c(s)$ and defining $d(t,x)$ by (15). $d(\cdot,\cdot)$ is continuous on $[o,T)\times[o,1]$. Using some properties of Green's function we obtain

$$d(t,x) = \int_o^1 G_M(t^*-t,x,\xi)d(t^*,\xi)d\xi \quad \text{for } t\in[o,t^*], \tag{16}$$

where $t_1\leq t^*<T$. For $x=1$, (14) gives

$$o = d(t,1) = \int_o^1 G_M(t^*-t,1,\xi)d(t^*,\xi)d\xi, \quad t\in[t_o,t_1]$$

and the analyticity of G_M in t for $t<t^*$ implies

$$\int_o^1 e_k(\xi)d(t^*,\xi)d\xi = o \qquad \text{for } k\in\mathbb{N}.$$

The completeness of $\{e_k(\cdot)\}$ leads to

$$d(t^*,\xi) = o \qquad \text{for } \xi\in[o,1]$$

and by (16) to

$$d(t,x) = o = c(t) \quad \text{for } t\in[o,t^*].$$

Since t^* was arbitrary chosen $t_1 \leq t^* < T$,

we obtain

$$c(t) = o \quad \text{for } t\in[o,T). \tag{17}$$

Hence (13) implies with (17)

$$\hat{1}(G_M(T-t,1,\;)) = o, \qquad t\in[o,T),$$

which is equivalent to

$$\sum_{k=1}^{\infty} \hat{1}(e_k)e_k(1)\exp(-\mu_k^2(T-t)) = o, \quad t\in[o,T).$$

Analyticity w.r.t. t and completeness of $\{e_k\}$ imply

$$\hat{1} = \Theta, \tag{18}$$

Since $\hat{1}$ is a subgradient of φ at $R\hat{u}-z$, (18) effects

$$\varphi(R\hat{u}-z) \leq \varphi(w) \qquad \text{for all } w\in C[o,1].$$

and

$$\varphi(R\hat{u}-z) \leq \varphi(Ru-z)$$

for all $u\in L_\infty[o,T]$ such that $u(t)\geq o$ a.e. on $[o,T]$.
This is a contradiction to the assumption, that the problem is well-posed.

In the case of the maximum-norm $\varphi(w) = \|w\|_\infty$ we can apply the theory since φ is convex and continuous.

References

[1] BUTKOVSKIY,A.G.,"Theory of Optimal Control of Distributed Parameter Systems", Elsevier, New York-Amsterdam-London, 1969.

[2] CARSLAW,H.S./JAEGER,J.C., "Conduction of heat in Solids", 2nd Edition, Oxford University Press, Oxford, 1959.

[3] GÖPFERT,A.,"Mathematische Optimierung in allgemeinen Vektorräumen", Teubner, Leipzig, 1973.

[4] EKELAND,I./TEMAM,R., "Convex Analysis and Variational Problems", North Holland, Amsterdam, 1976.

[5] LIONS,J.L.,Various Topics in the Theory of Optimal Control of Distributed Systems, in "Optimal Control Theory and its Applications" Eds.B.J.Kirby,part I, pp.166-3o9, Springer Lect.N.Ec.Math.Syst.1o5, 1974.

[6] SACHS,E., A Parabolic Control Problem with a Boundary Condition of the Stefan-Boltzmann type, to appear in ZAMM.

[7] v.WOLFERSDORF,L., Optimale Steuerung einer Klasse nichtlinearer Aufheizungsprozesse, ZAMM 55, 353-362 (1975).

[8] v.WOLFERSDORF,L., Optimale Steuerung bei Hammersteinschen Integralgleichungen mit schwach singulären Kernen, Math.Operationsforsch.u.Statistik 6,6o9-626,(1975).

[9] YVON,J.-P., Etude de quelques problèmes de controle pour les systèmes distribués, Thèse, Université Paris VI, 1973.

A VARIATIONAL INEQUALITY ASSOCIATED WITH A STEFAN PROBLEM
SIMULATION AND CONTROL

C. Saguez
IRIA-LABORIA
Domaine de Voluceau
78150 Le Chesnay

Abstract.

We study the simulation and the control of a Stefan problem with heat source along the free boundary. Such a problem can be associated with the continuous casting process taking the convection in the liquid into account. The initial problem is transformed into a variational inequality (V.I) in which the free boundary appears explicitly in the second member. We prove the existence of a solution of this V.I and we study some associated optimal control problems. Numerical methods are presented.

Introduction.

Many physical systems are characterized by the existence of a free boundary. For example, we can cite the classical Stefan problem and all systems with several phases. Here we study in the one dimensional case such a problem when the heat exchange along the free boundary is constitued by the latent heat and another term $f(t)$ only depending of t. Such a term can be interpreted as a term of convection during the continuous casting process. To study the optimal control of such a problem, we transform the initial system into a variational inequality (V.I) by a technique similar to the Baiocchi's method. In this case, we obtain a V.I where the free boundary appears explicitly on the second member. With convenient assumptions, the existence of a solution is proved. A numerical method is proposed.

Then we consider some associated optimal control problems when either the state or the free boundary are observed. We prove the existence of a solution and we obtain, for a semi-discretized problem, optimality conditions. Numerically, we use a gradient method.

I. Formulation of the problem :

Let $\theta(x,t)$ denotes the temperature. The equation of the free boundary is given by $x = s(t)$.

Then the problem is to find $\theta(x,t)$ and $s(t)$, solution of the following system :

(1.1) $$\frac{\partial\theta}{\partial t} - \frac{\partial^2\theta}{\partial x^2} + \theta = 0 \qquad 0 < x < s(t) \quad , \quad 0 < t < T$$

(1.2) $$\theta(0,t) = 0 \qquad 0 < t < T$$

(1.3) $$\begin{cases} \theta(s(t),t) = 0 & 0 < t < T \\ \dfrac{\partial\theta}{\partial x}(s(t),t) = - L\dfrac{ds}{dt} + f(t) & 0 < t < T \end{cases}$$

(1.4) $$\theta(x,0) = \theta_o(x) \qquad 0 < x < s_o \text{ given, } \theta_o(x) \geq 0$$

where $L \geq 0$ denotes the latent heat.

Remark 1 :

If a solution exists, then

$$- L\frac{ds}{dt} + f(t) \leq 0$$

II. The variational inequality :

We define $\Omega =]0,R[$ such that :

$$\{x|\ 0 \le x \le s(t)\} \subset]0,R[\qquad \forall\, t \in]0,T[$$

Let $\tilde{\theta}(x,t)$ denotes the extension of $\theta(x,t)$ to $Q = \Omega \times]0,T[$ by zero. We introduce the new variable :

$$y(x,t) = \int_x^R \tilde{\theta}(\xi,t)d\xi$$

After some calculations, we verify that $y(x,t)$ is solution of the following variational inequality :

To find
$$\begin{cases} y \in L^2(0,T;H^1(\Omega)) \\ \frac{dy}{dt} \in L^2(0,T;L^2(\Omega)) \\ s(t) \in H^1(0,T) \\ \Omega =]0,R[\end{cases} \quad \text{such that}$$

(2.1)
$$\left(\frac{\partial y}{\partial t},\varphi-y\right)+a(y,\varphi-y) \ge \left(-L\frac{ds}{dt}+f(t),\varphi-y\right)+v(t)(\varphi(o)-y(o,t))$$
$$\forall\, \varphi \in K = \{\varphi|\varphi \in H^1(\Omega),\ \varphi(R) = 0,\ \varphi \ge 0\}$$

(2.2)
$$y(x,0) = y_o(x) = \int_x^R \tilde{\theta}_o(\xi)d\xi$$

(2.3)
$$\begin{cases} s(t) = \operatorname{Inf}\{x|y(x,t) = 0\} \\ s(o) = s_o \end{cases}$$

(2.4)
$$s(t) < R$$

where $a(u,v) = \int_\Omega (\text{grad } u \text{ grad } v + uv)\, d\Omega$.

Remark 2 :

We see that the free boundary appears explicitly in the second member of this V.I. Another problem of this type has been studied by A. Friedman - D. Kinderlehrer [2].

To prove the existence of a solution, we do the following assumptions :

(2.5)
$$f(t) \in L^\infty(o,T)$$

(2.6)
$$v(t) \in L^\infty(o,T), \text{ increasing, positive}$$

(2.7)
$$y_o(x) \in H^3(\Omega) \text{ with } \begin{cases} \frac{dy_o}{dx} < 0 \qquad x \in]0,s_o[\\ \frac{dy_o}{dx}(s_o) = 0,\ y_o(x) = 0 \qquad x \ge s_o > 0 \\ \frac{d^3y_o}{dx^3} - \frac{dy_o}{dx} \le 0 \\ \int_o^{s_o} y_o(x)dx = c > 0,\ \frac{dy_o}{dx}(o) = -v_o \end{cases}$$

Then we have the proposition :

Proposition 1.

Under the assumptions (2.5)-(2.7), the problem (2.1)-(2.4) admits a solution (y,s) with :

$$y \in L^2(0,T;H^2(\Omega)) \qquad ; \quad s(t) \in H^1(\Omega)$$
$$\frac{dy}{dt} \in L^2(0,T;L^2(\Omega))$$

Proof :

We give only the principal steps of the demonstration (C. Saguez [6])

i) For R fixed, we define the semi-discretized problem :

$$\text{(2.8)} \quad \begin{cases} \left(\frac{y^{n+1}-y^n}{\Delta t},\varphi-y^{n+1}\right)+a(y^{n+1},\varphi-y^{n+1}) \geq \left(-L\,\frac{s^{n+1}-s^n}{\Delta t}+f^{n+1},\varphi-y^{n+1}\right) \\ \quad + v^{n+1}(\varphi(o)-y^{n+1}(o)) \qquad \forall\,\varphi \in K \\ \quad n = 0,\ldots,NT \\ y^o(x) = y_o(x) \qquad \text{(2.9)} \\ s^{n+1} = \operatorname{Inf}\{x \mid y^{n+1}(x) = 0\} \qquad s^o = s_o \qquad \text{(2.10)} \end{cases}$$

with $f^{n+1} = \frac{1}{\Delta t}\int_{n\Delta t}^{(n+1)\Delta t} f(t)dt \; ; \; v^{n+1} = \frac{1}{\Delta t}\int_{n\Delta t}^{(n+1)\Delta t} v(t)dt$

By using a fixed point theorem, we prove the existence of a solution for this semi-discretized problem.

ii) In a second part we prove that the free boundary associated with this semi-discretized problem is bounded independently of R and Δt. So we can choose R great enough such that $s^n < R \quad \forall\, n,\ \forall\, \Delta t$.

iii) Finally, with this choice of R, we obtain some a priori estimates and we prove the convergence of the solution of the semi-discretized problem to a solution of the initial V.I.

■

III. Numerical methods :

Let y_i^{n+1} denotes the solution of the following V.I :

$$\text{(3.1)} \quad \left(\frac{y_i^{n+1}-y^n}{\Delta t},\varphi-y_i^{n+1}\right)+a(y_i^{n+1},\varphi-y_i^{n+1}) \geq \left(-L\,\frac{s_i^{n+1}-s^n}{\Delta t}+f^{n+1},\varphi-y_i^{n+1}\right) + v^{n+1}(\varphi(o)-y_i^{n+1}(o)) \qquad \forall\,\varphi \in K$$

We define the following algorithm ;

1) Initialisation, $y^o = y_o(x)$, $n = 0$, $s^o = s_o$

2) $i = 0$, s_o^{n+1} given

3) Solve (3.1) with s_i^{n+1} given

4) $\tilde{s}_{i+1}^{n+1} = \operatorname{Inf}\{x \in \Omega \mid y_i^{n+1}(x) = 0\}$

5) $s_{i+1}^{n+1} = (1-\omega)s_i^{n+1} + \omega\,\tilde{s}_{i+1}^{n+1}$

6) Test of convergence :

7) if true $s^{n+1} = s^{n+1}_{i+1}$; $y^{n+1} = y^{n+1}_{i}$

else $i = i+1$, go to 3)

if n is equal to NT, stop.

else $n = n+1$, go to 2

Remark 3 :

Many modifications are possible for this algorithm.

Remark 4 :

The V.I (3.1) is solved by a relaxation-projection method. (R. Glowinski - J.L. Lions - R. Trémolières [3]).

Now we give a numerical example :

The initial problem is the following :

$$\begin{cases} \frac{\partial \theta}{\partial t} - \Delta\theta = 0 & (3.2) \\ \theta(o,t) = v(t) = t + \frac{1}{2} & (3.3) \\ \theta(s(t),t) = 0 & (3.4) \\ \frac{\partial \theta}{\partial x}(s(t),t) = -4\frac{ds}{dt} + \frac{2t}{\sqrt{3-2t}} & (3.5) \\ \theta(x,o) = (\frac{x^2}{2} - 2x + \frac{1}{2})^+ & (3.6) \end{cases}$$

So the exact solution is :

$$\begin{cases} \theta(x,t) = (\frac{x^2}{2} - 2x + t + \frac{1}{2})^+ \\ s(t) = 2 - \sqrt{3-2t} \end{cases}$$

With 30 points in space and 10 points in time, we obtain a good convergence after 20 iterations (see figures 1, 2), in 12s with IBM-370-168.

The numerical results are the following :

t	0	0,1	0,2	0,3	0,4	0,5	0,6	0,7	0,8	0,9	1
s(t) Exact	0,2679	0,3267	0;3875	0,4508	0,5168	0,5858	0,6584	0,7351	0,8168	0,9046	1,0CC
s(t) computed	0,2668	0,3235	0,3845	0,4468	0,5135	0,5832	0,6578	0,7363	0,8196	0,9112	0,9965

Another application has been carried out in collaboration with IRSID (Institute of Research of the french steel industry) to simulate the solidification of the steel in a continuous casting process taking the convection in the liquid into account (M. Larrecq - C. Saguez [4]).

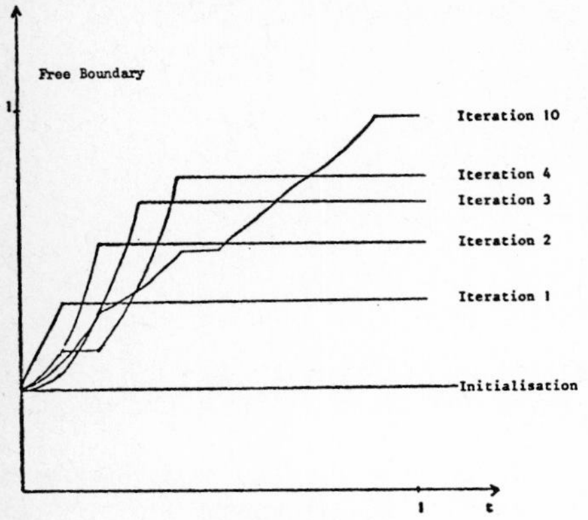

Figure 1

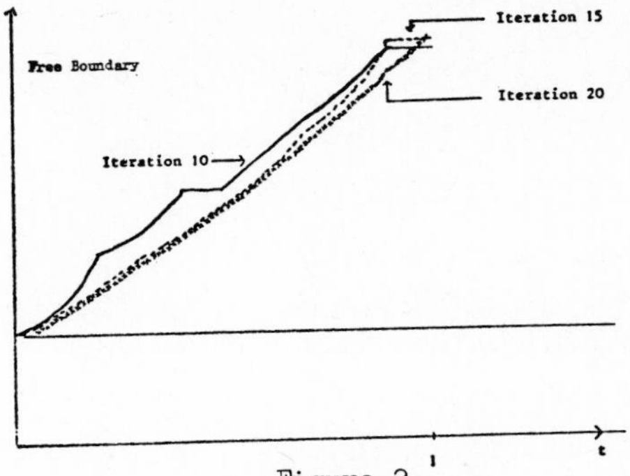

Figure 2

Convergence of the free boundary

IV. Optimal control problems :

Now we consider v(t) as a control variable.

Let $\mathcal{U}_{ad} = \{v \in L^2(o,T) \mid 0 < \alpha \leq v(t) \leq \beta,\ v(t) \text{ increasing}\}$ denotes the admissible set of control.

We choose R great enough such that $s_v(t) < R \quad \forall v \in \mathcal{U}_{ad}$.

We define the following functionals :

$$J_1(v) = \int_Q |y-z_d|^2 \, dxdt + \nu \| v \|^2_{L^2(o,T)} \tag{4.1}$$

$$J_2(v) = \int_0^T |s(t) - s_d(t)|^2 dt + \nu \| v \|^2_{L^2(o,T)} \tag{4.2}$$

$$J_3(v) = \int_Q |\chi_{]o,s(t)[} - \chi_d|^2 \, dxdt + \nu \| v \|^2_{L^2(o,T)} \tag{4.3}$$

where z_d, s_d, χ_d are given functions and ν a positive constant.

Then we consider the optimal control problems (P_i) :

$$\begin{cases} \text{To find } \bar{v} \in \mathcal{U}_{ad} \text{ such that :} \\ J_i(\bar{v}) = \inf\limits_{v \in \mathcal{U}_{ad}} J_i(v) \end{cases} \tag{4.4}$$

We have the result of existence :

Proposition 2 :

The problems (4.4) admit at least one solution.

To explain the numerical method, we consider only the functional J_1. The method and the results are similar for the other functionals.

We consider the semi-discretized problem associated to J_1 :

$$(4.5)\qquad \tilde{J}_1(v) = \sum_{i=1}^{NT+1} \| y^n - z_d^n \|^2_{L^2(\Omega)} \Delta t + \nu \sum_{i=1}^{n} |v^n|^2 \Delta t$$

We can prove the result of differentiability.

Proposition 3 :

The solution (y^n, s^n) of the semi-discretized problem is G-differentiable with respect to $\{v^n\}$.

If we introduce the following adjoint state :

$$\begin{cases} -\dfrac{p^{n+1} - p^n}{\Delta t} - \Delta p^n + p^n = 2\,\Delta t (y^{n+1} - z_d^{n+1}) & (4.6)\\ \dfrac{dp^n}{dx}(o) = 0 & (4.7)\\ \dfrac{dp^n}{dx}(s^{n+1}) = -\, q^n & (4.8)\\ p^{NT+1}(x) = 0 & (4.9)\\ \dfrac{L}{\Delta t}\displaystyle\int_o^{s^{n+1}} (p^n - p^{n+1}) = \left(-\, L\,\dfrac{s^{n+1} - s^n}{\Delta t} + f^{n+1}\right) p^n(s^{n+1}) & (4.10)\end{cases}$$

We obtain for the gradient of $\tilde{J}_1$ the expression :

$$(4.11)\qquad \tilde{J}_1'.h = \sum_{i=1}^{NT+1} h^n (2\nu\, \Delta t\, v^n + p^{n-1}(o))$$

Remark 5 :

This result is similar to the results obtained by F. Mignot for the elliptic variational inequality (F. Mignot [5]).

Numerically, we use a gradient method. We consider the same system (3.2) - (3.6) and we take :

$$z_d(x,t) = -\frac{x^3}{6} + x^2 - \left(t + \frac{1}{2}\right) x + (2 - \sqrt{3-2t})\left(-\frac{1}{3} + \frac{2t}{3} + \frac{\sqrt{3-2t}}{3}\right) \quad \text{if } x < 2 - \sqrt{3-2t}$$

$$z_d(x,t) = 0 \quad \text{else}$$

and $\nu = 0$.

So the exact solution is $v(t) = t + 1/2$.

We obtain the solution in 1mn with IBM 370-168. The results are presented figures 3, 4, 5.

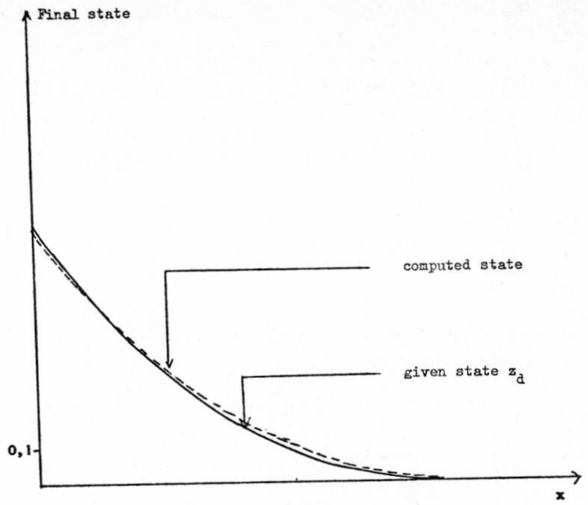

Final state

Figure 3

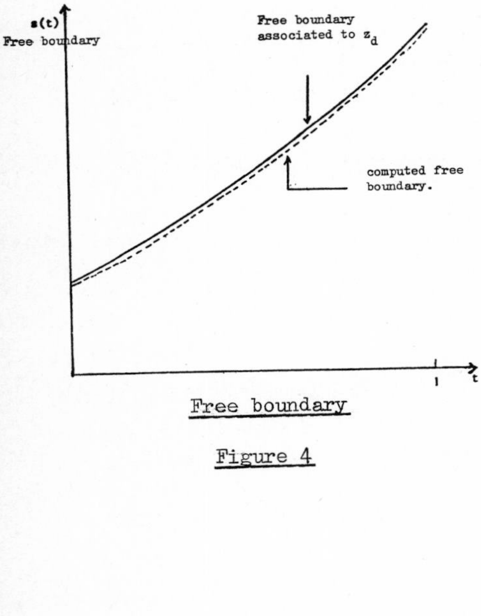

Free boundary

Figure 4

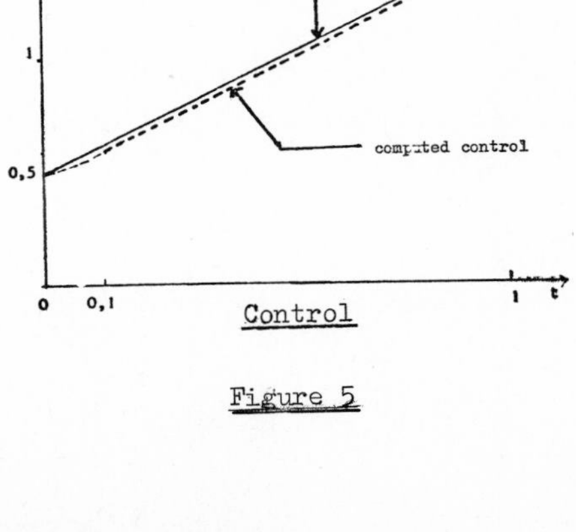

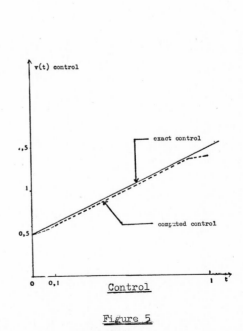

Control

Figure 5

Conclusion :

It appears that the transformation of the problem into a variational inequality is an efficient method to simulate and to control the system and that this method can be used for many free boundary problems.

Acknoledgment :

I am thankful to Prof. F. Mignot for his remarks and suggestions.

References :

[1] C. Baiocchi : Problèmes à frontière libre et inéquations variationnelles (C.R.A.S. t. 283, 12 Juillet 1976).

[2] A. Friedman, D. Kinderlehrer : A class of parabolic variational inequalities (J. of Diff. Eq. 21, 1976).

[3] R. Glowinski, J.L. Lions, R. Trémolières : Analyse numérique des inéquations variationnelles (Dunod, 1976).

[4] M. Larrecq, C. Saguez ; Modèle mathématique de la solidification en coulée continue tenant compte de la convection à l'interface liquide-solide (Journées Métallurgiques d'Automne, 1977).

[5] F. Mignot : Contrôle dans les inéquations variationnelles elliptiques (Thesis, Paris VI, 1975).

[6] C. Saguez : Un problème de Stefan avec source sur la frontière libre (Rapport Laboria, 1977).

NUMERICAL METHODS FOR A GENERALIZED OPTIMAL CONTROL PROBLEM

Claudia Lidia Simionescu
Department of Mathematics
University of Braşov
2200 Braşov, R.S.Romania

ABSTRACT

The present paper deals with an optimal control problem when the state of the system is given by an integro-differential equation. Methods regarding the numerical approximation for the solution are studied.

1. STATEMENT OF THE PROBLEM

Let V and H be two Hilbert spaces , $V \subset H$, V dense in H, the injection being continuous and let us consider , in the sense of scalar distributions , an integro-differential equation with variable coefficients:

$$\left(B_1(t)\frac{dy}{dt} + B_0(t)y(t) + B_{-1}(t)\int_0^\infty y(t-x)\,dx, \varphi\right)_H + a(t;y,\varphi) = (g+Bu,\varphi)_H , \varphi \in V \quad (1)$$

Here , $B_1(t)$, $B_0(t)$, $B_{-1}(t) \in \mathcal{E}_t(L(H,H))$, ($\mathcal{E}_t(L(H,H))$- the space of all linear continuous operators on H which are indefinit differentiable functions in t) and $B_1(t)$ is hermitic and satisfies the condition: $(\exists)\ b_1(t) \in \mathcal{E}_t$, $b_1(t) > 0$, such that

$$(B_1(t) \varphi, \varphi)_H \geq b_1(t) \| \varphi \|_H^2 \quad (\forall) \varphi \in H, \ t \in R \tag{2}$$

We denoted by $a(t;y,\varphi)$ a continuous sesquilinear form on VxV with the property that for any

$$\varphi, \psi \in V \longrightarrow a(t;\varphi,\psi) \in \mathcal{E}_t \tag{3}$$

and: $(\exists)\ \alpha(t) \in \mathcal{E}_t$, $\alpha(t) > 0$, $t \in R$ such that

$$\mathrm{Re}\ a(t;\varphi,\varphi) \geq \alpha(t) \| \varphi \|_V^2 \qquad \varphi \in V , \quad t \in R \tag{4}$$

From (3) and (4) it follows the existence of an operator $A(t) \in \mathcal{E}_t(L(V,V))$, such that

$$\mathrm{Re}\ (A(t) \varphi, \varphi)_V \geq \alpha(t) \| \varphi \|_V^2 \quad (\forall) \varphi \in V, \quad t \in R \tag{5}$$

In earlier papers ([3], [4], [5], [6]) we considered the optimal control problem:

<u>O.C.P.</u>: for a system governed by (1) , find $u \in U_{ad} \subset U$ (U the Hilbert space of controls and U_{ad} a closed convex subset of U) which minimizes a quadratic cost functional:

$$J(u) = \| Cy(u) - z_d \|_{\mathcal{H}}^2 + (Nu, u)_U \tag{6}$$

where C is the observation operator of the system , $z_d \in \mathcal{H}$ a given observation in $\mathcal{H}$, ($\mathcal{H}$ - a Hilbert space) and N a linear continuous operator on the control space U .

We established existence and uniqueness results for the above problem and in [5] we deald mainly with approximations for the optimal control via the Galerkin method (more general equations(1) have been taken into account)

This time , we shall be concerned with finit differential and penalty methods for solving (1)-(6).

Let us therefore start with the optimality system :

$$\left(B_1(t)\frac{dy}{dt} + B_0(t)y(t) + B_{-1}(t)\int_0^\infty y(t-x)\,dx, \varphi\right)_H + (A(t)y, \varphi)_V = (g + Bu^*, \varphi)_H \ , \ \varphi \in V$$

$$\left(-B_1(t)\frac{dp}{dt} + (B_0(t) - B_1'(t))p - \int_0^{\infty} B_{-1}(t-x)p(t-x)\,dx, \varphi\right)_H + (A^*(t)p, \varphi)_V =$$

$$= (C^*\Lambda(Cy(t,u^*) - z_d), \varphi)_V \qquad \varphi \in V \qquad (7)$$

$$(\Lambda_{\mathcal{U}}^{-1} B^* p(t,u^*) + Nu^*, u - u^*)_{\mathcal{U}} \geq 0 \qquad (\forall)\, u \in U_{ad} \subset U$$

where u^* is the optimal control , $p(t, u^*)$ is the adjoint state, B^x, C^x - the adjoints of B respectively C and Λ_U the canonical isomorphism $U \to U'$ and A^x is the adjoint for A . Without a loss of generality regarding the numerical methods we suppose $a(t; \varphi, \psi)$ -real hermitian , i.e. A(t) is autoadjoint, symmetric, and the space of observations $\mathcal{H}$,- the state space.

We know , [3] , that if $g + Bu^* \in \mathcal{D}^k(p; H)^{x)}$, B being a continuous linear operator from $U \to \mathcal{D}^k(p; H)$, then there is a unique $u^* \in U_{ad} \subset U$ and a unique couple $(y(t,u^*), p(t,u^*))$, satisfying (7) and

$$y \in \mathcal{D}^k(p + G; V) \cap \mathcal{D}^k(p, H)$$

$$p \in [\mathcal{D}^{-k}(-(p+G)+F; V) \cap \mathcal{D}^{-k}(-(p+G); H)] \cup [\mathcal{D}^{-k}(-p+G; V) \cap \mathcal{D}^{-k}(-p; H)]$$

where p,G,F are C^1 functions , such that if $p'(t) \geq p_o$, then also $(p+G)' \geq p_o$ and $F' - (p+G)' \geq p_o$, $(\forall)$ $t \in R$.

Since (7) holds for any $\varphi \in \mathcal{D}(V)$, we get in the weak sense:

$$B_1(t)\frac{dy}{dt} + B_0(t)y + B_{-1}(t)\int_0^{\infty} y(t-x)\,dx = Bu^*$$

$$y(t,u^*) = A^{-1}(t)Jg \qquad J \in L(V,V)$$

$$-B_1(t)\frac{dp}{dt} + (B_0(t) - B_1'(t))p - \int_0^{\infty} B_{-1}(t-x)p(t-x)\,dx = J^{-1}\Lambda\, y(t,u^*) \qquad (8)$$

$$p(t,u^*) = -A^{-1}(t)\Lambda\, z_d$$

$$(\Lambda_{\mathcal{U}}^{-1} B^* p(t,u^*) + Nu^*, u - u^*)_U \geq 0 \qquad (\forall)\; u \in U_{ad} \subset U$$

---

x) $\mathcal{D}^k(p; H)$ - Sobolev space with weight $p(t)$,for $k \in Z$,and $p \in C^1$, such that $p'(t) \geq p_o$, $t \in R$

2. FINITE DIFFERENCE METHOD

We shall refere us to the system (7) and shall approximate the optimal control u^* by means of an appropriate form of the wellknown finite difference method.

Let $\{N_L\}_{L\in\mathbb{N}}$ be a sequence of positive real-numbers converging to infinity and for every n let us take a step $h = \frac{N_L}{n}$. For $l= 0,1,\dots,n$ and $\xi_h^{\ell}\in[h(l-1), hl]$, we built the difference equations

$$B_1(lh)\frac{y_h^{\ell}-y_h^{\ell-1}}{h}+B_0(lh)y_h^{\ell}+B_{-1}(lh)\sum_{\ell=0}^{n-1}y(lh-\xi_h^{\ell})\cdot h,\varphi)_H+(A(lh)y_h^{\ell},\varphi)_V=$$

$$=(g(lh)+Bu^*_{lh},\varphi)_H \qquad \varphi\in V \tag{9}$$

$$-B_1(lh)\frac{p_h^{\ell}-p_h^{\ell-1}}{h}+(B_0(lh)-B_1'(lh))p_h^{\ell}-\sum_{\ell=0}^{n-1}B_{-1}(lh-\xi_h^{\ell})p(lh-\xi_h^{\ell})h,\varphi)_H+$$

$$+(A(lh)p_h^{\ell},\varphi)_V=(\Lambda(y_h^{\ell}-z_d),\varphi)_V \quad ,\ \varphi\in V \tag{10}$$

$$(\Lambda_U^{-1}B^*p_h^{\ell}+Nu^*_{lh},u-u^*_{lh})_U\geq 0 \qquad (\forall)\ u\in U_{ad}\subset U \tag{11}$$

Since B_{-1}, B_0, B_1, A, A^x have been supposed in $\mathcal{E}_t$, there has been possible to take their values in points $t= lh$ and not the mean value over an interval. We can start with

$$y_h^{-1}=A^{-1}(0)Jg(0)\ \in\ V$$

$$p_h^{-1}=-A^{-1}(0)\Lambda\, z_d\ \in\ V$$

In the equations(9) and (10) we must explicitate the term under the summation sign. In fact we can take

$$y(\,lh-\xi_h^{\ell})=y_h^1-\xi_h^{\ell}\frac{y_h^1-y_h^{1*}}{h}$$

where for y_h^{1*} we make an appropriate choice such that it corresponds to a value ξ , $1\frac{N_L}{n} - \xi_h^l < \xi < 1\frac{N_L}{n}$

The numerical computations we did shows us that we get pretty good approximations taking for $\xi_h^l = (l-1)h$ and for $y_h^{l*} = \frac{y_h^l - y_h^1}{2}$

If we do the same for (10) , then we obtaine an algebraic system of 2n+2 equations

$$B_1(lh)\frac{y_h^l - y_h^{l-1}}{h} + B_0(lh)y_h^l + B_{-1}(lh)\sum_{l=0}^{n-1}[y_h^l - \xi_h^l\frac{y_h^l - y_h^{l*}}{h}]h, \varphi)_H + \\ + (A(lh)y_h^l, \varphi)_V = (g(lh) + Bu_{lh}^*, \varphi)_H \qquad \varphi \in V \qquad (12)$$

$$-B_1(lh)\frac{p_h^l - p_h^{l-1}}{h} + (B_0(lh) - B_1'(lh))p_h^l - \sum_{l=0}^{n-1} B_{-1}(lh - \xi_h^l)[p_h^l - \xi_h^l\frac{p_h^l - p_h^{l*}}{h}]h, \varphi)_H + \\ + (A(lh)p_h^l, \varphi)_V = (\Lambda(y_h^l - z_d), \varphi)_V \qquad \varphi \in V \qquad (13)$$

Ordering the system by y_h^1 and p_h^1 , there is easy to be seen that because $g(lh) + B\,u_{lh}^* \in \mathcal{D}^K(p, H)$, the solution for (12)-(13) exists and is unique and belongs to the Sobolev space with weights $\mathcal{D}^K(p + G; V) \cap \mathcal{D}^K(p; H)$. Corresponding to p_h^1 , from (11) we get the approximate value u_{lh}^* for the optimal control. If the O.C.P. has no constrains , then

$$u_{lh}^* = -N^{-1}\Lambda_U^{-1}B^* p_h^1 \qquad (14)$$

and the whole problem reduces to solving (12) and (13) with (14) in the right hand side of (12).

This way , for any N_L we obtaine the sequences of states, adjoint states and optimal controls, for $t > 0$:

$$\begin{array}{l} y_{hN_L}^{-1}, y_{hN_L}^{0}, \ldots, y_{hN_L}^{n} \\ p_{hN_L}^{-1}, p_{hN_L}^{0}, \ldots, p_{hN_L}^{n} \\ u_{hN_L}^{-1*}, u_{hN_L}^{0*}, \ldots, u_{hN_L}^{n*} \end{array} \qquad (15)$$

For $t \in (-N_L, 0)$, we define $\bar{y}^{\,l}_{hN_L}$ by reflection, putting in (12) and (13) the coefficients $B_1(-lh)$, $B_o(-lh)$, $B_{-1}(-lh)$, $A(-lh)$ $g(-lh)$, $B_1'(-lh)$ and under the summation sign the values $y(-lh-\xi_h^l)$, $B_{-1}(-lh-\xi_h^l)p(-lh-\xi_h^l)$. Then, define:

$$y^n_{N_L} = \begin{cases} 0 & t \notin [-N_L, N_L] \\ \bar{y}^{\,l}_{hN_L} & t \in [-h(l-1), -hl) \\ y^l_{hN_L} & t \in [h(l-1), hl] \quad l = o, 1, \ldots, n \end{cases} \qquad (16)$$

By the same rule we built $(p^n_{N_L})$ and $(u^n_{N_L})$. This way, the approximations are defined over the real line. Also:

$$(B_1(lh)\varphi, \varphi)_H \geqslant b_1(lh)\|\varphi\|^2_H \qquad \varphi \in H \qquad (17)$$

$$(A(lh)\varphi, \varphi)_V \geqslant \alpha(lh)\|\varphi\|^2_V \qquad \varphi \in V \qquad (18)$$

and

$$J^n_{N_L} = J(u^{nx}_{N_L}) = \|y^n_{N_L} - z_d\|^2_{\mathcal{H}} + (Nu^{nx}_{N_L}, u^{nx}_{N_L})_U \qquad (19)$$

where $\mathcal{H} = \mathcal{D}^K(p+G; V) \cap \mathcal{D}^K(p; H)$, the state space being the observation space, $C = I$.

We must mentione that ([5]), with G_k large enough,

$$\|y^n_{N_L}\|^2_{V;p+G+G_k,k} + \|y^n_{N_L}\|^2_{H;p+G_k,k} \leqslant \operatorname{Re}(g(lh) + Bu^{nx}_{N_L}, y^n_{N_L})_{H;p+G_k,K} \qquad (20)$$

$$\|p^n_{N_L}\|^2_{V;p+G+G_k,k} + \|p^n_{N_L}\|^2_{H;p+G_k,K} \leqslant \operatorname{Re}(\Lambda(y^n_{N_L} - z_d), p^n_{N_L})_{V;p+G_k,K} \qquad (21)$$

Suppose then $B: U \longrightarrow \mathcal{D}^K(p; H)$, such that for any bounded sequence $(u_n)_{n\in\mathbb{N}}$ in U, $(Bu_n)_{n\in\mathbb{N}}$ is a Cauchy sequence in $\mathcal{D}^K(p; H)$ for at least a $k \in Z$ and $p'(t) \geqslant p_o$. For instance B might be a compact operator.

Let then $(u^{nx}_{N_L})_{n\in\mathbb{N}}$ be the bounded set of approximate optimal controls. For $n_i, n_j \geqslant M$, we have

$$\|y_{N_L}^{n_i} - y_{N_L}^{n_j}\|^2_{V;p+G+G_k,k} + \|y_{N_L}^{n_i} - y_{N_L}^{n_j}\|^2_{H;p+G_k,k} \le Re\left(Bu_{N_L}^{n_i*} - Bu_{N_L}^{n_j*}, y_{N_L}^{n_i} - y_{N_L}^{n_j}\right)_{H;p+G_k,k} \tag{22}$$

and $(y^n_{N_L})$ results Cauchy in $\mathcal{D}^k(V;p+G+G_k) \cap \mathcal{D}^k(H;\ p+G_k)$ and hence convergent to y_{N_L} in the norm of this space

$$\|y_{N_L}^{n} - y_{N_L}\| \xrightarrow[n\to\infty]{} 0$$

Then, for appropriate F_k and F, it follows

$$\|p_{N_L}^{n_i} - p_{N_L}^{n_j}\|^2_{V;p+F+F_k,k} + \|p_{N_L}^{n_i} - p_{N_L}^{n_j}\|^2_{H;p+F_k,k} \le \tag{23}$$

$$\le Re\left(\Lambda(y_{N_L}^{n_i} - y_{N_L}^{n_j}), p_{N_L}^{n_i} - p_{N_L}^{n_j}\right)_{V;p+F_k,k}$$

which again signifies that $\{p^n_{N_L}\}$ is a Cauchy sequence and concequently convergent to p_{N_L} in the norm of $\mathcal{D}^k(p+F+F_k;V) \cap \mathcal{D}^k(p+F_k;H)$,

$$\|p_{N_L}^{n} - p_{N_L}\| \xrightarrow[n\to\infty]{} 0$$

If we now take the limit in (11), we immediatly get (by a routine inequality knowing that $N \in L(\mathcal{U},\mathcal{U})$ and is positive), that $u_{N_L}^n \xrightarrow[n\to\infty]{} u_N$ in the norm of $\mathcal{U}$.

Let us observe that because $y \in \mathcal{D}^k(p+G;H)$ and $B_{-1} \in \mathcal{E}_t(L(H,H))$, then:

$$\left(B_{-1}(t)\int_0^{N_L} y(t-x)dx - B_{-1}(lh)\sum_{l=0}^{n-1}\left[y_h^l - \xi_h^l\,\frac{y_h^l - y_h^{l*}}{h}\right]h, \varphi\right)_H \xrightarrow[h\to 0]{} 0$$

$(\forall)\ \varphi \in V$ and

$$\left(\int_0^{N_L} B_{-1}(t-x)p(t-x)\,dx - \sum_{l=0}^{n-1} B_{-1}(lh-\xi_h^l)\left[p_h^l - \xi_h^l\,\frac{p_h^l - p_h^{l*}}{h}\right]h, \varphi\right)_H \xrightarrow[h\to 0]{} 0$$

for any $\varphi \in V$

Also, from (19) :

$$J(u_{N_L}^{n*}) \longrightarrow J(u_{N_L}^{*}) \tag{24}$$

for each finite N_L .

Now:

$$\int_{-\infty}^{\infty} \left(e^{-(p+G)} \mathcal{D}^k \left[B_{-1}(t) \int_0^{N_L} y(t-x)\,dx \right], e^{-(p+G)} \varphi \right)_{L^2(H)} dt \qquad \varphi \in V$$

and

$$\int_{-\infty}^{\infty} \left(e^{-(p+F)} \mathcal{D}^k \left[\int_0^{N_L} B_{-1}(t-x) p(t-x)\,dx, e^{-(p+F)} \varphi \right)_{L^2(H)} dt \qquad \varphi \in V$$

are convergent for all N_L and we can pass to the limit in these integrals when $N_L \to \infty$

Since we already know that y , p and u in (7) are unique , we have the result:

Theorem 21.

If we consider the system (7) that characterizes the unique solution for the stated O.C.P. , then the approximate solutions for the O.C.P. defined by (9), (10) and (11) are strongly convergent to the optimal solution in the corresponding spaces:

$$y^n_{N_L} \to y \qquad \text{in } \mathcal{D}^k(p+G;V) \cap \mathcal{D}^k(p;H)$$

$$p^n_{N_L} \to p \qquad \text{in } \mathcal{D}^k(p+F;V) \cap \mathcal{D}^k(p;H)$$

$$u^n_{N_L} \to u \qquad \text{in } \mathcal{U}$$

$$\mathcal{J}^n_{N_L} \to \mathcal{J}$$

$$n \to \infty$$
$$N_L \to \infty .$$

3. PENALTY METHOD

Returning to the system (7) we shall consider a penalty functional:

$$J_\varepsilon(y,v) = \|y(t,v) - z_d\|^2_{V;p+G,k} + \|y(t,v) - z_d\|^2_{H;p,k} + \gamma\|v\|^2_{\mathcal{U}} + \tag{25}$$
$$+\frac{1}{\varepsilon_1}\left\| B_1(t)\frac{dy}{dt} + B_0(t)y(t) + B_{-1}(t)\int_0^\infty y(t-x)\,dx - Bu^*\right\|^2_{H;p,k} + \frac{1}{\varepsilon_2}\|A(t)y - \mathcal{J}g\|^2_{V;p+G,k}$$

defined over $Y \times \mathcal{U}$, where

$$Y = \left\{ y \mid B_1(t)\frac{dy}{dt} + B_0(t)y(t) + B_{-1}(t)\int_0^\infty y(t-x)\,dx - Bu^* \in \mathcal{D}^k(p;H),\ A(t)y - \mathcal{J}g \in \in \mathcal{D}^k(p+G;V) \right\} \tag{26}$$

with the norm (denote $P(t,D) = B_1(t)\frac{dy}{dt} + B_0(t)y(t) + B_{-1}(t)\int_0^\infty y(t-x)\,dx$)

$$|||y|||^2_Y = \|y\|^2_{V;p+G,k} + \|y\|^2_{H;p,k} + \|P(t,D)y - Bu^*\|^2_{H;p,k} + \|A(t)y - \mathcal{J}g\|^2_{V,p+G,k}$$

The penalty problem consists in : find $(y_\varepsilon, u_\varepsilon)$ such that

$$J_\varepsilon(y_\varepsilon, v_\varepsilon) = \inf_{y\in Y,\, v\in \mathcal{U}_{ad}} J_\varepsilon(y,v) \tag{27}$$

The infimum exists and is unique.

Under the above assumptions - 2 - on P(t,D), A(t),B,B_{-1} B_1 and from the already classical result in [1] , we have:

<u>Theorem 3.1.</u>

Let $\omega_\varepsilon = (y_\varepsilon, v_\varepsilon)$ be such that

$$J_\varepsilon(\omega_\varepsilon) = \inf_{y\in Y,\, v\in \mathcal{U}_{ad}} J_\varepsilon(y,v)$$

If $\varepsilon_1, \varepsilon_2 \longrightarrow 0$, then $(y_\varepsilon, v_\varepsilon) \longrightarrow (y,v)$ strongly in

$$Y \times \mathcal{U}$$

and $J_\varepsilon(\omega_\varepsilon) \xrightarrow[\varepsilon\to 0]{} J(u)$.

Numerical computations have been done for both: finite difference and penalty method , but the comparative results together with graphs have been included in a different paper [6] ,under print.

The results in 2 and 3 can be extended without any difficulties to hyperbolic equations

x x x

REFERENCES.

1. P.Kenneth,M.Sibony,J.P.Yvon, La methode de penalization et ses application aux problemes de controle optimal, IRIA,Alg. num. d'optimization, cahier 2, mai 1970
2. J.L.Lions, Equations differentielles et problemes aux limites, Springer Verlag 1961
3. Cl. Simionescu, Optimal control problems in Sobolev spaces with weights, SIAM Journal on control and optimizations, 14-1, Ja 1976
4. Cl. Simionescu, Computational aspects for a control problem, Information processing 77, North Holand Publ. Comp. 1977
5. Cl. Simionescu, Optimal control problems in Sobolev spaces with weights, numerical approach, applications to plasma physics and time delay problems, Proc. of IFIP Confr. Nice, sept 1975, vol. 41. Springer Verlag, Lecture Notes in Computer Science
6. Cl. Simionescu, Metode numerice pentru o problemă de control frontieră, Bul. Univ. din Braşov (under print)

A COSINE OPERATOR APPROACH TO MODELLING BOUNDARY INPUT HYPERBOLIC SYSTEMS

Roberto Triggiani

Mathematics Department, Iowa State University, Ames, Iowa 50011

A semigroup approach to modelling Dirichlet boundary input problems for linear partial differential equations (p.d.e.) was given recently by Balakrishnan [B 1 - B 3] and Washburn [W 1 - 2]. It was motivated by, and applicable to, parabolic p.d.e. In fact, crucial use is made of the assumption that the semigroup be analytic (holomorphic). Therefore it does not apply to hyperbolic p.d.e. Here we propose to use an approach based on the cosine operator theory to model Dirichlet and mixed boundary input problems for linear hyperbolic p.d.e.

So far our results are satisfactorily complete for mixed boundary input problems on any space dimension and for Dirichlet boundary input problems only on one space dimension.

We purposely avoid giving here a presentation in full generality and rather confine ourselves to the canonical situation.

Acknowledgment: We are pleased to acknowledge that this note was inspired by the work of, and our conversations with, Balakrishnan and Washburn. It is not an accident, in fact, that it was initiated during the author's recent stay at UCLA (Spring '77).

1. Canonical example

Let Ω be a bounded open domain of R^n with boundary Γ. We consider a vibrating system based on Ω with input u applied on Γ, and zero initial data, i.e.

$$\frac{\partial^2 f}{\partial t^2} = \Delta f \quad \text{in } \Omega; \tag{1.1}$$

$$\text{(I.C.)} \qquad f\Big|_{t=0} = \frac{\partial f}{\partial t}\Big|_{t=0} = 0$$

(B.C.)

$$\text{either} \quad f\Big|_{\Gamma} = u \quad \text{on } \Gamma \tag{1.1.D}$$

$$\text{or} \quad \left[\frac{\partial f}{\partial \nu} + \sigma f\right]_{\Gamma} = u \tag{1.1.M}$$

In (1.1.M), σ is a real function defined on Γ and $\frac{\partial}{\partial \nu}$ is the normal (outward) derivative. The Neumann case obtains when $\sigma \equiv 0$. Following Nečas, the standing assumption we make throughout on the bounded domain Ω is

Assumption 1. We require Ω to be 'of class $\mathfrak{M}$' [see Nečas [N 1, p.245] for the technical definition]. This is a mild restriction which in particular allows Γ to have corners. For our purpose here, a main feature about $\Omega \in \mathfrak{M}$ is that the Dirichlet map $D : L_2(\Gamma) \to L_2(\Omega)$, defined at the beginning of section 3, is continuous [N 1, Thm. 1.2, p.250]. As for the mixed map M, defined at the beginning of section 4, it's even true that M is continuous from $H^{-1}(\Gamma) \to L_2(\Omega)$ [N 1, Thm. 2.3 p.257] but we will not make use of this.

The problem we consider is to find a direct input-solution formula giving f in $L_2(\Omega)$ in terms of u in $L_2(\Gamma)$. To this end we need the following results.

2. A few results for abstract second order equations in Banach space needed in the sequel

Let X be a Banach space and consider the following second order abstract Cauchy Problem (ACP)

$$\ddot{x} = Ax + h(t); \quad x(0) = x_0, \quad \dot{x}(0) = x_1 \tag{2.1}$$

For the homogeneous equation $(h \equiv 0)$, the ACP is uniformly well posed and of type $\leq \omega_0$ if and only if the linear operator A is (closed, with dense domain $\mathcal{D}(A)$ in X and range in X and) the infinitesimal generator of a strongly continuous cosine operator $\mathcal{C}(t)$ of bounded linear operators in X, $-\infty < t < \infty$. (Fattorini, [F 1]). The spectrum of A is contained in a parabolic sector defined by ω_0. When the X-valued function $h(\ )$ is Bochner integrable (locally L_1), the mild solution of (2.1) is by definition

$$x(t) = \mathcal{C}(t)x_0 + \int_0^t \mathcal{C}(\tau)x_1 d\tau + \int_0^t \left(\int_0^{t-\tau} \mathcal{C}(s)\, h(\tau) ds\right) d\tau \tag{2.2}$$

In particular, when $x_0, x_1 \in \mathcal{D}(A)$ and $h(t)$ is C^1, the mild solution is indeed the strict solution of (2.1) (twice strongly continuous differentiable) (Fattorini, [F 2]).

The following proposition will be crucially exploited in the sequel.

Proposition 2.1 [Sova, S 1] Let $\mathcal{C}(t)$ be a strongly continuous cosine operator on X, with generator A. Then

(i) $\int_0^t \sigma\, \mathcal{C}(t-\sigma)x\, d\sigma \in \mathcal{D}(A)$ for all x in X

(ii) $\mathcal{C}(t)\, x - x = A \int_0^t \sigma \mathcal{C}(t-\sigma)x\, d\sigma$, for all x in X

(iii) Moreover, if in particular $x \in \mathcal{D}(A)$, then A can be brought inside the integral sign.

3. Input-solution map: $u \to f$ for the Dirichlet case and smooth input.

Let D denote the 'Dirichlet map' for the elliptic problem corresponding to (1.1), (1.1.D) : $g = Dv$. Where

$$\Delta g = 0 \quad \text{in} \quad \Omega; \qquad g\Big|_{\Gamma} = v \quad \text{on} \quad \Gamma$$

We already remarked above that, for $\Omega \in \mathfrak{M}$, the Dirichlet map $D : L_2(\Gamma) \to L_2(\Omega)$ is continuous.

We first find a formula for the input-solution map of the Dirichlet problem (1.1), (1.1.D), in the case when the boundary input $u(t)$ belongs to a smooth class dense in $L_2[[0,T],L_2(\Gamma)]$.

<u>Theorem 3.1</u> Let the boundary input $u(t)$ of (1.1.D) be of class $C^2_{0^-}[[0,T],C(\Gamma)]$ (where here and hereafter the subscript 0^- means "compact support near the origin"). Then, under assumption 1, the solution $f(t) \in L_2(\Omega)$ of (1.1), (1.1D) is given by

$$f(t) = -A_D \int_0^t C(t-\beta) D(\int_0^\beta u(\tau)d\tau)d\beta \tag{3.1.A}$$

or, alternatively, by

$$f(t) = -A_D \int_0^t (\int_\tau^t C(t-\sigma)\ Du(\tau)d\sigma)d\tau \tag{3.1.B}$$

where A_D is the operator defined by

$$A_D g = \Delta g \quad \text{for} \quad g \in \mathcal{D}(A_D) = \{g \in L_2(\Omega): \ \Delta g \in L_2(\Omega) \ \text{and} \ g\Big|_{\Gamma} = 0\}$$

(the derivatives are taken in the sense of distributions) and $C(\cdot)$ is the corresponding cosine operator.

<u>Remark 3.1</u> In (3.1) the operator A_D cannot in general be brought inside the integral sign.

<u>Proof</u> It follows from assumption 1 that

$$Du(t) \in C^2_{0^-}[[0,T],\ C(\Omega)]$$

Consequently, (1.1) maybe rewritten as

$$\frac{\partial^2(f-Du)}{\partial t^2} = \Delta(f-Du) - \frac{\partial^2 Du}{\partial t^2} \quad \text{in} \ \Omega; \quad [f-Du]_\Gamma = 0 \ \text{in} \ \Gamma \tag{3.2}$$

$$[f-Du]_{t=0} = \frac{\partial(f-Du)}{\partial t}\Bigg|_{t=0} = 0$$

The operator A_D defined above is selfadjoint, with spectrum bounded above.

Therefore A_D generates a C_0- cosine operator $\mathcal{C}(t)$. By (2.2), the solution of the abstract version of (3.2) is

$$f(t) - Du(t) = -\int_0^t \left(\int_0^{t-\tau} \mathcal{C}(s)\, \frac{d^2 Du(\tau)}{d\tau^2}\, ds\right) d\tau$$

Integration by parts yields

$$f(t) - Du(t) = -\left\{ \left[\int_0^{t-\tau} \mathcal{C}(s) \frac{dDu(\tau)}{d\tau}\, ds\right]_{\tau=0}^{\tau=t} - \int_0^t \frac{d}{d\sigma}\left(\int_0^{t-\sigma} \mathcal{C}(s)\, \frac{dDu(\tau)}{d\tau} ds\right)\Bigg|_{\sigma=\tau} d\tau \right\}$$

$$= \text{zero} - \int_0^t \mathcal{C}(t-\tau)\, \frac{dDu(\tau)}{d\tau}\, d\tau$$

since $dDu/d\tau = 0$ at $\tau = 0$. We notice at this point that a further attempt to integrate by parts runs into difficulty because of domain problems. We circumvent this by means of Proposition 2.1, which we apply with $x = \frac{dDu(\tau)}{d\tau}$.
This yields

$$f(t) - Du(t) = -\int_0^t \left\{\frac{dDu(\tau)}{d\tau} + A_D \int_0^{t-\tau} \sigma \mathcal{C}(t-\tau-\sigma)\, \frac{dDu(\tau)}{d\tau}\, d\sigma\right\} d\tau$$

($\tau+\sigma=\beta$; also [H1 p.83-84]) $$= -Du(t) + \sout{Du(0)} - A_D \int_0^t \left(\int_\tau^t (\beta-\tau)\, \mathcal{C}(t-\beta) \frac{dDu(\tau)}{d\tau} d\beta\right) d\tau$$

(change order of integration) $$= -Du(t) - A_D \int_0^t \left(\int_0^\beta (\beta-\tau)\, \mathcal{C}(t-\beta) \frac{dDu(\tau)}{d\tau} d\tau\right) d\beta$$

$$= -Du(t) - A_D \int_0^t \mathcal{C}(t-\beta) \left(\int_0^\beta (\beta-\tau) \frac{dDu(\tau)}{d\tau}\, d\tau\right) d\beta$$

(by parts) $$= -Du(t) - A_D \int_0^t \mathcal{C}(t-\beta) \left[(\beta-\tau) Du(\tau)\Big|_{\tau=0}^{\tau=\beta} + \int_0^\beta Du(\tau) d\tau\right] d\beta$$

($Du(0) = 0$) $$= -Du(t) - A_D \int_0^t \mathcal{C}(t-\beta) D\left(\int_0^\beta u(\tau) d\tau\right) d\beta$$

and the proof of (3.1.A) is complete. Integrating by parts (3.1.A) yields (3.1.B) Q.E.D.

4. Input-solution map: $u \to f$ for the mixed case and smooth input

Let M denote the 'mixed map' for the elliptic problem corresponding to (1.1), (1.1.M) (N, in case $\sigma \equiv 0$): $g = Mv$ where

$$\Delta g = 0 \text{ in } \Omega; \quad \left[\frac{\partial g}{\partial \nu} + \sigma g\right]_\Gamma = v \text{ on } \Gamma$$

Let A_M denote the operator: $L_2(\Omega) \supset \mathcal{D}(A_M) \to L_2(\Omega)$ defined by

$$A_M g = \Delta g \quad \text{in } \Omega, \quad g \in \mathcal{D}(A_M) = \{g \in L_2(\Omega) : \left[\frac{\partial g}{\partial \nu} + \sigma g\right]_\Gamma = 0\} \tag{4.1}$$

Remark 4.1 When $\sigma \equiv 0$, the solution of the elliptic Neumann problem is unique only up to an additive constant which corresponds to the eigenvalue $\lambda = 0$. In this case, we stipulate to introduce a further criterion as to identify the constant uniquely. A natural convention, in the case where the method of separation of variables yields a solution in series form, is to set the constant equal to zero. This convention corresponds to eliminating the eigenvalue $\lambda = 0$ from the spectrum of the operator A_N defined in (5.3) and so A_N^{-1} becomes a well-defined bounded operator on $L_2(\Omega)$. As recalled before, M is a bounded operator from $H^{-1}(\Gamma)$ into $L_2(\Omega)$ for $\Omega \in \mathfrak{M}$. Another statement, more useful for our purposes, that the range of M is 'smoother' than the range of D is contained in the following Lemma and subsequent remark.

Lemma 4.1 Let $w \in L_2(\Omega)$. Then

$$M^* w = - A_M^{-1} w\Big|_\Gamma \tag{4.2}$$

In particular, if x_n is an eigenvector of A_M corresponding to a nonzero eigenvalue λ_n, (4.2) specializes to

$$M^* x_n = - \frac{1}{\lambda_n} x_n\Big|_\Gamma \tag{4.3}$$

Proof. Consider $A_M f = w$ $(f = A_M^{-1} w)$ and $g = Mv$,; i.e., consider the two elliptic systems

$$\begin{cases} \Delta f = w & \text{in } \Omega \\ \left[\frac{\partial f}{\partial \nu} + \sigma f\right]_\Gamma = 0 & \text{on } \Gamma \end{cases} \quad \text{and} \quad \begin{cases} \Delta g = 0 & \text{in } \Omega \\ \left[\frac{\partial g}{\partial \nu} + \sigma g\right]_\Gamma = v & \text{on } \Gamma \end{cases}$$

For domains Ω under consideration, Green's theorem holds:

$$\int_\Omega (f \Delta g - g \Delta f) d\xi = \int_\Gamma (f \frac{\partial g}{\partial \nu} - g \frac{\partial f}{\partial \nu}) d\sigma$$

Hence it follows that

$$(v, M^* w)_{L_2(\Gamma)} = (Mv, w)_{L_2(\Omega)} = - \int_\Gamma \{(A_M^{-1} w)[-\sigma(Mv) + v] - (Mv)(-\sigma A_M^{-1} w)\} d\sigma$$

$$= - \int_\Gamma (v A_M^{-1} w) d\sigma = (v, -A_M^{-1} w\Big|_\Gamma)_{L_2(\Gamma)} \qquad \text{Q.E.D.}$$

Remark 4.2 Eq. (4.2) should be contrasted with the analogous result for the Dirichlet map [W 1]

$$D^* x_n = \frac{1}{\lambda_n} \left. \frac{\partial x_n}{\partial \nu} \right|_\Gamma \qquad (4.4)$$

As will be apparent from the examples below, the difference of structure between $M^* x_n$ and $D^* x_n$ will be significant for the purposes of the saught after extension described in section 5. The analogous of Thm. 3.1 is then

Theorem 4.2 Let the boundary input u in (1.1.M) be of class $C^2_{0^-}[[0,T], C(\Gamma)]$. then, under assumption 1, the solution $f(t) \in L_2(\Omega)$ of (1.1), (1.1.M) is given by

$$f(t) = - A_M \int_0^t \mathcal{C}(t-\beta) M (\int_0^\beta u(\tau) d\tau) d\beta \qquad (4.5.A)$$

$$= - A_M \int_0^t (\int_\tau^t \mathcal{C}(t-\sigma) M u(\tau) d\sigma) d\tau \qquad (4.5.B)$$

where A_M is defined by (4.1) and $\mathcal{C}(\cdot)$ is the correspondent cosine operator.

Remark As one can check, in the examples below the operator A_M can be brought inside the first integral sign. Compare with Remark 3.1.

5. Extension of input-solution map to non-smooth input

Let $[0,T]$ be an arbitrary fixed finite time interval.

Motivated by theorem 3.1 and 4.2, consider the linear maps L_D and L_M defined by the right hand side of (3.1.B) and (4.5.B) respectively:

$$(L_D u)(t) = - A_D \int_0^t (\int_\tau^t \mathcal{C}(t-\sigma) D u(\tau) d\sigma) d\tau \qquad (5.1)$$

$$(L_M u)(t) = - A_M \int_0^t (\int_\tau^t \mathcal{C}(t-\sigma) M u(\tau) d\sigma) d\tau \qquad (5.2)$$

for any function $u(\cdot)$ in $L_2[[0,T], L_2(\Gamma)]$ for which they make sense, into $L_2[[0,T], L_2(\Omega)]$.

We first ask whether L_D and L_M admit continuous (bounded) extensions - denoted again by the same symbols - when viewed as operators from $L_2[[0,T], L_2(\Gamma)]$ into $L_2(\Omega)$; i.e., whether there exist constants d and m (depending only on T, but not on u) such that

$$\| (L_D u)(t) \|^2_{L_2(\Omega)} \leq d \int_0^T \| u(t) \|^2_{L_2(\Gamma)} dt \qquad (5.3)$$

$$\|(L_M u)(t)\|^2_{L_2(\Omega)} \leq m \int_0^T \|u(t)\|^2_{L_2(\Gamma)} dt \tag{5.4}$$

for all $t \in [0,T]$ and all $u \in L_2[[0,T], L_2(\Gamma)]$.

Should this be the case, it is then natural to call such extensions $(L_D u)(t)$ and $(L_M u)(t)$ "pointwise generalized solutions of (1.1), (1.1.D) and (1.1), (1.1.M)" respectively, for $L_2[[0,T], L_2(\Gamma)]$ - boundary inputs.

Less demandingly, we may ask whether L_D and L_M admit continuous (bounded) extensions - still denoted by the same symbols - when viewed as operators from $L_2[[0,T], L_2(\Gamma)]$ into $L_2[[0,T], L_2(\Omega)]$; i.e., whether there exist constants d' and m', depending only on T, but not on u, such that

$$\int_0^T \|(L_D u)(t)\|^2_{L_2(\Omega)} dt \leq d' \int_0^T \|u(t)\|^2_{L_2(\Gamma)} dt \tag{5.5}$$

$$\int_0^T \|(L_M u)(t)\|^2_{L_2(\Omega)} dt \leq m' \int_0^T \|u(t)\|^2_{L_2(\Gamma)} dt \tag{5.6}$$

In this latter case, it would only follow $(L_D u)(t) \in L_2(\Omega)$ and $(L_M u)(t) \in L_2(\Omega)$ a.e. in $[0,T]$ and a pointwise meaning of generalized solution is lost. However, it is still natural to define $\{(L_D u)(t), t \in [0,T]\}$ and $\{(L_M u)(t), t \in [0,T]\}$ as "generalized trajectory solutions of (1.1), (1.1D) and (1.1), (1.1.M)" respectively, for $L_2[[0,T], L_2(\Gamma)]$ - boundary inputs. Pointwise generalized solutions are needed for time optimal control problems or final value problems in general. However, generalized trajectory solutions are still sufficient for considering quadratic or more general integral cost functionals.

Our results so far indicate that:

(i) The continuous extension (5.3) for the Dirichlet case holds at least for one space dimension.

(ii) The continuous extension (5.4) for the mixed (in particular, Neumann) case holds for any space dimension.

For sake of clarity we illustrate next these results by means of examples which in fact display all the essential features.
To begin with, we state a general result covering all hyperbolic systems on bounded spatial domains and classical boundary conditions. Let A be a self-adjoint operator on some Hilbert space X with compact resolvent and spectrum bounded above. Then its corresponding cosine operator is

$$C(t)x = \sum_{n=1}^{\infty} \cos \sqrt{-\lambda_n}\, t[x, x_n]x_n, \qquad x \in X \tag{5.7}$$

where $\{x_n\}$ is an orthonormal basis of eigenvectors with corresponding (real) eigenvalues λ_n, $\lambda_n \to -\infty$.

As for our original Dirichlet and mixed problems, we then have from (5.7), (4.4), (5.1) and (5.7), (4.3), (5.2) respectively:

$$\mathcal{C}\,(t-\sigma)Du(\tau) = \sum_{n=1}^{\infty} \{\frac{1}{\lambda_n} \cos \sqrt{-\lambda_n}\,(t-\sigma)\,[u(\tau), \frac{\partial x_n}{\partial \nu}\Big|_\Gamma]_{L_2(\Gamma)}\}\, x_n$$

$$(L_D u)(t) = \sum_{n=1}^{\infty} \{\frac{1}{\sqrt{-\lambda_n}} \int_0^t \sin \sqrt{-\lambda_n}\,(t-\tau)\,[u(\tau), \frac{\partial x_n}{\partial \nu}\Big|_\Gamma]_{L_2(\Gamma)}\, d\tau\}\, x_n \tag{5.8}$$

$$\mathcal{C}\,(t-\sigma)Mu(\tau) = -\sum_{n=1}^{\infty} \{\frac{1}{\lambda_n} \cos \sqrt{-\lambda_n}\,(t-\sigma)\,[u(\tau), x_n\Big|_\Gamma]_{L_2(\Gamma)}\}\, x_n$$

$$(L_M u)(t) = -\sum_{n=1}^{\infty} \{\frac{1}{\sqrt{-\lambda_n}} \int_0^t \sin \sqrt{-\lambda_n}\,(t-\tau)\,[u(\tau), x_n\Big|_\Gamma]_{L_2(\Gamma)}\, d\tau\}\, x_n \tag{5.9}$$

We next apply Eqs. (5.8) and (5.9) to typical cases where the extensions (5.3) and (5.4) indeed hold

Example 5.1 (One dimensional Dirichlet case). Let $\Omega = (0,\pi)$ and $u(t) = [u_1(t), u_2(t)]$, with $u_1(t)$ and $u_2(t)$ scalar inputs applied at 0 and π, respectively. The eigenvalues are $\lambda_n = -n^2$, $n = 1,2,\ldots$ and the corresponding eigenfunctions are $x_n = \sqrt{\frac{2}{\pi}} \sin n\,\xi$ forming a complete orthonormal set in $X = L_2(0,\pi)$. Eq. (5.8) becomes

$$(L_D u)(t) = \sqrt{\frac{2}{\pi}} \sum_{n=1}^{\infty} \{\int_0^t \sin n(t-\tau)\,[u_1(\tau) + \cos n\pi\, u_2(\tau)\,]\}\, x_n \tag{5.10}$$

Call $L_D u_i$ the contribution of $L_D u$ in (5.10) due only to u_i, $i = 1,2$ and extend $u_i(\cdot)$ to vanish for negative argument (cf. Thm. 3.1). Then, (5.10) implies for $0 \le t \le \pi$, $\pi < t \le 2\pi$ etc.:

$$(L_D u_1)(t) = \sum_{n=1}^{\infty} \{\int_0^{\pi} \sqrt{\frac{2}{\pi}} \sin n\alpha\, u_1(t-\alpha)\, d\alpha\} \sqrt{\frac{2}{\pi}} \sin n\xi$$

$$= u_1(t-\xi) \quad , \quad \text{for} \quad 0 < t \le \pi; \tag{5.11.A}$$

$$(L_D u_1)(t) = u_1(t-\xi) - \sum_{n=1}^{\infty} \{\sqrt{\frac{2}{\pi}} \int_{\pi}^{t} \sin n\alpha\, u_1(t-\alpha) d\alpha\} \sqrt{\frac{2}{\pi}} \sin n\xi$$

$$(\alpha = 2\pi-\alpha) \quad = u_1(t-\xi) - \sum_{n=1}^{\infty} \{\int_0^{\pi} \sqrt{\frac{2}{\pi}} \sin n\sigma\, u_1(t-2\pi+\sigma) d\sigma\} \sqrt{\frac{2}{\pi}} \sin n\xi$$

$$= u_1(t-\xi) - u_1(t-2\pi+\xi) \quad , \quad \text{for} \quad \pi < t \le 2\pi; \tag{5.11.B}$$

etc., in agreement with the physical fact that the input u_1 applied at $\xi = 0$ travels with speed equal to 1, and is reflected at $\xi = \pi$. Eqs. (5.10), (5.11) etc. easily imply now the validity of inequality (5.3) Q.E.D.

Remark 5.1 The above computations exploit the fact that time and one-dimensional space variables are interchangeable.

Example 5.2 (One dimensional Neumann problem) Let $\Omega = (0,\pi)$. Because of the usual necessary condition for the Neumann problem, we must impose

$$\left.\frac{\partial f}{\partial \nu}\right|_{\xi=0} = -\left.\frac{df}{d\xi}\right|_{\xi=0} = u_1 \quad \text{and} \quad \left.\frac{\partial f}{\partial \nu}\right|_{\xi=\pi} = \left.\frac{df}{d\xi}\right|_{\xi=\pi} = u_2, \quad \text{with } u_1 = u_2$$

The eigenvalues of A_N are $= -n^2$, $n = 0,1,2,\ldots$ with corresponding eigenfunctions $x_n = \sqrt{\frac{2}{\pi}} \cos n\,\xi$. As explained in Remark 4.1, we discard the eigenvalue $\lambda = 0$ (i.e. $n = 0$) in order to uniquely identify the solution of the corresponding elliptic problem. Eq. (5.9) becomes

$$(L_N u)(t) = -\sum_{n=1}^{\infty} \{\sqrt{\tfrac{2}{\pi}}\ \tfrac{1}{n} \int_0^t \sin n(t-\tau)\ [u_1(\tau) + \cos n\pi\ u_2(\tau)]\ d\tau\}\ x_n \tag{5.12}$$

which should be constrasted with (5.10): the essential difference is the presence of $\sqrt{-\lambda_n} = n$ in the denominator of (5.12). Here a simple application of Schwartz inequality will do it:

$$\|(L_N u)(t)\|^2_{L_2(\Omega)} = \frac{2}{\pi} \sum_{\substack{n \\ \text{even}}}^{\infty} \frac{4}{n^2} \left| \int_0^t \sin n(t-\tau)\ u_1(\tau)\ d\tau \right|^2$$

$$\leq \frac{2}{\pi} \Big(\sum_{\substack{n \\ \text{even}}}^{\infty} \frac{2}{n^2} \Big) \cdot t \cdot \int_0^t [u_1^2(\tau) + u_2^2(\tau)]\, d\tau$$

and (5.4) is proved. Q.E.D.

Example 5.3 (Two dimensional Neumann problem on Ω = a square of side π: $0 \leq \xi, \zeta \leq \pi$). It is sufficient to consider the case

$$\left.\frac{\partial f}{\partial \xi}\right|_{\xi=0} = -u_1\ , \quad \text{while} \quad \left.\frac{\partial f}{\partial \xi}\right|_{\xi=\pi} = \left.\frac{\partial f}{\partial \zeta}\right|_{\zeta=0} = \left.\frac{\partial f}{\partial \zeta}\right|_{\zeta=\pi} = 0$$

and $\frac{\partial f}{\partial \nu}$ along the side $\{\xi = 0,\ 0 \leq \zeta \leq \pi\}$ is equal to $-\left.\frac{\partial f}{\partial \xi}\right|_{\xi=0}$. The eigenvalues of A_N are $\lambda_{nm} = -(n^2 + m^2)$ with corresponding eigenfunctions $\{x_{nm} = \frac{2}{\pi} \cos n\,\xi \cos m\,\zeta\}$, $n = m = 0,1,2,\ldots$ As explained in Remark 4.1, the zero eigenvalue $(n = m = 0)$ is discarded. Eq. (5.9) becomes

$$(L_N u)(t) = -\sum_{n,m=1}^{\infty} \{\frac{1}{n^2+m^2} \int_0^t \sin\sqrt{n^2+m^2}\ (t-\tau) g_m(\tau) d\tau\}\ x_{nm} \tag{5.13}$$

where $g_m(\cdot)$ is defined by

$$g_m(\tau) = \frac{2}{\pi}\,[u_1(\tau,\cdot),\ \cos m\ \cdot]_{L_2[0,\pi]} \in \ell_2$$

Since [J p.123, #124]

$$\sum_{n=1}^{\infty} \frac{1}{n^2+m^2} = \frac{\pi/2}{m} \coth \pi m - \frac{1/2}{m^2} = 0(\frac{1}{m})$$

Schwartz inequality yields (5.4):

$$\| (L_N u)(t) \|^2_{L_2(\Omega)} \leq t \sum_{m=1}^{\infty} [\int_0^t g^2_m(\tau)d\tau] (\sum_{n=1}^{\infty} \frac{1}{n^2+m^2}) \leq c \cdot t \cdot \int_0^t \| u_1(\tau) \|^2_{L_2[0,\pi]} d\tau$$

Q.E.D.

Example 5.4 (Three dimensional Neumann problem on $\Omega =$ a cube of side $\pi : 0 \leq \xi, \zeta, \eta \leq \pi$). It is sufficient to consider the case:

$\frac{\partial f}{\partial \nu}\Big|_{\xi=0} = - \frac{\partial f}{\partial \xi}\Big|_{\xi=0} = u_1$, while partials of f over other sides of Ω are zero

The eigenvalues of A_N are $\lambda_{nm\ell} = -(n^2 + m^2 + \ell^2)$, $n,m,\ell = 0,1,..$ with corresponding eigenfunctions $\{x_{nm\ell} = k \cos n \xi \cos m \zeta \cos \ell \eta\}$. As explained in Remark 4.1, the zero eigenvalue $(n = m = \ell = 0)$ is discarded. Eq. (5.9) becomes:

$$(L_N u)(t) = - \sum_{n,m,\ell=1}^{\infty} \frac{k}{\sqrt{n^2+m^2+\ell^2}} \{ \int_0^t \sin[\sqrt{n^2+m^2+\ell^2}\,(t-\tau)] \cdot g_{m\ell}(\tau)d\tau \} x_{nm\ell} \qquad (5.14)$$

where $g_{m\ell}(\tau)$ is defined by

$g_{m\ell}(\tau) = k[u_1(\tau,\cdot), \cos m \zeta \cos \ell \eta]_{L_2(S)}$ S: the square $0 \leq \zeta,\eta \leq \pi$.

Since [J, p.23]: $\sum_{n=1}^{\infty} \frac{1}{n^2+m^2+\ell^2} \to 0$ as $m,\ell \to \infty$, Schwartz inequality yields (5.4):

$$\| (L_N u)(t) \|^2_{L_2(\Omega)} \leq t \cdot \sum_{m,\ell=1}^{\infty} k \int_0^t g^2_{m\ell}(\tau)d\tau \; (\sum_{n=1}^{\infty} \frac{1}{n^2+m^2+\ell^2}) \leq C \cdot t \cdot \int_0^t \|u_1(\tau)\|^2_{L_2(S)} d\tau$$

Q.E.D.

Remark 5.2 It should be clear that the results on the canonical situation considered above can be extended to any second order self-adjoint operator on Ω whose eigenvalues behave asymptotically like n^2, $n^2 + m^2$, $n^2 + m^2 + \ell^2$, in one, two, three dimensions, respectively. See [C 1, Chapter VI, in particular §3.3 and §4].

References

[B 1] A. V. Balakrishnan, Applied Functional Analysis, Springer-Verlag, 1976.

[B 2] A. V. Balakrishnan, Filtering and Control Problems for Partial Differential Equations, Proceedings of the 2nd Kingston Conference on Differential Games and Control Theory, University of Rhode Island, Marcel Dekker Inc., New York, 1976.

[B 3] A. V. Balakrishnan, Boundary Control of Parabolic Equations: L - Q - R Theory, Proceedings of the Conference on Theory of Non Linear Operators, September 1977, Berlin, to be published by Akademie-Verlag, Berlin, 1978.

[C 1] R. Courant, D. Hilbert, Methods of Mathematical Physics Vol. I, Interscience 1953.

[F 1] H. O. Fattorini, Ordinary Differential Equations in Linear Topological Spaces, I and II, J. Differ. Equats. 5(1968), 72-105, and 6(1969), 50-70.

[F 2] H. O. Fattorini, Controllability of Higher Order Linear Systems, in Mathematical Theory of Control, Balakrishnan-Neustadeds, Acad. Press 1967.

[H 1] E. Hille and R. S. Phillips, Functional Analysis and Semigroups, American Mathematical Society, Providence,R. I., 1958.

[J 1] L. B. W. Jolley, Summation of Series, 2nd Revised Ed., Dover 1961.

[N 1] J. Nẹcas, Les Methodes Directes en Theorie des Equations Elliptiques, Masson et Cie, Editeurs, 1967.

[S 1] M. Sova, Cosine Operator Functions, Rozpr. Mat. XLIX (1966).

[W 1] D. C. Washburn, A semi-Group Approach to Time Optimal Boundary Control of Diffusion, Ph.D. Dissertation, 1974, UCLA.

[W 2] D. C. Washburn, A Semi-Group Theoretic Approach to Modeling of Boundary Input Problems, Proceedings of IFIP Working Conference, University of Rome, Lecture Notes in Control and Information Science, Springer-Verlag 1977.

Acknowledgment: Research partially supported by Air Force Office of Scientific Research under Grant AFOSR-77-3338.

THE LINEARIZATION OF THE QUADRATIC RESISTANCE TERM IN THE EQUATIONS OF MOTION FOR A PURE HARMONIC TIDE IN A CANAL AND THE IDENTIFICATION OF THE CHÉZY PARAMETER C

G. Volpi P. Sguazzero
IBM Venice Scientific Center
Dorsoduro 3228, 30123 Venice, Italy

Abstract

In the hydrodynamic equations, which govern wave propagation in a shallow water basin, the friction term plays a very important rôle [10]. In the applications, however, the coefficient of this term is often unknown. In this paper an identification method is described for the 1-dimensional case.

The procedure requires first the approximation of the shallow water semi-linear time-dependent partial differential system by means of a linear time-dependent one in such a way that the equivalence of the dissipated energies is obtained. The usual separation of variables transforms then the time-dependent linear system into a time-independent one.

On the last system the identification is performed introducing a parametrization of the friction coefficent, a set of observed data and the least square error criterion. The minimization is obtained making use of the gradient method and introducing an adjoint partial differential system in order to compute exactly the derivatives of the error functional.

Introduction

Wave propagation in a basin is described by the classical momentum and mass conservation equations in the water velocity and pressure [4], [8]. When the phenomenon is characterized by great wave length and small wave amplitude with respect to the basin depth, it is possible to integrate vertically the hydrodynamic equations and obtain the semi-linear shallow water system [4], [6], [7].

In the simple case of a canal closed at one end, with a forcing sinusoidal wave at the other end, taking into account the horizontal gradient forces, the inertness forces and the energy dissipation, the partial differential system is the following:

$$(1) \qquad \begin{cases} \dfrac{\partial q}{\partial t} + gs\dfrac{\partial \eta}{\partial x} + \dfrac{g}{C^2}\dfrac{|q|}{hs}q = 0 \\ b\dfrac{\partial \eta}{\partial t} + \dfrac{\partial q}{\partial x} = 0 \end{cases} \qquad (x,t) \in \Omega$$

$$(2) \qquad q(0,t) = 0 \quad ; \quad \eta(l,t) = a \cos \omega t \qquad t \in T$$

$$(3) \qquad q(x,t_0) = \varphi(x) \; ; \; \eta(x,t_0) = \psi(x) \qquad x \in L$$

where:

l is the canal length,

$h(x)$ is the canal depth,

$b(x)$ is the canal width,

$s(x)$ is the canal section,

$q(x,t)$ is the canal discharge,

$\eta(x,t)$ is the water level with respect to the equilibrium level,

$C(x)$ is the Chézy coefficient,

a is the forcing wave amplitude,

ω is the angular frequency,

$T =]t_0, t_1[$,

$L =]0, l[$,

$\Omega = T \times L$.

Several numerical methods have been developed to solve (1) with the conditions (2) and (3) [7], [9], [10]. It is the purpose of this work to contribute to the solution of a problem occurring in applications: the lack of direct measurements on the Chézy coefficient C [11]. Our expected goal is the introduction of a procedure of identification of the coefficient C with the aid of water level data at some observation points.

The hybrid system

The identification problem could be solved dealing directly with the time dependent system (1). However the huge computational effort required to follow such a procedure suggests us to modify the initial system in order to eliminate its dependence on time. The here proposed modification [3] consists in the replacement of the quadratic resistance term by a linear one, thus providing the system:

$$(4) \qquad \begin{cases} \dfrac{\partial q}{\partial t} + gs\dfrac{\partial \eta}{\partial x} + rq = 0 \\[2mm] b\dfrac{\partial \eta}{\partial t} + \dfrac{\partial q}{\partial x} = 0 \end{cases} \qquad (x,t) \in \Omega$$

(supplemented by the conditions (2) and (3)) where the new roughness coefficient $r = r(x)$ has to be linked in some way to the Chézy coefficient C and to the discharge q.

This linkage is obtained substituting the solution (q, η) of (4) into (1) and imposing the orthogonality in a given time interval T between the residual we get in (1):

$$\begin{pmatrix} \frac{g}{C^2 h s} |q| q - r q \\ 0 \end{pmatrix} \qquad (x,t) \in \Omega$$

and the solution of (4). The relation obtained this way:

$$(5) \qquad \frac{g}{C^2 h s} \int_T q^3 \, dt = r \int_T q^2 \, dt \qquad x \in L$$

expresses the equivalence in the mean of the energies dissipated by (1) and (4) in the time interval T and allows us to consider the solution of the hybrid (differential-algebraic) system (4), (5) as a solution in a weaker sense of the semi-linear system (1)

At this point it is possible to separate the variables in the new partial differential system (4). If we assume the following periodic functions:

$$(6) \qquad \begin{cases} q(x,t) = \mathbb{R}e \left[U(x) \exp(i\omega t) \right] \\ \eta(x,t) = \mathbb{R}e \left[E(x) \exp(i\omega t) \right] \end{cases} \qquad (x,t) \in \Omega$$

as solution of (4) (supplemented by the boundary condition (2) only) and (5), a new hybrid system is obtained:

$$(7) \qquad \begin{cases} \frac{d}{dx} \left[\frac{gs}{1 - i\frac{r}{\omega}} \frac{dE}{dx} \right] + \omega^2 E = 0 & x \in L \\ \frac{dE}{dx}(0) = 0 \quad ; \quad E(l) = 0 & \\ Q = - \frac{gs}{r + i\omega} \frac{dE}{dx} & x \in L \\ r = \frac{g}{C^2 h s} \frac{8}{3\pi} |Q| & x \in L \end{cases}$$

This non-linear, hybrid time independent system provides a solution "equivalent in the sense of the dissipated energies" to the solution of (1) (with the values of (6) at the time t_0 as initial conditions) and will be used in the following for the identification problem.

Comparison between the solutions of hybrid time-indipendent system and the time-dependent semi-linear system

In order to explain the preference accorded to the hybrid time-independent system (7) against the semi-linear time-dependent system (1) we report here some reasons motivating such a choice. To sum up we can say that (7) has a better computational performance than (1). Both systems have been discretized, more exactly:

a) the equations of (7) have been discretized by a finite difference method [2] on L. To solve the discretized hybrid system an iterative process has then been introduced. Starting from an initial guess of r, for instance $r^{(0)} \equiv 0$, the n - th step calls for the following actions:

1) $E^{(n)}$ is obtained from $r^{(n)}$ by the first equation of (7);

2) $Q^{(n)} = \dfrac{gs}{r+i\omega} \dfrac{dE^{(n)}}{dx} \qquad x \in L$

3) $r^{(n+1)} = \theta r^{(n)} + (1-\theta) \dfrac{g}{C^2 hs} \dfrac{8}{3\pi} |Q^{(n)}| \qquad x \in L$

The process is stopped when the increments $r^{(n+1)} - r^{(n)}$ are sufficiently small. At this point the expression (6) provides the solution. The numerical experiments show that the iterative algorithm is convergent if $0 < \theta < 1$.

b) The semi-linear system (1) has been discretized similarly by a second order explicit method space and time staggered [10]. The time step size Δt has been chosen as big as possible compatibly with the stability condition.

The comparison between the numerical solutions of (7) obtained with the following choice of the physical parameters:

$$(8) \begin{cases} l = 20 \text{ Km}; \quad h = 5 \text{ m}; \quad b = 1 \text{ km}; \\ T_a = \text{length of } T = 12 \times 3600 \text{ sec}; \\ a = 0.5 \text{ m}; \end{cases}$$

shows that:

a) the ratio between the computational efforts becomes more and more favourable to the hybrid system as the space step size Δx tends to 0 (see Fig. 1). (C is fixed equal to 60 $m^{1/2} \times \sec^{-1}$.).

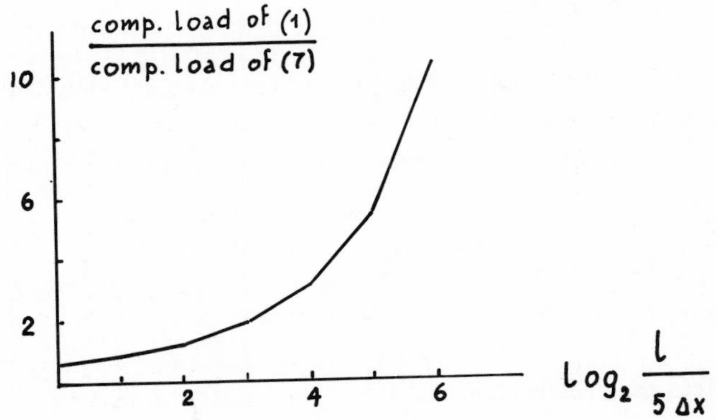

Fig. 1 Comparison between the computational efforts.

b) For fixed $\Delta x = l/40$ and denoting with:

(q_1, η_1) the numerical solution of (1) at t_1;

(q_2, η_2) the expression (6) at t_1 at the computational points;

$\| \cdot \|$ the discrete L^2 (L) norm;

the mean square normalized errors:

$$\varepsilon_q = \frac{\| q_1 - q_2 \|}{\sqrt{\| q_1 \| \, \| q_2 \|}} \quad ; \quad \varepsilon_\eta = \frac{\| \eta_1 - \eta_2 \|}{\sqrt{\| \eta_1 \| \, \| \eta_2 \|}}$$

tend to 0 as the amplitude a of the forcing wave tends to 0. In Fig. 2 and Fig. 3 the results for several values C are reported in logarithmic scale.

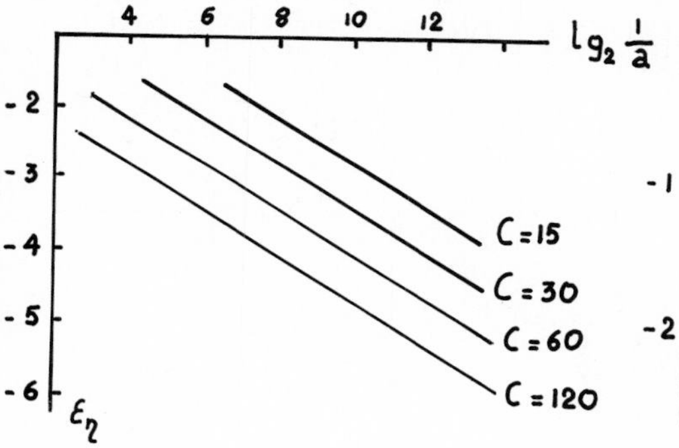

Fig. 2 Error ε_η

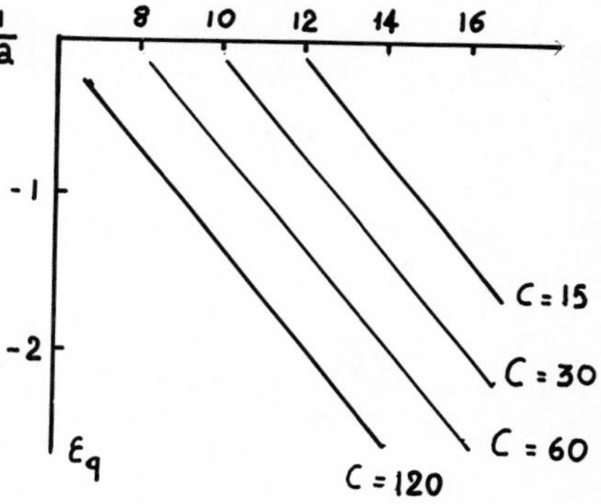

Fig. 3 Error ε_q

Chézy coefficent identification

As it can be found in the specialized literature [4], for the Chézy coefficient C it is possible to resort to some simple schematization; for instance to postulate a dependence of C on the depth and on other unknown parameters:

$$C(x) = C\left[h(x), \alpha_1, \alpha_2, \ldots, \alpha_M \right]$$

The identification of C is now equivalent to the identification of the parameters vector $\alpha = (\alpha_1, \alpha_2, \ldots, \alpha_M)$. In the following the goal will be to estimate α from some observed level oscillation amplitudes $\tilde{E}_k$, $k = 1,2, \ldots, K$ at the points $\tilde{X} = \{\tilde{x}_k, k = 1,2,\ldots,K\}$ along the canal. The chosen optimality criterion is the minimization of the mean square error between the computed values E at $\tilde{X}$ and the observed values $\tilde{E}_k$, $k = 1,2,\ldots,K$:

$$J(\alpha) = \sum_{K=1}^{K} \left| E(\tilde{x}_K, \alpha) - \tilde{E}_K \right|^2$$

A local minimum is obtained by the gradient method [1], [5]. Starting from an initial guess $\alpha^{(0)}$, at every step the new value of α is originated from the old one by the relation:

$$\alpha^{(n+1)} = \alpha^{(n)} + \Delta\alpha^{(n+1)}$$

where the increment $\Delta\alpha^{(n+1)} = \left\{ \Delta\alpha_m^{(n+1)}, m = 1,2,\ldots,M \right\}$ is the solution of a linear system obtained first setting:

$$E(x,\alpha^{(n+1)}) = E(x,\alpha^{(n)}) + \Delta\alpha^{(n+1)} \; \nabla E(x,\alpha^{(n)})$$

where:

$$\nabla E = \left\{ \frac{\partial E}{\partial \alpha_m} \; , \; m = 1,2,\ldots,M \right\}$$

and then minimizing J with respect to $E(x, \alpha^{(n+1)})$:

$$\left\{ \sum_{K=1}^{K} \mathbb{R}e \left[\frac{\partial E^*}{\partial \alpha_m} (\tilde{x}_K, \alpha^{(n)}) \quad \nabla E(\tilde{x}_K, \alpha^{(n)}) \right] \right\} \Delta\alpha^{(n+1)} =$$

$$= - \sum_{K=1}^{K} \mathbb{R}e \left\{ \left[E(\tilde{x}_K, \alpha^{(n)}) - \tilde{E}_K \right] \frac{\partial E^*}{\partial \alpha_m} (\tilde{x}_K, \alpha^{(n)}) \right\} \qquad m = 1,2,\ldots,M$$

In order to compute exactly $\nabla E(x_k, \alpha^{(n)})$ an adjoint system is introduced ($A_m = \frac{\partial E}{\partial \alpha_m}$, $m = 1,2, \ldots,M$):

$$(9) \quad \begin{cases} \dfrac{d}{dx}\left[\dfrac{gs}{1-i\frac{r}{\omega}} \dfrac{dA_m}{dx} \right] + \omega^2 A_m = - \dfrac{dP_m}{dx} & x \in L \\ \dfrac{dA_m}{dx}(0) = 0 \quad ; \quad A_m(l) = 0 & \\ P_m = \dfrac{Q}{1-i\frac{r}{\omega}} \dfrac{8}{3\pi sh} \left\{ -\dfrac{2g}{C^3} \dfrac{\partial C}{\partial \alpha_m} |Q| + \dfrac{g}{C^2} \mathbb{R}e\left[\dfrac{Q^*}{|Q|} \left(-gs \dfrac{\frac{dA_m}{dx}}{i\omega + r} + \dfrac{i}{\omega} P_m \right) \right] \right\} & x \in L \end{cases}$$

Each hybrid system (9) can be formally solved like (7) and each P_m, $m = 1,2,\ldots M$ plays in (9) the same rôle of r in (7) .

Numerical results

The results of a simple case are reported. In addition to the choice (8) of the physical quantities, the following assumptions have been made:

- $C = \sqrt{32\,g}\ \lg_{10}\ \alpha_1 h$;
- $\lambda = g/C^2$ is the new parameter to be identified;
- $\tilde{X}$ = is a set of equispaced points;
- $\tilde{E}_k = E(\tilde{x}_k, \lambda = \Lambda)$, $k = 1,2,\ldots,K$; $\Lambda = 2.725 \times 10^{-3}$;
- the relative tolerance on λ is 10^{-2}
- $\Delta x = 1/40$

Under these assumptions the parameter λ has been identified in 3 gradient iterations (starting from $\lambda = 0$) and with a relative error with respect to Λ less than 10^{-4}. This result is consistent with the polar graph of Fig. 4 reporting the adjoint variable $A_1(0)$ as function of λ: the values of A_1 are sufficiently far away from the origin for all the physically admissible values of λ.

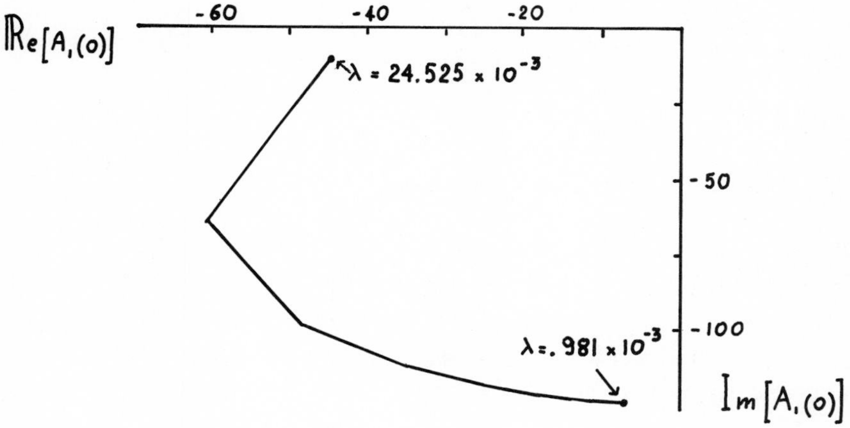

Fig. 4 Polar graph of the adjoint variable

The behaviour of A_1 at $x \neq 0$ is quite similar to the behaviour at $x = 0$.

A heuristic idea about the sensitivity of the identification problem can be drawn perturbing the original observed values and correlating the identification errors with the introduced perturbation.

In the example reported below 19 observation points are introduced and are perturbated by a multinormal error distribution whose standard deviation is 10% of the amplitude of the forcing wave; 100 runs have provided an error distribution on λ with mean almost equal to 0 and standard deviation equal approximately to the 22% of the observation error. The histogram of this distribution is reported in Fig. 5.

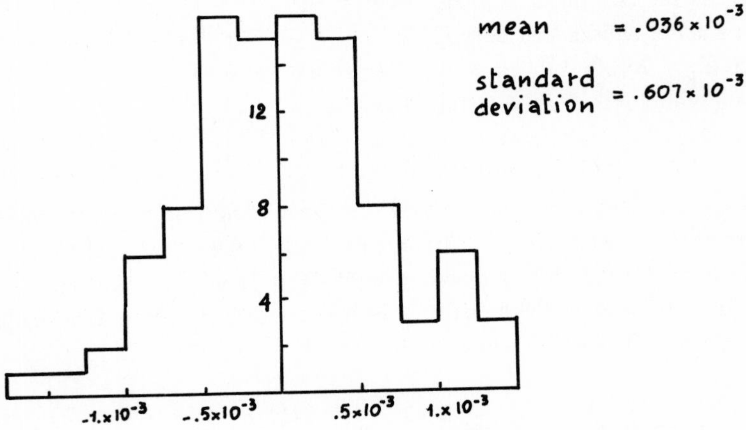

Fig. 5 Histogram of λ- error distribution

Conclusions

In order to solve the shallow water equations it is necessary to know the Chézy coefficient C. The very important rôle played by this coefficient on wave propagation requires a good estimation of it and therefore calls for efficient automatic identification procedures. The satisfactory numerical results obtained by the method presented in this paper encourage its extension to multicomponent wave propagation and to bidimensional domaines.

Bibliography

1. Canon D. et al.; Theory of Optimal Control and Mathematical Programming, McGraw Hill, New York, 1970.
2. Collatz L.; The Numerical Treatment of Differential Equations, Springer Verlag, Berlin, 1966.
3. Dronkers J.J.; The linearization of the quadratic resistence term in the equations of motion for a pure harmonic tide in a sea, Proc. Symp. Hydr. Meth. of Phys.

Oceanography, Institut für Meereskunde der Universität Hamburg, p. 195, 1971.

4. ; Tidal Computation in Rivers and Coastal Waters, North-Holland, Amsterdam, 1964.

5. Fletcher R.; Minimizing general functions subject to linear contraints, Numerical Methods for Non-linear Optimization ed.by F.A. Lootsma, Academic Press, London, p. 279, 1972.

6. Hansen W; Dent. Hydrog. Z. (Ergänzungsheft), 1,1 (1952).

7. ; Hydrodynamical methods applied to oceanographic problems, Proc. Symp. Math. Hydr. Meth. Phys. Oceanography, Institut für Meereskunde der Universität Hamburg p. 25, 1961.

8. Lamb H.; Hydrodynamics, Cambridge University Press, 1931.

9. Leendertsee J.J.; Aspects of a computational model for long period water-wave propagation, Rand Corporation, Santa Monica, California, RM-6230-RC, 1970.

10. Reid R.O. and Bodine B.R.; Numerical models for storm surges in Galveston Bay, J. Waterways Harbors Division, ASCE, 94,33 (1968).

11. Volpi G. and Sguazzero P.; La propagazione della marea nella laguna di Venezia; un modello di simulazione e il suo impiego nella regolazione delle bocche di porto, Rivista Italiana di Geofisica, 4 (1-2), 67 (1977).

STOCHASTIC SIMULATION OF SPACE-TIME DEPENDENT PREDATOR-PREY MODELS

D.M. DUBOIS and G. MONFORT
University of Liège
Institute of Mathematics
15, Avenue des Tilleuls
B-4000 LIEGE (BELGIUM).

ABSTRACT.

This paper deals with a new method for the stochastic simulation of a system of non-linear partial differential equations describing the dynamics of the spatial distribution of populations in predator-prey relationship.

This method belongs to Monte-Carlo techniques, used in Operation Research. By generation of random numbers, the horizontal distribution of prey and predator concentrations in a turbulent medium is simulated.

Remarkably, it is shown that stochastic fluctuations drive the system very far from equilibrium and induce a temporal dissipative structure as defined by I. Prigogine. One assists to the creation, propagation and annihilation of concentrations waves, and this, independently of initial and boundary conditions. In fact, stochastic fluctuations induce a bifurcation within the predator-prey system which is structurally unstable.

1. INTRODUCTION.

In 1975, Dubois proposed the following space (x)-time (t) dependent mathematical model of the Lotka-Volterra predator-prey system (written here in one spatial dimension) :

$$\frac{\partial N_1}{\partial t} = k_1 N_1 - k_2 N_1 N_2 + D \frac{\partial^2 N_1}{\partial x^2} - w \frac{\partial N_1}{\partial x} \qquad (1\text{-a})$$

$$\frac{\partial N_2}{\partial t} = - k_3 N_2 + k_4 N_1 N_2 + D \frac{\partial^2 N_2}{\partial x^2} - w \frac{\partial N_2}{\partial x} \qquad (1\text{-b})$$

where N_1 and N_2 are the prey and predator concentrations, respectively; D, a diffusion coefficient; w, an advection current velocity; k_1, the growth rate of the prey; k_3, the decay rate of the predator; k_2, the grazing rate of the predator; and $k_4 = \beta k_2$, where β is the utilisation factor.

The deterministic simulation of eqs. 1 shows the basic mechanism of a transient spatial structuration in the predator-prey populations obeying three fundamental laws (Dubois, 1975b) :
(i) the creation of predator-prey waves;
(ii) the propagation of these waves; and,
(iii) the annihilation of meeting waves.

This paper deals with stochastic versions of the above model.

2. FIRST STOCHASTIC SIMULATION.

Let us discretisize eqs. 1 under the form (for $w > o$; when $w < o$, w_{i-1} replaced by w_{i+1})

$$N_{1i}^{t+\Delta t} = N_{1i}^{t} + (k_1 N_{1i}^{t} - k_2 N_{1i}^{t} N_{2i}^{t} + \frac{D}{\Delta x^2}(N_{1i+1}^{t} + N_{1i-1}^{t} - 2N_{1i}^{t}) - \frac{w_{i-1}^{t} + w_{i}^{t}}{2\Delta x}(N_{2i}^{t} - N_{2i-1}^{t}))\Delta t \quad (2\text{-a})$$

$$N_{2i}^{t+\Delta t} = N_{2i}^{t} + (-k_3 N_{2i}^{t} + k_4 N_{1i}^{t} N_{2i}^{t} + \frac{D}{\Delta x^2}(N_{2i+1}^{t} + N_{2i-1}^{t} - 2N_{2i}^{t}) - \frac{w_{i-1}^{t} + w_{i}^{t}}{2\Delta x}(N_{2i}^{t} - N_{2i-1}^{t}))\Delta t \quad (2\text{-b})$$

where the spatial domain is divided in N cells numbered by the index i of equal length Δx. In the simulation of eqs. 2, the velocities w_i are randomly distributed following a Gaussian curve. Figure 1-a shows the creation, propagation of predator-prey waves in starting with a small patch of both populations as initial conditions. This simulation is quite similar to the deterministic simulation given in Dubois (1975b). In Figure 1-b, it is seen that the spatial repartition of both populations becomes more and more uniform. The final state will be a quasi-uniform spatial distribution depending on boundary conditions (see Dubois, 1975a).

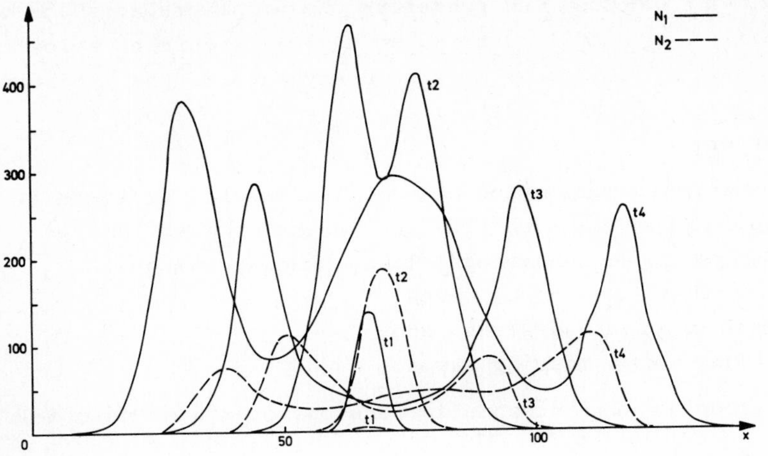

Figure 1-a

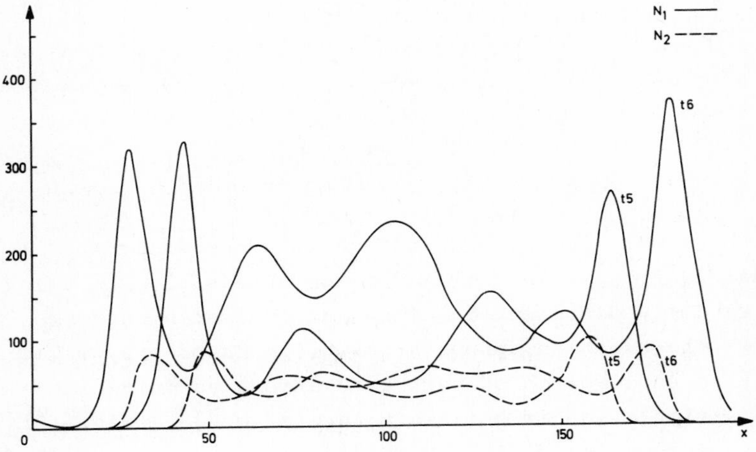

Figure 1-b

3. SECOND STOCHASTIC SIMULATION.

From eqs. 2, let us deduced transition probabilities inside a cell i for performing a pure stochastic simulation (similar to a Monte-Carlo method). In Table 1, these transition probabilities are clearly defined. For this simulation, we suppose only one transition during the interval of term Δt. For chosing the cell i where the transition will occur and what transition happens one adds the probabilities of all possibilities for each cell in adjusting the normalization constant K so that the sum of probabilities in all cells is equal to one. A random number (between 0 and 1) is generated and the cell number is thus determined as well as the type of transition. To calculate the interval of time Δt for the modification of one event in one cell, the following well-known formula is used

$$\Delta t = \frac{1}{K} \ln(1-R) \qquad (3)$$

where R is a random number (between 0 and 1) chosen in a uniform distribution. This technique is a generalization of the method of Bartlett (1960) who simulated the Lotka-Volterra time-dependent model system . Bartlett showed the instability of the system (one or two populations desappeared). Figure 2 gives the result of the space-time stochastic simulation (the velocity field is generated as in the first simulation) : let us notice the creation of very strong spatial heterogeneities. Initial conditions are given by a uniform distribution of both populations and boundary conditions are periodic. So, the spatial pattern is independent of both initial and boundary conditions. As in the Bartlett simulation, our system is also unstable : Figure 2-b shows the desappearance of the prey (at time t_8).

4. THIRD STOCHASTIC SIMULATION.

In view of stabilizing the system, a new assumption is taken into account. It is well-known that, when the prey concentration is very low, the predator does no more graze on it for efficiency raisons. It means, in practice, that the grazing rate k_2 (and $k_4 = \beta k_2$) is equal to zero for concentrations of N_1 below a certain threshold value.

TABLE 1

Cause	Probability		Effect	
Diffusion N_1	$K\lvert\frac{D}{\Delta x^2}(N_{1i+1}+N_{1i-1}-2N_{1i})\rvert$	$N_{1i}\to N_{1i}+1$	if the term is	> O
		$N_{1i}\to N_{1i-1}$	"	< O
Advection w>O; N_1	$K\lvert-\frac{w_{i+1}+w_i}{2\Delta x}(N_{1i}-N_{1i-1})\rvert$	$N_{1i}\to N_{1i}+1$	"	> O
		$N_{1i}\to N_{1i}-1$	"	< O
Advection w<O; N_1	$K\lvert-\frac{w_{i-1}+w_i}{2\Delta x}(N_{1i+1}-N_{1i})\rvert$	$N_{1i}\to N_{1i}+1$	"	> O
		$N_{1i}\to N_{1i}-1$	"	< O
Diffusion N_2	$K\lvert\frac{D}{\Delta x^2}(N_{2i+1}+N_{2i-1}-2N_{2i})\rvert$	$N_{2i}\to N_{2i}+1$	"	> O
		$N_{2i}\to N_{2i}-1$	"	< O
Advection w>O; N_2	$K\lvert-\frac{w_{i-1}+w_i}{2\Delta x}(N_{2i}-N_{2i-1})\rvert$	$N_{2i}\to N_{2i}+1$	"	> O
		$N_{2i}\to N_{2i}-1$	"	< O
Advection w<O; N_2	$K\lvert-\frac{w_{i+1}+w_i}{2\Delta x}(N_{2i+1}-N_{2i})\rvert$	$N_{2i}\to N_{2i}+1$	"	> O
		$N_{2i}\to N_{2i}-1$	"	< O
Growth of N_1	$K\ k_1\ N_{1i}$	$N_{1i}\to N_{1i}+1$		
Decay of N_2	$K\ k_3\ N_{2i}$	$N_{2i}\to N_{2i}-1$		
Predator-prey interaction	$K\ k_2\ N_{1i}\ N_{2i}$	$N_{1i}\to N_{1i}-1$		
"	$K\ k_4\ N_{1i}\ N_{2i}$	$N_{2i}\to N_{2i}+1$		

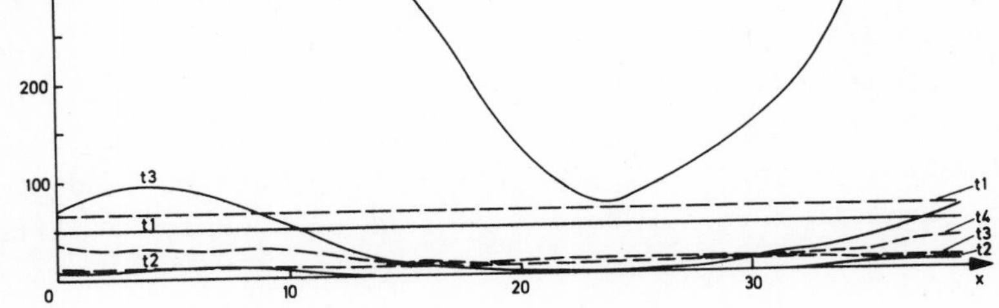

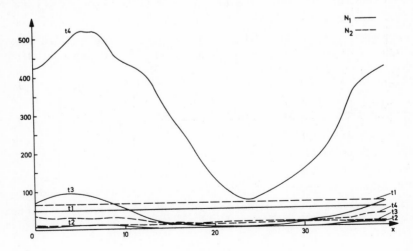

Figure 2a

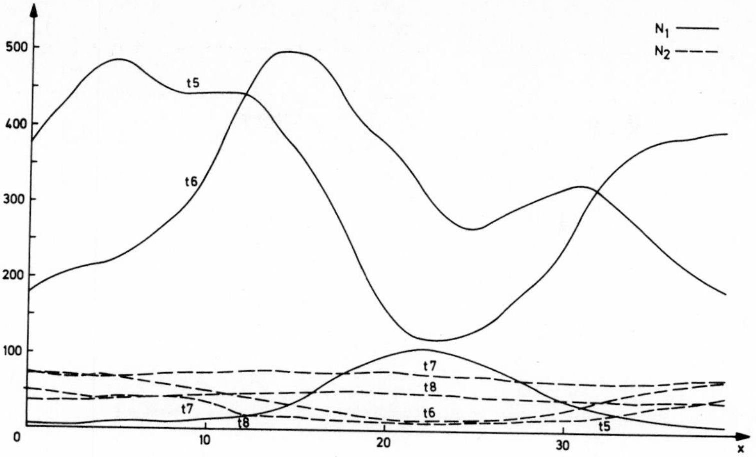

Figure 2-b

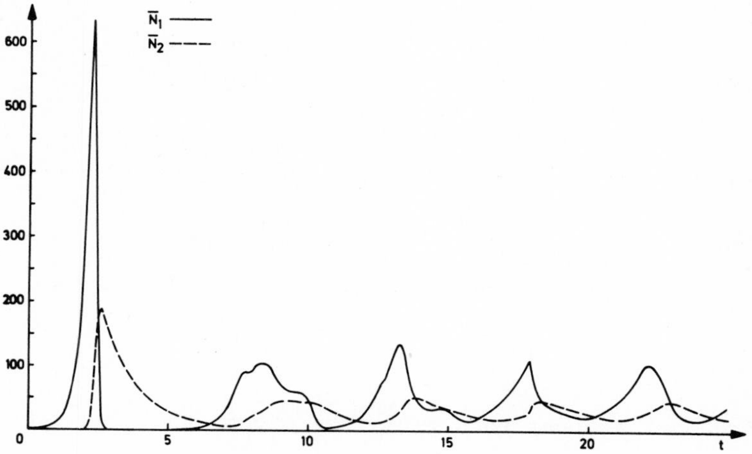

Figure 3

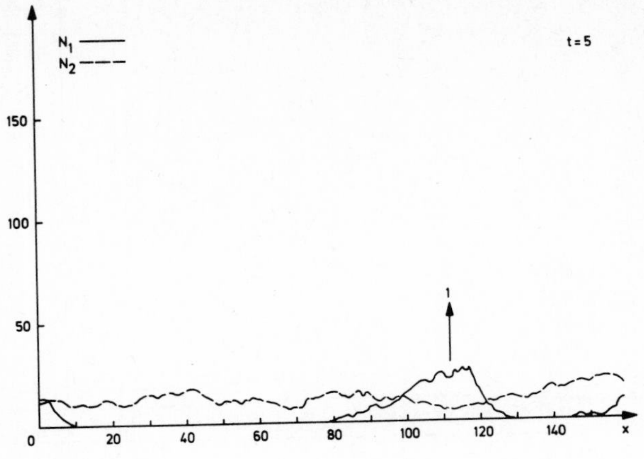

Figure 4-a

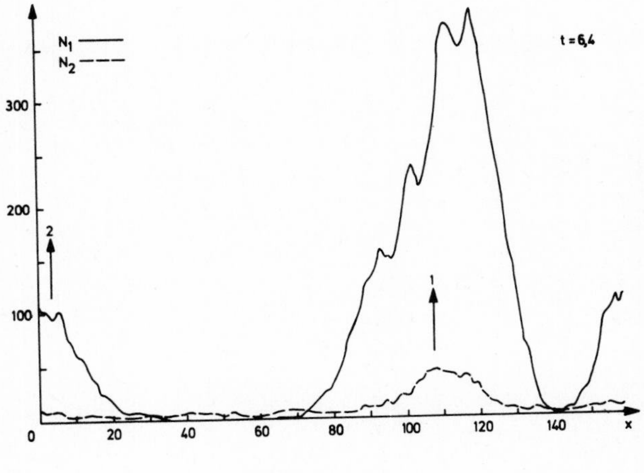

Figure 4-b

In taking into account this fact, the stochastic simulation (by the method given in the second simulation) shows the creation, propagation and annihilation of predator-prey waves. Figure 3 gives the time evolution of the spatial average of both population : one has a stable stochastic limit cycle. The space-time behavior of this simulation is resumed in Figures 4-a-b-c-d-e-f. Shoting at time t = o, with a uniform spatial distribution of both populations,

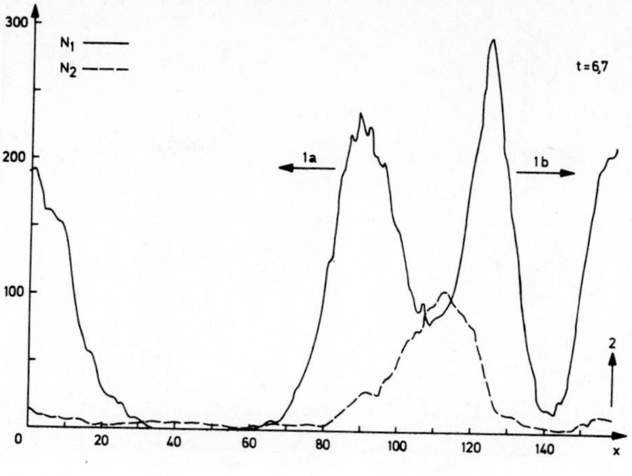

Figure 4-c

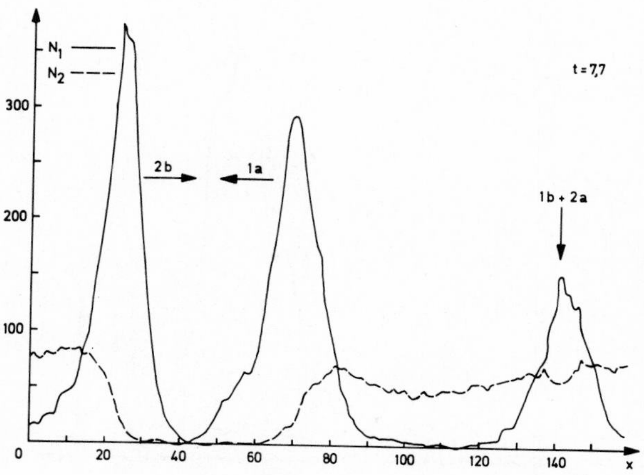

Figure 4-d

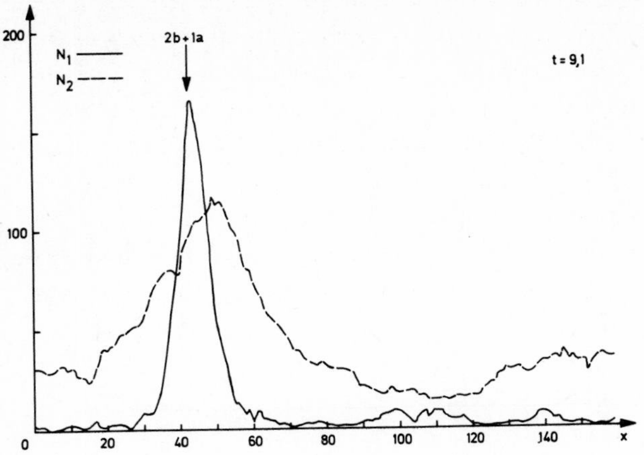

Figure 4-e

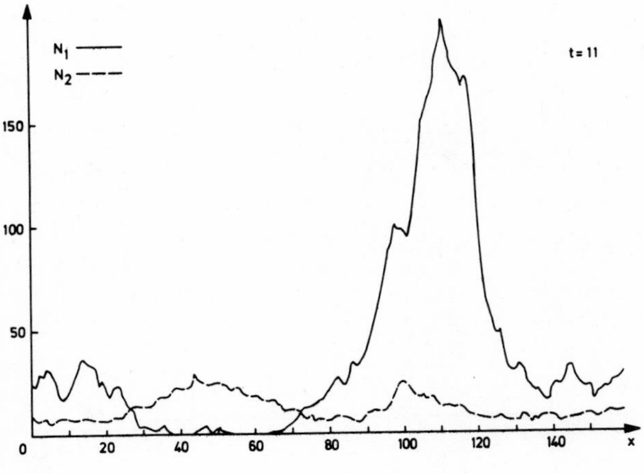

Figure 4-f

Figure 4-a shows , at time 5 (one time period later), the spontaneous emergence of one patch of prey. In Figure 4-b a high density prey patch is created (1) and a second one is beginning to create (2). A predator patch is also created (due to its growth with a time lag). The creation of two prey waves (1a and 1b) is well-shown in Figure 4-c due to the increase of predator populations in the center of the prey patch. Noticed also the development of the second prey patch. The first wave (1a) popagates to the left with a velocity given, approximately by (Dubois and Closset, 1975)

$$v \simeq 2 \sqrt{k_1 D} \tag{4}$$

Due to the same process, patch 2 splits also in two waves (2a and 2b). Figure 4d shows the propagation of waves 2b (to the right) and 1a (to the left) and the annihilation of the two meeting waves (1b and 2a). When waves 2b and 1a meet, one assists also to their annihilation, as given in Figure 4-e. Finally, Figure 4-f shows the creation of a new prey patch and the spatial pattern looks very similar to the spatial pattern of Figure 4-b. The same type of behavior is rediscovered in continuating the simulation : the process of waves creation, propagation and annihilation is quasi-periodic and belongs to the temporal dissipative structures class as defined by I. Prigogine. In fact, stochastic fluctuations induce a bifurcation in the system which is structurally unstable. Theoretical explanations about this sort of bifurcation are given in Dubois (1977c and 1978). More practical details about the computer programming of the stochastic simulation are described in Dubois and Monfort (1977).

REFERENCES

[1] BAILEY, N.T.J., (1964), "*The Elements of Stochastic Process*", London : Wiley.
(1967), "*The Mathematical Approach to Biology and Medicine*", London : Wiley.

[2] BARTLETT, M.S., (1960), "*Stochastic Population Models*", London : Methuen.

[3] DUBOIS, D.M., (1975a), "*The Influence of the Quality of Water on Ecological Systems*", in Modeling and Simulation of Water Resources Systems, North-Holland, ed. by G.C. Vansteenkiste, pp. 535-543.

[4] DUBOIS, D.M., (1975b), "*A Model of Patchniness for Prey-Predator Plankton Populations*", Ecol. Modelling, 1(1975) pp. 67-80.

[5] DUBOIS, D.M., (1977c), "*Limites à la Modélisation des Systèmes*", in Modélisation et Maîtrise des Systèmes, (AFCET), Ed. Hommes et Techniques, T.1., pp. 170-177.

[6] DUBOIS, D.M., (1978), "*Limits to Modeling and Stochastic Simulation of Biological Systems in Seas and Rivers*", in Modeling and Simulation of Land, Air and Water Resources Systems, North-Holland, Ed. by G.C. Vansteenkiste, in press.

[7] DUBOIS, D.M., CLOSSET, L., (1975), "*Patchiness in Primary and Secondary Production in the Southern Bight : a Mathematical Theory*", 10th European Symposium on Marine Biology, Ostend, Belgium, Vol. 2, pp. 211-229.

[8] DUBOIS, D.M., MONFORT, G., (1977), "*Simulation Stochastique de la Distribution Horizontale non-uniforme de Plancton Marin : Influence d'une Tache de Polluant*", Coll. des Publ. de la Fac. des Sc. Appl. de l'Univ. de Liège, n°68, pp. 3-45.

[9] LOTKA, A.J., (1925), "*Elements of Physical Biology*", William & Wilkins, Baltimore.

[10] PIELOU, E.C., (1969), "*An Introduction to Mathematical Ecology*", New York : Wiley.

[11] VOLTERRA, V., (1931), "*Leçon sur la Théorie Mathématique de la Lutte pour la Vie*", Paris, Gauthier-Villars.

OPTIMAL DERIVATION OF ANTIBODY DISTRIBUTION IN THE IMMUNE RESPONSE FROM NOISY DATA

C. Bruni
Istituto di Automatica, Università di Ancona, Italy

A. Germani
C.S.S.C.C.A.-C.N.R. Roma, Italy

G. Koch
Istituto di Matematica, Università di Lecce, Italy

INTRODUCTION

As several authors pointed out [1-5], in the immune response the antibodies produced are inhomogeneous with respect to their binding affinity with the stimulating antigen. Thus the problem arises (of analytically describing the antibody distribution with respect to the affinity, starting from the usually available experimental binding data. In particular, this problem constitutes a key point in the validation and identification of a mathematical model of the whole process [6,7]. Previous work related to this problem is contained in [1,3,4,8-15].

Of course, it would be highly desirable to describe the antibody distribution by a suitably chosen analytical expression depending only on few parameters. In this context, a research work was carried on [16,17] which led to select an analytical expression depending on four parameters. It appears to be able to interpret experimental data close enough during the whole maturation of the immune response and significantly generalizes the so called, widely adopted, Sips distribution [11].

Denoting by $\bar{K}$ the antibody site affinity with respect to a given antigen (K is supposed to range from 0 to ∞) and setting $x = \ln K$, this distribution has the form:

$$p(x) = \frac{1}{2\pi} \frac{\operatorname{sen}(\theta(x))}{\cos h\,[(x-x_o) - \ln M(x)] - \cos(\theta(x))} \tag{1.1}$$

$$\theta(x) = \varepsilon \cdot \arg\left\{ \cos h\,\left(\beta\left(x - x_o + \frac{\varepsilon}{2} + \eta\right)\right) + \cos h \frac{1}{2} \cdot \cos \beta\pi + j\left[\operatorname{sen} h \frac{1}{2}\ \operatorname{sen} \beta\pi\right]\right\} \tag{1.2}$$

$$M(x) = \left[e\ \frac{\cos \beta\pi + \cos h\left(\frac{1}{2} - \beta\left(x - x_o + \frac{\varepsilon}{2} + \eta\right)\right)}{\cos \beta\pi + \cos h\left(\frac{1}{2} + \beta\left(x - x_o + \frac{\varepsilon}{2} + \eta\right)\right)} \right]^{-\varepsilon/2} \tag{1.3}$$

and

$$\beta = \frac{1 - \varphi}{\varepsilon} \operatorname{cotgh} \frac{1}{4} \tag{1.4}$$

Clearly the previous distribution depends on the four parameters x_o, $\varepsilon, \eta, \varphi$. As shown in [17], eq.(1.1) describes a bimodal distribution, i.e. (losely speaking) a distribution with two dominant populations of antibodies; there is also a clear relation between the values of the four parameters and the main features of the distribution. To avoid inconsistent results such as negative valued or non unit area distributions, the set of parameters must be chosen according to the following constraints

$$\varepsilon \geq 0 \tag{1.5}$$

$$f_{2n+1}(\varepsilon) \leq \varphi \leq f_{2n}(\varepsilon) \quad \text{for a non negative integer } n \tag{1.6}$$

where

$$f_{2n+1}(\varepsilon) = \begin{cases} \max\{0, 1 - \dfrac{\varepsilon}{\operatorname{cotgh}\frac{1}{4}}(2n+1)\}, & 0 \leq \varepsilon \leq 1 \\ \max\{0, 1 - \dfrac{\varepsilon}{\operatorname{cotgh}\frac{1}{4}}\; 2n - \dfrac{2\varepsilon}{\pi \operatorname{cotgh}\frac{1}{4}} \operatorname{tg}^{-1}(\operatorname{cotgh}\frac{1}{4} \operatorname{tg}\frac{\pi}{2\varepsilon})\}, & \varepsilon \geq 1 \end{cases} \tag{1.7}$$

$$f_{2n}(\varepsilon) = 1 - \frac{\varepsilon}{\coth 1/4}\, 2n \tag{1.8}$$

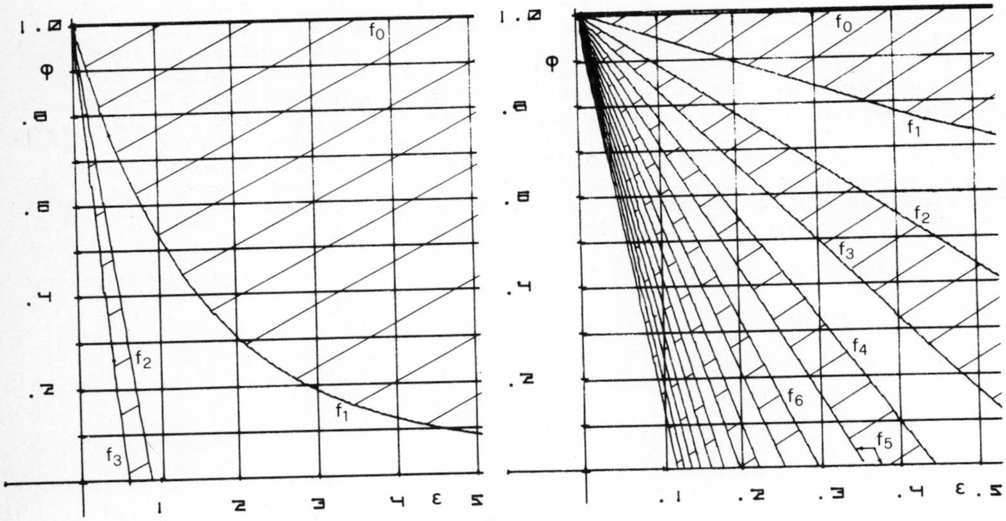

Fig. 1.

These constraints divide the (φ, ε) plane into a sequence of allowed and forbidden regions as shown in Fig. 1. In particular, we stress that the allowed regions are closed sets in $[0,1]\times[0,\infty)$ with piecewise linear boundaries (excepted $f_1(\varepsilon)$, $\varepsilon > 1$).

As well-known, denoting by H the free antigen concentration, by Ab_t the total concentration of antibody sites and by AbH the concentration of bound antibody sites, a sips plot consists of a representation of $\ln \frac{AbH}{Ab_t - AbH}$ against lnH. Now it turns out that from an experimental binding curve in this plot the four parameters $x_o, \varepsilon, \eta, \varphi$ can be easily determined by means of a direct geometrical construction [17].

A further remark is that the antibody population is not only characterized by its distribution but also by its Ab_t value. However, the Sips plot does not give information about Ab_t, although this value is required to draw the plot itself, which is very sensitive to it [18]. This motivates the convenience of another arrangement of binding data, namely the "double reciprocal plot" (1/AbH versus 1/H) [19]: not only are $x_o, \varepsilon, \eta, \varphi$ directly deducible from this plot also, but in addition $1/Ab_t$ is obtained as the limit value of $\frac{1}{AbH}$ as $\frac{1}{H} \to 0$ [18].

Thus, from the double reciprocal plot all the five parameters may be estimated: the estimates should satisfy constraints (1.5), (1.6) and in addition the obvious further condition

$$Ab_t \geq 0 \tag{1.9}$$

In [17] it is shown that assuming for the distribution the expression (1.1) is equivalent to assume the following analytical expression for 1/AbH as a function of 1/H:

$$\frac{1}{AbH} = e^{-\ln Ab_t}\left[1+\frac{1}{H}\,e^{-x_o}\left\{\frac{\sqrt{e}\,\cosh[\frac{1}{4}+\frac{1-\varphi}{2\varepsilon}(x_o-\frac{\varepsilon}{2}-\eta-\ln\frac{1}{4})\,\mathrm{cotgh}\,\frac{1}{4}]}{\cosh[\frac{1}{4}-\frac{1-\varphi}{2\varepsilon}(x_o-\frac{\varepsilon}{2}-\eta-\ln\frac{1}{4})\,\mathrm{cotgh}\,\frac{1}{4}]}\right\}^{\varepsilon}\right] \tag{1.10}$$

In the present work we deal with the optimal estimation problem for the five parameters from noisy experimental data arranged on a double reciprocal diagram. A statistical analysis of the measurement errors is carried out, and approximate formulas for their variances found. Subsequently the estimation problem is formally stated as a nonlinear constrained optimization problem. An algorithm is proposed, and its convergence properties are investigated.

2. STATISTICAL ANALYSIS OF MEASUREMENT ERRORS

We assume here that the situation procedure adopted is the Farr technique [18,19].
In it, 1/H and 1/AbH are respectively computed as:

$$1/H = \frac{1}{\%H \cdot H_R}$$
$$1/AbH = \frac{1}{(1-\%H)H_R} \tag{2.1}$$

where H_R is the hapten reacting concentration and %H is the fraction of H_R corresponding to the free hapten. It was shown [19] that these quantities are affected by essentially gaussian errors, whose mean values are zero and whose variances may be well approximated by the following formulas:

$$\sigma^2(1/H) = \frac{1}{H^2}\left[\frac{\sigma^2(\%H)}{(\%H)^2} + \frac{\sigma^2(H_R)}{H_R^2}\right] \tag{2.3}$$

$$\sigma^2(1/AbH) = \frac{1}{(AbH)^2}\left[\frac{\frac{\sigma^2(H_R)}{H_R^2} + (\%H)^2\left(\frac{\sigma^2(\%H)}{(\%H)^2} + \frac{\sigma^2(H_R)}{H_R^2}\right)}{(1-\%H)^2}\right] \tag{2.4}$$

$$\sigma(1/H, \frac{1}{AbH}) = \frac{1}{H\cdot AbH}\left[\frac{\sigma^2(H_R)}{H_R^2} - \frac{H}{AbH}\frac{\sigma^2(\%H)}{(\%H)^2}\right] \tag{2.5}$$

where σ^2 denotes error variance and $\sigma(1/H, \frac{1}{AbH})$ is the error covariance of $\frac{1}{H}$ and $\frac{1}{AbH}$. The values of $\sigma(\%H)/\%H$ and $\sigma(H_R)/H_R$ are known from experimental conditions and are fairly constant along each series of data: usual values are 2-3%.

Few remarks are listed below:

a) $\frac{\sigma(1/AbH)}{1/AbH}$ stays at low values (few percent) for low values of H, but quickly increases when 1/H approaches zero.

This circumstance heavily influences the feasibility of the Ab_t determination by the usual graphical extrapolation of data at low values of 1/H and constitutes one of the key motivations for choosing instead the value of Ab_t (along with the other four parameters) by means of an interpolation procedure over all the available data [19].

b) As may be easily seen from (2.3)-(2.5) [19] the correlation coefficient

$$\rho = \frac{\sigma(1/H, 1/AbH)}{\sigma(1/H)\cdot\sigma(1/AbH)} \tag{2.6}$$

monotonically decreases as 1/H increases and never takes the values ± 1 within any feasible experimental condition $(0 < H < \infty)$.

c) By the very way in which the titration procedure is arranged, no significant correlation exists between errors at different expetimental points.

3. STATEMENT OF THE PROBLEM

Let us assume to have N experimental points in the double reciprocal diagram and denote by z_H and z_B respectively the N-dimensional vectors of the measured values for $\frac{1}{H}$ and $\frac{1}{AbH}$. Let us further denote by $\gamma = [\gamma_1\gamma_2\gamma_3\gamma_4\gamma_5]^T = [\eta\ \varepsilon\ \varphi\ x_o\ \ln Ab_t]^T$ the unknown parameter vector and by a the N vector of the "true values" of $\frac{1}{H}$ (it was assumed as unknown parameter $\ln Ab_t$, instead of Ab_t, so as to take constraint (1.9) automatically into account and to have unknown parameters of the same order of magnitude). Then the N vector of the "true values" of 1/AbH is a function $h(a,\gamma)$ whose components h_i are given by eq. (1.10): each of them depends only on the corresponding component a_i of a, and of course on γ.

Due to the results of the previous section, the measured values are given by:

$$z_H = a+u \tag{3.1}$$

$$z_B = h(a,\gamma)+v \tag{3.2}$$

where u, v are the N-dimensional error vectors corrupting the a, h values which are gaussian distributed with zero mean and covariance matrices:

$$\Psi_u = \begin{bmatrix} \sigma^2_{u,1} & 0 \cdots\cdots 0 \\ 0 & \sigma^2_{u,2} \cdots 0 \\ \cdot & \cdot \quad \cdots \\ 0 & 0 \cdots \sigma^2_{u,N} \end{bmatrix} \tag{3.3}$$

$$\Psi_v = \begin{bmatrix} \sigma^2_{v,1} & 0 \cdots\cdots 0 \\ 0 & \sigma^2_{v,2} \cdots 0 \\ \cdot & \cdot \quad \cdots \\ 0 & 0 \cdots \sigma^2_{v,N} \end{bmatrix} \tag{3.4}$$

$$\Psi_{uv} = \begin{bmatrix} \sigma_{uv,1} & 0 \cdots\cdots 0 \\ 0 & \sigma_{uv,2} \cdots 0 \\ \cdot & \cdot \quad \cdots \\ 0 & 0 \cdots \sigma_{uv,N} \end{bmatrix} \tag{3.5}$$

Of course, the elements of the matrices (3.3)-(3.5) are given by the variances (2.3)-(2.5) at the various measurement points.

As an optimal estimate of the vectors a, γ we consider here the maximum likelihood estimate, for its well known properties. Due to gaussianity of errors, the maximum likelihood estimate is achieved by minimization of the functional:

$$J(a,\gamma) = \begin{bmatrix} z_H - a \\ z_B - h(a,\gamma) \end{bmatrix}^T Q^{-1} \begin{bmatrix} z_H - a \\ z_B - h(a,\gamma) \end{bmatrix} \tag{3.6}$$

with respect to a,γ; Q being the error covariance matrix:

$$Q = E\left(\begin{bmatrix} u \\ v \end{bmatrix} \begin{bmatrix} u^T & v^T \end{bmatrix} \right) \tag{3.7}$$

Obviously minimization of (3.6) is to be made taking constraints (1.5), (1.6) into account that is, γ ranges in $\Gamma \subset R^5$, where Γ denotes the set of the admissible values for the unknown parameters vector.

We now first solve the minimization problem with respect to a (and denote by $\hat{a}$ the optimal estimate), so to be left with a minimiza tion problem only with respect to γ.

Defining:

$$\left.\frac{\partial h(a,\gamma)}{\partial a}\right| \triangleq N(a,\gamma) \tag{3.8}$$

$$\left.\frac{\partial h(a,\gamma)}{\partial \gamma}\right| \triangleq M(a,\gamma) \tag{3.9}$$

and assuming that both these matrices are practically constant for $a = \alpha\hat{a} + (1-\alpha)z_H$, $\alpha \in [0,1]$ (which is reasonable to assume due to the small magnitude of the usual values of u), it is shown in [20] that the above mentioned problem reduces to minimizing the functional:

$$J(\gamma) = [h(z_H,\gamma) - z_B]^T \Psi^{-1}(z_H,\gamma)[h(z_H,\gamma) - z_B] \tag{3.10}$$

with respect to γ and taking constraints (1.5), (1.6) into account.The matrix $\Psi(z_H,\gamma)$ is given by:

$$\Psi(z_H,\gamma) = E\{[v - N(z_H,\gamma)u][v - N(z_H,\gamma)u]^T\} = \Psi_v - 2N(z_H,\gamma)\Psi_{uv} + N^2(z_H,\gamma)\Psi_u \tag{3.11}$$

It turns out to be positive definite [20] and such that:

$$\frac{\partial \Psi(z_H,\gamma)}{\partial \gamma} \simeq 0 \tag{3.12}$$

4. OPTIMIZATION PROCEDURE.

In order to evercome difficulties related to minimization of functional (3.10) with an infinite number of admissible regions (1.7), (1.8), the optimal estimate $\hat{\gamma}$ is looked for locally within a r_o-neighbourhood $S(\gamma^{(o)}, r_o)$ about a suitable initial guess $\gamma^{(o)}$, by a two-step procedure. In the first step the local unconstrained optimal estimate γ^* is achieved. If γ^* belongs to the admissible set, clearly $\gamma^* = \hat{\gamma}$. Otherwise, if γ^* belongs to the forbidden region, the second step of the procedure follows, in which $\hat{\gamma}$ is looked for in the intersection of the boundary of this region with $S(\gamma^{(o)}, r_o)$.

In both steps the Newton method (or the modified Newton method) can be adopted, for which "ad hoc" convergence conditions have been deduced from the general Kantorovich theorems [21]. Some theorems proved in [20], and reported in the following, substantiate this procedure.

The first step implies solution of the nonlinear algebraic equation:

$$P(\gamma) \triangleq \frac{1}{2} \frac{\partial J^T}{\partial \gamma} = M^T(z_H, \gamma) \Psi^{-1}(z_H, \gamma)[h(z_H, \gamma) - z_B] = 0 \tag{4.1}$$

The Newton method applied to (4.1) gives rise to the sequence:

$$\gamma^{(i+1)} = \gamma^{(i)} - [P'(\gamma^{(i)})]^{-1} P(\gamma^{(i)}), \quad i = 0, 1, \ldots \tag{4.2}$$

We first observe that, as shown in [20], for any set $D \subset R^5$ of the type:

$$D = \{\gamma : |\gamma_k - \gamma_k^{(o)}| < \delta_k, k = 1, 2, \ldots 5\} \tag{4.3}$$

a $\delta_I > 0$ exists (δ_I depending on D) such that $P(s) = P(\gamma + j\omega)$ is holomorphic in the set:

$$\Omega = \{s \in C^5 : \gamma \in D, |\omega_K| < \delta_I, k = 1, 2, \ldots 5\} \tag{4.4}$$

Now local existence and uniqueness of solution of (4.1) as well as convergence of sequence (4.2) are guaranteed by the following theorem:

THEOREM 1. For a given admissible $\gamma^{(o)}$, and D and Ω as in (4.3),(4.4), if:

a) $\left[P'(\gamma^{(o)})\right]^{-1}$ exists and B_o, η_o are such that:

$$B_o \geq \left\| \left[P'(\gamma^{(o)})\right]^{-1} \right\| \tag{4.5}$$

$$\eta_o = \|\gamma^{(1)} - \gamma^{(o)}\| \tag{4.6}$$

b) denoted by $m = \sup_{s \in \partial\Omega} \|P(s)\|$, it is

$$\ell_o \triangleq \eta_o B_o \frac{m}{\delta_I^2} \leq \frac{1}{4} \tag{4.7}$$

c) it is:

$$\min_k \delta_k \geq r_o \triangleq \frac{1-\sqrt{1-2\ell_o}}{\ell_o} \eta_o \tag{4.8}$$

then in $S(\gamma^{(o)}, r_o) = \{\gamma : \|\gamma - \gamma^{(o)}\| \leq r_o\}$ a unique solution γ^* of (4.1) exists which moreover is the absolute minimum of (3.10) in $S(\gamma^{(o)}, r_o)$, and the sequence (4.2) quadratically converges to γ^*:

$$\|\gamma^* - \gamma^{(i)}\| \leq \frac{1}{2^i} (2\ell_o)^{2^i} \frac{\eta_o}{\ell_o} \quad , \qquad i = 0,1,\ldots \tag{4.9}$$

◀

REMARK. Were (4.2) be substituted by the similar sequence implementing the modified Newton method, convergence to a unique γ^* is still guaranteed but not its speed.

At this point, if γ^* belongs to the admissible set, it constitutes the desired optimal constrained estimate in $S(\gamma^{(o)}, r_o)$. Should this be not the case, the optimal constrained estimate must be searched for on the intersection of $S(\gamma^{(o)}, r_o)$ with the boundary $\partial\Gamma$ of the forbidden region . The following theorem makes the search easier, limiting the portion of the boundary of the forbidden region which must be explored in the second step of the procedure [21].

For A a closed set in R^n and $z^* \notin A$, we denote by $P_{z^*}\{A\}$ the subset of the elements z of A such that: $\alpha z + (1-\alpha) z^* \notin A$, $\forall \alpha \in [0,1)$. Then the following theorem holds:

THEOREM 2. Under the same hypotheses of thm.1, if $\gamma^* \notin \Gamma$, the constrained absolute minimum $\hat{\gamma}$ of (3.10) in $S(\gamma^{(o)}, r_o)$ belongs to $P_{\gamma^*}\{S(\gamma^o, r_o) \cap \partial\Gamma\}$.

◀

The last result amounts to say that if $\gamma^* \notin \Gamma$, $\hat{\gamma}$ must be looked for in $S(\gamma^{(o)}, r_o)$ along the portions of hypersurfaces which constitute $P_{\gamma^*}\{S(\gamma^{(o)}, r_o) \cap \partial\Gamma\}$ and are defined by constraints of the type:

$$\gamma_3 = f_r(\gamma_2) \quad , \qquad r = 0,1,2,\ldots \tag{4.10}$$

In this second step of the procedure, Newton's method may again be exploited. After local linearization of the only nonlinear constraint

(4.10) for r = 1, in [20] conditions sufficient to guarantee local existence and uniqueness of a minimum point as well as quadratic convergence of the algorithm, have been deduced.

Specifically, these conditions are expressed in terms of the quantities B_o, m, ℓ_o, already defined in Thm. 1.

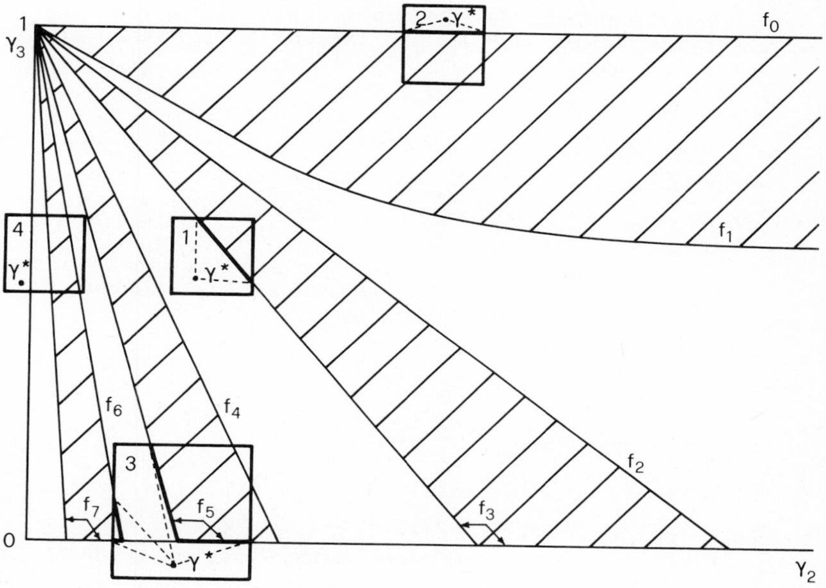

Fig. 2.

In Fig.2 same typical cases are represented: when $0 < \gamma_3^* < 1$, $\gamma_2^* > 0$ (case 1) one or two portions of straight contraints must be explored ; when $\gamma_3^* > 1$, $\gamma_2^* > 0$ (case 2) only one portion of the constraint $\gamma_3 = f_o(\gamma_2) = 1$ must be considered;when $\gamma_3^* < 0$, $\gamma_2^* > 0$ (case 3) several parts of constraints must be explored (included portions of the constraint $\gamma_3 = 0$). The case in which $\gamma_2^* < 0$ (case 4) is a degenerate one because, assuming for ε its limit value $\varepsilon = 0$, the distribution (1.1) turns out to be a Dirac function in x_o;therefore in this case, only two parameters (γ_4 and γ_5) have to be estimated and the solution may be easily achieved by means of standard linear estimation.

5. CONCLUDING REMARKS

In this paper the problem of the optimal estimate of the parameters of the antibody distribution in the immune response is rigorously settled and tackled. The estimation procedure exploits all available experimental data and is based on a statistical analysis of measurement errors.

An optimal estimation algorithm is formulated and its convergence properties are analyzed in detail.

Of course the check of the conditions of algorithm convergence in the different steps of the procedure, requires a relevant computational effort; moreover these conditions are only sufficient. Therefore, once the convergence and the right working of the procedure is verified at least in some typical conditions, it may be convenient to apply the procedure itself without the preliminary check of the above mentioned conditions. This can produce some trouble in the second step of the optimization due to the lack of information about $S(\gamma^{(o)}, r_o)$. A pragmatic way of operating is to substitute $P_{\gamma^*}\{S(\gamma^{(o)}, r_o) \cap \partial\Gamma\}$ with $P_{\gamma^*}\{\partial\Gamma\}$.

Surely the minimum thus found is not greater than the constrained minimum in $S(\gamma^{(o)}, r_o)$; it is true that the portions of constraints to explore are larger but, as a conterpart, the computational work for checking the convergence conditions is avoided.

In [20] examples of applications of the convergence conditions are shown and the related complete computing program is implemented; they are not reported here because of lack of space.

REFERENCES

[1] L.PAULING, D.PRESSMAN, A.L.GROSSBERG, J.Amer.chem.Soc.1944,66,784.

[2] H.N.EISEN, F.KARUSH, J.Amer. chem. Soc., 1949, 71, 363.

[3] F.KARUSH, J.Amer.chem.Soc., 1956, 78, 5519.

[4] A.NISONOFF, D.PRESSMAN, J. Immunol., 1958, 80, 417.

[5] H.N.EISEN, G.W.SISKIND, Biochemistry, 1964, 3, 996.

[6] C.BRUNI, M.A.GIOVENCO, G.KOCH, R.STROM, Math.Biosciences,1975,27,191.

[7] C.BRUNI, M.A.GIOVENCO, G.KOCH, R.STROM, In: Theoretical Immunology; G.I.Bell, G.H.Pimbley, A.S.Perelson Eds. M.Dekker, 1977.

[8] F.KARUSH, Adv. Immunol., 1962, 2, 1.

[9] R.N.PINCKARD, D.M.WEIR, Handbook of Exp. Immunol., D.M. Weir ed., Blackwell Scient. Pub., 1967, ch. 14.

[10] I.M.KLOTZ, In: The Proteins, H.Neurath, K.Bailey Eds., Academic Press, 1953, vol.I, Part B.

[11] R.SIPS, J.chem. phys., 1948, 16, 490.

[12] T.P.WERBLIN, G.W.SISKIND, Immunochemistry, 1972, 9, 987.

[13] T.P.WERBLIN, T.K.YOUNG, F.QUAGLIATA,G.W.SISKIND,Immunology,1973,24,477.

[14] J.D.BOWMAN, F.ALADJEM, J.theor. Biol., 1963, 4, 242.

[15] M.E.MEDOF, F.ALADJEM, Fedn. Proc., 1971, 30, 657.

[16] C.BRUNI,A.GERMANI,G.KOCH,Boll.Soc.It.Biol.Sper.,1974,50,1057.

[17] C.BRUNI, A.GERMANI, G.KOCH, R.STROM, J.theor.Biol.,1976,61,143.

[18] A.ORATORE, Rapp.Ist. Automatica, Università di Roma. To appear.

[19] G.KOCH,A.ORATORE,Rapp.Ist.Automatica,Università di Roma.To appear.

[20] C.BRUNI, A.GERMANI,Rapp.Ist.Automatica,Università di Roma.To appear.

[21] L.V.KANTOROVICH, G.P.AKILOV, Functional Analysis in Normed Spaces. Mc Millan, 1964.

COMPARTMENTAL CONTROL MODEL OF THE IMMUNE PROCESS

R. R. Mohler and C. F. Barton
Department of Electrical and Computer Engineering
Oregon State University
Corvallis, Oregon 97331 USA

ABSTRACT

A nonlinear model of the humoral immune process is derived with critical organs, such as spleen and lymph represented by compartments. B cells, T cells, macrophages and their progeny represent the mechanism for generating antibodies which in turn control antigen (alien material). The model is synthesized as a combination of bilinear subsystems. Simulation of the model is compared with experimental results.

INTRODUCTION

The purpose of this paper is to introduce a mathematical model of the humoral immune system which shows a distribution of control function through several important organs and transfer media. Analysis of such a model may lead eventually to a better understanding of the immune system, the role of various organs, the control of disease and transplant rejection or immune tolerance.

Obviously, such long-range goals are not accomplished in this brief paper, but would require a long concentrated research program conducted with an extensive experimental program. Some system research has been active since at least the 1966 preliminary analysis by Hege and Cole [1]. This was followed by a Poisson stochastic model by Jilek and Sterzl [2, 3]. Bell [4, 5, 6] derived a model which is more like the recent work such as reported here. This includes antigen-antibody binding according to their chemical affinities. The work reported here is closest to the work of Bruni, et. al. [7, 8], but none of these have considered the special distribution modes through the various organs which are immunologically significant. Bystryn, Schenkein and Uhr [9] present a non-mathematical model of functional compartments for immunogen and antibody secretion. Hammond [10] derived a spleen model with marginal zones and red and white pulp zones. The recent work of Marchuk [11] and Perelson [12] presents a good base for immune system analysis of disease control.

Briefly, it is the purpose of the immune system to recognize and to remove alien material (recognized chemically as antigens) such as viruses and pollutants from the body. For the humoral process, antibodies of certain chemical specificities are triggered by the presence of antigens of corresponding specific values. Antibodies prepare the alien for eventual ingestion and destruction by a class of cells called macrophages. The source of antibodies is a class of white blood cells called lymphocytes, and in particular, a subclass of sensitized lymphocytes called plasma cells is the main source. Lymphocytes, which have been sensitized according to a specific antigen, are further classified as B cells and T cells, the latter having been processed through the thymus and which play an important control role with some antigens. The model which is presented here as well as those of a mathematical nature considered throughout the literature assume only one class of such cells, viz B cells. An interactive T-B cell control model is presented by Mohler, Barton and Hsu [13]; and the B-cell model presented here is a compartmental generalization of that class [14].

SINGLE COMPARTMENTAL MODEL

Following Mohler, et. al. [13], it is readily seen from a birth-death conservation process that the humoral immune process with certain assumptions may be approximated by:

$$\frac{dx_1}{dt} = \alpha u_1 x_1 - x_1/\tau_1 + \beta_k \tag{1}$$

$$\frac{dx_2}{dt} = 2\alpha u_2 x_1 - x_2/\tau_2 \tag{2}$$

$$\frac{dx_3}{dt} = -c_k k x_5 x_3 - x_3/\tau_3 + \alpha' x_2 + c_k x_4 + \alpha'' x_1 \tag{3}$$

$$\frac{dx_4}{dt} = c_k k x_5 x_3 - (c_k + 1/\tau_4) x_4 \tag{4}$$

and

$$\frac{dx_5}{dt} = \dot{x}_{5i} - x_5/\tau_5 - k c_k x_5 x_3 + c_k x_4 \tag{5}$$

Here with the arguments (t, time and k, association constant) omitted for brevity:

x_1 is population density of immunocompetent cells (ICC), which are sensitized lymphocyte cells with particular surface receptors for antigen according to k. They may differentiate into plasma cells and divide into memory cells. The latter which may further divide and enter the pool of immunocompetent cells.

x_2 is the population density of plasma cells which are non-reproducing offspring of stimulated immunocompetent cells.

x_3 is population density of "antibody sites."

x_4 is population density of immune complex which individually include antibody sites and antigen.

x_5 is antigen concentration which triggers the response mechanism.

u_1 $= p_s(1 - 2p_d)$

u_2 $= p_s p_d$

α is birth-rate constant of stimulated immunocompetent cells.

p_d is probability that an immunocompetent cell differentiates into a plasma cell.

p_s is probability that antigen stimulates cell.

β_k is the rate of generation of new immunocompetent cells (from bone marrow).

τ_1 is the mean lifetime of immunocompetent cells.

τ_2, τ_3, τ_4 are appropriate life times.

α' is plasma-cell antibody production rate.

α'' is ICC antibody production.

c_k, kc_k are dissociation rate and association rate coefficients respectively of immune complex (k dependent).

It is structurally interesting that this system may be decoupled as a feedback combination of bilinear systems. Mohler [15, 16], Mohler and Ruberti [17] and Ruberti and Mohler [18] provide a summary of significant theoretical developments of bilinear systems and their application to biological processes in particular.

With $r_b(t, k)$ and $r_f(t, k)$, the number of bound and free receptors on B-cell surface per unit volume respectively, the affinity k and probability p_r that receptory of affinity k is occupied, respectively are

$$k = \frac{r_b}{r_f \cdot x_5} \quad \text{and} \quad p_r = \frac{r_b}{r_b + r_f} = \frac{kx_5}{1 + kx_5} .$$

It is shown by Bruni, et. al. [7] that

$$p_d = \frac{kx_5}{1 + kx_5} \tag{6}$$

and

$$p_s = \begin{cases} 1, & \text{for some sensitive interval, } \gamma_1 \le kx_5 < \gamma_2 \\ 0, & \text{for other } kx_5 \end{cases} \tag{7}$$

MULTI-COMPARTMENTAL MODEL

Figure 1 presents a compartmentation of the humoral immune system according to the most relevant organs and transfer media. These include bone marrow, thymus, spleen, blood, lymph, lymph nodes and gut associated lymphoid tissue termed GALT. The latter location of antigen-antibody reaction includes a lumping of tonsils, appendix and Peyer's patches. Bone marrow is the source of stem cells or precursor cells for the process. T cells mature in the spleen but are neglected in the model which is derived here. Spleen and lymph nodes are important locations of antigen-antibody reactions. Blood and lymph are important transport media but also represent significant storage of cells and molecules.

$$\frac{dx_{2s}}{dt} = 2\alpha u_{2s}x_{1s} - x_{2s}/\tau_2 - a_{20s}x_{2s} \qquad (11)$$

$$\frac{dx_{21}}{dt} = 2\alpha u_{21}x_{11} - x_{21}/\tau_2 - a_{201}x_{21} \qquad (12)$$

$$\frac{dx_{3b}}{dt} = a_{3b1}x_{31} - (a_{31b} + a_{30b})x_{3b} \qquad (13)$$

$$\frac{dx_{3s}}{dt} = \alpha' x_{2s} - c_k k x_{4s}x_{3s} - (a_{31s} + a_{30s})x_{3s} \qquad (14)$$

$$\frac{dx_{31}}{dt} = \alpha' x_{21} - c_k k x_{41}x_{31} + a_{31b}x_{3b} + a_{31s}x_{3s} - (a_{3b1} + a_{301})x_{31} \qquad (15)$$

$$\frac{dx_{4s}}{dt} = -a_{40s}x_{4s} - c_k k x_{3s}x_{4s} + \dot{x}_{4si} + x_{4s}/\tau_4 \qquad (16)$$

$$\frac{dx_{41}}{dt} = -a_{401}x_{41} - c_k k x_{31}x_{41} + \dot{x}_{41i} + x_{41}/\tau_4 \qquad (17)$$

where a_{kji} indicates the transfer rate coefficient of material k to jth compartment from ith compartment; u_{11}, u_{1s}, u_{21}, u_{2s} are computed as for the single compartment with the second subscript referring to the compartment (lymph and spleen) from which the state variables are used in the computation.

Figure 2 and 3 provide a computer simulation of this model with typical numerical values taken from the literature, for a response to sheep red blood cells, a commonly used antigen in experiments. Here, the average k is 10^7, and the latter two terms in equations (16) and (17) are regulated with the antigen introduced by initial conditions, $x_{4s}(0)$ and $x_{41}(0)$. Also, it is assumed that complex dissociation rate and death rate, are small compared to association rate and the two, tending to cancel one another, are neglected. The comparison of simulated response of antibodies in blood is in fair agreement with experimental results as shown in Figure 3.

CONCLUSIONS

As a consequence of the nature of population equations and chemical reactions, the humoral immune system is modeled by a feedback combination of bilinear systems which makes the total model quadratic plus other nonlinearities which arise from

As shown in Figure 1, stem cells migrate from bone marrow to thymus and spleen, back via blood to GALT and lymph nodes, from both to lymph and back to blood again. During this migration, the process dynamics to varying degrees may be approximated throughout the compartments similar to equations (1) to (5). From the immunological literature, however, it is seen that plasma cells are concentrated in the spleen, lymph nodes and GALT, and that mature plasma cells, due to their large size, do not circulate. Antigen, on the other hand, gets trapped in the spleen, lymph nodes and GALT after a very short circulation time which is neglected here. Antibodies are produced mainly by plasma cells in spleen, lymph nodes and GALT, and are too large to enter the blood stream directly from tissue, but are transferred to the blood via lymphatic channels. Antibody reacts with trapped antigen mainly in spleen, lymph nodes, and GALT to form complexes which in turn are desctroyed by macrophages. The model presented here assumes that macrophages and T cells are present in sufficient quantity but their dynamics are neglected. Furthermore, lymph nodes and lymph are lumped together into one compartment since further experimental data could hardly be collected anyway. Also, GALT is appropriately lumped with other compartment since its only unique role would be for the generation of a particular antibody (IgA) which is regulated here.

The state equations for the multi-compartmental model may be derived as for the lumped model except that the above discussion is incorporated with a conservation of cells or molecules in traversing the compartment.

Here, the first subscript on the state variables refers to the lumped model (i.e., 1 for immunocompetent cells, etc.) with a second subscript added according to b for blood, s for spleen, 1 for lymph and lymph nodes. Also, it is assumed that dissociation rate is negligible compared to association rate and the rate of destruction by macrophages. As noted by Bell [19], this assumption is reasonable if it is assumed that most antibody-antigen bonds are multivalent and that single binding is instantaneous. Consequently, it is not necessary to define a complex density and x_{4s}, for example, refers to antigen concentration in spleen.

Then the system is approximated by:

$$\frac{db_{1b}}{dt} = \beta_k + a_{1bs}x_{1s} + a_{1b1}x_{11} - (a_{1sb} + a_{11b} + a_{10b})x_{1b} \tag{8}$$

$$\frac{dx_{1s}}{dt} = \alpha u_{1s}x_{1s} - x_{1s}/\tau_1 + a_{1sb}x_{1b} - (a_{1bs} + a_{10s})x_{1s} \tag{9}$$

$$dx_{11} = u_{11}x_{11} - x_{11}/\tau_1 + a_{11b}x_{1b} - (a_{1b1} + a_{101})x_{11} \tag{10}$$

the approximation of stimulation and differentiation probabilities in terms of state variables. The compartmental model presented is only a first attempt to better describe the humoral immune control process. Further refinements, sensitvity analysis and tracer experiments to further verify the model are needed. Previous results on required tracer accessibility for minimal compartmental realization should be useful for the experimental program [16, 20].

ACKNOWLEDGEMENT

The authors wish to acknowledge the controbutions of K. Lawrence particularly to the simulation reported here. Also, they are grateful for the support of National Science Foundation through Grant No. ENG74-15530 A01.

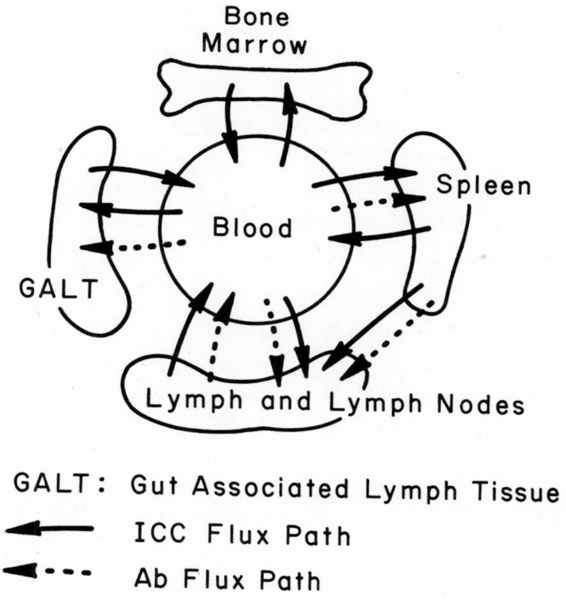

Figure 1. Immunological Model Compartments.

BIBLIOGRAPHY

1. Hege, J. S. and J. L. Cole, "Mathematical Model Relating Circulating Antibody and Antibody Forming Cells," J. Immunol. 94, 34-40, 1966.

2. Jilek, M., "Immune Response and Its Stochastic Theory," Proc. IFAC Symposium on System Identification, The Hague, 209-212, 1973.

3. Jilek, M. and J. Sterzl, "Modeling of the Immune Response," in Morphological and Functional Aspects of Immunity, 333-349, Plenum Press, New York, 1971.

4. Bell, G. I., "Mathematical Model of Clonal Selection and Antibody Production," J. Theoret. Biol. 29, 191-232, 1970.

5. Ibid., Part II, J. Theoret. Biol. 33, 339-378, 1971.

6. Ibid., Part III, "The Cellular Basis of Immunological Paralysis," 379-398, 1971.

7. Bruni, C., M. A. Giovenco, G. Koch and R. Strom, "A Dynamical Model of Immune Response," in Variable Structure Systems with Application to Biology and Economics, (Ruberti and Mohler, eds.), Springer-Verlag, New York, 1975.

8. Bruni, C., A. Germani and G. Koch, "Optimal Derivation of Antibody Distribution in the Immune Response from Noisy Data," in this volume.

9. Bystryn, J. C., I Schenkein and J. Uhr, "A Model for the Regulation of Antibody Synthesis by Serum Antibody," in Progress in Immunology, (B. Amos, ed.), 627-636, Academic Press, New York, 1971.

10. Hammond, B. J., "A Compartmental Analysis of Circulatory Lymphocytes in the Spleen," Cell Tissue Kinetics 8, 155-169, 1975.

11. Marchuk, G. I., "An Immunological Model of Virus and Bacterial Diseases," in this volume.

12. Perelson, A. S., "The IgM-IgG Switch Looked at From a Control Theoretic Viewpoint," in this volume.

13. Mohler, R. R., C. F. Barton and C. S. Hsu, "T and B Cell Models in the Immune System," in Theoretical Immunology, (G. Bell, et. al., eds.), Marcel Dekker, New York, 1978.

14. Barton, C. F., R. R. Mohler and C. S. Hsu, "System Theoretic Control in Immunology," Proceedings IFIP Symposium on Optimization, Nice, Springer-Verlag, 1976.

15 Mohler, R. R., Bilinear Control Processes, Academic Press, New York, 1973.

16. Mohler, R. R., "Biological Modeling with Variable Compartmental Structures," IEEE Trans. Auto. Control AC 19, 922-926, 1974.

17. Mohler, R. R. and A. Ruberti, Theory and Application of Variable Structure Systems, Academic Press, New York, 1972.

18. Ruberti, A. and R. R. Mohler, Variable Structure Systems with Application to Biology and Economics, Springer-Verlag, New York, 1975.

19. Bell, G. I., "Model for the Binding of Multivalent Antigen to Cells," Nature 248, 430-431, 1974.

20. Smith, W. D. and R. R. Mohler, "Necessary and Sufficient Conditions in the Tracer Determination of Compartmental Order," J. Theoret. Biol. 57, 1-21, 1976.

21. Alder, F. L., "Studies on Mouse Antibodies, I. The Response to Sheep Red Blood Cells," J. Immunology 95, 26-38, 1965.

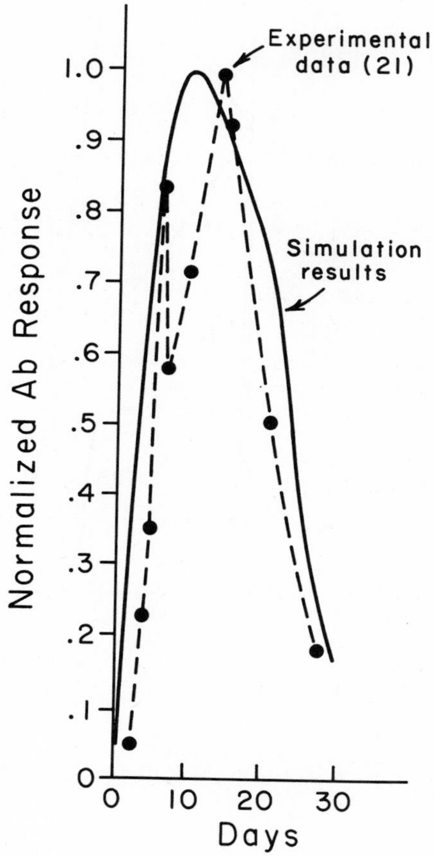

Figure 2. Comparison of Experimental and Simulation Time Responses of Blood Antibody.

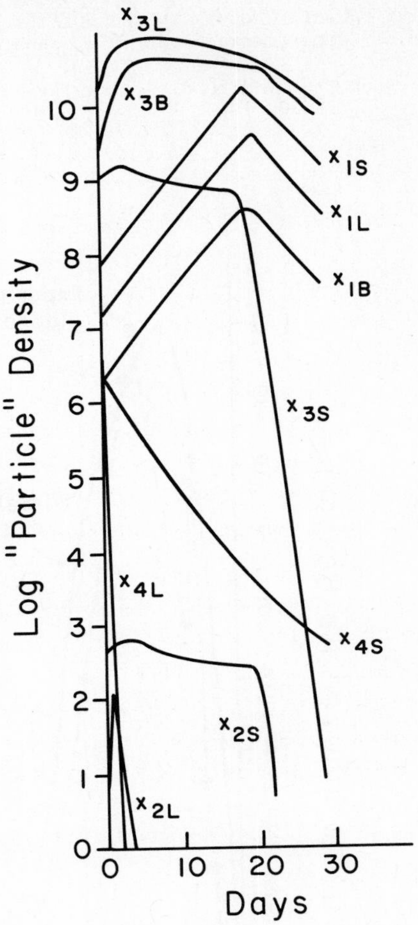

Figure 3. Simulated Compartmental States.

THE IgM-IgG SWITCH LOOKED AT FROM A CONTROL THEORETIC VIEWPOINT

Alan S. Perelson
Theoretical Division
University of California
Los Alamos Scientific Laboratory

Los Alamos, NM 87545/USA

The vertebrate immune system is a collection of molecules and cells designed to defend an animal from disease causing agents. When an *antigen* (a foreign molecule either in solution or on a cell) is introduced into an animal it stimulates a class of white blood cells, *B lymphocytes*, to proliferate and to produce protein molecules known as *antibodies*. Antibodies specifically bind to the antigen and lead to its elimination from the animal. One curious feature of the antibody response to an animal's first encounter with antigen (the primary immune response) is that under many circumstances two distinct types of antibody molecule are made, immunoglobulin M (IgM) and immunoglobulin G (IgG). If the amounts of IgM and IgG in the blood serum of an animal are measured as a function of time after injection of antigen, one finds that IgM appears in the blood serum first and IgG appears after some delay. Other analyses which I shall not discuss here have shown that single B lymphocytes first make IgM and then switch to the production of IgG. One further noteworthy feature of this switch is that both the IgM and IgG made by a single cell have the same specificity for antigen (see Figure 1).

In this paper I shall address the question: why should a cell make two different types of antibody with the same specificity for antigen? Further, why should a cell first make one type of antibody (IgM) and then switch to the production of another type (IgG) later in the immune response? Since biological systems are the result of millions of years of evolution by natural selection one might hypothesize that the IgM-IgG switch provides some advantage to an animal. In order to examine this possibility I, in collaboration with Byron Goldstein of Los Alamos Scientific Laboratory and Sol Rocklin of the University of California at Berkeley, have devised a model of the interaction of the immune system with a growing antigen (e.g., pathogen) and have attempted to optimize the performance of the immune system with respect to its antibody production. Although there are many pitfalls in using optimization arguments to predict the course of evolution (cf Perelson, Mirmirani

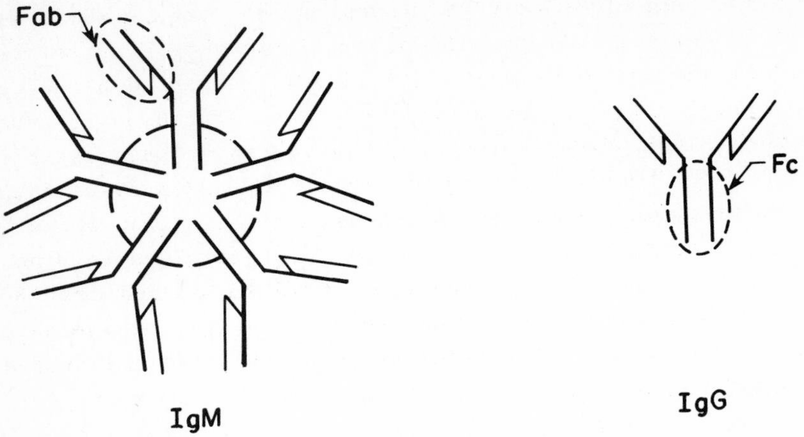

Figure 1. The structure of IgM and IgG. IgM contains five subunits joined by disulfide bridges (circular dotted line) and a J chain (not shown). Each subunit is similar to a single IgG. When a cell switches from IgM to IgG production, both immunoglobulins are believed to contain identical antigen binding fragments (Fab), but different complement binding fragments (Fc).

and Oster, 1976; Oster and Wilson, 1978), the evolutionary stability of the immune system and the magnitude of the selective forces acting on it give one some confidence in the predictions of optimization procedures.

There are many host defenses that act against disease causing organisms (cf Mims, 1976). Here I shall only discuss one defense mechanism, complement dependent lysis, which I have singled out because it relies on both IgM and IgG. Besides directly interacting with antigen to form large antibody-antigen aggregates, antibody also acts as a tag, marking cells as foreign. Once a cell is so marked it may be engulfed by a large migratory phagocytic cell such as a neutrophil or macrophage or it may be attacked by a series of eleven serum glycoproteins known as *complement*. The complement components literally drill a hole in the cell membrane leading to the death of the cell by osmotic forces (cf Mayer, 1973). Not all cells are susceptible to complement dependent lysis (some cells may be able to repair the damage to their membranes), but gram-negative bacteria and virus infected cells are among those which succumb to complement (Osler, 1976; Porter, 1971).

The cascade of binding reactions which eventually can lead to cell lysis is initiated by the first complement component, C1, binding to one IgM or a pair (or higher multiples) of IgG molecules in close proximity on the cell membrane (Borsos and Rapp, 1965a,b). Studies by Humphrey and Dourmashkin (1965) and Humphrey (1967) showed that for a red blood cell about 800 IgG molecules would be required to attach at random for there to be an even chance that two such molecules would be at adjacent sites. Thus, at least for red cells, it would at first sight seem best if the immune system secreted only IgM. For a pathogen smaller than a red blood cell, c, the critical number of IgG molecules required to bind in order to initiate the complement reaction would be smaller than 800, and should scale roughly as the ratio of the surface area of the pathogen to that of the red blood cell (assuming equal densities of antigenic determinants).

Although IgM and IgG have the same specificity for antigen they need not bind to a cell with equal efficiencies. As shown in Figure 1, IgG has two binding sites, while IgM being a pentamer has ten binding sites. Studies of the dynamics of red cell lysis by IgM and IgG in the hemolytic plaque assay have indicated that IgM rapidly binds and dissociates from the cell surface with an equilibrium constant indicative of single site interaction (Goldstein and Perelson, 1976; DeLisi, 1975a,b), whereas IgG is known to bind bivalently to surfaces (Hornick and Karush, 1972). If one assumes that IgG binds bivalently and IgM binds monovalently then the equilibrium constant for IgG binding may be as much as 10^6 times as great as that of IgM, implying that a cell in a solution containing equal concentrations of IgM and IgG would be much more likely to have 800 IgG molecules than 1 IgM molecule on its surface. When there exists such large differences in the binding constants of IgM and IgG it would seem advantageous if the immune system secreted only IgG. However, during the initial stages of an immune response there may not be enough antibody to put 800 molecules of IgG on each pathogen's surface and, in fact, with large infections this may take some time. During this initial period only IgM can lead to cell lysis and thus first producing IgM and the producing IgG may in fact be an optimal strategy.

Model

Consider two populations of lymphocytes, one of which, L_M, secretes only IgM, and the other of which, L_G, secretes only IgG. At any time t, I shall assume that a fraction u(t) of the L_M cells are proliferating with per capita rate b and the remaining fraction of L_M cells are differentiating with per capita rate d into L_G cells. In order not to bias the model, I shall also assume that a fraction, v(t), of L_G cells are proliferating at per capita rate b and the remaining fraction, 1 - v(t), are differentiating with per capita rate d into L_M cells (Figure 2).

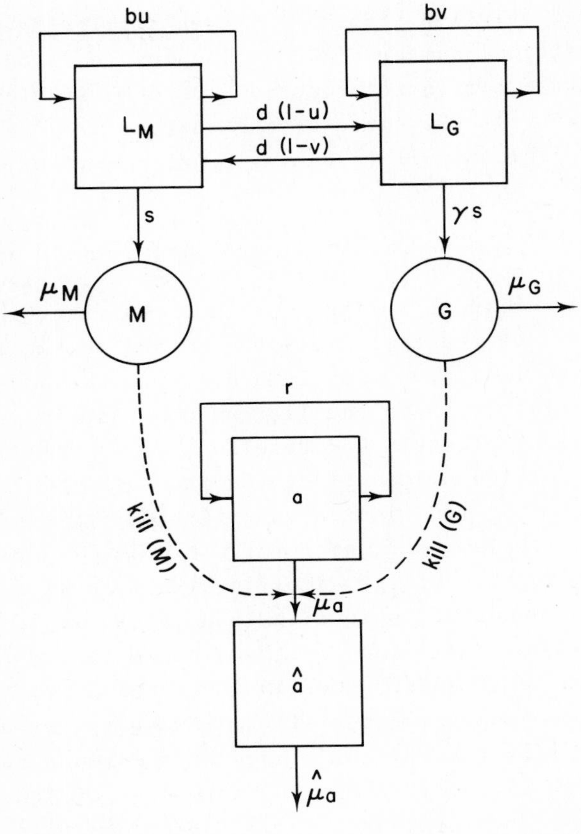

Figure 2. Block diagram of complement dependent killing. The dotted lines indicate portions of the model involving the binding of IgM and IgG to the antigen and the subsequent cell killing. Also the loss of IgM and IgG by the elimination of dead antigen is not shown.

I further assume L_M cells secrete IgM at rate s and L_G cells secrete IgG at a rate γs, where $\gamma \geq 1$ is introduced to account for the difference in size and complexity of the two antibodies. (For example, since IgM is a pentamer it may be possible to secrete 5 times as many IgG molecules as IgM molecules.) The antigen, a, I assume can grow at a per capita net rate r in the absence of antibody and complement and is killed at a per capita rate μ_a multiplied by p_{kill}, the probability of lysis by complement. Once complement has acted to kill a cellular antigen, the dead cell, $\hat{a}$, is assumed to remain in the system until it is removed, say by phagocytosis, at per capita rate $\hat{\mu}_a$.

To complete the model I need to specify how antibody binds to cells and the probability of a cell being killed by complement dependent lysis. I shall assume that each antigen (cell) has a total of $\tilde{\rho}_0$ sites at which antibody can bind, and the concentrations of sites bound by IgM and IgG on live and dead antigens are $\rho_M, \rho_G, \hat{\rho}_M$ and $\hat{\rho}_G$, respectively. As the antigen grows the total number of available sites in the system, $a\tilde{\rho}_0$, increases, but as dead cells are removed so are sites and antibodies bound to the dead cells. IgM is assumed to bind with only one of its ten sites while IgG is assumed to bind bivalently. The forward and reverse rate constants for the binding reactions of IgM and IgG are k_M, k'_M, k_G and k'_G, respectively. With this set of assumptions one obtains the following state equations:

$$\dot{L}_M = bu(t)L_M - d[1 - u(t)]L_M + d[1 - v(t)]L_G \tag{1}$$

$$\dot{L}_G = bv(t)L_G - d[1 - v(t)]L_G + d[1 - u(t)L_M \tag{2}$$

$$\dot{M} = sL_M - \mu_M M - \hat{\mu}_a\hat{\rho}_M \tag{3}$$

$$\dot{G} = \gamma sL_G - \mu_G G - \hat{\mu}_a\hat{\rho}_G/2 \tag{4}$$

$$\dot{a} = ra - \mu_a a\ p_{kill} \tag{5}$$

$$\dot{\hat{a}} = \mu_a a\ p_{kill} - \hat{\mu}_a\hat{a} \tag{6}$$

$$\dot{\rho}_M = 10k_M(M - \rho_M - \hat{\rho}_M)(\tilde{\rho}_0 a - \rho_M - \rho_G) - k'_M\rho_M - \mu_a\rho_M\ p_{kill} \tag{7}$$

$$\dot{\hat{\rho}}_M = 10k_M(M - \rho_M - \hat{\rho}_M)(\tilde{\rho}_0\hat{a} - \hat{\rho}_M - \hat{\rho}_G) - k'_M\hat{\rho}_M + \mu_a\rho_M\ p_{kill} - \hat{\mu}_a\hat{\rho}_M \tag{8}$$

$$\dot{\rho}_G = k_G(2G - \rho_G - \hat{\rho}_G)(\tilde{\rho}_0 a - \rho_M - \rho_G) - k'_G\rho_G - \mu_a\rho_G\ p_{kill} \tag{9}$$

$$\dot{\hat{\rho}}_G = k_G(2G - \rho_G - \hat{\rho}_G)(\tilde{\rho}_0\hat{a} - \hat{\rho}_M - \hat{\rho}_G) - k'_G\hat{\rho}_G + \mu_a\rho_G\ p_{kill} - \hat{\mu}_a\hat{\rho}_G \tag{10}$$

and probability of being killed by complement, p_{kill}, is given by

$$p_{kill} = p_M + p_G - p_M p_G \tag{11}$$

Here $p_M = p_M(\rho_M, a)$ and $p_G = p_G(\rho_G, a)$ are the probabilities of an antigen being killed with IgM and IgG, respectively. Since $\tilde{\rho}_M \triangleq \rho_M/a$ is mean number of IgM molecules bound per antigen, p_M, i.e., the probability of having at least one IgM bound to each antigen, is given by

$$p_M(\rho_M, a) = \sum_{i=1}^{\tilde{\rho}_0} \binom{\tilde{\rho}_0}{i} \left(\frac{\tilde{\rho}_M}{\tilde{\rho}_0}\right)^i \left(1 - \frac{\tilde{\rho}_M}{\tilde{\rho}_0}\right)^{\tilde{\rho}_0 - i} \approx 1 - \exp(-\tilde{\rho}_M) \tag{12}$$

while the probability of having at least c IgG molecules bound per antigen is

$$p_G(\rho_G, a) = \sum_{i=2c}^{\tilde{\rho}_0} \binom{\tilde{\rho}_0}{i} \left(\frac{\tilde{\rho}_G}{\tilde{\rho}_0}\right)^i \left(1 - \frac{\tilde{\rho}_G}{\tilde{\rho}_0}\right)^{\tilde{\rho}_0 - i} \tag{13}$$

where $\tilde{\rho}_G \triangleq \rho_G/a$ and $\tilde{\rho}_G \geq 2c$ for c IgG molecules to be bound. Using the De-Moivre Laplace theorem one can show

$$p_G \simeq \frac{1}{2} \operatorname{erf}\left(\frac{\tilde{\rho}_0 - \tilde{\rho}_G}{\sqrt{2\tilde{\rho}_G(1-\tilde{\rho}_G/\tilde{\rho}_0)}}\right) - \operatorname{erf}\left(\frac{2c - \tilde{\rho}_G}{\sqrt{2\tilde{\rho}_G(1-\tilde{\rho}_G/\tilde{\rho}_0)}}\right) \tag{14}$$

In deriving (12) and (14) I have relied on the fact that $\tilde{\rho}_0$ is typically 10^5 and hence much greater than $\tilde{\rho}_M$ or $\tilde{\rho}_G$ on live cells.

The optimization problem I wish to consider is minimize the time, i.e.,

$$\min_{u(\cdot),v(\cdot)} \int_0^T dt \tag{15}$$

to go from the initial state:

$$L_M(0) = L_{M0}, \quad L_G(0) = L_{G0}, \; M(0) = G(0) = 0, \quad a(0) = a_0$$
$$\hat{a}(0) = \rho_M(0) = \hat{\rho}_M(0) = \rho_G(0) = \hat{\rho}_G(0) = 0 \tag{16}$$

to the final manifold:

$$a(T) = a^* \qquad (\text{e.g.,} \quad a^* \leq 1 \text{ antigen/animal}) \tag{17}$$

subject to the dynamic constraints of Eqs. (1) - (10) and the static constraints

$$0 \leq u(t) \leq 1, \quad 0 \leq v(t) \leq 1, \quad t \in [0,T] \tag{18}$$

Results

If a_0 is too large the antigen grows without bound and the final manifold cannot be reached. For smaller values of a_0, using numerical techniques, I have compared the times needed to reach a* for the following strategies: 1) secrete only IgM, 2) secrete IgM and then switch to IgG secretion, 3) secrete only IgG, 4) secrete IgG and then switch to IgM secretion. Here I shall only report results for a typical set of biologically reasonable parameter values. A more complete discussion of results, including a study of parameter sensitivity, will be published elsewhere.

As a typical parameter set I have chosen $b = d = 0.1\ h^{-1}$, $s = 3.6 \times 10^6$ antibodies h^{-1}, $\gamma = 5$, $\mu_M = 0.03\ h^{-1}$, $\mu_G = 0.006\ h^{-1}$, $r = 0.5\ h^{-1}$, $\mu_a = 2.0\ h^{-1}$, $\hat{\mu}_a = 0.693\ h^{-1}$, $k_M = 3.6 \times 10^{-11}\ cm^3$ molecule^{-1} h^{-1}, $k_M' = 4.32 \times 10^5\ h^{-1}$, $k_G = 3.6 \times 10^{-11}\ cm^3$ molecule^{-1} h^{-1}, $k_G' = 1.44\ h^{-1}$, $c = 32$ and $\tilde{\rho}_0 = 4 \times 10^3$. Concentrations are expressed in molecules or cells per cm^3. I have assumed the immune response is occurring in a mouse with a serum volume of $1.25\ cm^3$. Further, c and ρ_0 have been chosen to represent a pathogen, such as a bacteria, with a surface area 1/25 that of a red blood cell. Using these parameters I show in Figure 3 how the final time T varies with the time t_s at which the control switches from $u = 1$, $v = 0$ to $u = 0$, $v = 1$ (IgM to IgG switch) for varying initial concentrations of a_0, with $L_{M0} = 1 \times 10^4$ cells/cm^3 and $L_{G0} = 0$. For $a_0 \geqslant 2 \times 10^{11}$ cells/cm^3 the antigen grows without bound.

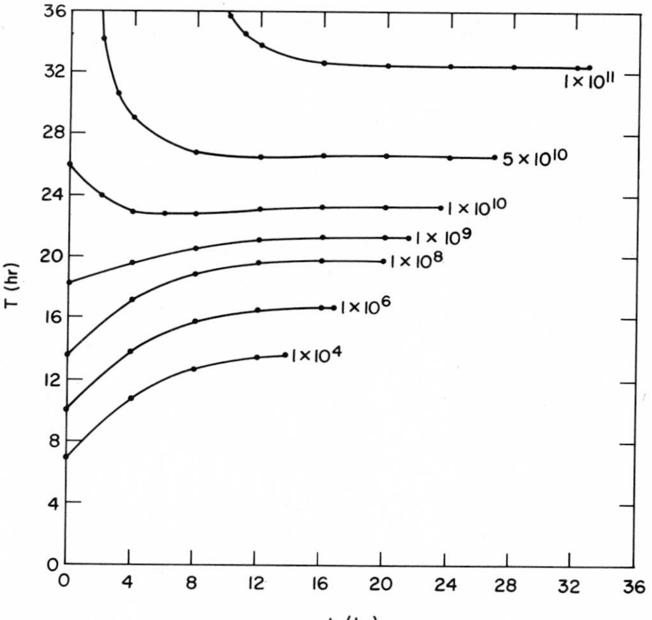

Figure 3. The final time vs the switching time for an immune response employing an IgM-IgG switch. The case of no switch, i.e., IgM production only, corresponds to the last point on each curve where $t_s = T$. Initially, $L_{M0} = 1 \times 10^4$ cells/cm^3, $L_{G0} = 0$, and a_0 was varied between 1×10^{11} cells/cm^3 to 1×10^4 cells/cm^3. The value of a_0 is indicated next to each curve in the figure.

With somewhat smaller values of a_0 the antigen can be destroyed only if the switch to IgG production is delayed beyond some critical time (e.g., approximately 1.6 h for $a_0 = 5 \times 10^{10}$ cells/cm^3). Thus early IgM production is crucial. For each antigen concentration there is some optimal time to switch to IgG production; which minimizes the total response time T. When $a_0 \leqslant 1 \times 10^9$ cells/cm^3 the optimal switching time is zero, whereas for $a_0 = 1 \times 10^{10}$, 5×10^{10}, and 1×10^{11} cells/cm^3 the optimal switching times are roughly 6 h, 12 h, and 20 h, respectively.

In Figure 4 I illustrate the effects of beginning an immune response with cells that secrete IgG and then switching at time t_s to the production of L_M cells. If $a_0 \geqslant 5 \times 10^{10}$ cells/cm^3 (not shown) then for a pure IgG response or for any choice of switching time the antigen grows without bound. For $a_0 = 6 \times 10^9$ cells/cm^3 the antigen can be controlled only if a switch to IgM production is made very early. When $a_0 \leqslant 1 \times 10^9$ cells/cm^3 IgG is sufficiently effective that switching to IgM production has no effect on the total response time. Another type

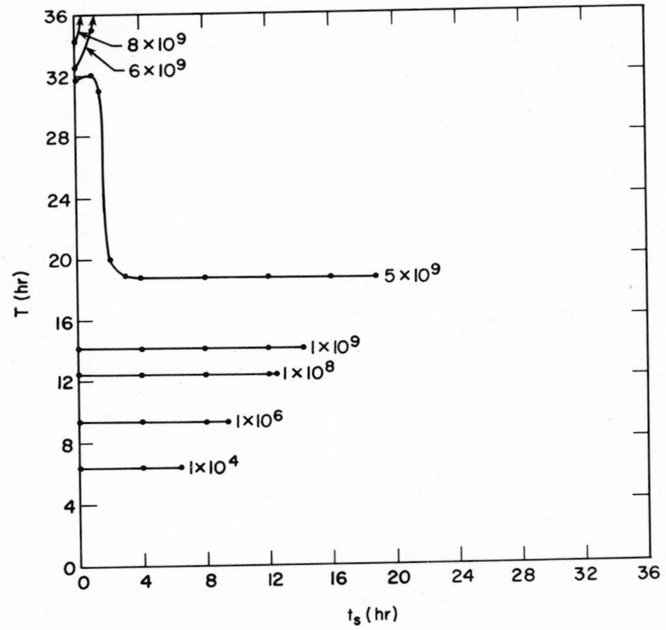

Figure 4. The final time vs the switching time for an immune response employing an IgG-IgM switch [i.e., $u(t) = 0$, $v(t) = 1$, $0 \leqslant t < t_s$; $u(t) = 1$, $v(t) = 0$, $t_s < t \leqslant T$] with $L_{M0} = 0$. $L_{G0} = 1 \times 10^4$ cells/cm^3 and various a_0 as indicated next to each curve on the figure.

of behavior occurs when $a_0 = 5 \times 10^9$ cells/cm³. Here a switch to IgM production at t = 0 considerably lengthens the response; the L_G population is being depleted, so killing by IgG is initially ineffective and there is a long delay ($\sim$ 12 h) before the L_M population is sufficiently large to prevent the antigen population from increasing. However, if the switch is delayed or if only IgG is secreted, then enough IgG is produced to quickly control the antigen.

Comparing Figures 3 and 4 one notices that for $a_0 < 1 \times 10^9$ cells/cm³ the total response time T is less for pure IgG immune responses than for responses which employ an IgM-IgG switch. Thus for "low" antigen doses it is better to employ a pure IgG response while for "high" antigen doses an IgM-IgG switch is better. In fact, employing a pure IgG response at "high" doses can be a fatal mistake. Here "high" and "low" doses are defined relative to the initial lymphocyte populations, L_{M0} and L_{G0}, since the ratio of bound antibodies to antigens is the crucial parameter in determining cell lysis. Thus if L_{G0} is large enough one would expect that a pure IgG response would be effective against all realizeable antigen concentrations and consequently would be a good strategy. However, if the initial lymphocyte population is low then it would seem best to employ an IgM-IgG switch since an animal may be confronted with a "high" antigen dose. In fact, this divergence in strategies is observed biologically. When the same antigen is encountered by an animal for a second time (the secondary response) the immune system has ready a large population of lymphocytes able to react with the antigen and the immune response is observed to be almost a total IgG response. In contrast, when an antigen is encountered by an animal for the first time (the primary response) a much smaller number of lymphocytes are able to react with the antigen and a switch in the type of antibody from IgM to IgG is usually observed.

Conclusions

For $a_0 \leqslant 10^{11}$ cells/cm³ and the other biologically reasonable parameter values used to generate Figures 3 and 4 one can draw the following conclusions:

1) It is better to begin an immune response with L_M cells rather than L_G cells if the antigen concentration is high ($a_0 > 1 \times 10^9$ cells/cm³).

2) Beginning with only L_M cells one can always reduce the time needed to eliminate the antigen by switching to IgG production at an appropriate time.

3) At high antigen doses switching from IgM to IgG production too early can allow the antigen to grow unbounded, switching too late or not at all only lengthens the response time.

4) At low antigen doses it is always better to begin the immune response with L_G cells. Switching these L_G cells into L_M cells provides no advantage to the animal.

If complement dependent killing of pathogenic organisms were an important defense strategy over evolutionary time, then it seems reasonable that natural selection would have led to the development of an IgM-IgG switch for the primary immune response, and an all IgG secondary response. Whether more complicated switching strategies or singular control would lead to an even more efficient response is not yet known.

Acknowledgements

This work was performed under the auspices of the U. S. Department of Energy.

References

Borsos, T. and Rapp, H. J. (1965a). J. Immunol. 95, 559-566.

Borsos, T. and Rapp, H. J. (1965b). Science 150, 505-506.

DeLisi, C. (1975a). J. Theor. Biol. 52, 419-440.

DeLisi, C. (1975b). J. Math. Biol. 2, 317-331.

Goldstein, B. and Perelson, A. S. (1976). Biophysical Chem. 4, 349-362.

Hornick, C. L. and Karush, F. (1972). Immunochem. 9, 325-340.

Humphrey, J. H. (1967). Nature 216, 1295-1296.

Humphrey, J. H. and Dourmashkin, R. R. (1965). In Ciba Found. Symp. Complement (eds. Wolstenholme, G.E.W. and Knight, J.) Churchill, London, pp. 175-189.

Mayer, M. (1973). Sci. Am. 229 (No. 5), 54-66.

Mims, C. A. (1976). The Pathogenesis of Infectious Disease, Academic Press, London.

Osler, A. G. (1976). Complement: Mechanisms and Function, Prentice Hall, Englewood Cliffs, New Jersey.

Oster, G. F. and Wilson, E. O. (1978). Ecology and Evolution of Castes in Social Insects, Princeton University Press, Princeton, NJ.

Perelson, A., Mirmirani, M. and Oster, G. (1976). J. Math. Biol. 3, 325-367.

Porter, D. D. (1971). Ann. Rev. Microbiol. 25, 283-290.

TWO-LEVEL OPTIMIZATION TECHNIQUES IN ELECTRIC POWER SYSTEMS

M.BIELLI, G.CALICCHIO, M.CINI, F.NICOLO'

Centro di Studio dei Sistemi di Controllo e Calcolo Automatici, National Research Council; Automatica Institute University, Via Eudossiana, 18
00184 Rome, Italy

ABSTRACT

In this paper a multi-area computer hierarchy particularly adapted to handle the problem of real-time intervention on electric power systems is proposed.

Concerning the active power subproblem, the economic dispatching for a given configuration of voltages and loads is examined. The overall optimization problem is formulated according to Lagrange, introducing the interaction constraints among the areas in the objective function. The resulting problem has a decomposable structure and can be solved by a two-level iterative computation, i.e. by goal coordination or interaction balance method.

For a given value of the Lagrange multiplier, local constrained quadratic optimizations are performed using the Beale algorithm. The supervisor computer in turn iteratively up-dates the multiplier to guarantee the interaction balance, using gradient like algorithms. The conditions of applicability and coordinability of the method are recalled and particularized for the economic dispatching problem.

To verify the effective applicability of the methodologies examined in this work, a worked out computation program is tested on a real electric network and the results for several coordination algorithms are discussed and compared.

1. INTRODUCTION

In electric power systems a number of decentralized hierarchical structures have been developed for state estimation for voltage and reactive power control and for economic dispatching. These hierarchical structures provide for the use of regional or departmental or area computers which are coordinated by a supervisor computer.

Essentially the advantages of this decentralization consist in the reduced amount of information to be centralized to the supervisor computer, in the possibility to use different calculation algorithms for the departmental computers independently of the coordination and in the reduction of calculation time for all the on-line operations.

Indeed it is true for the electrical power system the general remark that multi-level hierarchy is useful when a decentralized multicomputer configuration is used, while a pure mathematical decomposition of the problem on a centralized computer may not justify.

In this framework in the following we will consider the problem of optimal active power set point assignment with economic objective function and network steadystate constraints.

On the basis of a suitable network decomposition and a convenient choice of the interconnection variables between the subnetworks, the dispatching problem is decomposed into subproblems which are coordinated with the goal coordination method.

2. TWO-LEVEL ECONOMIC DISPATCHING

We suppose here that the reactive production problem has been solved on the basis of the reactive power balance equations. Therefore our preliminary hypothesis is that the voltage magnitudes at nodes are given values. This hypothesis is acceptable in on-line control, when the model parameters are computed at each control time interval on the basis of small variations of the operating conditions. Indeed the equations of active power balance are not sensitive too much to the voltage magnitude variations at the nodes and the reactive power equations depend weakly on the phase angles of the voltages (Billinton, 1971).

Let us consider the global network as an aggregation of nodes and lines with a power injection (positive or negative) in each node. We choose as intermediate variables the phase angles of the voltages at the nodes (see Fig. 1a).

Now we assume that a load flow problem has been solved for the network with constant and forecasted active loads so that we have a working point in the space of injections and phase angles. Then the optimization problem is to compute power injection variations such that an economic cost function is minimized with technical constraints on power generations and on power flows in the lines.

In order to solve the problem via a hierarchical structure we decompose the network into R subnetworks overlapped on interconnection tie-lines as shown in Fig. 1b.

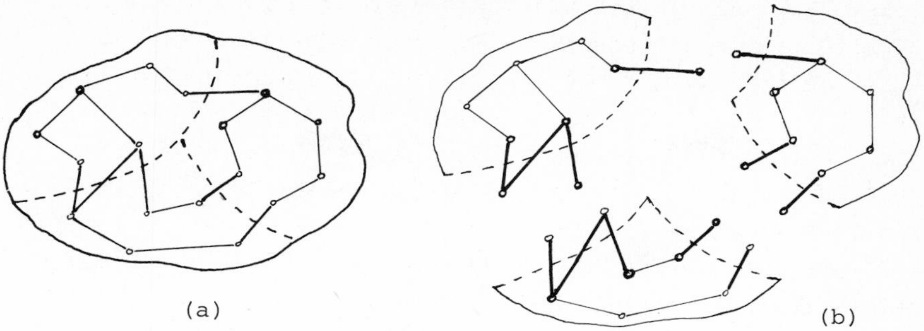

(a) (b)

Fig.1 Network decomposition

Then we associate to each area subnetwork a subsystem which has vector c^r, $(r = 1,\ldots,R)$ of the incremental power generations as control input, vector u^r $(r = 1,\ldots,R-1)$ of incremental phase angles at the external nodes of the tie-lines as interconnection input and vector y^r $(r = 1,\ldots,R)$ of incremental phase angles at the internal nodes of the tie-lines as output. The components of each vector c^r correspond to the generation nodes, of the relative area r, only. It is well known that a reference for the phase angle values must be chosen by fixing at zero the value of the phase at a node (whatever) defined slack. We choose the slack node of the network at one of the external nodes of area R, which is an internal node of area R-1, therefore the interconnection input u^R, for area R, is defined as the aggregation of the phase angle increments at the external nodes unless the slack node [Bielli et alii, 1977].

In view of applying the method to on-line control with short optimization time-interval, we can assume that the starting working point gives quasi-optimal operating conditions and a linearized steady state models can be used for the subsystems:

$$y^r = Q_r\, c^r + B_r\, u^r \qquad r = 1,\ldots,R \tag{1}$$

while technical constraints, via linearized expressions of intermediate variables, are given by linear inequalities:

$$S_r\, c^r + T_r\, u^r \leq d^r \qquad r = 1,\ldots,R \tag{2}$$

In formulas (1) and (2) Q_r, B_r, S_r, T_r and d^r are constant matrices and a constant vector respectively, of proper dimensions.

The above choice of the slack node introduces an equality constraint in area R-1:

$$y_{slack}^{R-1} = q_s^{R-1} c^{R-1} + b_s^{R-1} u^{R-1} = 0 \qquad (3)$$

where q_s^{R-1} and b_s^{R-1} are proper rows of Q_{R-1} and B_{R-1}.

Constraints (2) and (3) define bounded polyhedrons D^r in the space of (c^r, u^r) for $r = R-1, R$.

Further technical constraints can be obtained with global physical considerations such that polyhedrons D^r are bounded for $r=1,\ldots,R-2$ also, when $R > 2$. Then in the following we assume D^r bounded.

At last define vectors c,u,y as the aggregations of c^r, u^r, y^r respectively and D as the Cartesian product of D^r $(r = 1,\ldots,R)$.

The interconnection equation between areas (see example in fig.2) is:

$$u = My \qquad (4)$$

where M is a selection matrix with entries 0 or 1. Equation (4) is decomposed in:

$$u^r = \sum_{k=1}^{R} M_{rk}\, y^k \qquad (5)$$

and by (1) can be written as

$$u = M(Qc + Bu) \qquad (6)$$

where matrices Q and B are the aggregations of Q_r and B_r for $r=1,..,R$.

The global objective function to be minimized is:

$$f(c) = \sum_{r=1}^{R} f^r(c^r) \qquad (7)$$

where each f^r is the area cost generation function which we assume, as usual, to be quadratic.

The global dispatching problem is to minimize (7) on D with constraint (6).

In order to obtain a two-level iterative computation structure by goal coordination or interaction balance method (MESAROVIC 1970, TITLI 1972, FINDEISEN 1974) we build up a Lagrangian which takes into account of constraints (4) or (6):

$$L(c,u;\lambda) = \sum_{r=1}^{R} f^r(c^r) + \lambda[u - My] = \sum_{r=1}^{R} f^r(c^r) + \lambda[u - M(Qc+Bu)] \triangleq$$
$$\triangleq \sum_{r=1}^{R} L^r(c^r, u^r; \lambda) \qquad (8)$$

In our hypothesis values $(\hat{c},\hat{u})$ solution of the primal problem of minimizing (7) on $D = D^1 \times D^2 \times \ldots \times D^R$ and with constraint (6) coincide with those of the saddle point $(\hat{c},\hat{u},\hat{\lambda})$ of Lagrangian (8) restricted on D.

Indeed the saddle point exists because the primal problem has solution, f^r are convex functions on convex sets D^r and all constraints satisfy the reverse convex constraint qualification (Mangasarian , 1969), owing to their linearity. The job of the first level is to minimize (8) on D^r $(r=1,\ldots,R)$ for a given value of λ and the coordination job of the second level is to find λ such that the function:

$$h(\lambda) = \min_{(c,u)\in D} L(c,u;\lambda) \tag{9}$$

is maximized. The existence of the saddle-point guarantees the applicability of this kind of price coordination method.

In order to perform the task of the first level each area computer needs only to know the local parameters of the network, the values of the multiplier λ and the interaction selection matrix M.

3. COORDINABILITY OF THE DISPATCHING PROBLEM.

With the above formulation our global optimization problem can be splitted in the lower level problems:

$$\min_{(u^r,c^r)\in D^r} L^r(c^r,u^r;\lambda) \qquad r = 1,\ldots,R \tag{10}$$

and the upper level problem:

$$\max_{\lambda} \left[h(\lambda) = \min_{(u,r)\in D} L(c,u;\lambda) \right] \tag{11}$$

We say that the problem is price-coordinable if the dual function $h(\lambda)$ is differentiable and its gradient coincides with the grandient of L with respect to λ.

If the problem is coordinable the job of second level can be tackled by gradient type algorithms; the gradient of $h(\lambda)$ is $[u-My]$ and information on the values of u and y is the only needed from the first level local optimizations.

In our case coordinability does not hold unless for very special configuration of area interconnection: values of λ exist for which local problems (10) have not unique solution and $[u-My]$ is not a constant vector over the set of these solutions, then conditions for differentiability of $h(\lambda)$ are violated (Lasdon, 1972). Non uniqueness is due to the linearity of L^r with respect to u^r together with either the linearity of bounds of D^r or the existence of stationary values of L^r independent from a few components of u^r for some values of λ. Moreover an inspection of the problem shows that in general at the saddle point $h(\lambda)$ is not differentiable, then coordinability does not hold (see example below). In particular cases this remark may suggest to search maximum of $h(\lambda)$ by exploring only subspaces of the space of λ.

In the general case , convergence by gradient type techniques may be gained by introducing slight local cost on each interaction input u^r; then we have new local problems with strictly convex Lagrangians L^r_ε:

$$\min_{(c^r,u^r)\in D^r} L^r_\varepsilon(c^r,u^r;\lambda) = \min_{(c^r,u^r)\in D^r} [L^r(c^r,u^r;\lambda)+\varepsilon\|u^r\|^2] \quad r=(1,\ldots,R) \qquad (12)$$

This ε-convexification guarantees (Foord, 1975) in our conditions that the sequence of solutions of max min problem (the saddle point), in correspondence to a sequence of ε_K $(K = 1,2,\ldots)$ converging to zero, converges to the solution of max min problem without convexification.

The above discussion can be illustrated by a simple example. Let us have two areas with tie-lines and one power generation node in each area (Fig.2). In this case the area subproblems give:

$$h^r(\lambda) \triangleq \min_{(c^r,u^r)\in D^r} L^r(c^r,u^r;\lambda) = \min_{(c^r,u^r)\in D^r} \left[f^r(c^r)+\tilde{b}^r(\lambda)u^r_1+\tilde{q}^r(\lambda)c^r\right] \qquad (13)$$
$$r = 1,2$$

where D^r are of the type of fig.3 and $\tilde{b}^r(\lambda), q^r(\lambda)$ are linear omogeneous.

Thus for the values of λ^k such that $\tilde{b}^r(\lambda^k) = 0$, $h^r(\lambda^k)$ and then h^1+h^2, is not differentiable because local optimal $u^r(\lambda^k)$ is between $\bar{u}^r(\lambda^k)$ and $\underline{u}^r(\lambda^k)$ for each local optimal $c^r(\lambda^k)$ (with k=1,2, see Fig.3) and vector u-My has not constant value when u^r $(r = 1,2)$ ranges these intervals, while other variables are constant.

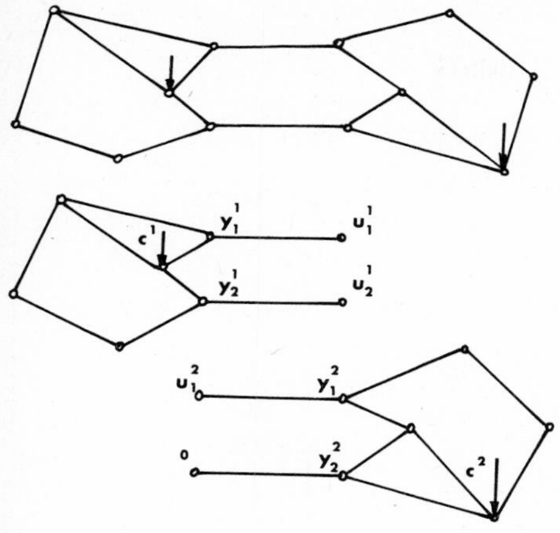

Fig.2 Example . Two area interconnection variables.

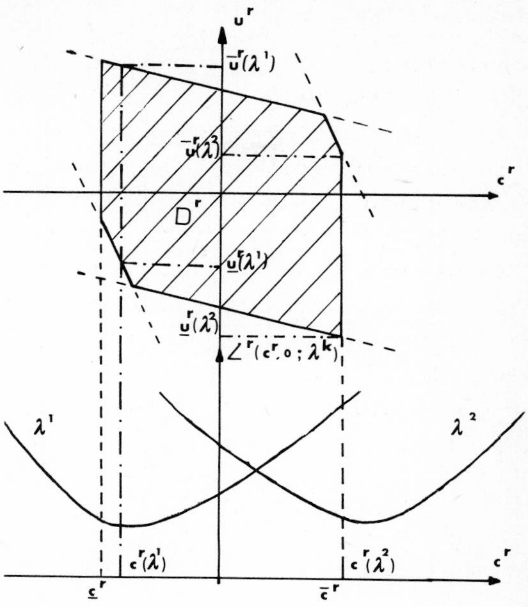

Fig.3 Example . Non differentiability of dual function $h(\lambda)$.

4. CASE STUDY

We applied the method to a real test network of 36 nodes, 49 lines and 4 transformers derived from a subnetwork of the Italian electrical distribution system.

Then we decomposed the network into two areas interconnected by two tie-lines. The first area is composed by 11 passive nodes and 2 active power generations, the second one by 20 and 3 respectively.

On the basis of data that in the actual implementation would be measured via local state estimators of the two subnetworks (i.e. operating conditions, forecasted loads and electrical parameters) matrices Q_r and B_r [see(1)] have been computed.

Following the notations introduced in previous section in Table I and II all the formulas specified for the case under study are written down.

Then we worked out a computer package for a two-level optimization procedure simulating the decentralized structure described above; the flow chart is in Fig.4.

The first level optimizers use Beale algorithm for quadratic programming (Land, 1973).

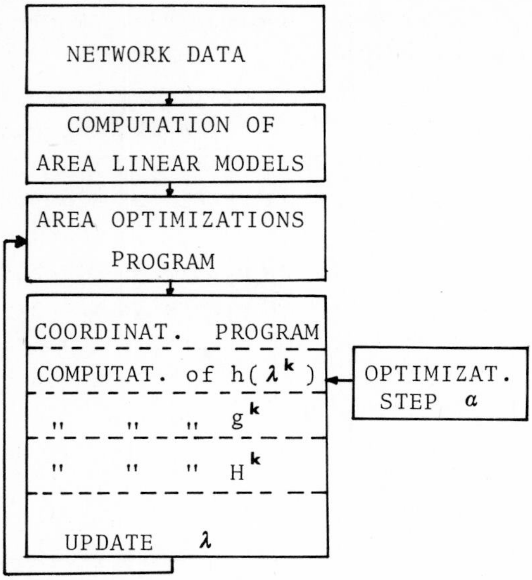

Fig.4 Program flow chart of case study .

The coordinability achieved by ε-convexification allows to use gradient algorithms in the second level optimization (Fletcher,1969). After testing several gradiental or quasi-Newton techniques we found the following algorithm suitable for our second level problem.

We shown that the gradient of the dual function $h(\lambda)$ is:

$$g(\lambda) = [\tilde{u}(\lambda) - M\tilde{y}(\lambda)] \tag{14}$$

where $\tilde{u},\tilde{y}$ are the solutions of lower problems for a given λ. Then at the iteraction k the new value of λ is computed by:

$$\lambda^{k+1} = \lambda^k + \alpha^k H^k g^k \tag{15}$$

where $g^k = g(\lambda^k)$ and α^k is computed minimizing with respect to scalar α the function

$$\Phi^k(\alpha) = h(\lambda^k + \alpha H^k g^k) \tag{16}$$

The recurrence formula that updates the inverse Hessian estimate H^k is:

$$H^{k+1} = \begin{cases} H^k - \dfrac{H^k \gamma^k \gamma^{k^T} H^k}{\gamma^{k^T} H^k \gamma^k} & \text{for } k \neq n \dim \lambda \\ & n = 1,2,\ldots \\ E & \text{for } k = n \dim \lambda \end{cases} \tag{17}$$

where $\gamma^k = g^{k+1} - g^k$, and E is the identity matrix.

In Figs. 5 and 6 the behaviour of dual function $h(\lambda)$, of Euclidean norm of the gradient $\|g\|$ and of components of multiplier λ, versus iterations of upper level algorithm is shown.

First we applied the method without the ε-convexification of cost functions, starting from zero values of Lagrange multipliers ($\lambda_1 = \lambda_2 = \lambda_3 = 0$), i.e. with zero price on the unbalance of the interconnection (Fig.5).

As expected these values proved to be a very bad estimate of optimal $\hat{\lambda}$, so that, during the first iterations the value of $h(\lambda)$ is very low and that of $\|g\|$ very high. In spite of this bad estimate, the algorithm in a few iterations leads to values of λ and $h(\lambda)$ near the optimal ones but to values of $\|g\|$ still far from zero.

In the following iterations the algorithm improves $h(\lambda)$ slowly while $\|g\|$ decreases fastly, showing a great sensitivity with respect to λ.

At the end the algorithm finds values of λ such that the coefficient of variable u_1^2 is zero, corresponding to a point of non differenziability of $h(\lambda)$, and is unable to head straight for optimum.

Afterwards we tested the method with the ε-convexification with values $\varepsilon_1 = 10^{-2}$, $\varepsilon_2 = 10^{-3}$, $\varepsilon_3 = 10^{-4}$ and $\varepsilon = 10^{-5}$. The convexification is meaningful when a reasonable estimate of λ is got, so in fig.6 (for $\varepsilon = 10^{-3}$) plot is shown for the last iterations only.

Inspection of plot shows that convergence is achieved (with $\|g\| \leq 10^{-6}$ degrees). As expected ε_k-optimal solutions converge as ε goes to zero, i.e. solution with $\varepsilon = 10^{-4}$ coincides with that shown for $\varepsilon = 10^{-3}$, while lower values of ε give quasi convergence situations wimilar to that with $\varepsilon = 0$, because of computation errors.

At last some remarks can be done about the other second-level algorithms tested. Regarding Fletcher-Reeves and Kelley-Myers algotithms we found their convergence extremely slow, so they seem not convenient for this problem. The DFP and Rank 1 algorithms have convergence properties similar to those of the algorithm proposed in (17), i.e. they attain the optimum in about the same number of iterations. Their disadvantage consists in the difficulties of optimizing step α^k, because its optimal value ranges on a broad interval during the iterations and, as a consequence, a large number of evaluations of function (16) is need.

On the contrary in the proposed algorithm optimal step α^k shows smaller variations which facilitates the linear search for its optimization.

REFERENCES

M.BIELLI, G.CALICCHIO, M.CINI (1977). *Application of multilevel techniques to economic despatch of electrical power systems.* IFAC Symposium on Automatic Control and protection of electric power Systems, Melbourne.

W.FINDEISEN (1974). *Wielopozionowe uklady sterowania.* Warszawa.

R.FLETCHER (1969). *A review of methods for unconstrained optimization.* Optimization. E.R.Fletcher, Academic Press, London.

A.G.FOORD (1975). *Eliminating instabilities in price coordination or balance methods.* Workshop Discussion on Multilevel Control. Technical University, Warsaw.

A.H.LAND, S.POWELL (1973). *Fortran Codes for Mathematical Programming.* Jhon Wiley & Sons.

L.S.LASDON (1972). *Optimization Theory for Large Systems.* MacMillan C. N.Y.

O.L.MANGASARIAN (1969). *Non linear programming.* McGraw-Hill Publ. C., New Delhi.

M.D.MESAROVIC, D.MACKO, Y.TAKAHARA (1970). *Theory of Hierarchical, Multilevel, Systems.* Academic Press, N.Y. & London.

A.TITLI (1972). *Contribution a l'ètude des structures de Commande Hiérarchisées en une de l'optimisation des processus complexes.* Thèse Université de Toulose.

$$\left| u^1 \right| = \begin{vmatrix} u_1^1 \\ u_2^1 \end{vmatrix} \qquad \left| c^1 \right| = \begin{vmatrix} c_1^1 \\ c_2^1 \\ c_3^1 \end{vmatrix} \qquad \left| y^1 \right| = \begin{vmatrix} y_1^1 \\ y_2^1 \end{vmatrix}$$

Linear model around the working point

$$I_1^1 = 233.339 + c_1^1$$
$$I_2^1 = 575.083 + c_2^1 \qquad \text{total injection in megawatt}$$
$$I_3^1 = 112.514 + c_3^1$$

$$\psi_1^1 = 5.266 + y_1^1$$
$$\psi_2^1 = 1.559 + y_2^1 \qquad \text{total angle in degrees}$$

$$\begin{vmatrix} y_1^1 \\ y_2^1 \end{vmatrix} = \begin{vmatrix} .01760448 & .01742711 & .0318928 \\ .01829959 & .01807809 & .01338381 \end{vmatrix} \left| c^1 \right| + \begin{vmatrix} .424054 & .575943 \\ .177954 & .622043 \end{vmatrix} \left| u^1 \right|$$

$$f^1(c^1) = 0.5 \left| c^1 \right|^T \begin{vmatrix} .66 & .00 & .00 \\ .00 & .73 & .00 \\ .00 & .00 & 4.20 \end{vmatrix} \left| c^1 \right| + \begin{vmatrix} 18\,000 & 0 & 0 \\ 0 & 15\,700 & 0 \\ 0 & 0 & 16\,000 \end{vmatrix} \left| c^1 \right| + 232\,842$$

initial cost value = 1'900'358 it. lire x hour

$$\mathcal{L}^1(c^1, u^1; \lambda) = f^1(c^1) + .5\,\varepsilon \left\| u^1 \right\|^2 - \lambda_3 y_1^1 + \lambda_1 u_1^1 + \lambda_2 u_2^1$$

technical constraints on power generations	1	0	0	0	0	c_1^1	≤	300
	-1	0	0	0	0		≤	- 60
	0	1	0	0	0	c_2^1	≤	588
	0	-1	0	0	0		≤	-240
	0	0	1	0	0	c_3^1	≤	140
	0	0	-1	0	0		≤	- 70
slack node	.017604	.017427	.031893	.42405	.57595		=	0
power transm. constr. on tie-lines	.017604	.017427	.031893	-.57595	.57595	u_1^1	≤	14.907
	-.017604	-.017427	-.031893	.57595	-.57595		≤	-5.093
	.018299	.018078	.013384	.17795	-.17795	u_2^1	≤	2.796
	-.018299	-.018078	-.013384	-.17795	.17795		≤	1.204

Table 1: - Area 1

$$|u^2| = |u_1^2| \qquad |c^2| = \begin{vmatrix} c_1^2 \\ c_2^2 \end{vmatrix} \qquad |y^2| = \begin{vmatrix} y_1^2 \\ y_2^2 \end{vmatrix}$$

Linear model around the working point

$$I_1^2 = 215.468 + c_1^2 \qquad \text{total injection in megawatt} \qquad \psi_1^2 = 10.173 + y_1^2 \qquad \text{total angle in degrees}$$

$$I_2^2 = 63.892 + c_2^2 \qquad \psi_2^2 = 2.355 + y_2^2$$

$$\begin{vmatrix} y_1^2 \\ y_2^2 \end{vmatrix} = \begin{vmatrix} .04585006 & .00912333 \\ .00870289 & .02094112 \end{vmatrix} |c^2| + \begin{vmatrix} .384792 \\ .883225 \end{vmatrix} |u^2|$$

$$f^2(c^2) = .5 \, |c^2|^T \begin{vmatrix} .90 & .00 \\ .00 & 2.80 \end{vmatrix} |c^2| + \begin{vmatrix} 1.500 & 0 \\ 0 & 2.100 \end{vmatrix} |c^2| + 113.340$$

initial cost value = 597.322 it. lire x hour

$$\mathcal{L}^2(c^2, u^2; \lambda) = f^2(c^2) + .5\,\varepsilon \,\|u_1^2\|^2 + \lambda_3 u_1^2 - \lambda_1 y_1^2 - \lambda_2 y_2^2$$

technical constraints on power generations	-1	0	0	c_1^2	≤	- 60
	1	0	0		≤	330
	0	-1	0	c_2^2	≤	- 20
	0	1	0		≤	95
power transm. constr. on tie-lines	-.045850	-.009123	.38479	u_1^2	≤	-14.907
	.045850	.009123	-.38479		≤	5.093
	.008703	.020941	.11677		≤	1.204
	-.008703	-.020941	-.11677		≤	- 2.796

Table 2: - Area 2.

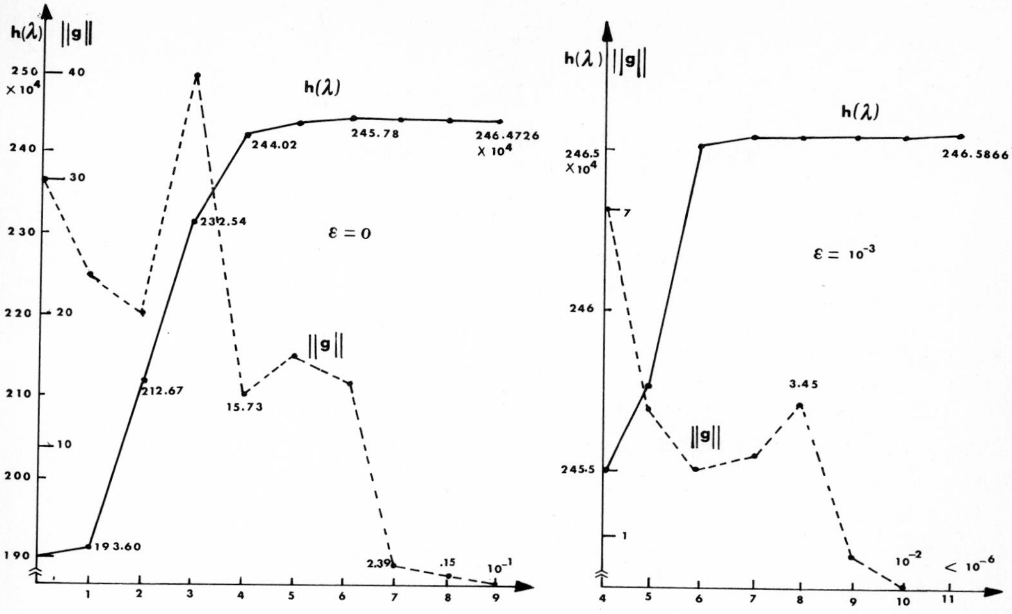

final solutions in megawatt	
$I_1^1 = 231.63$ $I_2^1 = 496.07$ $I_3^1 = 74.50$	$I_1^1 = 256.02$ $I_2^1 = 517.99$ $I_3^1 = 79.22$
$I_1^2 = 330.00$ $I_2^2 = 20.00$	$I_1^2 = 330.00$ $I_2^2 = 20.00$

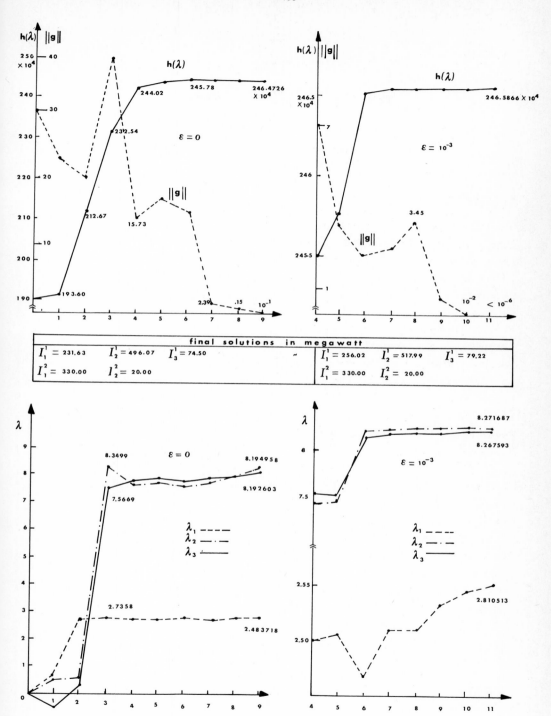

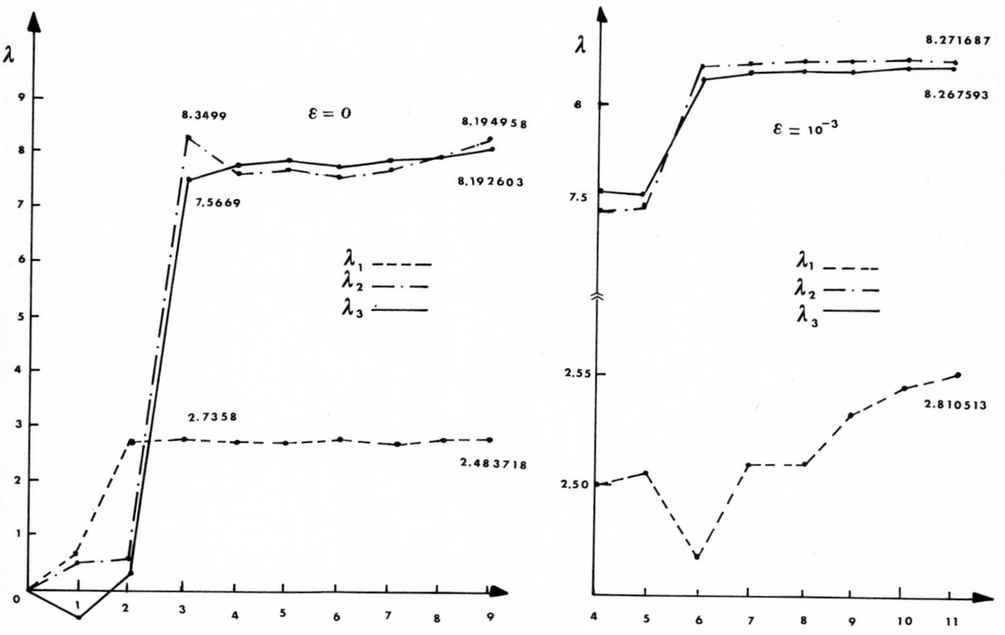

Fig. 5) and 6) - Iteration-by-iteration plots.

Multiobjective Programming and Siting of Industrial Plants

G. Halbritter

Abteilung für Angewandte Systemanalyse
Kernforschungszentrum Karlsruhe, Postfach 3640
7500 Karlsruhe

Introduction

Investigations of the outdoor pollutant concentration within the Upper Rhine Region raised the question for a more appropriate, perhaps an optimal distribution of sites for industrial plants. Which criteria, however, are the relevant ones to define such an optimum? Economical and ecological aspects should at least be taken into account.

Regarding the economical aspects minimum costs could be the determining factor and as far as the ecological aspects are concerned a minimum outdoor pollutant concentration on the population can be taken as the main objective. So we have a problem of multiobjective optimization. For these problems only so-called Pareto-optimum-solutions will be obtained, none of them clearly preferable to the others.

The calculations for "optimal" siting were done for the example of power stations within the Upper Rhine Region, i.e. the Upper Rhine Valley from Mannheim to Kehl. It is assumed that the energy is generated by means of fossil fuels implying the emission of the sulphur dioxide (SO_2). The pollutant concentrations caused by the emission sources are considered in a field point grid extending 60 km in the west-east and 120 km in the north-south directions. The source point grid, containing the eligible, preestablished sites, lies within the field point grid. There are 108 source points. Care was taken that the main regions exposed to pollutant concentration from sources located at the periphery of the source point grid do not lie outside the field point grid. A standard power station unit of 100 MWe is taken as a basis. The assumption that the burnt fuel oil (S) contains 2 wt.% of sulphur yields a sulphur dioxide (SO_2) emission of about 0.9 t/h and a heat emission from the stack of about 3 x 10^3 kcal/sec. The stack height is taken to be 150 m. A total energy of 5 GW_e is to be generated within the region. The environmental standard for normal areas is 140 μg SO_2/m^3 in the FRG, this standard is to be observed.

Cost Minimization Model

With the help of the Cost Minimization Model we search for the lowest costs distribution of power station units and heating power station units in the source point grid. In a first approximation the following costs can be considered as site specific:

1) Costs for secondary energy transport systems to the nearest centers of consumption (e.g., transmission lines, distant heat transport lines, pipelines).

2) Costs for a cooling water transport system to the nearest main canal.

The site-specific costs for the site j are obtained by multiplication of the cost function per unit distance $f_1(x_j)$ and $f_2(x_j)$, respectively, by the respective distance from the nearest center of consumption (cost fraction 1) and the nearest main canal (cost fraction 2), respectively. The degressive development of costs raises a problem of non-linear programming:

$$\min \left\{ \sum_{l=1}^{p} \sum_{j=1}^{n} D_{lj} f_1(x_j) + \sum_{j=1}^{n} E_j f_2(x_j) \right\}$$

subject to the constraints

$$\sum_{j=1}^{n} T_{ij} \cdot x_j \leq b_i \qquad i = 1,\ldots, m$$

$$\sum_{j=1}^{n} Tl_{lj} \cdot x_j \geq bl_l \qquad l = 1,\ldots, p$$

$$x_j \geq 0 \qquad j = 1,\ldots, n$$

where

x_j occupation number of the source point j by standard power stations and standard heating power stations, respectively

$f_1(x_j)$ cost function for the secondary energy transport per unit distance of the site j with x_j standard units installed

$f_2(x_j)$ cost function for cooling water transport per unit distance of the site j with x_j standard units installed

D_{lj} matrix element expressing the distance from the location of energy generation j to the center of consumption l (l=1,...,p; j=1,..., n)

E_j vector component expressing the distance of the point of energy generation j from the nearest main canal (j=1,..., n)

T_{ij} element of the environmental transfer matrix T(m x n), describing the influence of a specific emission (emission per occupation number x_j) at the point j on the outdoor pollutant concentration at the field point i

$T1_{lj}$ element of the technical transfer matrix T1(p x n) describing the possible contribution of a standard power station and standard heating power station, respectively, to the total supply of electricity and heat, respectively, of the center of consumption l

b_i environmental quality standard to be observed at the field point i

bl_l minimum production of electricity and heat, respectively, for the center of consumption l.

The first m constraints ensure compliance with the environmental quality standards. The following p constraints ensure the minimum energy generation for the p centers of consumption in terms of electric current and heat. The environmental transfer matrix T is determined by means of diffusion calculations. The elements T_{ij} of this transfer matrix describe the influence of a standard source at the point j of the source point grid on the point i of the field point grid. The elements of the technical transfer matrix $T1_{lj}$ decribe the contribution of a standard power station and a standard heating power station, respectively, at the point j of the source point grid to the current and heat generation, respectively, of the center of consumption 1.

Minimum Impact to Population Model

We search for the occupations of power stations and heating power stations in the given source point grid, which involves the lowest impact of outdoor pollutant concentration on the population.

In this model the siting should achieve a minimization of the weighted pollutant concentrations $p_i \cdot x_i$ at the field points i of the region while complying with the environmental standards b_i and with a minimum production level bl_l in the subregion l. Weighting is done proportional to the density of population. The minimization of the impact to population in addition to compliance with the environmental standards can be justified as follows: Although in the environmental standards the findings of industrial medicine are considered, these standards are, on the whole, the result of political privisions in which also economical requirements play a role. Environmental standards are no threshold for the non-occurence of damage. Therefore, besides the observation of standards for individual persons, minimizing of the total risk for the population should be achieved.

The following problem arises:

$$\min \left(\sum_{i=1}^{m} p_i \cdot x_i \right)$$

$$x_i = \sum_{j=1}^{n} T_{ij} \cdot x_j$$

thus

$$\min \left(\sum_{i=1}^{m} p_i \cdot \sum_{j=1}^{n} T_{ij} \cdot x_j \right)$$

subject to the following constraints

$$\sum_{j=1}^{n} T_{ij} \cdot x_j \leq b_i \qquad \text{for all values } i=1,\dots, m$$

$$\sum_{j=1}^{n} Tl_{lj} \cdot x_j \geq bl_l \qquad \text{for all values } l=1,\dots, p$$

$x_j \geq 0$ for all values j=1,..., n

where

p_i weighting of the field point i according to the density of population

x_i outdoor pollutant concentration at the field point i.

Model for Calculating Compromise Solutions

The occupation vector $\underline{x}$ must be found for a given site grid, i.e., the number of standardized facilities (e.g. 100 MW_e power stations) at given grid points with the best possible achievement of the following objective concepts:

1) minimum costs for the facilities,

2) mimimum impact by pollutants to the population.

These objective concepts shall be optimized subject to the following constraints:

a) The environmental standards (long-term and short-term standards) have to be observed at all points of the region,

b) a minimum production level, e.g., of energy generation in the region, must be maintained.

The problem known as the vector maximum problem can be represented as follows:

Def. 1:

$$\left\{ \begin{array}{c} c_1(\underline{x}) \\ c_2(\underline{x}) \\ \vdots \\ c_k(\underline{x}) \end{array} \quad \underline{x} \in X \right\} \quad k = 1,\ldots, K$$

with $X = \{\underline{x} | A \cdot \underline{x} \leq \underline{b}, \underline{x} \geq 0\}$ convex polyhedron in R^n

$$\underline{x} = \begin{pmatrix} x_1 \\ \vdots \\ x_n \end{pmatrix}$$ vector of strategies (occupation numbers)

$c_1(\underline{x}),\ c_2(\underline{x}) \ldots,\ c_k(\underline{x})$ objective functions

For the general case of conflicting objectives no vector of strategies $\underline{x}$ will be found reaching all goals at the same time. Therefore, so-called efficient objective vectors $\underline{c}(\underline{x})$ will be searched, these objective vectors being Pareto-optimum meaning that for a transition from $\underline{c}(\underline{x})$ to another admissible objective vector $\underline{c}(\underline{x}')$ never holds $\underline{c}(\underline{x}') \geq \underline{c}(\underline{x})$ in other words, starting from an efficient objective vector, no higher level can be attained, for all objective functions at the same time. Vectors of strategies $\underline{x}$ yielding the efficient objective vectors are called functionally efficient which means that there is no vector x' having the property $\underline{c}(\underline{x}') \geq \underline{c}(\underline{x})$. The amount of all functionally efficient vectors of a vector maximum problem is called the complete solution of the vector maximum problem /DINKELBACH (1969)/. For practical problems the complete solution of the vector maximum problem can mostly not be determined. For the given problem, i.e., the siting of large-scale technical facilities, two methods were used in order to obtain efficient solutions:

(1) Maximization of the sum of goal achievements of the individual objectives and
(2) maximization of a common minimum goal achievement for all objectives.

The conflicting objective is not solved by a uniform model but by the following single steps using the models already described.

1) Determination of scaling for goal achievement by the individual objectives in the space of solutions.

2) Determination of compromise solutions for goal achievement by the individual objectives using different approaches.

3) Evaluation of compromise results obtained.

Scaling (single step 1) results from the determination of the most favorable and most unfavorable solutions for each objective function in the space of solutions defined by the system of constraints. The solutions of the problems of minimization described in model 1 and 2 yield the most favorable solutions, the so-called scalar maxima $c_1(\underline{x}^*) = f^{01}$ and $c_2(\underline{x}^*) = f^{02}$.

The most unfavorable solutions, the so-called scalar minima $c_1(\underline{\bar{x}}_1) = f_{01}$ and $c_2(\underline{\bar{x}}_2) = f_{02}$ are obtained from the respective maximum problems of models 1 and 2. The differences between the scalar maximum and the scalar minimum are mapped to the interval /0, 1/ and the goal achievements for each solution can be found on the scales so obtained.

The second single step comprises application of the two procedures to find compromise solutions.

Method (2) yields the greatest possible goal achievement which can be obtained simultaneously for both objective concepts. This minimum goal achievement will be exactly applicable, at least for one objective concept, and can be exceeded for further objective concepts. Thus, the method (2) is an achievement of individual goals having equal weights. The following expressions are obtained for the problem of siting:

(1) $\max\ (v_1 + v_2)$

subject to the following constraints

$$c_1(\underline{x}) - (c_1(\underline{x}_1^*) - c_1(\overline{\underline{x}})) \cdot v_1 \leq c_1(\overline{\underline{x}}_1)$$

$$c_2(\underline{x}) - (c_2(\underline{x}_2^*) - c_2(\overline{\underline{x}})) \cdot v_2 \leq c_2(\overline{\underline{x}}_2)$$

$\underline{v} = (v_1, v_2)^T$ vector of individual goal achievements

(2) $\max\ v$

subject to the following constraints

$$c_1(\underline{x}) - (c_1(\underline{x}^*) - c_1(\overline{\underline{x}})) \cdot v \leq c_1(\overline{\underline{x}}_1)$$

$$c_2(\underline{x}) - (c_2(\underline{x}^*) - c_2(\overline{\underline{x}})) \cdot v \leq c_2(\overline{\underline{x}}_2)$$

v minimum goal achievement

where

$\underline{x} \in X,\ X = \{\underline{x} | T \cdot \underline{x} \leq \underline{b} \wedge T1 \cdot \underline{x} \geq \underline{b1} \wedge \underline{x} \geq \underline{0} \wedge v$ bzw. $\underline{v} \geq 0\}$

In the first lines of (1) and (2) the individual goal achievements and the common minimum goal achievement, respectively, are maximized.

Method 1 corresponds to an equal weight addition of both value functions (single goal achievements) to form a common benefit function. However, the criterion of optimality does not ensure that very different single goal achievements might be obtained.

Method 2 does not correspond to a direct aggregation of individual objectives. The approach according to the theory of games - choice of strategy without knowing the strategies of the opponent-guarantees equal consideration of individual objective concepts.

Compromise Solutions for Siting

The first part of the evaluation problem is solved by setting up evaluation scales for the goal achievements by the individual objectives. Within the system of constraints the most favorable and most unfavorable solutions can be calculated for each objective. Using these solutions scales are defined. To be able to compare goal achievements on these scales, they are standardized which means that the scales are mapped to the interval /0, 1/. The scale value 1 is always correlated by the most favorable and the scale value 0 by the most unfavorable value of solution.

Figs. 1 to 4 show the most favorable and most unfavorable sites whose cost and impact values fix the two scales. The solution which is most favorable in terms of costs (Fig. 1) yields power station sites in the vicinity of rivers. By contrast, the most adverse case in terms of costs (Fig. 2) would be sites very much distant from rivers. In both cases accumulations of 100 MW_e power units are obtained. Figs. 3 and 4 show the results of calculations in case of the most favorable and most unfavorable impact to the population from pollutant concentration. It is characteristic of the most favorable impact that the eastern column of the source point grid is preferred (Fig. 3). In case of the most unfavorable burden the sites are located south-west of the regions more densely populated (Fig. 4). These sites are determined by the main wind direction which is from south-west. In both cases accumulations of 100 MW_e nuclear power units are again obtained.

Figs. 5 and 6 show the outdoor pollutant concentration for the cases of the most favorable cost and the most favorable impact solution. The improvement of the impact for the densely populated area of MANNHEIM can be seen in Fig. 6 compared with the impact of the best cost solution in Fig. 5.

With the help of the two methods selected

1) Maximization of the sum of individual goal achievements;
2) maximization of the common minimum goal achievement for both obejective functions,

possible compromise solutions are searched.

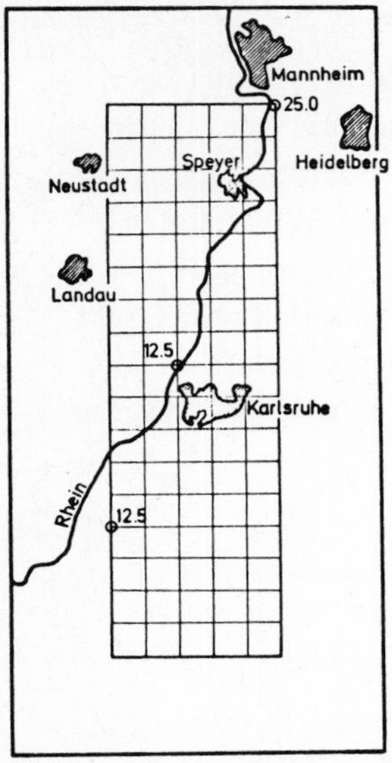

Fig.1: Distribution of occupation-numbers for 100 MWe-power stations in the source-point grid for minimum costs.

Energy generation capacity: 5 GWe
Environmental standard: 140 μg SO_2/m^3

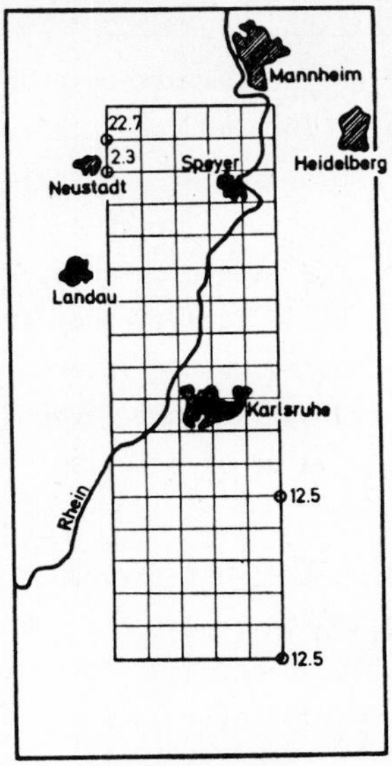

Fig.2: Distribution of occupation-numbers for 100 MWe-power stations in the source-point grid for maximum costs.

Energy generation capacity: 5 GWe
Environmental standard: 140 μg SO_2/m^3

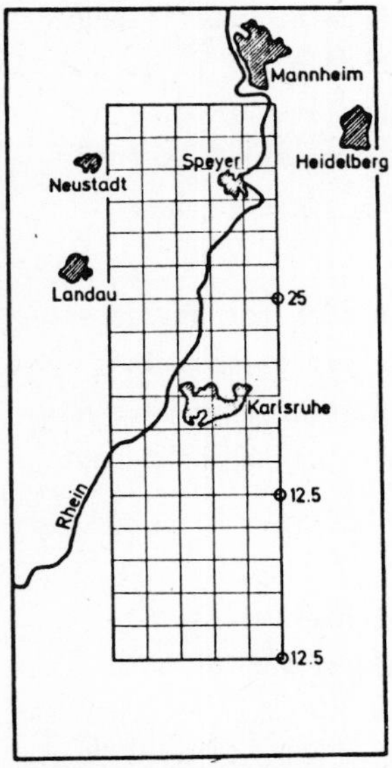

Fig.3: Distribution of occupation-numbers for 100 MWe-power stations in the source-point grid for minimum impact on the population.
Energy generation capacity: 5 GWe
Environmental standard: 140 μg/SO_2/m^3

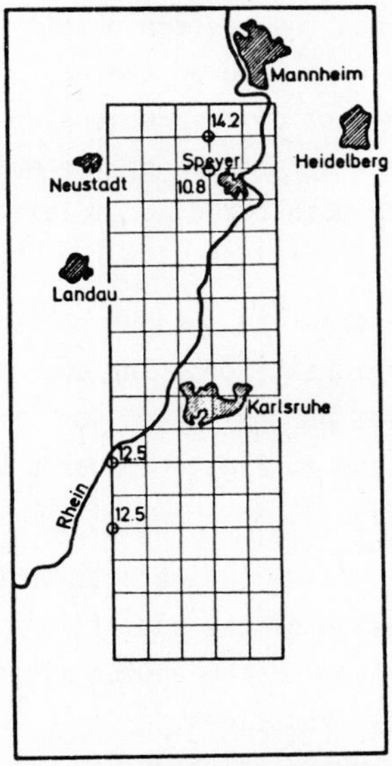

Fig.4: Distribution of occupation-numbers for 100 MWe-power stations in the source-point grid for maximum impact on the population

Energy generation capacity: 5 GWe
Environmental standard: 140 μg SO_2/m^3

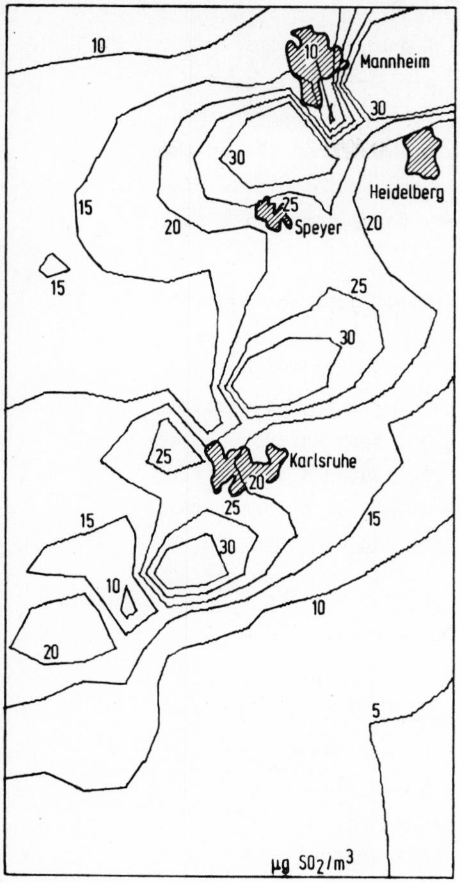

Fig.5: Ambient pollutant concentration (SO_2) for the minimum costs solution (Fig.1)

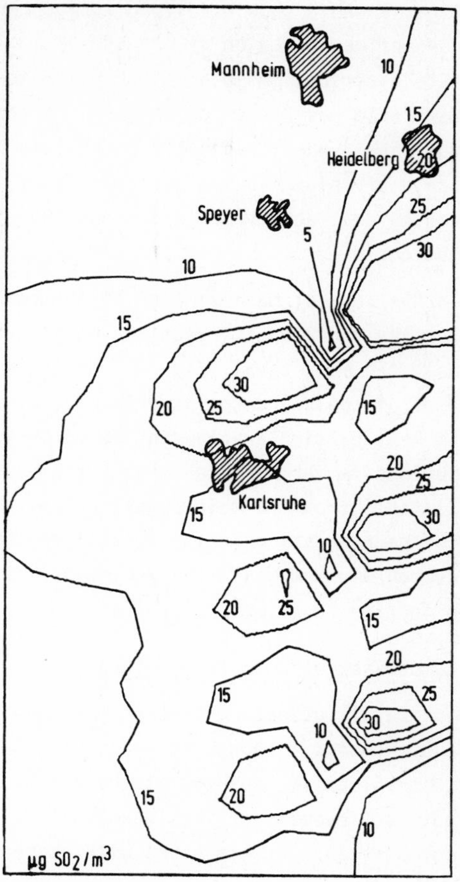

Fig.6: Ambient pollutant concentration (SO_2) for the minimum impact solution (Fig.2)

Figs. 7 and 8 show results of the compromise calculations. In the case of maximization of the sum of individual goal achievements a site distribution is obtained which is very similar to the most favorable cost solution (Fig. 7). Accordingly, the values for the goal achievement are 89% on the scale of cost values and 47% on the scale of values indicating the impact to the population. These differing goal achievements are not satisfactory for a compromise solution.

In the case of maximization of the common minimum goal achievement for the individual objectives only some of the sites are located near the main canal (Fig. 8). No sites can be found in the northern part of the source point grid so that the burden to the population is kept particulary low in the northern region. Very unfavorable solutions in terms of costs are obtained for the southern sites (far distance from the main canal) which, however, entails a lower impact to the population in the central region of the field point grid. A common minimum goal achievement of 67% in total is obtained for both objective functions.

Comparison of Results Obtained

The result of method 1 - maximization of the sum of individual goal achievements - show that under this method the goal achievement of cost values is improved at the expense of a deterioration of the goal achievement of the impact values. This confirms the critical comments already expresses with respect to this approach.

In conclusion it can be stated with respect to the practical problem of siting for technical-scale facilities and considering the lack of knowledge of exogenous preference for single objective concepts, the method 2 - maximization of a common minimum goal achievement for both goals - leads to appropriate compromise solutions. Equal weight additions of the single goal achievements for the objective functions (method 1) does not ensure an equal result for the goal achievement. This is clear from Fig. 9 which represents the objective function values from all scaling computations and from all vector valued optimization calculations within the space mapped of both objective functions. The so-called efficient borderline of possible objective function values will lie between the points (1) - best cost solution - and (3) - best solution for the impact to the population. All values on this borderline are characterized by Pareto optimality. In most of the practical calculations the full course of this borderline

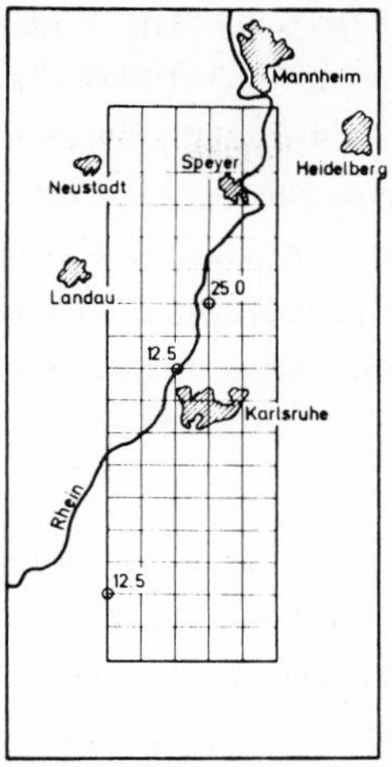

Fig.7: Distribution of occupation-numbers for 100 MWe-power stations in the source point grid for maximal sum of objective attainments (objective attainment best costs: 89 %; objective attainment best impact: 47 %)
Energy generation capacity: 5 GWe
Environmental standard: 140 μg SO_2/m^3

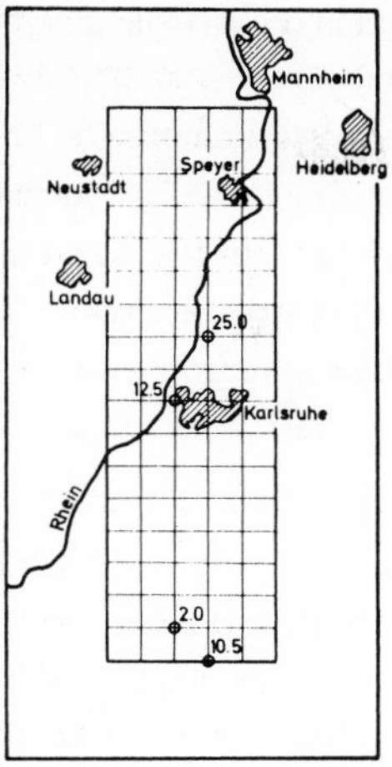

Fig.8: Distribution of occupation-numbers for 100 MWe-power stations in the source-point grid for best minimum objective attainment (minimum objective attainment: 67 %).

Energy generation capacity: 5 GWe
Environmental standard: 140 μg SO_2/m^3

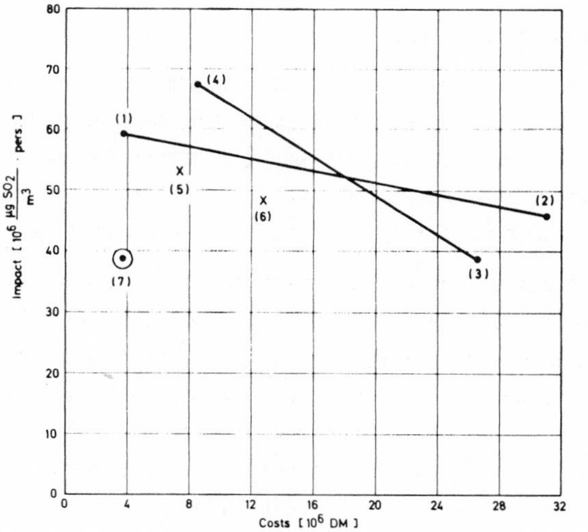

(1) minimum costs solution

(2) maximum costs solution

(3) minimum impact solution

(4) maximum impact solution

(5) compromise solution - maximal sum of objective attainments

(6) compromise solution - best minimum objective attainment

(7) ideal solution

Fig.9 Values of objective functions

will remain unknown and only some points can be obtained. These results will not always offer a satisfactory solution, despite Pareto optimality. Therefore, it is necessary to fix, either by appropriate solution finding procedures or by further constraints imposed by external preferences, the possible solutions in such a way that the minimum level of aspiration is attained. The exclusive guarantee that functionally efficient solutions are obtained is not sufficient to solve the problem under consideration. This limitation will apply to the majority of practical problems.

REFERENCES

ALLGAIER, R. (1974), Zur Lösung von Zielkonflikten, Dissert., TU Karlsruhe

DINKELBACH, W. (1969), Entscheidungen bei mehrfacher Zielsetzung und die Problematik der Zielgewichtung, In: BUSSE v.KOLBE, W., MEYER-DOHM, P., Unternehmerische Planung und Entscheidung, Bertelsmann Universitätsverlag, Bielefeld.

HADLEY, G. (1969), Nichtlineare und dynamische Programmierung, Physica Würzburg-Wien.

JÜTTLER, H. (1968), Ein Modell zur Berücksichtigung mehrerer Zielfunktionen bei Aufgabenstellungen der mathematischen Optimierung, In: Math. Modelle und Verfahren der Unternehmensforschung, Köln, S. 11-31.

KÖRTH, H. (1969), Zur Berücksichtigung mehrerer Zielfunktionen bei der Optimierung von Produktionsplänen, In: Mathematik und Wirtschaft, Band 6, Berlin, S. 184-201.

PASQUILL, F. (1962), Atmospheric Diffusion. D.van NOSTRAND Company Ltd.

SLADE, D.H. (1968), Meteorology and Atomic Energy. U.S. Atomic Energy Commission, Division of Technical Information.

Economic Operation of Electric Power System
under Environmental Impacts
by
Peter Georg Harhammer

1,0 Introduction

The main goal of this paper is to describe a method for committing thermal generation of fossil-fuel fired units and scheduling their output economically so as to comply with environmental objectives, such as limitations on emissions, groundlevel concentrations and wasted heat. The model presented here is formulated generally, making it possible to extract a series of special models. These special models reflect both all limitations of the power system under investigation and its operational objective as well which may vary problem dependent.

2,0 Optimization Method

The operational model of an electric power system with thermal generation requires an optimization method capable to handle discrete (e.g. valve points of thermal units), discontinuous (e.g. absolute cost curves between operating limits) and non-linear (e.g. absolute cost curves with large changes of gradient) functions. Moreover, decision variables with values 0 or 1 are necessary to describe different operating conditions (e.g. unit up or down).

Therefore, a Branch and Bound based Mixed-Integer Programming technique ((1)) is used for optimization purposes. This method offers in each case the global optimum independent of the solution space's form (convex, non convex, non connex).

3,0 Optimization Model

The following Mixed-Integer Programming model represents the mathematical description of the operation of an electric power system with fossil-fired thermal generating power plants taking into account their environmental impacts. The inclusion of these environmental impacts into operation planning models should be possible to meet legislative regulations or to improve the public relations of the Electric Power Industry.

In order to better understand the primary concern of this paper, only those model elements of thermal units were formulated that must be seen in a direct connection with the solution of the economic-environmental problem. Other necessary parts of the objective function and respective constraints, such as start-up costs, limited number of start-up procedures, power changing velocity, limitations of up-time and down-time of units etc., can be found in other publications of the author ((2,3,4,5)). These model elements are not treated in this paper although they are necessary when formulating real-life models especially for shortterm optimization (day, weekend).

Certain assumptions concerning the pollution of thermal generating plants are necessary in order to include their environmental impacts into the model.

- The amount of sulphur contaminations (e.g. SOx) is directly proportional to the type and amount of fuel burned and is therefore direct proportional to the fuel input - power output curve.

- The same is valid for solid particulates.

- The amount of nitrogen oxide (NOx) products released is related to the fuel type, the air-fuel ratio and the temperature of combustion. Therefore, NOx curves must be obtained by testing of individual units. They can be expressed as a function of the power output by applying regression analysis to the measured data ((6)).

- The discharge of thermal energy (wasted heat) into watercourses is defined as a function of the power output. Here the incremental that rate is used.

Models for ground concentration of gaseous emissions (SOx, NOx) with respect to the location (surface coordinates x,y representing the distance from the power plant) are not treated in this paper. The environmental-oriented literature offers different proposals for respective models in deterministic ((7)) and stochastic formulation ((7,8)).

Operations planning of an electric power system is a continuously proceeding process taking into account planning horizons from minutes to years. Although the model presented here is mainly oriented towards short term optimization (day, weekend) of electric power systems' operation it can also be used on-line and incorporated into larger models for medium-term and long-term operations planning too. This requires the inclusion of other model elements ((3)) taking into account all company-specific requirements for shorter (minutes) or longer planning horizons (months, years) respectively.

When speaking of plants, units hereof are modelled.

Indices used:

i generation unit
t time step (t_t length of time step)
T number of time steps within the planning horizon (T_T length of planning horizon)

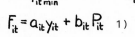

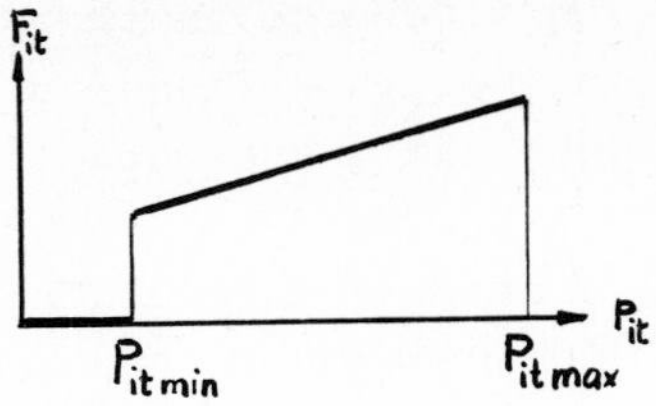

THERMAL POWER PLANTS

- Fuel Input - Power Output Curve
  - o Continuous Operation $F_{it} = a_{it} y_{it} + b_{it} P_{it}$ 1)
  - o Operation by Valve Point Loading $F_{it} = \sum_{m=1}^{M} F_{mit}(P_{mit}) y_{mit}$
- Absolute Fuel Costs $C_{Fit} = c_{Fit} F_{it}(P_{it})$
- Fuel Costs Constraints $C_{Fit} \leq C_{Fit\,max}$
- Operations Limitations
  - o Continous Operations $y_{it} P_{it\,min} \leq P_{it} \leq y_{it} P_{it\,max}$
  - o Operation by Valve Point Loading $P_{it} = \sum_{m=1}^{M} y_{mit} P_{mit}$
- Operations Indicators (OFF/ON)
  - o Continous Operation $y_{it} = 0 \text{ or } 1$
  - o Operation by Valve Point Loading $y_{it} = \sum_{m=1}^{M} y_{mit}$, $y_{it} = 0 \text{ or } 1, y_{mit} = 0 \text{ or } 1$

ENVIRONMENTAL FUNCTIONS

- Emission (e.g. SOx, solid particulates - E_P-Emission) (direct proportional to fuel input-power output curve) $E_{Pit} = e_{it} F_{it}(P_{it})$
  - o Continuous Operation $E_{Pit} = e_{it}(a_{it} y_{it} + b_{it} P_{it})$
  - o Operations by Valve Point Loading $E_{Pit} = e_{it} \sum_{m=1}^{M} F_{mit}(P_{mit}) y_{mit}$

1) Here F_{it} is assumed to be a linear function of P_{it}. Other types of functions are represented by pice-wise linear approximations ((3,4)).

- Emission (NO - E_C-Emission)
  (function of combustion, to be measured as a function of power output) $E_{Cit} = E_{Cit}(P_{it})$

  - o Continous Operation $E_{Cit} = g_{it}y_{it} + h_{it}P_{it}$ 1)

  - o Operation by Valve Point Loading $E_{Cit} = \sum_{m=1}^{M} E_{Cmit}(P_{mit})y_{mit}$

- Ground Level
  Concentration of pollutants (e.g. SO_2)
  (function of distance x_i, y_i from power plant) $G_{it} = G_{it}(x_i, y_i)$

- Wasted Heat
  (temperature increase of cooling water) $\Delta T_{it} = \Delta T_{it}(P_{it})$

  - o Continuous Operation $\Delta T_{it} = o_{it}y_{it} + q_{it}P_{it}$ 1)

  - o Operation by Valve Point Loading $\Delta T_{it} = \sum_{m=1}^{M} \Delta T_{mit}(P_{mit})y_{mit}$

- Emission Costs (Taxes)

  - o E_P-Emission $C_{EPit} = c_{EPit}E_{Pit}$

  - o E_C-Emission $C_{ECit} = c_{ECit}E_{Cit}$

- Constraints
  (index g refers to a geographical region with environment polluting thermal power plants)

  - o E_P-Emission $E_{Pit} \leq E_{Pit\,max}, \sum_{i \in i_g} E_{Pit} \leq E_{Pg\,t\,max}$

  - o E_C-Emission $E_{Cit} \leq E_{Cit\,max}, \sum_{n \in i_g} E_{Cit} \leq E_{Cg\,t\,max}$

1) Here E_{Cit} and T_{it} are assumed to be linear functions of P_{it}. Other types of functions are represented by piece-wise linear approximations ((7,8)).

o Ground-Level Concentration $G_{it} \leq G_{it\,max}$

o Wasted Heat $\Delta T_{it} \leq \Delta T_{it\,max}$

LOAD SUBMODEL

- Energy Balance Equation $P_{Lt} + P_{Losst} = \sum_{i=1}^{I} P_{it}$

- Power System Losses $P_{Losst} = 0{,}01 a_{Lt} P_{Lt}, \; 0 \leq a_{Lt} \leq 15\%$

4,0 Model Classification

This paper intends is to present a generally formulated model optimizing an operational electric power system with thermal generation including environmental impacts. The adjacent Mixed-Integer Programming model can be used with different objectives to solve a specific problem.

General Objective Function:

$$C = \sum_{t=1}^{T} \sum_{i=1}^{I} \left[C_{Fit} + C_{EPit} + C_{ECit} + C_{\Delta Tit} \right] t_t \longrightarrow \text{MIN}$$

$$C_{Fit} = c_{Fit} F_{it}(P_{it}), \; C_{EPit} = c_{EPit} E_{Pit}[F_{it}(P_{it})], \; C_{ECit} = c_{ECit} E_{Cit}(P_{it}), \; C_{\Delta Tit} = c_{\Delta Tit} \Delta T_{it}(P_{it})$$

Two categories of models can be distinguished when classifying them taking into account economic-environmental aspects.

- Economic dispatch models ($c_{Fit} \neq 0$)
  with environmental constraints and

- Environmental dispatch models ($c_{EPit} \neq 0$, $c_{ECit} \neq 0$, $c_{\Delta Tit} \neq 0$)
  with cost constraints.

4.1 Economic Dispatch Models with Environmental Constraints

PROBLEM'S OBJECTIVES	Objective Function Coefficients				Constraints (Limits of Variables)				
					MAX	MAX	MAX	MAX	MAX
	c_{Fit}	c_{EPit}	c_{ECit}	$c_{\Delta Tit}$	E_{Pit}	E_{Cit}	G_{it}	ΔT_{it}	C_{Fit}
Economic Dispatch	≠0	0	0	0	0	0	0	0	0
Economic Dispatch Constrained E_P-Emission	≠0	0	0	0	yes	0	0	0	0
Economic Dispatch Constrained E_C-Emission	≠0	0	0	0	0	yes	0	0	0
Economic Dispatch Constrained GL-Concentration	≠0	0	0	0	0	0	yes	0	0
Economic Dispatch Constrained Wasted Heat	≠0	0	0	0	0	0	0	yes	0
Economic Dispatch Constrained Emission	≠0	0	0	0	yes	yes	0	0	0
Economic Dispatch Constrained Emission & GL-Concentration (AIR-oriented)	≠0	0	0	0	yes	yes	yes	0	0
Economic Dispatch Constrained Emission & GL-Concentration Wasted Heat (AIR- & WATER-oriented)	≠0	0	0	0	yes	yes	yes	yes	0
Economic-Environmental Dispatch (TAX-Model)	≠0	≠0	≠0	≠0	0	0	0	0	0

4.2 Environmental Dispatch Models with Cost Constraints

PROBLEM'S OBJECTIVES	Objective Function Coefficients C_{Fit}	C_{EPit}	C_{ECit}	$C_{\Delta Tit}$	Constraints (Limits of Variables) MAX E_{Pit}	MAX E_{Cit}	MAX G_{it}	MAX ΔT_{it}	MAX C_{Fit}
Environmental Dispatch (A) E_P-oriented	O	1	O	O	O	O	O	O	O
Environmental Dispatch (B) E_C-oriented	O	O	1	O	O	O	O	O	O
Environmental Dispatch (C) T-oriented	O	O	O	1	O	O	O	O	O
Environmental Dispatch (A) Constrained Cost	O	1	O	O	O	O	O	O	yes
Environmental Dispatch (B) Constrained Cost	O	O	1	O	O	O	O	O	yes
Environmental Dispatch (C) Constrained Cost	O	O	O	1	O	O	O	O	yes
Environmental (Air) Dispatch (TAX)	O	1	1	O	O	O	O	O	O
Environmental (Air) Dispatch (TAX) Constrained Wasted Heat	O	1	1	O	O	O	O	yes	O
Environmental (Air&Water) Dispatch (TAX)	O	1	1	1	O	O	O	O	O

In both cases the main operational requirement is met by covering the power system's load. This is secured by the energy balance equation which is necessary in each model independent of the respective objective function.

When minimizing an objective function subject to different constraints it is obvious that the objective function's value will increase with an increasing number of constraints. I.e. the models under discussion of the first group (4.1) will lead to 2 to 4 % higher operating costs ((6,9)). Those of the second group (4.2) emphasize environmental aspects expressed by an appropriate objective function. Costs are taken into account by constraints only. This way of modelling results normally in higher operating costs too when compared to the objective function's value of the mere economic dispatch strategy. Tax models (SO_2 emission) are reported ((10)) to result in 15 to 35 % average increase of operating costs.

5,0 Conclusion

This paper presents a generally formulated Mixed-Integer Programming model in order to optimize economic-environmental oriented operational objectives of an electric power system with thermal generation. Different company-specific requirements in the context of environmental impacts can be considered in different models. Modelling is done by extracting the necessary model elements from the general model presented herein in order to describe a real-world power system. These models are clearly structured and easily to understand by both practical experienced energy-engineers and dispatching personnel as well. These models are primarily oriented towards short-term optimization (day, weekend) and on-line application but may be applied to every other planning horizon (from years to minutes) as well. The solutions of economic-environmental optimization problems should be the proper basis for decision making in the real-life practice of responsible power system dispatching.

6,0 References

((1)) IBM-Program Reference Manual:
Mathematical Programming System Extended /370 (MPSX/370), Mixed-Integer Programming /370 (MIP/370), 1974

((2)) P. G. Harhammer: 1)
Wirtschaftliche Lastaufteilung auf Basis der Gemischt-Ganzzahligen Planungsrechnung
Dissertation, Technical University of Vienna, May 1974

((3)) P. G. Harhammer:
Wirtschaftliche Lastaufteilung auf Basis der Gemischt-Ganzzahligen Planungsrechnung
ÖZE, 1976, No. 3, Page 87 - 94

((4)) P. G. Harhammer
Economic Operation of Electric Power Systems
Paper presented at the IXth International Symposium on Mathematical Programming, Budapest, August 1976

((5)) P. G. Harhammer
Wirtschaftlicher Verbundbetrieb
EuM, Heft 7, 1977

((6)) Mr. R. Gent, J. W. Lamont:
Minimum-Emission Dispatch
IEEE-Transactions on Power Apparatus and Systems
Vol. PAS-90, 1971, No. 6, Page 2650 - 2660

1) This study has been awarded the State-Price 1975 for Energy Research by the Austrian Federal Ministry for Science and Research.

((7)) F. Schweppe et al
Supplementary Control Systems - A Demonstration
IEEE-Transactions on Power Apparatus and Systems
Vol. PAS-96, 1976, No. 2, Page 309 - 317

((8)) R. L. Sullivan, D. W. Hilson
Computer Aided Ambient Air Quality Assessment for Generation System Planning
1975 PICA-Conference Proceedings, Page 247 - 251

((9)) J. K. Deloon
Controlled Emission Dispatch
IEEE-Transactions on Power Apparatus and Systems
Vol. PAS-93, 1974, No. 5, Page 1359 - 1366

((10)) J. B. Cadogan, L. Eisenberg
Sulphur Oxide Emissions Management for Electric Power Systems
IEEE-Transactions on Power Apparatus and Systems,
Vol. PAS-96, No. 2, 1977, Page 393 - 403

Dipl.-Ing. Dr. Peter G. HARHAMMER
IBM Österreich
Obere Donaustraße 95
A - 1020 Wien
Austria

AN OPTIMUM OPERATION OF PUMP AND RESERVOIR IN WATER SUPPLY SYSTEM

I. Nakahori[†], I. Sakaguchi[††] and J. Ozawa[*]

†,* Mitsubishi Electric Corp., Central Res. Lab., Amagasaki, Japan

†† Osaka Prefectural Gov., Water Works Div., Hirakata, Japan

ABSTRACT

The operation problem in water supply systems is considered. The objective of the problem is to minimize pump operation cost under constraints of flow volumes of pump stations and water levels of reservoirs.
Especially, the flow volumes of pump stations are assumed to have discrete values. The problem is described in a form of Linear Programming. The discrete flow volumes of pump stations are obtained by the modified Branch and Bound method. Finally, the feasibility of the proposed method is confirmed by a numerical example of the water supply system of Osaka Prefectural Government.

1. INTRODUCTION

Recently, a water demand in urban areas shows a remarkable tendency to increase in Japan. Therefore, a large amount of water is often transported in the long distance between sources and urban areas. There already appear some water supply systems whose lengths are over 50 Km, and flow volumes are over 1.0 million tons per day. These big water supply systems need to be operated economically and reliably.

Water supply systems which connect sources with dimands, are composed of pipe lines, reservoirs, pump stations and demands. Economy of the operation depends mostly on the pumping cost. Reliability of the operation is achieved by keeping the water level of reservoirs high. Therefore, the optimum operation of water supply systems should meet both requirements for economy and reliability.

F.Fallside and P.Perry presented this problem as a nonlinear optimum control problem.[1,2] They proposed to linearlize the original problem and to apply two-level goal coordination algorithm in order to obtain the solution. But the method seems not to be practical in the sense of computability, if the system becomes large. On the other hand, in many practical systems operators used to obtain desirable solutions combining experiences and try and error methods. But these methods are also troublesome in the case of large systems.

This paper is concerned with a simple description and some algorithmic approach for the operation problem of water supply systems. The objective of the operation problem is to minimize the pump operation cost under constraints of flow volumes of pump stations and water levels of reservoirs. Especially, the flow volumes of pump stations are assumed to have discrete values.

The dynamics of water levels of reservoirs are represented by linear combinations of flow volumes of pump stations and demands. The objective function of the problem is also described by a linear form of flow volumes of pump stations. As a result, the problem is described in a form of Linear Programming. Further, the discrete flow volumes of pump stations are obtained by the Branch and Bound method, i.e., using the above Linear Programming, iteratively. The feasibility of the proposed method is confirmed by a numerical example of the water supply system of Osaka Prefectural Government.

2. REPRESENTATION OF WATER SUPPLY SYSTEMS

The main components of water supply systems, pipe lines, reservoirs, pump stations and demands are shown in Fig. 1. In comparison with water distribution systems, the features of water supply systems are represented as follows;

i) Reservoirs are connected with each other through pump stations,

ii) Demands are connected with reservoirs directly.

Note that each demand of water supply systems usually corresponds to the total demand of a water distribution system.

The connecting points of reservoirs and their demands correspond to nodes, on the other hand, pump stations correspond to branches. Consequently, water supply systems are represented by directed graphs as shown in Fig. 2.

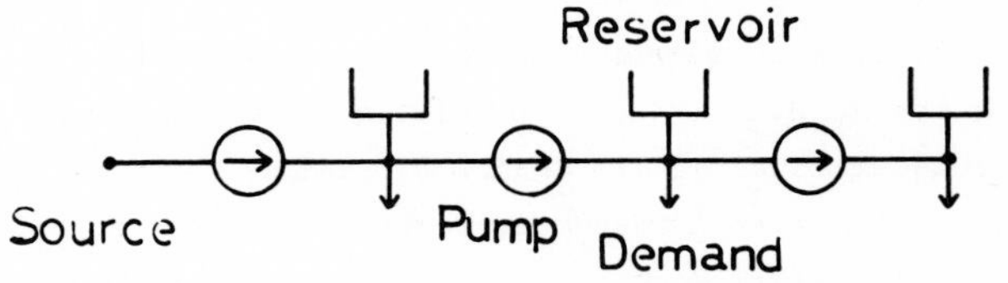

Fig.1. Schematic figure of a water supply system.

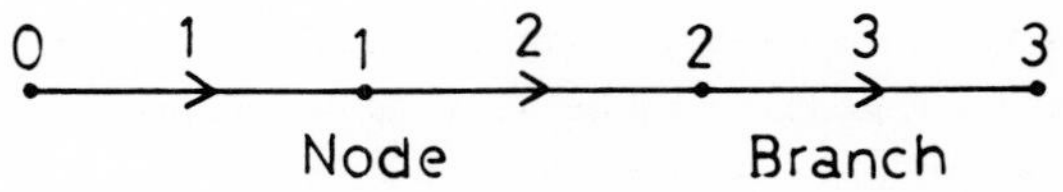

Fig.2. Directed graph representing the structure of a water supply system.

Excluding a base node which usually corresponds to the source, we obtain the incidence matrix F as follows.

$$F = \text{node} \begin{array}{c} \\ 1 \\ 2 \\ 3 \end{array} \overset{\text{branch}}{\begin{array}{ccc} 1 & 2 & 3 \\ \end{array}} \begin{bmatrix} -1 & 1 & \\ & -1 & 1 \\ & & -1 \end{bmatrix} \tag{1}$$

Using the incidence matrix F, the law of mass conservation in a system is represented by the following matrix form.

$$Z\ \{r(k)-r(k-1)\} = F\ p(k)-I\ d(k) \tag{2}$$

where, $r(k)$: a vector of water level of reservoirs at time kT

$p(k)$: a vector of pump flow (flow volume of pump stations) at time interval kT ($(k-1)T \leqq t \leqq kT$)

$d(k)$: a vector of demand flow (flow volume of demands) at time interval kT.

Further, Z is a diagonal matrix representing the size of reservoirs and I is the identity matrix.

3. OBJECTIVE FUNCTION OF WATER SUPPLY SYSTEMS

When considering the operation of water supply systems, it is important to harmonize requirements from both sides of pump stations and reservoirs. The side of pump stations requires (1) minimizing pumping cost, (2) minimizing operations especially at night. The side of reservoirs requires (1) keeping water level high, (2) minimizing the operation range of water level.

Note that requirements from both sides clash each other. Roughly speaking, the side of pump stations requires a constant pump flow causing big fluctuations of water level of reservoirs. On the other hand, the side of reservoirs requires a constant water level which leads to frequent changes of pump flow.

Here, for simplification only the requirements of the side of pump stations are taken for the objectives and the requirements of the side of reservoirs are taken for the constraints.

Let's assume that the time when pump flow changes is given and during the time interval the pump flow is constant. When the flow volume of pump stations is controlled by the number of pumps, the pumping cost is almost proportional to the pump flow. And the maintenance cost depends on the number of times changing pump flow. Therefore, we obtain the next objective function.

$$J = \sum_{i=1}^{N} \sum_{k=1}^{K} w_{1i}(k) p_i(k) + \sum_{i=1}^{N} \sum_{k=1}^{K} w_{2i}(k) \cdot |p_i(k) - p_i(k-1)| \tag{3}$$

Where the first term of the right hand side represents the pumping cost and the second term represents the maintenance cost. N and K are the number of pump stations and the number of time intervals, $w_{1i}(k)$ and $w_{2i}(k)$ represent the weighting coefficients.

In order to eliminate the absolute values of Eq.(3) we can use the common technique which uses the following auxiliary variables x_1 and x_2.

$$x_{1i}(k) - x_{2i}(k) = p_i(k) - p_i(k-1) \quad ; \quad x_{1i}(k),\ x_{2i}(k) \geqq 0 \tag{4}$$

From Eq.(4), p is rewitten as follows;

$$\begin{bmatrix} p(1) \\ \vdots \\ p(K) \end{bmatrix} = \begin{bmatrix} 1 & & 0 \\ \vdots & 1 & \\ 1 & \cdots & 1 \end{bmatrix} \begin{bmatrix} x_1(1) \\ \vdots \\ x_1(K) \end{bmatrix} - \begin{bmatrix} 1 & & 0 \\ \vdots & 1 & \\ 1 & \cdots & 1 \end{bmatrix} \begin{bmatrix} x_2(1) \\ \vdots \\ x_2(K) \end{bmatrix} + \begin{bmatrix} 1 \\ \vdots \\ 1 \end{bmatrix} p(0) \tag{5}$$

Substituting Eq.(5) into Eq.(2)., the objective function is arranged as follows;

$$J = \sum_{i=1}^{N} \sum_{k=1}^{K} w_{1i}(k) \{p_i(0) + \sum_{j=1}^{k} (x_{1i}(j) - x_{2i}(j))\} + \sum_{i=1}^{N} \sum_{k=1}^{K} w_{2i}(k) \{x_{1i}(k) + x_{2i}(k)\} \tag{6}$$

4. CONSTRAINTS OF WATER SUPPLY SYSTEMS

As described in the previous chapter, the requirements from the side of reservoirs are included in the constraints. The condition is represented by the feasible range of water level for a reservoir as follows;

$$r_i(k)_{min} \leqq r_i(k) \leqq r_i(k)_{max} \tag{7}$$

Equation (7) means that each reservoir has its own range of water level at each time. Each pump station has also its own range of flow volume for each time interval.

$$p_i(k)_{min} \leqq p_i(k) \leqq p_i(k)_{max} \tag{8}$$

Note that Eq. (8) plays a fundamental role for the Branch and Bound method which will be described later.
Eqs. (7) and (8) are represented in vector form as follows;

$$r(k)_{min} \leqq r(k) \leqq r(k)_{max} \tag{9}$$

$$p(k)_{min} \leqq p(k) \leqq p(k)_{max} \tag{10}$$

For the convenience of descriptions, let's introduce the following aggregated vectors.

$$r \triangleq \begin{bmatrix} r(1) \\ r(2) \\ \vdots \\ r(K) \end{bmatrix},\ p \triangleq \begin{bmatrix} p(1) \\ p(2) \\ \vdots \\ p(K) \end{bmatrix},\ d \triangleq \begin{bmatrix} d(1) \\ d(2) \\ \vdots \\ d(K) \end{bmatrix},\ x \triangleq \begin{bmatrix} x_1 \\ x_2 \end{bmatrix},\ x_1 \triangleq \begin{bmatrix} x_1(1) \\ \vdots \\ x_1(K) \end{bmatrix},\ x_2 \triangleq \begin{bmatrix} x_2(1) \\ \vdots \\ x_2(K) \end{bmatrix} \tag{11}$$

Other vectors such as r_{min}, r_{max}, p_{min} and p_{max} are also defined similarly.

Denote that Eq. (9) is represented in the form of constraints for pump flow by using Eq. (1). Therefore, Eqs (9) and (10) are combined into the following inquallity using the aggregated vectors in Eq. (11).

$$\begin{bmatrix} D_1 \\ D_2 \\ D_3 \\ D_4 \end{bmatrix} p \leqq \begin{bmatrix} e_1 \\ e_2 \\ e_3 \\ e_4 \end{bmatrix} \tag{12}$$

where,

$$D_1 = -D_2 \triangleq \begin{bmatrix} Z^{-1}F & & 0 \\ Z^{-1}F & Z^{-1}F & \\ \vdots & & \ddots \\ Z^{-1}F & \cdots & Z^{-1}F \end{bmatrix},\ D_3 = -D_4 \triangleq I,\ e_1 \triangleq r_{max} - \begin{bmatrix} Z^{-1} & & 0 \\ \vdots & \ddots & \\ Z^{-1} & \cdots & Z^{-1} \end{bmatrix} d - \begin{bmatrix} I \\ \vdots \\ I \end{bmatrix} r(0)$$

$$e_2 \triangleq r_{min} + \begin{bmatrix} Z^{-1} & & 0 \\ \vdots & \ddots & \\ Z^{-1} & \cdots & Z^{-1} \end{bmatrix} d + \begin{bmatrix} I \\ \vdots \\ I \end{bmatrix} r(0) \qquad e_3 \triangleq p_{max}, \; e_4 \triangleq -p_{min}$$

Furthermore, substituting Eq. (5) into Eq. (12), we finally obtain the inequality in matrix form using variables x_1 and x_2 as follows;

$$\begin{bmatrix} A_{11} & A_{12} \\ A_{21} & A_{22} \\ A_{31} & A_{32} \\ A_{41} & A_{42} \end{bmatrix} \begin{bmatrix} x_1 \\ x_2 \end{bmatrix} \leq \begin{bmatrix} b_1 \\ b_2 \\ b_3 \\ b_4 \end{bmatrix} \tag{13}$$

where,

$$A_{11} = -A_{12} = -A_{21} = A_{22} \triangleq \begin{bmatrix} Z^{-1}F & & 0 \\ 2Z^{-1}F & Z^{-1}F & \\ \vdots & & \ddots \\ KZ^{-1}F & \cdots & Z^{-1}F \end{bmatrix}$$

$$A_{31} = -A_{32} = -A_{41} = A_{42} = \begin{bmatrix} I & & 0 \\ \vdots & I & \\ I & \cdots & I \end{bmatrix}$$

$$b_1 \triangleq e_1 - \begin{bmatrix} Z^{-1}F \\ 2Z^{-1}F \\ \vdots \\ KZ^{-1}F \end{bmatrix} p(0) \quad , \quad b_2 \triangleq e_2 + \begin{bmatrix} Z^{-1}F \\ 2Z^{-1}F \\ \vdots \\ KZ^{-1}F \end{bmatrix} p(0)$$

$$b_3 \triangleq e_3 - \begin{bmatrix} I \\ \vdots \\ I \end{bmatrix} p(0) \quad , \quad b_4 \triangleq e_4 + \begin{bmatrix} I \\ \vdots \\ I \end{bmatrix} p(0)$$

Equation (13) will be the constraints of the problem described later.

5. DESCRIPTION OF THE PROBLEM AND ALGORITHM

As described in chapters 3 and 4, the primal problem is reduced to the Linear Programming problem whose objective is Eq. (6) and constraint is Eq. (13). The problem is settled in the following arranged form.

[Problem 1] Find an optimum solution x^* which minimizes $J = cx$, subject to $Ax \leqq b$, $x \geqq 0$.

In problem 1, A,b and c represent the corresponding matrix and vectors in Eqs. (6) and (13). The optimum pump flow p^* is obtained by substituting x^* into Eq. (5).

When pump stations are controlled by the number of pumps with constant speed, we must add the condition that the corresponding pump flow p^* has the given discrete values, to Problem 1. Then the problem is settled in the following form.

[Problem 2] Find an optimum solution x^* which minimizes $J = cx$, subject to $Ax \leqq b$, $x \geqq 0$, under the condition that the corresponding pump flow p^* has the given discrete values.

The Branch and Bound method of Integer Programming is applicable in solving Problem 2. In the procedure, Problem 1 which differs from the original form in the range of pump flow should be solved iteratively. Using this characteristics, the sensitivity analysis method is successfully applied to solve the Problem 1.

The Branch and Bound method needs the numerous iterations when the system is large. Therefore, fast methods to obtain the integer solution is necessary for the practical use. For this purpose, we reform the objective function to the next one,

$$J = cx - \alpha n \tag{14}$$

Where n means the number of the components in p which has already had the given discrete values in the previous procedure. Positive α is an accelerated coefficient. As α increases, the solution differs from the original one, but the number of iterations reduces. Note that when α equals zero, this problem coincides with Problem 2.

6. NUMERICAL EXAMPLES

The effectiveness of the method is now shown by some numerical examples. Fig. 3 shows a schematic figure of the water supply system of Osaka Prefectural Government. The length of its trunk main is about 70 km. The total amount of the water demand is from 0.5 to 1.5 million tons per day. Each demand corresponds to the water distribution system of the surrounding cities of Osaka.

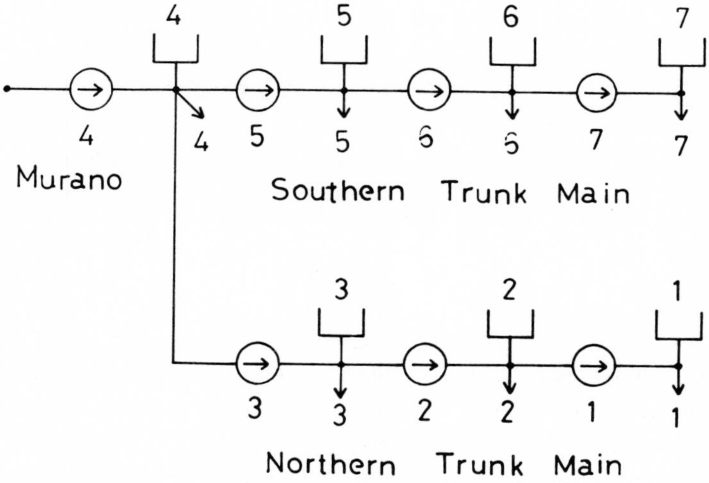

Fig. 3. The water supply system of Osaka Prefectual Government in Japan.

One day is divided into four time intervals, each of which is six hours. Total amount of the demand, 1.0 million tons, is distributed as shown in Table 1. Each pump stations have ten discrete values as shown in Table 2. The conditions of the reservoirs are shown in Table 3.

Table 1. Demand distribution in the example

Step \ No	1	2	3	4	5	6	7
1	3.22	1.79	0.36	3.76	1.61	1.97	5.20
2	5.02	2.79	0.56	5.86	2.51	3.07	8.10
3	5.20	2.89	0.58	6.06	2.60	3.18	8.37
4	4.63	2.57	0.51	5.40	2.31	2.83	7.46

Table 2. Given flow volume of pump stations in the example

No. \ flow	1	2	3	4	5	6	7	8	9	10
1	1.2	2.4	3.6	4.8	6.0	7.2	8.4	9.6	10.8	13.2
2	2.4	3.6	4.8	6.0	7.2	8.4	9.6	10.8	12.0	13.2
3	1.8	-	-	-	-	-	-	-	-	18.0
4	3.6	7.2	10.8	14.4	18.0	21.6	25.2	28.8	32.4	36.0
5	6.0	7.2	9.0	10.8	12.0	13.8	15.0	16.8	18.0	19.8
6	1.8	3.6	5.4	7.2	9.0	10.8	12.6	14.4	14.8	13.8
7	1.2	2.4	3.6	4.8	6.0	7.2	8.4	9.6	10.8	12.0

Table 3. Conditions of reservoirs in the example

No	$r_{(o)}$	r_{max}	r_{min}	z
1	1.0	3.7	1.0	1.5
2	1.0	3.7	1.0	0.5
3	1.0	3.7	1.0	0.8
4	1.0	3.7	1.0	0.8
5	1.0	3.7	1.0	0.8
6	1.0	3.7	1.0	0.5
7	1.0	3.7	1.0	1.5

Table 4 shows the optimum solution of Problem 2 obtained bv the common Branch and Bound method. The solution obtained by the accelerated Branch and Bound method, differs from the optimum one only in the part which is shown by parentheses in Table 4. Difference of iteration numbers between these two methods is shown in Fig. 4. Figure 4 shows the effectiveness of the accelerated method, especially when the number of time intervals increases.

Table 4. The optimum solution in the example Pump Flow

Pump Flow

Step \ No.	1	2	3	4	5	6	7
1	4.80	7.20	8.45	25.20	12.00	9.00	6.00
2	4.80	7.20	8.45	28.80	13.80	10.80	8.40
3	4.80	8.40	8.45	28.80	13.80	12.60	9.60
4	4.80	7.20	(7.80)	25.20	(12.00)	(9.00)	(6.00)

Reservoir Level

Step \ No.	1	2	2	4	5	6	7
0	1.00	1.00	1.00	1.00	1.00	1.00	1.00
1	2.05	2.22	2.11	2.24	2.74	3.06	1.53
2	1.91	1.44	2.98	2.10	3.35	1.72	1.73
3	1.64	2.86	2.31	3.70	1.60	1.36	2.55
4	1.75	2.52	(2.43)	(3.70)	2.46	(1.70)	(1.58)

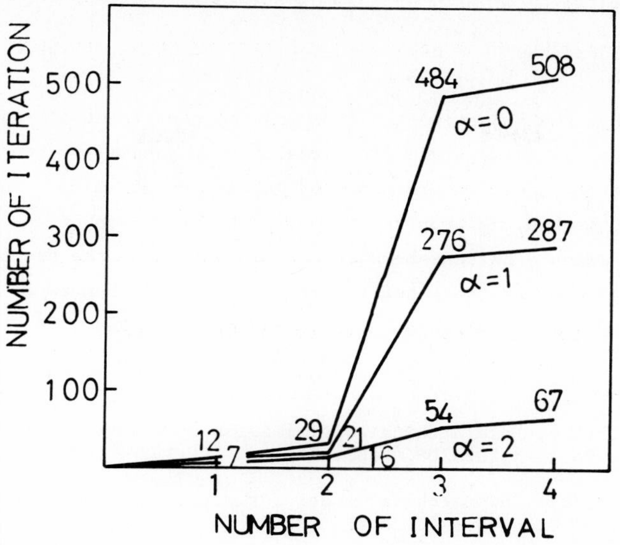

Fig. 4. Number of iterations in the example.

7. CONCLUDING REMARKS

A systematic procedure has been developed for the operation of water supply systems.

It is clarified that the objective function and the constraint of the system are able to be described in linear algebraic forms. Correspondingly, two problems are derived in the form of Linear Programming and Integer Programming. An accelerated Branch and Bound method, is proposed to obtain the solution of the Integer Programming. A simple example tested shows the effectiveness of the proposed method.

The authors wish to express their heartly thanks to Dr. K. Uemura of Mitsubishi Electric Corp. for his guidance and encouragement during this study.

REFERENCES

[1] F.Fallside and P.F.Perry, "Hierarchical optimization of a water supply network", PROC. IEE, Vol.122, No.2, pp202-208, Feb., 1975

[2] F.Fallside and P.F.Perry, "Hierarchical model for water supply system control", PROC. IEE, Vol.122, No.4, pp441-443, Apr., 1975

OPTIMAL EXPANSION OF GENERATING CAPACITY IN NATIONAL ELECTRIC POWER ENERGY SYSTEM

Katsuya OGINO
Dept. of Applied Mathematics & Physics
Faculty of Engineering, Kyoto University
Kyoto 606 JAPAN

ABSTRACT

The present paper deals with a long-range expansion policy of generating capacity in national electric power energy system. In order to take into account the originally regional feature of environmental problems, the national system is divided into several regional subsystems throughout the paper, and the overall national decision on capacity expansion is investigated as a cooperation between national decision unit and regional decision units.

By the material balance approach, a capacity balance model is firstly developed to represent generating capacity balance in each time period. A power plant evaluation model is secondly developed, where an environmental impreference measure is introduced to provide an order in regional environmental preference to each power plant type. By interfacing the models under electric demand restrictions, a least cost capacity expansion model is clearly developed and grasped as a hierarchical multiobjective system in such a way that regional subsystems hold rights in capacity expansion planning to select power plant type from the point of view of environmental preference, while nationalsystem dominates regional selection through allocation of available energy resource to the subsystems. The parametric linear programming plays a fundamental role in this modelling.

1. Introduction

A diverse range of consideration of growing electric demand, limited supply of primary energy resource and nation's environmental concern is essential in electric power generating capacity expansion policy. In an integrated framework of recent energy, resource and environment problems, the present paper represents the author's research attempts to apply systems optimization techniques to evaluation of alternative long-range expansion policies of generating capacity in national electric power energy system(Figure 1).

Since the oil embargo in 1973, various energy-related research projects have been conducted at national, institutional and university levels [1-4]. Regarding the capacity expansion in generating electric power energy, Farrar and Woodruff [5] adopted the comprehensive approach, where the linear programming model with an environmental simulation model plays a fundamental role. Beglari and Laughton [6] proposed to apply the Z-Substitute method to reduce the number of constraints in linear programming capacity expansion model, while Peterson [7] developed a dynamic programming approach and proposed a technique, applicable to a large class of capital budgeting problems under uncertainty. Following the 8th IFIP Conference on Optimization Techniques in Würzburg, the author presented a two-objective programming capacity expansion model [8] with special emphases on the cost-environment trade-off in electric power energy system planning, where, in selecting expansion mix of electric power plants, the nation's

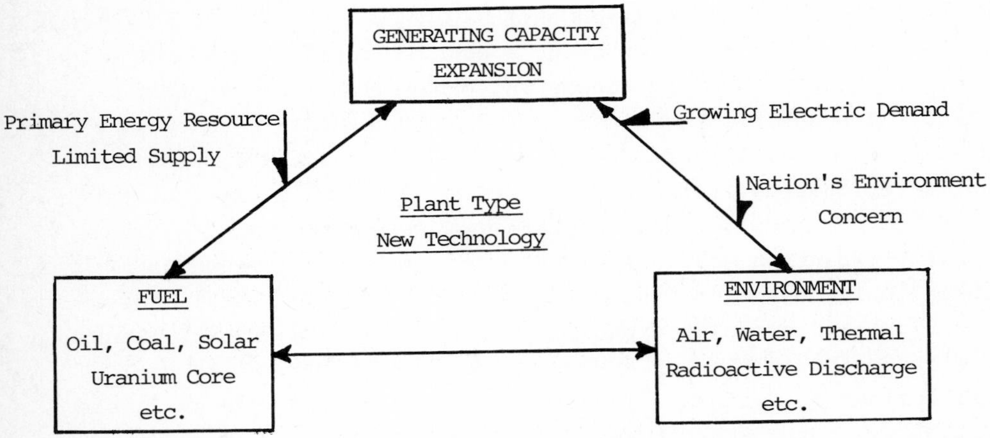

Fig. 1 National Electric Power Energy System

environmental concern is quantitatively represented by one objective function in integrating in a whole national frame the nation's environmental impreference to each type of power plants.

Taking account of limited supply of primary energy resource, the present paper proposes a hierarchical multiobjective approach to the expansion problem of generating capacity in national electric power energy system, where overall national system is divided into several regional subsystems in order to account environmental preference of regional residents in selecting alternative types of electric power plant in capacity expansion. A generating capacity balance model is firstly developed in a form of linear difference equation to represent generating capacity balance in each region over the study period. The introduced environmental impreference measure as a power plant evaluation model gives an order in regional preference to each alternative power plant types. By interfacing the models under restrictions of electric demand and primary energy resource supply, the capacity expansion problem in a least cost sense is clearly formulated as a hierarchical multiobjective model, where the parametric linear programming plays an essential role. The special feature of the proposed capacity expansion model is that rights of environmental selection of power plant types in national electric power energy system are left to regional residents, while national system dominates regional selections through primary energy resource supply.

2. Generating Capacity Balance Model

Given a long range time horizon, restrictions of growing electric demand and limited supply of primary energy resource ,and quantitative measure of nation's environmental concern, the final goal of present research is to evaluate national policies of selecting alternative power plant mix in expanding electric power generating capacity. The overall national electric power energy system is divided into L regional subsystems throughout the present paper in order to take a realistic standpoint that environmental problems are originally concerned with regional residents. The present section develops a generating capacity balance model essential for expansion policies evaluation, where every possible mix of power plant types are needed to be investigated.

Consider a time horizon [o, T], which is broken down into n periods of equal time interval dt (i.e. nxdt=T). The generating capacity in the k-th period in region L is equal to the effective amount of capacity available from the previous period plus the expanded amount of capacity in period k in region L. Thus, the generating capacity balance in each period in region L can be represented by a linear difference equation

$$c_{kL} = (I - A_{kL})c_{k-1L} + e_{kL}, \quad (k=1,2,..,n \; ; \; L=1,2,..,L), \tag{1}$$

$$c_{0L} \; ; \text{ given}, \tag{2}$$

with

$$c_{kL} = \text{col.}(c_{1kL}, c_{2kL}, .., c_{mkL}), \tag{3}$$

$$c_{0L} = \text{col.}(c_{10L}, c_{20L}, .., c_{m0L}), \tag{4}$$

$$e_{kL} = \text{col.}(e_{1kL}, e_{2kL}, .., e_{mkL}), \tag{5}$$

$$I \; ; \; m\text{x}m\text{-unit matrix}, \tag{6}$$

$$A_{kL} = \begin{bmatrix} a_{1kL} & & & O \\ & a_{2kL} & & \\ & & \ddots & \\ O & & & a_{mkL} \end{bmatrix}, \tag{7}$$

where

c_{ikL} ; generating capacity by the i-th type electric power plant at the end of period k in region L,

a_{ikL} ; attrition rate of the i-th type electric power plant in period k in region L,

e_{ikL} ; expanded amount of capacity of the i-th type electric power plant during period k in region L,

m ; number of types of electric power plants under consideration.

As the generating capacity balance system (1) is linear, it can be rewritten as

$$c_{kL} = \prod_{i=1}^{k} D_{iL} c_{0L} + \prod_{i=2}^{k} D_{iL} e_{1L} + \dots + \prod_{i=j+1}^{k} D_{iL} e_{jL} + .. + e_{kL}$$

$$= G_{1L} c_{0L} + G_{2L} e_{1L} + \dots + G_{k+1L} e_{kL}, \tag{8}$$

with

$$D_{kL} = I - A_{kL}, \quad (k=1,2,..,n \; ; \; L=1,2,..,L). \tag{9}$$

The above model (8) with (9) is hereafter referred to as capacity balance model.

It is to be noted here that for a new i-th type electric power plant and/or a conventional i-th type power plant with new technology, the initial value of the capacity c_{i0L}, the expanded amount e_{ikL} and the attrition rate a_{ikL} are restricted as

$$c_{i0L} = 0, \quad (L=1,2,..,L), \tag{10}$$

$$e_{ikL} = 0, \quad (L=1,2,..,L \; ; \; k < v_i), \tag{11}$$

$$a_{ikL} = 0, \quad (L=1,2,..,L \; ; \; k < v_i), \tag{12}$$

where

v_i ; period in which development of new i-th type power plant and/or new technology is completed and its commercial operation can be commenced.

Regarding the existing conventional i-th type power plant, it is assumed that the new technology is installed in the power plant after its development and its installation cost is accounted by increasing the operating and maintenance cost of the power plant thereafter. Thus, every possible mix of electric power plants could be examined for investigation of capacity expansion policy.

3. Power Plant Environmental Evaluation Model

Considerations of limited supply of primary energy resource and environmental impacts are essential in electric generating capacity expansion, where the need of comprehensive technical and economic data of good quality is a matter of course. The aim of the power plant evaluation model in this section is twofold: The first is to evaluate the resource-consumptive, environmental and economic characteristics of each type of electric power plants over the entire study period. The second is to process and adjust the data in proper form for their effective use in numerical optimization. Thus, the model could be referred to as an intensive computer-aided data base. The model specifies the following alternatives; type of generating electric power plants; capital and operating & maintenance costs of power plants; type and amount of pollutions discharged from power plants; type and consumption amount of fuels, and etc. As environmental problems are rather regional ones and resource supply problems are to be grasped at a national level, this section mainly investigates power plant environmental evaluations.

Assume that the total absolute amount of the j-th type pollution (j=1,2,...,q) in the period k is represented in an adjusted matrix form as

$$P_k = \begin{bmatrix} p^1_{1k} \; p^1_{2k} \; \cdots \; p^1_{mk} \\ \cdots\cdots\cdots\cdots \\ \cdots\cdots\cdots\cdots \\ p^q_{1k} \; p^q_{2k} \; \cdots \; p^q_{mk} \end{bmatrix}, \quad (k=1,2,\ldots,n), \tag{13}$$

where

p^j_{ik} ; discharge amount of pollution j per unit power output of the electric power plant i in the period k.

Then, the nation's environmental concern can be quantified as

$$imp_{kL} = w_{kL} P_k, \quad (k=1,2,..,n \; ; \; L=1,2,..,L), \tag{14}$$

$$w_{kL} = (w^1_{kL}, w^2_{kL}, .., w^q_{kL}), \quad imp_{kL} = (imp_{1kL}, .., imp_{mkL}), \tag{15}$$

where

w_{kL} ; comparative measure of impreference of region L residents to pollutions j(between j=1,..,q) in period k.

The environmental impreference measure imp_{kL} to each plant type , which could correspond to the environmental standard and/or tax, gives an order in regional preference to each alternative plant type and provides the basis for comparative standard in selecting the types of power plants to be expanded in region L. Hereafter, the above model in (14) with (13) and (15) is referred to as plant environmental evaluation model.

4. Generating Capacity Expansion Model

Based on the capacity balance model and the plant environmental evaluation model developed so far, this section formulates the capacity expansion problem as a hierarchical multiobjective program to investigate the national policies of selecting in a least cost sense an expansion mix of power plant alternative types in such a way to meet a long range electric demand, to satisfy primary energy resource supply restriction and to account nation's environmental concern in the most realistic way.

Cost Function

The cost function to be minimized consists of the present worth of capital cost and operating & maintenance cost over the study period. Thus, the cost function in region L is written as

$$J_{0L} = \sum_{k=1}^{n} (cp_{kL} e_{kL} + om_{kL} CPF_{kL} c_{kL} dt), \quad (L=1,2,..,L) \tag{16}$$

with

$$cp_{kL} = (cp_{1kL}, cp_{2kL}, .., cp_{mkL}), \quad (17)$$

$$om_{kL} = (om_{1kL}, om_{2kL}, .., om_{mkL}), \quad (18)$$

$$CPF_{kL} = \begin{bmatrix} cpf_{1kL} & & 0 \\ & \ddots & \\ 0 & & cpf_{mkL} \end{bmatrix}, \quad (19)$$

where

cp_{ikL} ; present worth of capital cost per unit capacity of the i-th type electric power plant in period k in region L,

om_{ikL} ; present worth of operating and maintenance cost per unit power output of the i-th type electric power plant in period k in region L,

cpf_{ikL} ; capacity factor of the i-th type electric power plant in period k in region L.

Electric Demand Restrictions

The available installed capacity in region L should exceeds regional peak electric demand in each period, and the electric power output is to be greater than the regional electric energy demand in each period. Thus, the electric demand restrictions can be given by

$$uv \cdot c_{kL} \geq (1 + em_{kL}) pd_{kL}, \quad (k=1,2,..,n \; ; \; L=1,2,..,L), \quad (20)$$

$$cpf_{kL} c_{kL} dt \geq ed_{kL}, \quad (k=1,2,..,n \; ; \; L=1,2,..,L) \quad (21)$$

with

$$uv = (1,1,..,1), \quad (22)$$

$$cpf_{kL} = (cpf_{1kL}, cpf_{2kL}, .., cpf_{mkL}), \quad (23)$$

where

em_{kL} ; capacity excess margin in period k in region L,

pd_{kL} ; peak demand in period k in region L,

ed_{kL} ; electric power energy demand in period k in region L.

Primary Energy Resource Restrictions

With the assumption that the available primary energy resource in region L should exceed the amount of the regional electric power output in Btu equivalence in each period, restrictions on primary energy resource supply can be given by

$$3412 CPF_{kL} c_{kL} dt \leq pr_{kL}, \quad (k=1,2,..,n \; ; \; L=1,2,..,L) \quad (24)$$

with

$$pr_{kL} = \text{col.}(pr_{1kL},..,pr_{mkL}), \tag{25}$$

where

pr_{ikL} ; available amount of primary energy resource in Btu for the i-th type power plant in period k in region L.

Environmental Index

When the environmental selection of alternative plant types in capacity expansion is permitted at each region, based on regional environmental preference, the environmental impreference measure of power plants imp_{kL} in (14) is utilized to form a L-regional environmental index as

$$J_{2L} = \sum_{k=1}^{n} imp_{kL} e_{kL}, \quad (L=1,2,..,L) \tag{26}$$

to be minimized.

The optimal capacity expansion problem can now be described in the most general form as "For the capacity balance model (8) with (9), find an optimal expansion mix of power plants over a given period [0, T] such that the cost function (16) and the environmental index (26) are minimized under electric demand restrictions (20) and (21) and the resource supply restriction (24)." This dynamic problem can be converted to a static one by defining ee_L as

$$ee_L = \text{col.}(e_{1L}, e_{2L},..,e_{nL}), \tag{27}$$

and by applying (8) into (16), (20), (21), (24) and (26), which are finally rewritten as follows:

$$J_{0L} = a_{0L} + a_{1L} ee_L = a_{0L} + J_{1L}, \tag{28}$$

$$A_{1L} ee_L \geq b_{1L}, \tag{29}$$

$$A_{2L} ee_L \geq b_{2L}, \tag{30}$$

$$A_{3L} ee_L \leq b_{3L}, \tag{31}$$

$$J_{2L} = a_{2L} ee_L, \tag{32}$$

where

$$a_{0L} = \sum_{k=1}^{n} om_{kL} CPF_{kL} G_{1L} c_{0L} dt, \tag{33}$$

$$a_{1L} = (1,1,..,1) \times \tag{34}$$

$$\begin{bmatrix} cp_{1L}+om_{1L}CPF_{1L}G_{2L}dt & & & \\ \cdot & \cdot & & 0 \\ \cdot & & \cdot & \\ om_{nL}CPF_{nL}G_{2L}dt & \cdots & om_{nL}CPF_{nL}G_{nL}dt & cp_{nL}+om_{nL}CPF_{nL}G_{n+1L}dt \end{bmatrix},$$

$$A_{1L} = \begin{bmatrix} uvG_{2L} & & \\ \vdots & \ddots & O \\ uvG_{2L} & \cdots & uvG_{n+1L} \end{bmatrix}, \tag{35}$$

$$b_{1L} = \text{col.}((1+em_{1L})pd_{1L}-uvG_{1L}c_{0L} \cdots (1+em_{nL})pd_{nL}-uvG_{1L}c_{0L}), \tag{36}$$

$$A_{2L} = \begin{bmatrix} cpf_{1L}G_{2L}dt & & \\ \vdots & \ddots & O \\ cpf_{nL}G_{2L}dt & \cdots & cpf_{nL}G_{n+1L}dt \end{bmatrix}, \tag{37}$$

$$b_{2L} = \text{col.}(ed_{1L}-cpf_{1L}G_{1L}c_{0L}dt \cdots ed_{nL}-cpf_{nL}G_{1L}c_{0L}dt), \tag{38}$$

$$A_{3L} = \begin{bmatrix} 3412dtCPF_{1L}G_{2L} & & O \\ \vdots & \ddots & \\ 3412dtCPF_{nL}G_{2L} & \cdots & 3412dtCPF_{nL}G_{n+1L} \end{bmatrix}, \tag{39}$$

$$b_{3L} = \text{col.}(pr_{1L}-3412dtCPF_{1L}G_{1L}c_{0L} \cdots pr_{nL}-3412dtCPF_{nL}G_{1L}c_{oL}), \tag{40}$$

$$a_{2L} = (imp_{1L} \cdots imp_{nL}). \tag{41}$$

Hierarchical Multiobjective Capacity Expansion Model

In view of the regional feature of environmental problems and rather national feature of primary energy resource supply, present paper proposes to grasp the capacity expansion problem as a hierarchical multiobjective system [9] such that both national system and regional subsystems have decision units respectively in allocation phase of primary energy resource and environmental selection phase of power plant mix, and national and regional decisions cooperate to perform overall expansion policy. In this national-regional framework, the national system pursues national profit through its objective function (overall cost function), which reflects the regional subsystems expansion policies implicitly, while dominating subsystems through allocation of available primary energy resource. Regional subsystems hold their own rights to select alternative power plant mix in capacity expansion in region L by regional environmental preference measure under restrictions of regional electric demand and primary energy resource supply, imposed by the national system (Figure 2).

The capacity expansion problem can thus be clearly represented as a hierarchical multiobjective program as

"National Decision Unit;

* minimize overall cost function

$$J = \sum_{L=1}^{L} J_{1L} + \sum_{k=1}^{n} \sum_{L=1}^{L} cpr_k pr_{kL} \tag{42}$$

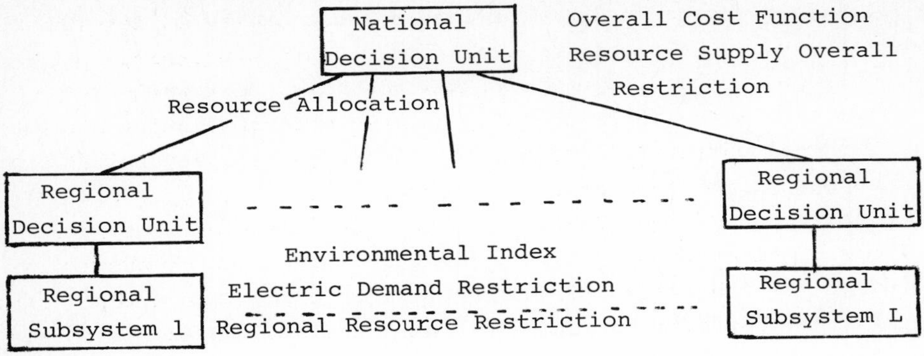

Fig. 2 Hierarchical Multiobjective Capacity Expansion Model

under primary energy resource supply overall restriction

$$\sum_{L=1}^{L} pr_{kL} \leq prs_k, \quad (k=1,2,..,n) \tag{43}$$

with

$$cpr_k = (cpr_{1k},..,cpr_{mk}) \; ; \; prs_k = col.(prs_{1k},..,prs_{mk}), \tag{44}$$

where

cpr_{ik} ; present worth of primary energy resource cost per unit amount for the i-th type power plant in period k,

prs_{ik} ; overall supply restriction of primary energy resource for the i-th type power plant in period k.

<u>Regional Decision Units(L=1,..,L);</u>

* minimize regional environmental index J_{2L} in (32) under electric demand restrictions (29) and (30) and primary energy resource restrictions (31)."

In the above problem formulation, pr_{kL} and ee_L are decision variables of the national system and regional subsystems. It is to be noted that the optimal solution of the regional decision units ee_L^* is parametric solution with respect to pr_{kL}, as easily understood from (40). The parametric solution $ee_L^*(pr_{kL})$ implements the national decision on energy resource allocation to regional subsystems. Thus, the parametric linear program becomes essential in the above capacity expansion problem.

5. Conclusions

From the point of view of recent energy, resource and environment problems, an expansion model of generating capacity is developed for national long range electric power energy systems planning. By material balance approach, a capacity balance model is firstly developed to represent generating capacity balance over the entire study period. The environmental impreference measure is secondly introduced to provides basis for comparative standard in selecting types of power plants to be expanded. The least cost capacity expansion model under restrictions of electric demand and primary energy resource supply is then developed as a hierarchical multiobjective program such that both national system and regional subsystems have decision units respectively in resource supply and environmental selection of power plant type in overall national expansion policy. The parametric linear programming plays an important role in the model. Computer simulation study is under development.

References

1. The nation's energy future a report to R.M. Nixon President of the United States, submitted by Dr. D.L. Ray, Wash-1281, US Government Printing Office, Washington, D.C., 1973.
2. Summary Report of the Cornell Workshop on energy and the environment, US Government Printing Office, Washington, D.C., 1972.
3. A report to the energy policy project of the Ford Foundation, Ballinger Publishing Company, Cambridge, Mass., 1974.
4. Project Independence, US Government Printing Office, Washington, D.C., 1974.
5. D.L. Farrar and F. Woodruff, Jr. : A model for the determination of optimal electric generating system expansion patterns, NTIS PB-223 995, Springfield, Va., 1973.
6. F. Beglari and M.A. Laughton : Model building with particular reference to power system planning : the improved Z-substitutes method, in Energy modelling, IPC Business Press Ltd., 1974.
7. E.R. Petersen : A dynamic programming model for the expansion of electric power systems, Management Science, Vol.20, No.4, pp.656-664, 1973.
8. K. Ogino and Y. Sawaragi : Capacity expansion in generating electric power energy system, Proc. of the International Conference on Cybernetics and Society, Washington, D.C., Sept. 1977.
9. K. Shimizu and Y. Anzai(in Japanese) : Optimization for a hierarchical system with independent local objectives, Trans. SICE, Vol. 10, No.1, pp.63-70, 1975.

PIPELINE NETWORK OPTIMIZATION-
AN APPLICATION TO SLURRY PIPELINES

Dr. A. Gündüz Ulusoy
Department of Industrial Engineering
Boğaziçi University, Istanbul - Turkey

Dr. David M. Miller
IEOR Department, VPI and SU
Blacksburg, Virginia 24061 - USA

ABSTRACT

The problem considered here is that of obtaining the least cost design of a pipeline network supplying demand points with known demand functions for dry solids from a single source over a finite time horizon. The problem is decomposed into a network problem and a subproblem which is the optimum design of a single arc (a pipeline). The subproblem is solved by a technique developed which exploits the monotonicity relationships of the decision variables in the objective function and in the constraints. For the network problem, it is shown that the objective function is quasi-concave and that it takes its minimum at at least one of the extreme points. The extreme points are shown to correspond to arborescences. An adjacent extreme point method searching over adjacent arborescences is developed. The method makes extensive use of the subproblem. A local optimum is defined and the solution procedure is shown to converge to this local optimum in a finite number of iterations.

INTRODUCTION

Pipeline network optimization has been one of the more frequently attacked problems of engineering optimization. A reason for this interest is that of scale. Pipeline networks typically require intensive capital investment and even small improvements in design can add up to considerable sums. Another reason is that pipelines can carry a wide spectrum of goods.

Among pipeline networks, those carrying slurry are of specific interest.

In most general terms, slurry pipelining is the transportation of solid material through a pipeline using a liquid as the transporting medium. In almost all theoretical and practical work accomplished in slurry pipelining, water has been accepted as the only transporting medium. Today, different materials such as coal, gold tailings, copper, iron, gilsonite, and limestone are being transported through slurry pipelines [1]. The largest slurry pipeline operating today is the Black Mesa coal slurry pipeline which is an 18 in., 273 mile pipeline with a capacity of 5.5 million tons of coal per year. It provides the fuel requirements for two 550 megawatt power generation units in southern Nevada [1].

The spreading use of slurry pipelines and the introduction of coal slurry pipelines as a promising means of coal transportation requires that an optimal design method be developed for slurry pipelines and slurry pipeline networks. This study is aimed at developing such a design method.

DESCRIPTION OF THE PROBLEM

In this study, the analysis will be restricted to homogeneous slurries. The path of the pipeline between any two nodes is assumed to be predetermined and all node locations are fixed. Specifically, the following decision problem is to be analyzed :

> Given (n-1) demand nodes and a source with sufficient capacity to satisfy the known static demands at the demand nodes, design the least annual discounted cost pipeline network satisfying dry solid requirements at each demand node over a finite time horizon.

The decision variables **are** the pipe diameter and the volumetric concentration of the solids in the slurry for each link, the network configuration and the level of dry solids flowing through each link of the network. Due to lack of sufficient experimentation for estabilishing functional relationships, the particles' size distribution is not included as a variable of optimization.

A slurry pipeline system consists in general of the following four components : (1) slurry preparation plants; (2) pumping stations; (3) pipe lines; and (4) dewatering and drying facilities. Since slurry preparation, dewatering and drying activities strongly depend on the particles' size distribution, they will not be considered there.

FLUID MECHANICAL ASPECTS OF THE PROBLEM

Types of Slurry Flow. One distinguishes between four types of slurry flow : 1) homogeneous flow; 2) heterogeneous flow; 3) intermediate flow, and 4) saltation flow.

In homogeneous flow, the particles are fine and are distributed evenly across the pipe cross-section. This is mainly due to low settling velocity of the solid particles and the fact that only a small turbulence is needed to distribute solids uniformly in the conduit. (The slurry is treated as a 'true' fluid with its density and viscosity being those of the slurry).

Heterogeneous slurries have a concentration gradient across the pipe cross-section but no deposition of the particles at the bottom of the pipe is allowed. In intermediate flow, homogeneous and heterogeneous flows exist simultaneously. In saltation flow, the solid particles settle to the bottom of the pipe and move in discontinuous jumps, sliding and rolling.

Critical Velocity and Pressure Drop. Investigators have modelled the homogeneous slurry flow as a non-Newtonian single phase fluid. The reason is that the homogeneous slurries act like a single phase fluid with its density and viscosity being those of the suspension. Two models employed are the Bingham plastic and pseudoplastic models. Faddick [2] has observed that the model best describing the flow changes from Newtonian at low volumetric concentrations, C_V, to pseudoplastic and further to yield-pseudoplastic and at higher C_V (above 40%) to Bingham plastic. Here, critical velocity and pressure drop expressions are obtained for pseudoplastic model.

Unless a very fine solution exists, a turbulent flow is required to keep the solid particles in suspension. Durand [3] recommends the selection of the flow velocity to be substantially larger than V_T. An expression for V_T is given in [4].

Pressure drop consists of the components due to friction loss and elevation difference. Pressure drop together with flow rate determines the horspower requirement of the pumps. Brakehorsepower expressions for pumps (BHPP) and for pump motors (BHPM) are derived in [4] based on the recommendation by Metzner and Reed [5] that the Newtonian relationship for the friction factor f in rough pipes can be used as a conservative measure for non-Newtonian fluids.

Sectioning. The elevation difference between the ends of the pipeline might not be representative of the entire elevation gradient needed for determining the pump size required. For the purpose of accounting for all types of elevation profile, four classes of profiles are defined. In type 1, the elevation profile is an ascending one and stays below the level of the terminal point. In type 3, the terminal point is again at a higher level than the starting point, but the elevation profile at some portions of the line lies above the level of the terminal point. If no crest of the elevation profile is higher than the terminal point except the starting point, then the elevation profile is of type 2. Any other descending profile belongs to type 4. After the elevation profile of interest is classified as one of the four types, a sectioning procedure can be applied to obtain the number of sections between node i and node j, m_{ij}, the length of each section, L_{ijk}, and the elevation difference of each section, $\Delta hijk$. These are used as input data for solving the design problem.

PROBLEM FORMULATION

The objective function consists of pipe, pump and pump motor costs and their operation and maintenance costs. It is expressed as an annual discounted cost. Pipe cost is taken to be a linear function of its weight [6]. Positive displacement pumps and electric motors are assumed to be employed in the pipeline. Pump cost is taken to be a linear function of BHPP and pump motor cost a concave piecewise linear function of BHPM [6] . They are assumed not to have any capacity constraint. The availability of the system is assumed to be less than

100%. The maintenance cost is expressed as a fixed proportion of the initial investment cost. Costs associated with pumps and pump motors are expressed in terms of BHPP and BHPM. An expression for BHPP is given by the enegy equation constraint (11) and BHPM is related to BHPP as : BHPP=$(1-C_V)$ (eff. of pump) BHPM.

It is assumed that concentration and dry solid flow rate over any arc stay constant and the pumps are located at the starting points of the pipelines when needed.

The following notation is used : x is volumetric solids concentration, y pipe inside diameter, z dry solid flow rate, ρ water density, g gravitational acceleration, s safety increment, p pressure, ρ_c solid density, p_V vapor pressure p_{min} minimum pressure requirement, V_{max} maximum allowable velocity, e demand for dry solid, D set of commercially available steel pipes, t pipe thickness and δ pressure difference.

$$\min_{\underline{x},\underline{y},\underline{z},\underline{u}} \; Z = \sum_{i=1}^{n} \sum_{j=1}^{n} \Bigg\{ \sum_{k=1}^{m_{ij}} A_1 (y_{ijk} + t_{ijk})\, t_{ijk}\, L_{ijk} + \sum_{k=1}^{m_{ij}} A_2 (BHPP)_{ijk} + \{(1-u_{ij}) A_3 + u_{ij} A_4\} \sum_{k=1}^{m_{ij}} (BHPM)_{ijk} + u_{ij} A_5 + \sum_{k=1}^{m_{ij}} A_6 (BHPM)_{ijk} \Bigg\} \qquad (M1)$$

<u>Subject to</u> :

(1) $\sum_{j=1}^{n} Z_{1j} = \sum_{j=2}^{n} e_j$

(2) $\sum_{i=1}^{n} Z_{ij} - \sum_{i=2}^{n} Z_{ij} = e_j \qquad \forall j \neq 1$

(3) $Z_{jj} = 0 \qquad \forall j \neq 1$

(4) $Z_{j1} = 0 \qquad \forall j$

(5) $y_{ijk} < 24\, (Z_{ij} / \pi g \rho_c X_{ij} V_T)^{1/2} \qquad \forall i,j,k$

(6) $y_{ijk} \geq 24\, (Z_{ij} / \pi g \rho_c X_{ij} V_{max})^{1/2} \qquad \forall i,j,k$

(7) $y_{ij} \in D \qquad \forall ijk$

(8) $x_{ij} \leq C_{vmax} \qquad \forall ij$

(9) $p_{ijm_{ij}} \geq (p_{min})_j \qquad \forall ij$

(10) $p_{ijk} \geq (p_V)_{ij} - s \qquad \forall ijk$

(11) $550 (BHPP)_{ijk}\, g\, \rho_c\, X_{ij}\, Z_{ij}^{-1} = \delta_{ijk} + (\Delta h)_{ijk}\, g\, \{\rho_c\, x_{ij} + \rho(1-x_{ij})\}$

$+ 32\, Z_{ij}^2 \{\rho_c\, x_{ij} + \rho(1-x_{ij})\} (\pi \rho_c\, g\, x_{ij})^{-2} (12/y_{ijk})^5 L_{ijk}\, f_{ijk} \qquad \forall ijk$

(12) $u_{ij} = \begin{cases} 1, & \text{if } (BHPM)_{ij} > 800 \text{ HP} \\ 0, & \text{otherwise} \end{cases} \qquad \forall ij$

(13) $Z_{ij} \geqslant 0 \quad \forall_{ij}$ (14) $X_{ij} > 0 \quad \forall_{ij}$

The problem above is a nonconvex minimization problem with nonlinear and linear constraints. Since there are no well established techniques to solve this problem, the special structure of the problem will be exploited to get a solution.

DECOMPOSITON OF THE PROBLEM

Observe that for a given network to bea a least cost network, it should not only have the least cost compared to all other feasible networks but it must also have at least cost for that specific network structure and the resulting flow rate assignment. Observe further that the linking variables for the arcs in the network are the flow rate in each arc and the pressure at the end of each arc. But since the capacity of pump at each arc is assumed to be continuous, then the constraint sets (9) and (10) will be satisfied as equalities. In view of this, the constraint sets (1), (2), (3), (4) and (13) of (M1) are 'independent' from the rest of the constraint sets in the sense that the above constraint sets are related to the network structure and the remaining ones are imposed on the flow in each arc. The mathematical formulation of the overall problem will be rewritten, this time reflecting the concept of decomposition in this context.

$$\min_{\underline{z}} \left\{ \sum_i \sum_j \min_{x,\underline{y},u} \left[\sum_k Z_{ijk} \mid (5), (6), (7), (8), (9), (10), (11), (12), (14) \right] \mid (1), (2), (3), (4), (13) \right\} \qquad \text{(Eq. 1)}$$

The decomposition as expressed above results in the introduction of a subproblem which is the optimization of an arc with a given level of dry solids flow rate. This subproblem is called the single line optimization problem.

SINGLE LINE OPTIMIZATION

Single line optimization problem (M2) is the inside minimization problem of (Eq.1). The objective function of (M2) is denoted by Z^1.

Also, let $y^1 \triangleq \{\min y \mid y \in D \text{ and } y \geqslant y_{min}\}$ and $y^2 \triangleq \{\max y \mid y \in D \text{ and } y < y_{max}\}$ where y_{max} and y_{min} are calculated from constraints (5) and (6) respectively.

Two different cases will be distinguished when attempting to solve (M2). These are the nonnegative and negative elevation differences. The reason for such a classification is the structure of Z^1. The terms in Z^1 which are not a function of the elevation difference Δh are always positive for any peasible value of the decision variables. But the terms which are function of Δh can be either negative zero or positive, depending on the value of Δh. In characterizing the monotonicity of the objective function, the sign of each term in the objective function as well as constraints becomes important. Therefore, basing the classification of the solution approach on the sign of Δh is justified.

The results leading to a very simple and fast solution method are summarized in Table I.

TABLE I

Case	Criterion	Characteristic	Solution
I	$\Delta h > 0$	Pumping required.	$x^* = C_{Vmax}$; y^* is obtained by searching over all feasible pipe diameters (result 1).
II	$\Delta h < 0$	No pumping required.	$x^* = C_{Vmax}$; $y^* = y^1$ (result 2).
		a) Pumping required for all feasible pipe diameters.	$x^* = C_{Vmax}$; y^* is obtained by searching over all feasible pipe diameters (result 3).
		b) Pumping required only for a range of pipe diameters.	$x^* = C_{Vmax}$; y^* is obtained by searching over all feasible pipe diameters (result 4).

THE NETWORK ALGORITHM

It is shown 4 that the outer minimization problem of (Eq.1) corresponds to the minimization of a quasi-concave function over a linear constraint set S. Such functions are shown to take their minimum at at least one of the extreme points of S and that the extreme points of S correspond to arborescences.

The algorithm starts with an initial arborescence. Adjacent arborescences are generated by dropping an arc and adding an arc such that the network structure still spans all the nodes. At each iteration, a node is considered for a change of its current supply node. If there are one or more adjacent arborescences with lower cost then the one with minimum cost is selected. Otherwise, the current arborescence is preserved. In essence then, the algorithm is an adjacent extreme point method. Considering every node for a change of its supply node is called a PASS. When during a PASS no change occurs then the algorithm stops. Nodes are considered in a PASS in decreasing order of unit cost of flow. When considering a node, say node m, for a possible change of its supply node, the set of candidate nodes CN(m) is comprised of all the remaining nodes except the current supply node. For an efficient algorithm, a set of criteria has to be developed for reducing CN(m) without going through any optimization process. These rules are called reduction properties. Following definitions are needed for stating those properties.

The _path_ P(j) of node j is the ordered set of nodes covered when starting from node 1 and going to node j.

A _branch_ B is a collection of all those paths such that the second element of each path included is the same.

Reduction Property 1 (RP1) :The node j, if $m \in P(j)$, has to be eliminated from CN(m).

RP1 prevents the creation of infeasible flow. The remaining reduction properties are based on the idea of economies of scale. RP2, RP3 and RP4 are applicable only starting with PASS 2.

Reduction Property (RP2).Suppose the flow in the arcs of the path P(m) have not decreased after node m has been considered for a change of its supply node in the previous PASS. Then all the elements of other branches which have exprienced no changes, or only loss of nodes to other branches since the previous PASS, are eliminated from CN(m). Furthermore, the elements of the branch that node m belongs to should be eliminated from CN(m) if this branch has not changed since the previous PASS.

Reduction Property 3 (RP3). Suppose there has been no decrease of flow in the arcs of P(m) since the previous PASS when node m has been considered for a change of its supply node. Then, eliminate node k which belongs to the same branch as node m, from the candidate supply node set CN(m) if the flow in the arcs of P(k) not belonging to arcs of P(m) does not increase.

Reduction Property 4 (RP4). Suppose there has been no decrease of flow in the arcs of P(m) since the previous PASS when node m has been considered for a change of its supply node. Then, node k an element of a different branch is eliminated from CN(m) if there has been no increase of flow in the arcs of P(k)since the previous PASS.

Reduction Property 5 (RP5). Suppose node k is the element of CN(m) which has the least distance from node m. Then, all the nodes which belong to CN(m) and are downstream of node k are eliminated from CN(m).

Reduction Property 6 (RP6). Suppose node m has as its supply node the source node. All the nodes of CN (m) are eliminated if the closest node to node m is the source node.

THE CONVERGENCE OF THE ALGORITHM

The convergence of algorithm to a local optimum will be estabilished by first proving that the algorithm produces a solution in a finite number of steps, and then showing that the solution obtained is indeed a local optimum.

For proving that the algorithm produces a solution in a finite number of steps, observe that no arborescence is repated during the procedure. This follows from the fact that for an arborescence to become the new network structure, the objective function value has to strictly decrease according to improvement criterion. That is, once an arborescence has been eliminated when compared with the current network structure, or an arborescence has been replaced by another arborescence as the new network structure, then it can never again be accepted as the new network structure. Moreover, since the number of arborescences that can be generated from a finite set of nodes with a specified source node is finite, it follows that the algorithm arrives at a solution in a finite number of steps.

A local optimum is defined as follows :

Definition : A network flow is said to be a local minimum if it is an arborescence flow and no adjacent arborescence has a smaller objective function value.

Recall that a PASS corresponds to a search over adjacent arborescences and the algorithm stops when no change occurs in a PASS, i.e. no better adjacent arborescence is found. It then follows that the resulting arborescence is indeed a local optimum as defined above.

CONCLUSIONS

Several example problems have been solved using the coal slurry data reported by Faddick [2]. As a result, it has been concluded that the volumetric solids concentration will always be at its upper bound when transporting homogeneous slurries. Furthermore, it is found that in ascending section of the pipeline the pipe diameter is chosen to be relatively larger in order to reduce the pumping requirement. These results are found in agreement with conclusions reached by Bain and Bonnington [7].

Problems with 5 to 12 nodes have been solved using the solution procedure suggested. In almost all problems, the algorithm has produced a local optimum in two or three passes. Specifically, in a five node problem, a local optimum has been achieved in two passes. Out of the possible 125 arborescences, 22 have been considered and for only 3 of these the annual total cost has been calculated. The subproblem has been solved 17 times. Furthermore, the local optimum found has been identified via complete enumeration to be the global optimum.

In conclusion, it appears that a promising solution technique has been developed for the single source tree shape pipeline networks. A further contribution of this study has been the development of a very fast optimization method for the homogeneous slurry pipelines which can be extended to other non-Newtonian fluids and heterogeneous and intermediate slurries.

References

1. E.J. Wasp, T.L. Thompson and P.E. Snoek, "The Era of Slurry Pipelines", Chem. Tech., Vol. 1, No. 9, pp. 552-562, Sept. 1971.

2. R.R. Faddick, "Flow Properties of Coal-Water Slurries", Proc., Third Int. Conf. on the Hydraulic Transport of Solids in Pipes, May 1974.

3. R. Durand, "The Hydraulic Transport Coal and Other Solid Material in Pipes", Colloquium of the National Coal Board, London, Nov. 5, 1952.

4. A. G. Ulusoy and D.M. Miller, "Pipeline Network Optimization- An Application to Slurry Pipelines", RN-7706, Dept. of Ind. Engg, Bogaziçi Uni., Istanbul, Turkey, 1977.

5. A.B. Metzner and J.C. Reed, "Flow of Non-Newtonian Fluids-Correlation of the Laminar, Transition and Turbulent Flow Regions", A.I.Ch.E. Journal, Vol. 1, pp.434-440, 1965

6. J.M. Link, N.J. Lavingia and R.R. Faddick, "The Economic Selection of a Slurry Pipeline", Proc., Third Int. Conf. on the Hydraulic Transport of Solids in Pipes, May 1974.

7. A.G. Bain and S.T. Bonnington, "The Hydraulic Transport of Solids by Pipeline", Pergamon Press, New York, 1970.

LINEAR FITTING OF NON-LINEAR FUNCTIONS IN OPTIMIZATION. A CASE STUDY: AIR POLLUTION PROBLEMS.

L.F. Escudero
A.M. Vazquez-Muñiz
IBM Madrid - Scientific Center
Madrid - 1 Spain

SUMMARY

A problem on mathematical programming is the linear approximation of nonlinearities in the constraints or in the objective function of a linear programming problem. In this paper, we compare the representation of nonlinear functions of a single argument by approximations based on piecewise constant, piecewise adjacent, piecewise non-adjacent additional and piecewise non-adjacent segmented functions. In each modelization we show the problem size and the results of the following techniques: separable programming, mixed integer programming with Special Order Sets of type 1, linear programming with Special Order Sets of type 2 and mixed integer programming.
The considerations involved in these alternative modelizations are illustrated using an air pollution abatement model. Also we study the possibilities of the sensitivity analysis in each type of modelization and optimizing strategy.

1. INTRODUCTION

This paper is an end part of [6] but the results have also a wide applicability. We analyze several ways of representation and subsequent optimization of single argument non-linear functions in mathematical programming models. The study is based on a model developed for the evaluation and selection of pollutant emission control policies and standards. To this purpose, the concepts, notation, formulations and some optimizing techniques are the same used in [6].
The problem description is presented in sec. 2. The main feature of this work is the comparative analysis of these methods from the aspect of the representation and optimization of the functions on a sophisticated mathematical programming model, Sec. 3 describes the modelization in the piecewise constant function approximation [6]. Secs. 4, 5 and 6 present the piecewise non-adjacent additional, piecewise non-adjacent segmented and piecewise adjacent representations (table 1). The structure of secs. 3 to 6 is as follows: The formal presentation of the

problem, several ways of optimization, and problem dimensions. Sec. 7 discusses computational experience obtained from the problem size optimization results and in each one of the following techniques: separable programming, mixed integer programming with Special Order Sets of type 1, linear programming with Special Order Sets of type 2, and mixed integer programming. Finally, sec. 8 discuss the comparative analysis of these methods.

2. PROBLEM DESCRIPTION

The air pollution reduction model described in 6 uses probabilistic limits of pollutant concentration represented by the global probability that real concentration, exceeds a concentration of reference. The problem is to find the best linear approximation of non-linear functions in the probability constraints of the model.
Notation and terminology are based on ordinary mathematical programming usage and some special notation and terminology of air pollution abatement is used. Escudero and Vázquez-Muñiz [6] make a disecription of the concepts described herein.
We use the subscript e for the influent emitters (NE total number) the subscript r for the polluted receptors (NR total number), the subscript m for the meteorological situations that are significant in the pollution effects (M total number), the subscript rm for a receptor r, polluted under the meteorological situation m (RM total number).
MP_m is the ocurrence probability of the meteorological situation m for the period under study. AQ_e is the emission of the emitter e for the period considered, this emission will be reduced if it is necessary but no more than MQ_e. Due to the effect caused, the emission coming from e is penalized by W_e in the sense of "more influence, more reduction". K_{rem} is the unit influence of emitter e upon receptor r in the meteorological situation m, this value is given by a diffusion model. This diffusion model also gives the estimation of the probability distribution of the pollutant concentration in the receptor r for each meteorological situation m. Representative values of this distribution are the value TC_{rm} (i.e. the mean), and the probability of exceed the concentration limit (AL), say PL_{rm}. Thus the global probability of exceed this limit will be:

$$PL_r = MP_m \; PL_{rm} \leq RL \tag{1}$$

less than a fixed maximum RL. If the value AL is fixed and we reduce the pollution in r, this reduction may be represented by the values XC_{rm}, PL_{rm}. These values give a descent function in PL_{rm} values.

There are several ways of representing the non-linear function PL_{rm} of a single argument XC_{rm} in mathematical programming terms. We may replace this function by a piecewise linear approximation based on a finite number of points (NP_{rm}). Let the co-ordinates of these points be $UC_{rm}^{(p)}$ and $PAL_{rm}^{(p)}$ where $p=1,2,\ldots,NP_{rm}$. Escudero and Vázquez-Muñiz [6] the method used to obtain NP_{rm}, and $(UC^{(p)}, PAL^{(p)})$.
Upper bound $UC_{rm}^{(NP)}$ is the concentration TC_{rm} corresponding to the emissions set without any reduction.
The objective function is the minimization of the emissions reduction,

$$\text{Min } AP = \sum_{e=1}^{NE} \frac{1}{W_e} XQ_e \tag{2}$$

such that priority is given to the emission from the emitter that pollutes most, subject to the constraints (1).
Table 1 gives several ways of linear approximation of non-linear functions and four types of optimizing strategies.

Table 1. Type of modelization and optimizing strategy (mod-opt)

Type of modelization \ Optimizing strategy	BB (a)	S1 (b)	S2 (c)	SEP (d)
Piecewise constant (1)	x	x		
Piecewise non-adjacent additional (2)	x	x		
Piecewise non-adjacent segmented (3)	x			x
Piecewise adjacent (4)	x	x	x	x

The notation of Table 1 is the following: (a) BB, Branch and bound; (b) S1, mixed integer programming (MIP) with Special Order Sets of Type 1; (c) S2, linear programming with Special Order Sets of type 2; (d) SEP, separable programming.
The optimizing strategy BB, considers that are binary variables. It uses the posibilities described in [6 secs. 6.2 and 7].
The optimizing strategies S2 and S2 (mod-opt 1b,2b,4b and 4c) are based on the Special Order Sets concept introduced by Beale and Tomlin [2] and developed by Tomlin [15] and Beale and Forrester [4]. See also [3, 10]. Special Order Sets of type 1, are sets of variables of which only one member may be non-zero in the final solution. Special Order Sets of type 2, are sets of variables of which, one or two members may be non-zero in the final solution, with the condition that if there are two, they must be adjacent. In our model we impose the condition that the sum of variables of the set be one, and in the strategy S1, the

condition that the variables are binary. Hence in the strategy S1, only one variable must be 1, and in the strategy S2, only the sum of at most two adjacent variables is 1.

The optimizing strategies BB and S1 use the MPSX system [9]. The optimizing strategy S2 uses a special algorithm of linear programming included in the SCICONIC system [14].

The optimizing strategy SEP (mod-opt 3d and 2d) uses separable programming [1, 8, 12], which assumes that all non-linear expressions can be separated into sums of linear functions of single argument that form a set of special variables called separable set. In our model, each separable set corresponds to each rm situation. $(UC_{rm}^{(p)}, PAL_{rm}^{(p)})$ (p=1,2, ..., NP_{rm}). In sucessive references these are the limits of p except if other are stated.

If PL_{rm} is convex 11 local optimum is always global optimum, but the probability function PL_{rm} is non-convex when the increment of probabilities $PAL_{rm}^{(p)} - PAL_{rm}^{(p-1)}$ is not always decreasing.

Table 2. Model dimensions in the mod-opt types of Table 1

Mod-opt type	Ordinary rows	special rows SOS/SEP	continuous variables	binary/SOS or separable variables
1a	2+2*RM+NC	-	NE	NT(BB)
1b	2+RM+NC	RM	NE	NT(S1)
2a	2+2*RM+NC+NT	-	NE+NT	NT(BB)
2b	2+RM+NC+NT	RM	NE+NT	NT(S1)
3a	2+2*NT-RM+NC	-	NE+NT	NT+RM(BB)
3d	2+RM+NC	RM(SEP)	NE+NT	NT(SEP)
4a	2+3*RM+NC+NT	-	NE+NT+RM	NT(BB)
4b	2+2*RM+NC+NT	RM	NE+NT+RM	NT(S1)
4c	2+RM+NC	RM	NE	NT+RM(S2)
4d	2+RM+NC	RM(SEP)	NE	NT+RM(SEP)

3. MODELIZATION PIECEWISE CONSTANT

This type [13 p. 110] is set in Escudero and Vázquez-Muñiz [6]. The basis employed for each significant situation rm is the binomial $(UC_{rm}^{(p)}, PAL_{rm}^{(p)})$.

We consider that XC_{rm} represents the concentration corresponding to the new emissions, then

$$XC_{rm} = TC_{rm} - \sum_{e=1}^{NE} K_{rem}(AQ-XQ_e) \quad \forall rm \in RM \tag{3}$$

and the probability that the concentration is greater than AL is the value $PAL_{rm}^{(p)}$ corresponding to the upper bound $UC_{rm}^{(p)}$ immediately above XC_{rm}. This solution no offers great exactness, since the linealization is achieved by steps, estimating a greater value for PL_{rm}.
The model is the following:

$$\text{Min } AP = \sum_{e=1}^{NE} \frac{1}{W_e} XQ_e \tag{4}$$

subject to

$$\sum_{p=1}^{NP_{rm}} UC_{rm}^{(p)} Y_{rm}^{(p)} + \sum_{e=1}^{NE} K_{rem} XQ_e \geq TC_{rm} \qquad \forall rm \varepsilon RM \tag{5}$$

$$\sum_{p=1}^{NP_{rm}} Y_{rm}^{(p)} = 1 \qquad \forall rm \varepsilon RM \tag{6}$$

where

$$\sum_{m}^{M} MP_m \sum_{p=1}^{NP_{rm}} PAL_{rm}^{(p)} Y_{rm}^{(p)} \leq RL \qquad \forall rm \varepsilon RM \tag{7}$$

$$0 \leq XQ_e \leq MQ_e \qquad \forall e \varepsilon NE \tag{8}$$

$$Y_{rm}^{(p)} \varepsilon \{0;1\} \quad \text{for } p=1,2,\ldots, NP_{rm} \text{ and} \qquad \forall rm \varepsilon RM \tag{9}$$

There are continuous variables (XQ_e) and binary variables ($Y_{rm}^{(p)}$). Hence, the optimization of this model should be carried out using mixed integer programming by strategies BB and S1, respectively mod-opt 1a and 1b.
If variable $Y_{rm}^{(p)}$ is considered binary, the optimizing strategy BB is applied. On the other hand, given the constraint (6) may be considered that variables $Y_{rm}^{(p)}$ belong to a S1 set in which case the optimizing strategy S1 could be applied.
When in a node, the next branching variable selection of the BB strategy only considers those variables $Y_{rm}^{(p)}$ which are not integer and it chooses the variable with greatest associated probability $MP_m PAL_{rm}^{(NP_{rm})}$.
The branching process uses a variable with strong influence in a constraint difficult to satisfy making a greater deterioration of the functional value.
The S1 strategy (mod-opt 1b) considers that the variables (6) form a S1 set and use the strategy described in [6 secs. 6 and 7].
The importance of each variable $Y_{rm}^{(p)}$ in each S1 set is measured by a weight by the constraint (7).
Table 2 shows the dimensions of these mod-opt types.

4. MODELIZATION PIECEWISE NON-ADJACENT ADDITIONAL

This type obtains a real linearization of the probability PL_{rm}. The piecewise non-adjacent additional type [1 p. 140] considers the values of XC_{rm} and PL_{rm} to be respectively the concentration $UC_{rm}^{(p)}$ and probability $PAL_{rm}^{(p)}$ inmediately below plus the corresponding difference $L_{rm} \leq UC_{rm}^{(p+1)} - UC_{rm}^{(p)}$ and the linear proportion $G_{rm}L_{rm}$, where $G_{rm}^{p} = (PAL_{rm}^{(p+1)} - PAL_{rm}^{(p)})/(UC_{rm}^{(p+1)} - UC_{rm}^{(p)})$.

The model will be the following:

$$\text{Min. } AP \sum_{e=1}^{NE} \frac{1}{W_e} XQ_e \tag{10}$$

subject to

$$\sum_{p=1}^{NP_{rm}} UC_{rm}^{(p)} Y_{rm}^{(p)} + \sum_{p=1}^{NP_{rm}} L_{rm}^{(p)} + \sum_{e=1}^{NE} K_{rem} XQ_e \leq TC_{rm} \qquad \forall rm \varepsilon RM \tag{11}$$

$$\sum_{p=1}^{NP_{rm}} Y_{rm}^{(p)} = 1 \qquad \forall rm \varepsilon RM \tag{12}$$

$$\sum_{m=1}^{M} \sum_{p=1}^{NP_{rm}} MP_m \, PAL_{rm}^{(p)} Y_{rm}^{(p)} + \sum_{m=1}^{M} \sum_{p=1}^{NP_{rm}} MP_m \, G_{rm}^{(p)} L_{rm}^{(p)} \leq RL \qquad \forall r \varepsilon NC \tag{13}$$

$$0 \leq L_{rm}^{(p)} \leq (UC_{rm}^{(p+1)} - UC_{rm}^{(p)}) \, Y_{rm}^{(p)} \quad \text{for } p=1,2,\ldots,NP_{rm} \quad \forall rm \varepsilon RM \tag{14}$$

$$0 \leq XQ_e \leq MQ_e \qquad \forall e \varepsilon NE \tag{15}$$

$$Y_{rm}^{(p)} \varepsilon \{0;1\} \text{ and continuous for } p=0,1,2,\ldots,NP_{rm} \qquad \forall rm \varepsilon RM \tag{16}$$

In this model there are three types of variables: XQ_e (continuous), $Y_{rm}^{(p)}$ (binary), which determines the value p and $L_{rm}^{(p)}$ (continuous).
If we consider the variables $Y_{rm}^{(p)}$ binary. The strategy BB(mod-opt 2a) is applied. On the other hand, given the constraint (12) and considering that these variables form a S1 set, the strategy S1 (mod-opt 2b) may also be applied.
The optimizing strategy BB(mod-opt 2a) follows the same procedure and criteria set for the optimizing strategy BB(mod-opt 1a). The optimizing strategy S1(mod-opt 2b) use the same strategy as for mod-opt 1b. The only difference lies in that the weight is not in the constraints, but in the reference row

$$W_{rm} = \sum_{p=0}^{NP_{rm}} W_{rm}^{(p)} Y_{rm}^{(p)} \qquad \forall rm \varepsilon RM \tag{17}$$

where $W_{rm}^{(p)} = p$. Escudero and Vazquez-Muñiz [6 sec. 6] indicate the branching procedure on the rm set and the method for obtaining the branching index l ($W_{rm}^{l} \leq W_{rm} \leq W_{rm}^{l+1}$). Table 2 shows the model dimensions for the strategies BB and S1.

5. MODELIZATION PIECEWISE NON ADJACENT SEGMENTED

This type also brings a real linealization of the probability PL_{rm}. It has the same type as that of the preceding section and gives identical results.

This type of modelization [9 chap. 6] considers that the binomials required are $(UC_{rm}^{(p)} - UC_{rm}^{(p-1)},\ PAL_{rm}^{(p)} - PAL_{rm}^{(p-1)})$. Let us consider that values for XC_{rm} and PL_{rm} are respectively:

$$XC_{rm} = UC_{rm}^{(p-1)} + (UC_{rm}^{(p)} - UC_{rm}^{(p-1)})\ Y_{rm}^{(p)} \tag{18}$$

$$PL_{rm} = PAL_{rm}^{(p-1)} + (PAL_{rm}^{(p)} - PAL_{rm}^{(p-1)})\ Y_{rm}^{(p)} \tag{19}$$

for $\forall rm \varepsilon RM$, where $0 \le Y_{rm}^{(p)} \le 1$ and $p=1,2,\ldots,NP_{rm}$. (20)

This type of modelization is known as segmented because the values for XC_{rm} and PL_{rm} are obtained by adding the respective increases in the points of linealization.

The model is the following:

$$\text{Min} \quad AP = \sum_{e=1}^{NE} \frac{1}{W_e} XQ_e \tag{21}$$

subject to

$$\sum_{p}^{NP_{rm}} (UC_{rm}^{(p)} - UC_{rm}^{(p-1)})\ Y_{rm}^{(p)} + \sum_{e=1}^{NE} K_{rem} XQ_e \le TC_{rm} \qquad \forall rm \varepsilon RM \tag{22}$$

$$\sum_{m=1}^{M} \sum_{p=1}^{NP_{rm}} MP_m\ (PAL_{rm}^{(p)} - PAL_{rm}^{(p-1)})\, Y_{rm}^{(p)} \le RL \qquad \forall r \varepsilon NC \tag{23}$$

$$0 \le XQ_e \le MQ_e \qquad \forall e \varepsilon NE \tag{24}$$

$$0 \le Y_{rm}^{(p)} \le 1 \text{ and continuous } \quad \forall p \varepsilon NP_{rm},\ rm \varepsilon RM \tag{25}$$

In this model, all variables are continuous (XQ_e and $Y_{rm}^{(p)}$) with the additional condition that if $Y_{rm}^{(p)} \ne 0 \rightarrow Y_{rm}^{(1)} = \ldots = Y_{rm}^{(p-1)} = 1$ and $Y_{rm}^{(p+1)} = \ldots = Y_{rm}^{(NP_{rm})} = 0 \quad \forall rm \varepsilon RM$.

With the application of strategy SEP [22 chap. 6] it is not necessary to include any other constraint or variable, but we must take into account that if the function PL_{rm} is non convex (sec. 2), the local optimum may possibly not be a global optimum and that case may produce unexpected results.

The disadvantage that the local minimun may possibly not be a global minimum may be avoided if we use the optimizing strategy BB (mod-opt 3a). In this case it is necessary to add explicity to the model (21) to (25)

the condition of segmentability $Y_{rm}^{(p)} \neq 0 \rightarrow Y_{rm}^{(H)} = 1$ for H=1,2,...,p-1 and $Y_{rm}^{(L)} = 0$ for $L=p+1,...,NP_{rm}$, $\forall rm \varepsilon RM$.
For this it is necessary to introduce the binary variable $_{rm}^{(p)}$ associated to each variable $Y_{rm}^{(p)}$ in such a manner that they satisfies the condition of segmentatibility. This formally results in

$$Y_{rm}^{(p)} \geq \delta_{rm}^{(p)} \quad \text{for } p=1,2,...,Np_{rm}-1 \tag{26}$$

$$Y_{rm}^{(p)} \leq \delta_{rm}^{(p-1)} \quad \text{for } p=2,3,...,NP_{rm} \tag{27}$$

$$\delta_{rm}^{(p)} \ \varepsilon\{0;1\} \quad \text{for } p=1,2,..., NP_{rm}-1 \tag{28}$$

The optimizing strategy BB(mod-opt 3a) has the same procedure and criteria as mod-opt 1a and 2b. Thus, the method of choosing the binary variable $\delta_{rm}^{(p)}$ is the same as for the variable $Y_{rm}^{(p)}$ in mod-opt 1a, using the concepts in a similar manner as well as the parameters PC_{rm} and $PAL_{rm}^{(p)}$ of the affected variable $Y_{rm}^{(p)}$ and strategies employed as before.

6. MODELIZATION PIECEWISE ADJACENT

Although this type also produces a real linealization of function PL_{rm}, thus giving the same results, its treatment is different since for the level p, XC_{rm} and PL_{rm} take the proportional value depending on the distance to each adjacent bound.

$$XC_{rm} = UC_{rm}^{(p-1)} \ Y_{rm}^{(p-1)} + UC_{rm}^{(p)} Y^{(p)} \tag{29}$$

$$Y_{rm}^{(p-1)} + Y_{rm}^{(p)} = 1 \qquad 0 \leq Y^{(p)} \leq 1 \quad \forall p \tag{30}$$

$$PL_{rm} = PAL_{rm}^{(p-1)} \ Y_{rm}^{(p-1)} + UC_{rm}^{(p)} \ Y^{(p)} \tag{31}$$

This type of piecewise is known as adjacent or consecutive because the values of XC_{rm} and PL_{rm} are obtained by a weighting of the inmediately upper and lower binomials ($UC_{rm}^{(p)}$, $PAL_{rm}^{(p)}$) [1 pp. 138-139, 5 p. 542, 7 p. 13, 13 pp. 110-111, 16 pp. 451-454].
The model is the following

$$\text{Min AP} = \sum_{e=1}^{NE} \frac{1}{W_e} XQ_e \tag{32}$$

subject to

$$\sum_{p=1}^{NP_{rm}} UC_{rm}^{(p)} \ Y_{rm}^{(p)} + \sum_{e=1}^{NE} K_{rem} \ XQ_e \leq TC_{rm} \qquad \forall rm \varepsilon RM \tag{33}$$

$$\sum_{p=0}^{NP_{rm}} Y_{rm}^{(p)} = 1 \qquad \forall rm \varepsilon RM \tag{34}$$

$$\sum_{m=1}^{M}\sum_{p=1}^{NP_{rm}} MP_{rm}\ PAL_{rm}^{(p)}\ Y_{rm}^{(p)} \leq RL \qquad \forall r \varepsilon NC \tag{35}$$

$$0 \leq Y_{rm}^{(p)} \leq 1 \quad \text{and continuous for } p=0,1,\ldots,NP_{rm}, \quad rm \varepsilon RM \tag{36}$$

All the variables in the model are continuous (XQ_e and $Y_{rm}^{(p)}$) with the additional condition that, in each rm situation, a maximum of two variables $Y_{rm}^{(p)} \neq 0$ are allowed, and in this case they should be adjacent. Table 1 shows that the optimizing strategies of this type of modelization are BB(mod-opt 4a), S1(mod-opt 4b), S2(mod-opt 4c) and SEP(mod-opt 4d) applying respectively mixed integer programming (MIP), MIP with SOS rows, linear programming with S2 sets and separable programming. With the strategy SEP(mod-opt 4d) [1, 8, 12], no formal inclusion is necessary of any other constraint or variable. The simplex method is used restricting 1 p. 125 the set of candidates for entering the basis. If two variables from a separate set (57) are in the basis, no third is allowed to enter. If only one is present, only its adjacent variables are considered. This modification is known as restricted basis entry. Also this modification is not needed if the function approximate is convex.

As indicated in sec. 5, we should consider that if the function PL_{rm} is not convex the local optimum does not guarantee the global optimum in separable programming. Table 2 shows the problem dimensions.

The local minimum disadvantage is evited by the S2 strategy (mod-opt 4c). For this, there is no need to add to the model any constraint. On the other hand, in the optimizing method we should take into account that the constraint (34) forms a S2 set [2, 15]. Thus, it should be allowed that for each rm situation there are at most two variables $Y_{rm}^{(p)} \neq 0$ for $p=0,1,2,\ldots,NP_{rm}$, and these should be adjacent. The method to be used (see references above) is very similar to the mixed integer programming method used by the optimizing strategy S1. This strategy has the same criteria as described by Escudero and Vázquez-Muñiz [6 secs. 6 and 7].

In the strategy S2, a weighted condition similar to (17) is always associated with each S2 set. In this case, the condition may be the following, not explicity expressed in the model,

$$PCY_{rm} = \sum_{p=1}^{NP_{rm}} MP_m\ PAL_{rm}^{(p)}\ Y_{rm}^{(p)} \tag{37}$$

The weights PCY_{rm} represent the importance given to each variable $Y_{rm}^{(p)}$. In the case of S2 sets, the dichotomy for the branching index (l) is obtained as follows: in a node all variables which precede $Y^{(l)}$ should be 0. We observe that the only variable common to the two nodes is precisely $Y_{rm}^{(1)}$. As we advance the number of variables "free" becomes lesser, and it is considered that the node is satisfied when no more than two variables $Y_{rm}^{(p)} \neq 0$ are adjacent. Beale and Forrest [4] describe a more sophisticated process. See also [3, 10]. Table 2 shows the model dimensions of mod-opt 4c.

The SCICONIC system 14 supports the strategy S2 according to the procedure described in [3, 4].

The strategies BB and S1 (respectively mod-opt 4a and 4b) need the inclusion in the model (32) to (36) of the binary variables $\delta_{rm}^{(p)}$, in such a manner that for each rm situation only one variable $\delta_{rm}^{(p)}$ may have the value of 1.

$$\sum_{p=1}^{NP_{rm}} \delta_{rm}^{(p)} = 1 \quad \forall rm \varepsilon RM \tag{38}$$

$$Y_{rm}^{(p)} \leq \delta_{rm}^{(p)} + \delta_{rm}^{(p+1)} \quad \text{for } p=1,2,\ldots,NP_{rm}-1; \quad \delta_{rm}^{(0)} = 0 \tag{39}$$

$$Y_{rm}^{(NP_{rm})} \leq \delta_{rm}^{(NP_{rm})} \quad \forall rm \varepsilon RM \tag{40}$$

$$\delta_{rm}^{(p)} \varepsilon \{0;1\} \quad \text{for } p=1,2,\ldots,NP_{rm},\ rm \varepsilon RM \tag{41}$$

It can be seen from (32) to (36) and (38) to (41) that there are three types of variables: XQ_e (continuous) $0 \leq Y_{rm}^{(p)} \leq 1$ (continuous) which gives the weighting of binomials when obtaining XC_{rm} and PL_{rm}, and $\delta_{rm}^{(p)}$ (binary) which indicates the pair of binomials to weight.

If we consider that in the problem the variables $\delta_{rm}^{(p)}$ are binary, strategy BB(mod-opt 4a) is applied. On the other hand, according to the constraints (38), it may be considered that for each rm situation, these variables form a S1 set and therefore the strategy S1(mod-opt 4b) may also be applied.

The optimizing strategy BB(mod-opt 4a) follows the process and criteria set out for mod-opt 1a, 2a and 3a. In this sense, the process for choosing the binary variable $_{rm}^{(p)}$ is the same as the applied for the variable $Y_{rm}^{(p)}$ in the case 1a. Table 2 shows the model dimensions for mod-opt 4a(BB).

The optimizing strategy S1(mod-opt 4b) considers that binary variables $\delta_{rm}^{(p)}$ form a S1 set (38) and thus, substituting $Y_{rm}^{(p)}$ for $\delta_{rm}^{(p)}$, uses the strategy described by Escudero and Vazquez-Muñiz [6 sec. 7]. Table 2 shows the problem dimensions. Very similar results may also be obtained by using the same strategy of mod-opt 2a and 2b in mod-opt 4a and 4b considering that for each variable $\delta_{rm}^{(p)}$ there is associated the weight $W_{rm}^{(p)} = p$. Hence, everything indicated in sec. 4 in this direction is valid for mod-opt 4a and 4b, substituting the variable $Y_{rm}^{(p)}$ by the variable $\delta_{rm}^{(p)}$, in such a manner that the reference row (17) would be

$$W_{rm} = \sum_{p=1}^{NP_{rm}} W_{rm}^{(p)} \delta_{rm}^{(p)} \tag{42}$$

In this way, the importance of variables $\delta_{rm}^{(p)}$ is measured by the value of $\underline{p}$ in direct relation to $PAL_{rm}^{(p)}$.

7. NUMERICAL ANALYSIS

Due to the lack of space, we summarize only the results of the five tests of each one of the eight strategies (mod-opt 1a, 1b, 2a, 2b, 3a, 3d, 4a, 4b; see table 1). The S2 modelling option has not been tested because the SCICONIC system is not available in our computer. All tests have been done in a IBM-370/158 using the MPSX-MIP/370 program for optimization, using the ECL option under PL/I control. The operating system was OS/VS 2 Rel 3 and the core average was 600 K-bytes of virtual memory.

The range of the dimensions (see table 2) was: NE from 30 to 40, NC from 8 to 20, M from 10 to 12, RM from 70 to 100 and NC_{rm} (mean) around 6.

In problem dimensions and matrix density, 1a, 1b and 3d have the lower dimensions and the strategies, 4a and 4b have bigger size than the 2a and 2b strategies.

The strategies 1a and 1b, give a 30% less accuracy in curve fitting than the rest of strategies, also have less non-zero elements and less CPU time in optimization. The rest of strategies have the same objective function and curve fitting results. The optimizing mode S1 (1b, 2b, 4b) has lower CPU time than the BB mode (1a, 2a, 3a, 4a), and the mode SEP (3d) has the best average CPU time in the five cases tested.

In the parametrization of the RL value, 1a and 1b gives a non-significative parametrization, 3d (SEP) offer a complete one, and the rest admits a partial parametrization of the RL value in the range

$\sum_{m=1}^{M} MP_m \; PAL_{rm}^{(p-1)} \leq RL \leq \sum_{m}^{M} MP_m \; PAL_{rm}^{p} \quad \forall r \varepsilon NC$. This range is determined fixing the binary variables in the optimal node solution. In contrast all strategies except 3d (SEP) give alternative solutions near the optimum one.

8. CONCLUSION. COMPARATIVE RESULTS

From the former chapter is obtained that the strategy 3d (piecewise non-adjacent segmented with SEP) is the simpler and faster one, having a complete precision in curve fitting. Also offers a complete parametrization of the RL value. If curves are non-convex Ch.2, this does not guarantee the global optimum and for alternative solutions strategy 2b will be preferred.

REFERENCES

[1] E.M.L. Beale, Mathematical Programming in Practice (Pitman, London, 1968).
[2] E.M.L. Beale and J.A. Tomlin, Special facilities in a general mathematical programming system for non-convex problems using ordered sets of variables, in: J. Lawrence (ed.), Op. Res. '69 (Tavistock Publ., London, 1969) 447-454.
[3] E.M.L. Beale, Branch and bound methods for mathematical systems, Discrete Optimization Conference DO'77 Vancouver, 1977.
[4] E.M.L. Beale and J.J.H. Forrest, Global optimization using Special Ordered Sets, Math. Prog. 10 (1976) 52-69.
[5] G.B. Dantzig, Recent advances in linear programming, Mgmt. Sci.2 (1956) 131-144.
[6] L.F. Escudero and A. Vazquez-Muñiz, The air pollution abatement. A mathematical model, in M. Roubens (ed.), Adv. in Op. Res. (North-Holland, 1977), 659-667.
[7] R.S. Garfinkel and G.L. Nemhauser, Integer programming (J. Wiley, N.Y., 1972).
[8] G. Hadley, Non-linear and Dynamic Programming (Addison-Wealey, London, 1964).
[9] IBM, Mathematical Programming Systems Extended/370:MPSX/370, SH19-1095 and SH19-1099, N.Y., 1974.
[10] A. Land and S. Powell, Computer codes for problems of integer programming, Discrete Optimization Conference DO'77, Vancouver, 1977.
[11] O.L. Mangasarian, Non-linear Programming (McGraw Hill, N.Y., 1969) chap. 4.
[12] C.E. Miller, The simplex method for local separable programming, in R.L. Graves and P. Wolfe (eds.), Rec. Adv. in Mat. Prog. (McGraw Hill, N.Y., 1963) 311-317.
[13] G. Mitra, Theory and Application of Mathematical Programming (Academic Press, London, 1976).
[14] SCICON, Scientific Control in Core: SCICONIC, Milton Keynes, England, 1977.
[15] J.A. Tomlin, Branch and bound method for integer and non-convex programming, in J.A. Abadie (ed.), Integer and Non-linear programming (North-Holland, Amsterdam, 1970) 437-450.
[16] G. Zoutendijk, Mathematical Programming Methods (North-Holland, Amsterdam, 1976) 451-454.

LIST OF AUTHORS	Volume/Page
Apostolescu, V.	2/22
Archetti, F.	2/380
d'Atri, G.	2/190
Bagchi, A.	1/149
Balakrishnan, A.V.	1/1
Bali, S.	2/59
Baras, J.S.*	
Bartholomew-Biggs, M.C.	1/92
Barton, C.F.	1/421
Benveniste, A.	1/128
Betrò, B.	2/380
Bianco, L.	2/443
Bielli, M.	1/441
Bräuninger, J.	2/33
Breakwell, J.V.*	
Bruni, C.	1/411
Bulirsch, R.*	1/-
Calicchio, G.	1/441
Camoin, B.	2/454
Carvalho Oliveira, R.	2/316
Chavent, G.	1/282
Cini, M.	1/441
Cirina, M.	2/389
Cohen, G.	1/282
Colding-Jørgensen, C.	2/422
Costa Lobo, M.L.	2/316
Cromme, L.J.*	
Cumani, A.	2/327
Curry, S.*	
DeClaris, N.*	
Del Bello, R.	2/327
Deuflhard, P.	2/22
Dickie, A.A.	1/117
Diekhoff, H.-J.*	1/-

*paper not received

Lecture Notes
in Control
nd Information
ciences

Volume 1

Distributed Parameter Systems: Modelling and Identification

Proceedings of the IFIP Working Conference, Rome, Italy, June 21–24, 1976
Edited by A. Ruberti
1977. Approx. 500 pages
ISBN 3-540-08405-3

Volume 2

New Trends in Systems Analysis

International Symposium, Versailles, December 13–17, 1976
Edited by A. Bensoussan and J.L. Lions
1977. 104 figures, 32 tables.
VII, 759 pages (101 pages in French)
ISBN 3-540-08406-1

Volume 3

Differential Games and Applications

Proceedings of a Workshop, Enschede, Netherlands, March 16–25, 1977
Edited by P. Hagedorn, H.W. Knobloch and G.J. Olsder
1977. 60 figures, 6 tables.
XII, 236 pages
ISBN 3-540-08407-X

Volume 4

M.A. Crane, A.J. Lemoine

An Introduction to the Regenerative Method for Simulation Analysis

1977. 4 figures, 10 tables.
VII, 111 pages
ISBN 3-540-08408-8

Volume 5

D.J. Clements, B.D.O. Anderson

Singular Optimal Control

The Linear-Quadratic Problem
1978. V, 93 pages
ISBN 3-540-08694-3

Volume 6

Optimization Techniques

Proceedings of the 8th IFIP Conference on Optimization Techniques, Würzburg, September 5–9, 1977
Part 1
Editor: J. Stoer
1978. 115 figures.
XIII, 528 pages
ISBN 3-540-08707-9

Volume 7

Optimization Techniques

Proceedings of the 8th IFIP Conference on Optizimation Techniques, Würzburg, September 5–9, 1977
Part 2
Editor: J. Stoer
1978. 82 figures.
XIII, 512 pages
ISBN 3-540-08708-7

Springer-Verlag
Berlin
Heidelberg
New York

Lecture Notes in Computer Science

Editors: G. Goos, J. Hartmanis

Volume 3
Fifth Conference on Optimization Techniques
Part 1
Editors: A. Ruberti, R. Conti
1973. Numerous figures.
XIII, 565 pages (34 pages in French)
ISBN 3-540-06583-0

Volume 4
Fifth Conference on Optimization Techniques
Part 2
Editors: A. Ruberti, R. Conti
1973. Numerous figures.
XIII, 389 pages (12 pages in French)
ISBN 3-540-06600-4

Volume 6
Matrix Eigensystem Routines-EISPACK Guide
By B. T. Smith, J. M. Boyle, J. J. Dongarra, B. S. Garbow, Y. Ikebe, V. C. Klema, C. B. Moler
2nd editon 1976. XI, 551 pages
ISBN 3-540-07546-1

Volume 25
Category Theory Applied to Computation and Control
Proceedings of the First International Symposium, San Francisco, February 25–26, 1974
Editor: E. G. Manes
1975. X, 245 pages
ISBN 3-540-07142-3

Volume 27
Optimization Techniques IFIP Technical Conference
Novosibirsk, July 1–7, 1974
Editor: G. I. Marchuk
1975. VIII, 507 pages
ISBN 3-540-07165-2

Volume 40
Optimization Tech
Modeling and Optimiz
the Service of Man 1
Proceedings, 7th IFIF
rence, Nice, Septeml
1975
Editor: J. Cea
1976. 204 figures, 3
XIV, 854 pages (113
French)
ISBN 3-540-07622-0

Volume 41
Optimization Techniq
Modeling and Optimiz
the Service of Man 2
Proceedings, 7th IFIP Co
rence, Nice, September 8–
1975
Editor: J. Cea
1976. XIII, 852 pages
ISBN 3-540-07623-9

Volume 51
Matrix Eigensystem Routines-EISPACK Guide Extension
By B. S. Garbow, J. M. Boyle, J. J. Dongarra, C. B. Moler
1977. VIII, 343 pages
ISBN 3-540-08254-9

Springer-Verlag
Berlin
Heidelberg
New York

a
S

Edit
and M.